**Chemoinformatics
in Drug Discovery**

*Edited by
Tudor I. Oprea*

Methods and Principles in Medicinal Chemistry

Edited by R. Mannhold, H. Kubinyi, G. Folkers

Editorial Board
H.-D. Höltje, H. Timmerman, J. Vacca, H. van de Waterbeemd, T. Wieland

Recently Published Volumes:

T. Lengauer (ed.)

**Bioinformatics –
From Genomes to Drugs**
Vol. 14

2001, ISBN 3-527-29988-2

J. K. Seydel, M. Wiese

Drug-Membrane Interactions
Vol. 15

2002, ISBN 3-527-30427-4

O. Zerbe (ed.)

BioNMR in Drug Research
Vol. 16

2002, ISBN 3-527-30465-7

P. Carloni, F. Alber (eds.)

Quantum Medicinal Chemistry
Vol. 17

2003, ISBN 3-527-30456-8

H. van de Waterbeemd,
H. Lennernäs, P. Artursson (eds.)

Drug Bioavailability
Vol. 18

2003, ISBN 3-527-30438-X

H.-J. Böhm, G. Schneider (eds.)

Protein-Ligand Interactions
Vol. 19

2003, ISBN 3-527-30521-1

R. E. Babine, S. S. Abdel-Meguid (eds.)

**Protein Crystallography
in Drug Discovery**
Vol. 20

2004, ISBN 3-527-30678-1

Th. Dingermann, D. Steinhilber,
G. Folkers (eds.)

**Molecular Biology
in Medicinal Chemistry**
Vol. 21

2004, ISBN 3-527-30431-2

H. Kubinyi, G. Müller (eds.)

**Chemogenomics in
Drug Discovery**
Vol. 22

2004, ISBN 3-527-30987-X

Chemoinformatics in Drug Discovery

Edited by
Tudor I. Oprea

WILEY-VCH

WILEY-VCH Verlag GmbH & Co. KGaA

Series Editors:

Prof. Dr. Raimund Mannhold
Biomedical Research Center
Molecular Drug Research Group
Heinrich-Heine-Universität
Universitätsstrasse 1
40225 Düsseldorf
Germany
raimund.mannhold@uni-duesseldorf.de

Prof. Dr. Hugo Kubinyi
Donnersbergstrasse 9
67256 Weisenheim and Sand
Germany
kubinyi@t-online.de

Prof. Dr. Gerd Folkers
Department of Applied Biosciences
ETH Zürich
Winterthurerstrasse 19
8057 Zürich
Switzerland
folkers@pharma.anbi.ethz.ch

Volume Editor:

Prof. Dr. Tudor I. Oprea
Division of Biocomputing
MSC08 4560
University of New Mexico
School of Medicine
Albuquerque, NM 87131
USA

Library of Congress Card No.: Applied for
British Library Cataloging-in-Publication Data A catalogue record for this book is available from the British Library.

Bibliographic information published by Die Deutsche Bibliothek
Die Deutsche Bibliothek lists this publication in the Deutsche Nationalbibliografie; detailed bibliographic data is available in the internet at http://dnb.ddb.de.

© 2005 WILEY-VCH Verlag GmbH & Co. KGaA Weinheim

Printed in the Federal Republic of Germany
Printed on acid-free paper

Composition Laserwords Private Ltd, Chennai, India
Printing betz-druck GmbH, Darmstadt
Bookbinding Litges & Dopf Buchbinderei GmbH, Heppenheim

ISBN-13: 978-3-527-30753-1
ISBN-10: 3-527-30753-2

Contents

A Personal Foreword *XV*

Preface *XVII*

List of Contributors *XIX*

1 **Introduction to Chemoinformatics in Drug Discovery –**
A Personal View *1*
Garland R. Marshall

1.1 Introduction *1*
1.2 Historical Evolution *4*
1.3 Known versus Unknown Targets *5*
1.4 Graph Theory and Molecular Numerology *6*
1.5 Pharmacophore *7*
1.6 Active-Analog Approach *8*
1.7 Active-Site Modeling *9*
1.8 Validation of the Active-Analog Approach and Active-Site Modeling *10*
1.9 PLS/CoMFA *11*
1.10 Prediction of Affinity *12*
1.11 Protein Structure Prediction *13*
1.12 Structure-Based Drug Design *15*
1.13 Real World Pharmaceutical Issues *15*
1.14 Combinatorial Chemistry and High-throughput Screens *16*
1.15 Diversity and Similarity *16*
1.16 Prediction of ADME *17*
1.17 Failures to Accurately Predict *17*
1.18 Summary *18*
References *19*

Chemoinformatics in Drug Discovery. Edited by Tudor I. Oprea
Copyright © 2004 WILEY-VCH Verlag GmbH & Co. KGaA, Weinheim
ISBN: 3-527-30753-2

Part I   Virtual Screening *23*

2   Chemoinformatics in Lead Discovery *25*
         Tudor I. Oprea

2.1   Chemoinformatics in the Context of Pharmaceutical Research *25*
2.2   Leads in the Drug Discovery Paradigm *27*
2.3   Is There a Trend for High Activity Molecules? *29*
2.4   The Concept of Leadlikeness *32*
2.5   Conclusions *37*
         References *38*

**3   Computational Chemistry, Molecular Complexity and Screening
         Set Design** *43*
         Michael M. Hann, Andrew R. Leach, and Darren V.S. Green

3.1   Introduction *43*
3.2   Background Concepts: the Virtual, Tangible and Real Worlds of
         Compounds, the "Knowledge Plot" and Target Tractability *44*
3.3   The Construction of High Throughput Screening Sets *45*
3.4   Compound Filters *47*
3.5   "Leadlike" Screening Sets *48*
3.6   Focused and Biased Set Design *54*
3.7   Conclusion *55*
         References *56*

4   Algorithmic Engines in Virtual Screening *59*
         Matthias Rarey, Christian Lemmen, and Hans Matter

4.1   Introduction *59*
4.2   Software Tools for Virtual Screening *61*
4.3   Physicochemical Models in Virtual Screening *62*
4.3.1   Intermolecular Forces in Protein–Ligand Interactions *63*
4.3.2   Scoring Functions for Protein–Ligand Recognition *66*
4.3.3   Covering Conformational Space *67*
4.3.4   Scoring Structural Alignments *68*
4.4   Algorithmic Engines in Virtual Screening *69*
4.4.1   Mathematical Concepts *69*
4.4.2   Algorithmic Concepts *76*
4.4.3   Descriptor Technology *81*
4.4.4   Global Search Algorithms *85*
4.5   Entering the Real World: Virtual Screening Applications *89*
4.5.1   Practical Considerations on Virtual Screening *89*
4.5.2   Successful Applications of Virtual Screening *91*
4.6   Practical Virtual Screening: Some Final Remarks *99*
         References *101*

5 **Strengths and Limitations of Pharmacophore-Based Virtual Screening** *117*
Dragos Horvath, Boryeu Mao, Rafael Gozalbes, Frédérique Barbosa, and Sherry L. Rogalski

5.1 Introduction *117*
5.2 The "Pharmacophore" Concept: Pharmacophore Features *117*
5.3 Pharmacophore Models: Managing Pharmacophore-related Information *118*
5.4 The Main Topic of This Paper *119*
5.5 The Cox2 Data Set *119*
5.6 Pharmacophore Fingerprints and Similarity Searches *120*
5.7 Molecular Field Analysis (MFA)-Based Pharmacophore Information *123*
5.8 QSAR Models *125*
5.9 Hypothesis Models *125*
5.10 The Minimalist Overlay-Independent QSAR Model *126*
5.11 Minimalist and Consensus Overlay-Based QSAR Models *128*
5.12 Diversity Analysis of the Cox2 Compound Set *131*
5.13 Do Hypothesis Models Actually Tell Us More Than Similarity Models About the Structural Reasons of Activity? *131*
5.14 Why Did Hypothesis Models Fail to Unveil the Key Cox2 Site–Ligand Interactions? *134*
5.15 Conclusions *136*
References 137

Part II **Hit and Lead Discovery** *141*

6 **Enhancing Hit Quality and Diversity Within Assay Throughput Constraints** *143*
Iain McFadyen, Gary Walker, and Juan Alvarez

6.1 Introduction *143*
6.1.1 What Makes a Good Lead Molecule? *144*
6.1.2 Compound Collections – Suitability as Leads *144*
6.1.3 Compound Collections – Diversity *145*
6.1.4 Data Reliability *146*
6.1.5 Selection Methods *149*
6.1.6 Enhancing Quality and Diversity of Actives *153*
6.2 Methods *154*
6.2.1 Screening Library *155*
6.2.2 Determination of Activity Threshold *156*
6.2.3 Filtering *156*
6.2.4 High-Throughput Screen Clustering Algorithm (HTSCA) *157*
6.2.5 Diversity Analysis *160*

6.2.6   Data Visualization   *161*
6.3   Results   *162*
6.3.1   Peptide Hydrolase   *162*
6.3.2   Protein Kinase   *167*
6.3.3   Protein–Protein Interaction   *168*
6.4   Discussion and Conclusion   *169*
         References   172

7   Molecular Diversity in Lead Discovery: From Quantity to Quality   *175*
         Cullen L. Cavallaro, Dora M. Schnur, and Andrew J. Tebben

7.1   Introduction   *175*
7.2   Large Libraries and Collections   *176*
7.2.1   Methods and Examples for Large Library Diversity Calculations   *177*
7.3   Medium-sized/Target-class Libraries and Collections   *181*
7.3.1   Computational Methods for Medium- and Target-class Libraries and
         Collections   *183*
7.4   Small Focused Libraries   *189*
7.4.1   Computational Methods for Small and Focused Libraries   *190*
7.5   Summary/Conclusion   *191*
         References   192

8   *In Silico* Lead Optimization   *199*
         Chris M.W. Ho

8.1   Introduction   *199*
8.2   The Rise of Computer-aided Drug Refinement   *200*
8.3   RACHEL Software Package   *201*
8.4   Extraction of Building Blocks from Corporate Databases   *201*
8.5   Intelligent Component Selection System   *203*
8.6   Development of a Component Specification Language   *205*
8.7   Filtration of Components Using Constraints   *207*
8.8   Template-driven Structure Generation   *208*
8.9   Scoring Functions – Methods to Estimate Ligand–Receptor
         Binding   *209*
8.10   Target Functions   *212*
8.11   Ligand Optimization Example   *214*
         References   219

Part III   Databases and Libraries   *221*

9   WOMBAT: World of Molecular Bioactivity   *223*
         Marius Olah, Maria Mracec, Liliana Ostopovici, Ramona Rad, Alina Bora,
         Nicoleta Hadaruga, Ionela Olah, Magdalena Banda, Zeno Simon,
         Mircea Mracec, and Tudor I. Oprea

9.1   Introduction – Brief History of the WOMBAT Project   *223*
9.2   WOMBAT 2004.1 Overview   *224*
9.3   WOMBAT Database Structure   *227*
9.4   WOMBAT Quality Control   *228*
9.5   Uncovering Errors from Literature   *231*
9.6   Data Mining with WOMBAT   *234*
9.7   Conclusions and Future Challenges   *235*
         References   *237*

10   Cabinet – Chemical and Biological Informatics Network   *241*
         Vera Povolna, Scott Dixon, and David Weininger

10.1   Introduction   *241*
10.1.1   Integration Efforts, WWW as Information Resource and
         Limitations   *241*
10.1.2   Goals   *243*
10.2   Merits of Federation Rather than Unification   *243*
10.2.1   The Merits of Unification   *244*
10.2.2   The Merits of Federation   *244*
10.2.3   Unifying Disparate Data Models is Difficult, Federating them
         is Easy   *245*
10.2.4   Language is a Natural Key   *246*
10.3   HTTP is Appropriate Communication Technology   *248*
10.3.1   HTTP is Specifically Designed for Collaborative Computing   *248*
10.3.2   HTTP is the Dominant Communication Protocol Today   *248*
10.3.3   HTML Provides a Universally Accessible GUI   *249*
10.3.4   MIME "Text/Plain" and "Application/Octet-Stream" are Important
         Catch-alls   *249*
10.3.5   Other MIME Types are Useful   *250*
10.3.6   One Significant HTTP Work-around is Required   *250*
10.4   Implementation   *251*
10.4.1   Daylight HTTP Toolkit   *251*
10.4.2   Metaphorics' Cabinet Library   *252*
10.5   Specific Examples of Federated Services   *252*
10.5.1   Empath – Metabolic Pathway Chart   *253*
10.5.2   Planet – Protein–ligand Association Network   *254*
10.5.3   EC Book – Enzyme Commission Codebook   *254*
10.5.4   WDI – World Drug Index   *254*

10.5.5 WOMBAT – World of Molecular Bioactivity *255*
10.5.6 TCM (Traditional Chinese Medicines), DCM (Dictionary of Chinese Medicine), PARK (Photo ARKive) and zi4 *255*
10.5.7 Cabinet "Download" Service *256*
10.5.8 Cabinet Usage Example *256*
10.6 Deployment and Refinement *262*
10.6.1 Local Deployment *264*
10.6.2 Intranet Deployment *264*
10.6.3 Internet Deployment *265*
10.6.4 Online Deployment *266*
10.7 Conclusions *266*
References *268*

11 Structure Modification in Chemical Databases *271*
Peter W. Kenny and Jens Sadowski

11.1 Introduction *271*
11.2 Permute *274*
11.2.1 Protonation and Formal Charges *274*
11.2.2 Tautomerism *275*
11.2.3 Nitrogen Configurations *276*
11.2.4 Duplicate Removal *276*
11.2.5 Nested Loop *276*
11.2.6 Application Statistics *277*
11.2.7 Impact on Docking *277*
11.3 Leatherface *279*
11.3.1 Protonation and Formal Charges *279*
11.3.2 Tautomerism *280*
11.3.3 Ionization and Tautomer Model *281*
11.3.4 Relationships between Structures *282*
11.3.5 Substructural Searching and Analysis *283*
11.4 Concluding Remarks *283*
References *284*

12 Rational Design of GPCR-specific Combinational Libraries Based on the Concept of Privileged Substructures *287*
Nikolay P. Savchuk, Sergey E. Tkachenko, and Konstantin V. Balakin

12.1 Introduction – Combinatorial Chemistry and Rational Drug Design *287*
12.2 Rational Selection of Building Blocks Based on Privileged Structural Motifs *288*
12.2.1 Privileged Structures and Substructures in the Design of Pharmacologically Relevant Combinatorial Libraries *288*

12.2.2   Analysis of Privileged Structural Motifs: Structure Dissection Rules   *291*
12.2.3   Knowledge Database   *293*
12.2.4   Target-specific Differences in Distribution of Molecular Fragments   *295*
12.2.5   Privileged versus Peripheral Retrosynthetic Fragments   *296*
12.2.6   Peripheral Retrosynthetic Fragments: How to Measure the
          Target-specific Differences?   *297*
12.2.7   Selection of Building Blocks   *300*
12.2.8   Product-based Approach: Limiting the Space of Virtual Libraries   *305*
12.2.9   Alternative Strategy: Property-based Approach   *306*
12.2.10   Kohonen Self-organizing Maps   *307*
12.3   Conclusions   *309*
          References   *311*

Part IV   Chemoinformatics Applications   *315*

13   A Practical Strategy for Directed Compound Acquisition   *317*
          Gerald M. Maggiora, Veerabahu Shanmugasundaram, Michael S. Lajiness,
          Tom N. Doman, and Martin W. Schultz

13.1   Introduction   *317*
13.2   A Historical Perspective   *319*
13.3   Practical Issues   *320*
13.4   Compound Acquisition Scheme   *322*
13.4.1   Preprocessing Compound Files   *322*
13.4.2   Initial Compound Selection and Diversity Assessment   *325*
13.4.3   Compound Reviews   *327*
13.4.4   Final Selection and Compound Purchase   *328*
13.5   Conclusions   *328*
13.6   Methodologies   *329*
13.6.1   Preprocessing Filters   *329*
13.6.2   Diverse Solutions (DVS)   *330*
13.6.3   Dfragall   *330*
13.6.4   Ring Analysis   *331*
          References   *331*

**14   Efficient Strategies for Lead Optimization by Simultaneously Addressing
          Affinity, Selectivity and Pharmacokinetic Parameters**   *333*
          Karl-Heinz Baringhaus and Hans Matter

14.1   Introduction   *333*
14.2   The Origin of Lead Structures   *336*
14.3   Optimization for Affinity and Selectivity   *338*
14.3.1   Lead Optimization as a Challenge in Drug Discovery   *338*
14.3.2   Use and Limitation of Structure-based Design Approaches   *339*

14.3.3 Integration of Ligand- and Structure-based Design Concepts *340*
14.3.4 The Selectivity Challenge from the Ligands' Perspective *342*
14.3.5 Selectivity Approaches Considering Binding Site Topologies *344*
14.4 Addressing Pharmacokinetic Problems *347*
14.4.1 Prediction of Physicochemical Properties *347*
14.4.2 Prediction of ADME Properties *348*
14.4.3 Prediction of Toxicity *349*
14.4.4 Physicochemical and ADMET Property-based Design *350*
14.5 ADME/Antitarget Models for Lead Optimization *350*
14.5.1 Global ADME Models for Intestinal Absorption and Protein Binding *350*
14.5.2 Selected Examples to Address ADME/Toxicology Antitargets *354*
14.6 Integrated Approach *357*
14.6.1 Strategy and Risk Assessment *357*
14.6.2 Integration *359*
14.6.3 Literature and Aventis Examples on Aspects of Multidimensional Optimization *360*
14.7 Conclusions *366*
References *367*

15 **Chemoinformatic Tools for Library Design and the Hit-to-Lead Process: A User's Perspective** *381*
Robert Alan Goodnow, Jr., Paul Gillespie, and Konrad Bleicher

15.1 The Need for Leads: The Sources of Leads and the Challenge to Find Them *381*
15.2 Property Predictions *383*
15.3 Prediction of Solubility *384*
15.4 Druglikeness *390*
15.4.1 Are There Differences between Drugs and Nondrugs? *390*
15.4.2 Is the Problem Tractable within a Single Program? *391*
15.4.3 Do We Have a Training Set that Will Allow Us to Address the Issue? *392*
15.4.4 Approaches to the Prediction of Druglikeness *392*
15.5 Frequent Hitters *394*
15.6 Identification of a Lead Series *395*
15.7 The Hit-to-lead Process *397*
15.7.1 Prioritization of Hits *397*
15.7.2 Identification of Analogs *402*
15.7.3 Additional Assays *403*
15.8 Leads from Libraries: General Principles, Practical Considerations *404*
15.9 Druglikeness in Small-molecule Libraries *406*
15.10 Data Reduction and Viewing for Virtual Library Design *407*
15.11 Druglikeness *408*

15.12 Complexity and Andrews' Binding Energy   *408*
15.13 Solubility   *411*
15.14 Polar Surface Area   *411*
15.15 Number of Rotatable Bonds   *412*
15.16 hERG Channel Binding   *413*
15.17 Chemoinformatic Analysis of the Predicted Hansch Substituent Constants of the Diversity Reagents for Design of Vector Exploration Libraries   *415*
15.18 Targeting Libraries by Virtual Screening   *416*
15.19 Combinatorial Design Based on Biostructural Information   *418*
15.20 Ligand-based Combinatorial Design: The RADDAR Approach   *419*
15.21 Virtual Screening of Small-molecule Library with Peptide-derived Pharmacophores   *421*
15.22 Chemoinformatic Tools and Strategies to Visualize Active Libraries   *423*
15.23 Visualization of Library Designs during Hit-to-lead Efforts   *423*
15.24 Summary and Outlook for Chemoinformatically Driven Lead Generation   *425*
References   *426*

16 **Application of Predictive QSAR Models to Database Mining**   *437*
Alexander Tropsha

16.1 Introduction   *437*
16.2 Building Predictive QSAR Models: The Importance of Validation   *438*
16.3 Defining Model Applicability Domain   *441*
16.4 Validated QSAR Modeling as an Empirical Data-modeling Approach: Combinatorial QSAR   *443*
16.5 Validated QSAR Models as Virtual Screening Tools   *445*
16.6 Conclusions and Outlook   *452*
References   *453*

17 **Drug Discovery in Academia – A Case Study**   *457*
Donald J. Abraham

17.1 Introduction   *457*
17.2 Linking the University with Business and Drug Discovery   *457*
17.2.1 Start-up Companies   *457*
17.2.2 Licensing   *458*
17.3 Research Parks   *459*
17.4 Conflict of Interest Issues for Academicians   *459*
17.5 Drug Discovery in Academia   *461*
17.5.1 Clinical Trials in Academia   *461*

17.6　　 Case Study: The Discovery and Development of Allosteric
　　　　 Effectors of Hemoglobin　　*462*
17.6.1　 Geduld (Patience)　　*463*
17.6.2　 Glück (Luck)　　*463*
17.6.3　 Geschick (Skill)　　*464*
17.6.4　 Geld (Money)　　*471*
　　　　 References　481

Subject Index　　*485*

A Personal Foreword

This volume brings together contributions from academic and industrial scientists who develop and apply chemoinformatics strategies and tools in drug discovery. From chemical inventory and compound registration to candidate drug nomination, chemoinformatics integrates data via computer-assisted manipulation of chemical structures. Linked with computational chemistry, physical (organic) chemistry, pharmacodynamics and pharmacokinetics, chemoinformatics provides unique capabilities in the areas of lead and drug discovery. This book aims to offer knowledge and practical insights into the use of chemoinformatics in preclinical research.

Divided in four sections, the book opens with a first-hand account from Garland Marshall, spanning four decades of chemoinformatics and pharmaceutical research and development. Part one sets the stage for virtual screening and lead discovery. Hit and lead discovery via *in silico* technologies are highlighted in part two. In part three, data collection and mining using chemical databases are discussed in the context of chemical libraries. Specific applications and examples are collected in part four, which brings together industrial and academic perspectives. The book concludes with another personal account by Don Abraham, who presents drug discovery from an academic perspective.

The progression hit identification → lead generation → lead optimization → candidate drug nomination is served by a variety of chemoinformatics tools and strategies, most of them supporting the decision-making process. Key procedures and steps, from virtual screening to *in silico* lead optimization and from compound acquisition to library design, underscore our progress in grasping the preclinical drug discovery process, its needs for novel technologies and for integrated informatics support. We now have the ability to identify novel chemotypes in a rational manner, and *in silico* methods are deep-rooted in the process of systematic discovery. Our increased knowledge in a variety of seemingly unrelated phenomena, from atomic level issues related to drug–receptor binding to bulk properties of drugs and pharmacokinetics profiling, is likely to lead us on a better path for the discovery of orally bioavailable drugs, at the same time paving the way for novel, unexpected therapeutics.

I want to acknowledge all the contributors who made this book possible. Their insights, examples and personal accounts move beyond the sometimes dry language of science, turning this volume into an interesting and fascinating book to read.

Finally, I thank Frank Weinreich and Hugo Kubinyi for their encouragement and timely pressure to prepare this book on time.

Albuquerque, January 2005   *Tudor I. Oprea*

Chemoinformatics in Drug Discovery. Edited by Tudor I. Oprea
Copyright © 2004 WILEY-VCH Verlag GmbH & Co. KGaA, Weinheim
ISBN: 3-527-30753-2

Preface

The term "chemoinformatics" was introduced in 1998 by Dr. Frank K. Brown in the Annual Reports of Medicinal Chemistry. In his article "Chemoinformatics: What is it and How does it Impact Drug Discovery", he defines chemoinformatics as follows: *"The use of information technology and management has become a critical part of the drug discovery process. Chemoinformatics is the mixing of those information resources to transform data into information and information into knowledge for the intended purpose of making better decisions faster in the area of drug lead identification and organization".*

In fact, Chemoinformatics is a generic term that encompasses the design, creation, organization, management, retrieval, analysis, dissemination, visualization and use of chemical information. Related terms of chemoinformatics are cheminformatics, chemi-informatics, chemometrics, computational chemistry, chemical informatics, and chemical information management/science.

Reflecting the above given definitions, the present volume on "Chemoinformatics in Drug Discovery"covers its most important aspects within four main sections. After an introduction to chemoinformatics in drug discovery by Garland Marshall, the first section is focused on *Virtual Screening*. T. Oprea describes the use of "Chemoinformatics in Lead Discovery" and M.M. Hann et al. deal with "Computational Chemistry, Molecular Complexity and Screening Set Design". Then, M. Rarey et al. review "Algorithmic Engines in Virtual Screening" and D. Horvath et al. review the "Strengths and Limitations of Pharmacophore-Based Virtual Screening". The next section is dedicated to *Hit and Lead Discovery* with chapters of I.J. McFadyen et al. on "Enhancing Hit Quality and Diversity Within Assay Throughput Constraints", of C.L. Cavallaro et al. on "Molecular Diversity in Lead Discovery", and of C. Ho on "In Silico Lead Optimization". Topics of the third section refer to *Databases and Libraries*. They include chapters on "WOMBAT: World of Molecular Bioactivity" by M. Olah et al., on "Cabinet – Chemical and Biological Informatics Network" by V. Povolna et al., on "Structure Modification in Chemical Databases" by P.W. Kenney and J. Sadowski, and on the "Rational Design of GPCR-specific Combinatorial Libraries Based on the Concept of Privileged Substructures" by N.P. Savchuk et al.

According to our intention, to provide in this series on "Methods and Principles in Medicinal Chemistry" practice-oriented monographs, the book closes with a section on *Chemoinformatics Applications*. These are exemplified by G.M. Maggiora et al. in a chapter on "A Practical Strategy for Directed Compound Acquisition", by

Chemoinformatics in Drug Discovery. Edited by Tudor I. Oprea
Copyright © 2004 WILEY-VCH Verlag GmbH & Co. KGaA, Weinheim
ISBN: 3-527-30753-2

K.-H. Baringhaus and H. Matter on "Efficient Strategies for Lead Optimization by Simultaneously Addressing Affinity, Selectivity and Pharmacokinetic Parameters", by R.A. Goodnow et al. on "Chemoinformatic Tools for Library Design and the Hit-to-Lead Process" and by A. Tropsha on the "Application of Predictive QSAR Models to Database Mining". The section is concluded by a chapter of D.J. Abraham on "Drug Discovery from an Academic Perspective".

The series editors would like to thank Tudor Oprea for his enthusiasm to organize this volume and to work with such a fine selection of authors. We also want to express our gratitude to Frank Weinreich from Wiley-VCH for his valuable contributions to this project.

September 2004

Raimund Mannhold, Düsseldorf
Hugo Kubinyi, Weisenheim am Sand
Gerd Folkers, Zürich

List of Contributors

DONALD J. ABRAHAM
Department of Medicinal Chemistry
Virginia Commonwealth University
800 E. Leigh Street
Richmond, VA 23219-1540
USA

JUAN C. ALVAREZ
Chemical and Screening Sciences
Wyeth Research
200 Cambridge Park Drive
Cambridge, MA 02140
USA

KONSTANTIN V. BALAKIN
Chemical Diversity Labs, Inc.
Computational and Medicinal Chemistry
11575 Sorrento Valley Road
San Diego, CA 92121
USA

MAGDALENA BANDA
Romanian Academy
Institute of Chemistry
"Coriolan Dragulescu"
Bv. Mihai Viteazul No. 24
300223 Timisoara
Romania

FRÉDÉRIQUE BARBOSA
Cerep S. A.
128, Rue Danton
92506 Rueil-Malmaison
France

KARL-HEINZ BARINGHAUS
Aventis Pharma Deutschland GmbH
DI&A Chemistry
Computational Chemistry
Industriepark Höchst, Bldg. G 878
65926 Frankfurt am Main
Germany

KONRAD BLEICHER
F. Hoffmann-La Roche Ltd.
PRBD-C
4070 Basel
Switzerland

ALINA BORA
Romanian Academy
Institute of Chemistry
"Coriolan Dragulescu"
Bv. Mihai Viteazul No. 24
300223 Timisoara
Romania

CULLEN L. CAVALLARO
Pharmaceutical Research Institute
Bristol-Myers Squibb Co.
Province Line Road
Princeton, NJ 08540
USA

SCOTT DIXON
Metaphorics, LLC
441 Greg Avenue
Sante Fe, NM 87501
USA

Chemoinformatics in Drug Discovery. Edited by Tudor I. Oprea
Copyright © 2004 WILEY-VCH Verlag GmbH & Co. KGaA, Weinheim
ISBN: 3-527-30753-2

THOMPSON N. DOMAN
Lilly Research Laboratories
Structural and Computational Sciences
Indianapolis, IN 46285
USA

PAUL GILLESPIE
Hoffmann-La Roche Inc.
Discovery Chemistry
340 Kingsland Street
Nutley, NJ 07110
USA

ROBERT A. GOODNOW, Jr.
Hoffmann-La Roche Inc.
New Leads Chemistry Initiative
340 Kingsland Street
Nutley, NJ 07110
USA

RAFAEL GOZALBES
Cerep S. A.
128, Rue Danton
92506 Rueil-Malmaison
France

DARREN V. S. GREEN
GlaxoSmithKline Research
and Development
Gunnels Wood Road
Stevenage, SG1 2NY
United Kingdom

NICOLETA HADARUGA
Romanian Academy
Institute of Chemistry
"Coriolan Dragulescu"
Bv. Mihai Viteazul No. 24
300223 Timisoara
Romania

MICHAEL M. HANN
GlaxoSmithKline Research
and Development
Gunnels Wood Road
Stevenage, SG1 2NY
United Kingdom

CHRIS HO
Drug Design Methodologies, LLC
4355 Maryland Ave.
St. Louis, MO 63108
USA

DRAGOS HORVATH
UMR8525-CNRS
Institut de Biologie de Lille
1, rue Calmette
59027 Lille
France

PETER W. KENNY
AstraZeneca Mereside
Alderley Park
Macclesfield, SK10 4TG
United Kingdom

MICHAEL S. LAJINESS
Lilly Research Laboratories
Structural and Computational Sciences
Indianapolis, IN 46285
USA

ANDREW R. LEACH
GlaxoSmithKline Research
and Development
Gunnels Wood Road
Stevenage, SG1 2NY
United Kingdom

CHRISTIAN LEMMEN
BioSolveIT GmbH
An der Ziegelei 75
53757 Sankt Augustin
Germany

GERALD M. MAGGIORA
Dept. of Pharmacology and Toxicology
University of Arizona
College of Pharmacy
Tucson, AZ 85271
USA

BORYEU MAO
Cerep Inc.
15318 NE 95th Street
Redmond, WA 98052
USA

GARLAND R. MARSHALL
Center for Computational Biology
Washington University
School of Medicine
660 S. Euclid Ave.
St. Louis, MO 63110
USA

HANS MATTER
Aventis Pharma Deutschland GmbH
DI&A Chemistry
Computational Chemistry
Industriepark Höchst, Bldg. G 878
65926 Frankfurt am Main
Germany

IAIN McFADYEN
Chemical and Screening Sciences
Wyeth Research
200 Cambridge Park Drive
Cambridge, MA 02140
USA

MARIA MRACEC
Romanian Academy
Institute of Chemistry
"Coriolan Dragulescu"
Bv. Mihai Viteazul No. 24
300223 Timisoara
Romania

MIRCEA MRACEC
Romanian Academy
Institute of Chemistry
"Coriolan Dragulescu"
Bv. Mihai Viteazul No. 24
300223 Timisoara
Romania

IONELA OLAH
Romanian Academy
Institute of Chemistry
"Coriolan Dragulescu"
Bv. Mihai Viteazul No. 24
300223 Timisoara
Romania

MARIUS OLAH
Division of Biocomputing, MSC08 4560
University of New Mexico
School of Medicine
Albuquerque, NM 87131
USA

TUDOR I. OPREA
Division of Biocomputing, MSC08 4560
University of New Mexico
School of Medicine
Albuquerque, NM 87131
USA

LILIANA OSTOPOVICI
Romanian Academy
Institute of Chemistry
"Coriolan Dragulescu"
Bv. Mihai Viteazul No. 24
300223 Timisoara
Romania

VERA POVOLNA
Metaphorics, LLC
441 Greg Avenue
Sante Fe, NM 87501
USA

RAMONA RAD
Romanian Academy
Institute of Chemistry
"Coriolan Dragulescu"
Bv. Mihai Viteazul No. 24
300223 Timisoara
Romania

MATTHIAS RAREY
Center for Bioinformatics (ZBH)
University of Hamburg
Bundesstrasse 43
20146 Hamburg
Germany

SHERRY L. ROGALSKI
Cerep Inc. 15318 NE 95th Street
Redmond, WA 98052
USA

JENS SADOWSKI
AstraZeneca R&D Mölndal
Structural Chemistry Laboratory
SC264
43183 Mölndal
Sweden

NIKOLAY P. SAVCHUK
Chemical Diversity Labs, Inc.
Chemoinformatics
11575 Sorrento Valley Road
San Diego, CA 92121
USA

DORA M. SCHNUR
Pharmaceutical Research Institute
Bristol-Myers Squibb Co.
Province Line Road
Princeton, NJ 08540
USA

MARTIN W. SCHULTZ
Pfizer Global Research and Development
301 Henrietta Street
Kalamazoo, MI 49007
USA

VEERABAHU SHANMUGASUNDARAM
Computer-Assisted Drug Discovery
Pfizer Global Research and Development
2800 Plymouth Road
Ann Arbor, MI 48105
USA

ZENO SIMON
Romanian Academy
Institute of Chemistry
"Coriolan Dragulescu"
Bv. Mihai Viteazul No. 24
300223 Timisoara
Romania

ANDREW J. TEBBEN
Pharmaceutical Research Institute
Bristol-Myers Squibb Co.
Province Line Road
Princeton, NJ 08540
USA

SERGEY E. TKACHENKO
Chemical Diversity Labs, Inc.
Computational and Medicinal Chemistry
11575 Sorrento Valley Road
San Diego, CA 92121
USA

ALEXANDER TROPSHA
Laboratory for Molecular Modeling
University of North Carolina
Chapel Hill, NC 27599
USA

GARY WALKER
Chemical and Screening Sciences
Wyeth Research
401 N. Middletown Road
Pearl River, NY 10965
USA

DAVID WEININGER
Metaphorics, LLC
441 Greg Avenue
Sante Fe, NM 87501
USA

1

Introduction to Chemoinformatics in Drug Discovery – A Personal View

Garland R. Marshall

1.1
Introduction

The first issue to be discussed is the definition of the topic. What is chemoinformatics and why should you care? There is no clear definition, although a consensus view appears to be emerging. "Chemoinformatics is the mixing of those information resources to transform data into information and information into knowledge for the intended purpose of making better decisions faster in the area of drug lead identification and organization" according to one view [1]. Hann and Green suggest that chemoinformatics is simply a new name for an old problem [2], a viewpoint I share. There are sufficient reviews [3–6] and even a book by Leach and Gillet [7] with the topic as their focus that there is little doubt what is meant, despite the absence of a precise definition that is generally accepted.

One aspect of a new emphasis is the sheer magnitude of chemical information that must be processed. For example, Chemical Abstracts Service adds over three-quarters of a million new compounds to its database annually, for which large amounts of physical and chemical property data are available. Some groups generate hundreds of thousands to millions of compounds on a regular basis through combinatorial chemistry that are screened for biological activity. Even more compounds are generated and screened *in silico* in the search for a magic bullet for a given disease. Either one of the two processes for generating information about chemistry has its own limitations. Experimental approaches have practical limitations despite automation; each *in vitro* bioassay utilizes a finite amount of reagents including valuable cloned and expressed receptors. Computational chemistry has to establish relevant criteria by which to select compounds of interest for synthesis and testing. The accuracy of prediction of affinities with current methodology is just now approaching sufficient accuracy to be of utility.

Let me emphasize the magnitude of the problem with a simple example. I was once asked to estimate the number of compounds covered by a typical issued patent for a drug of commercial interest. The patent that I selected to analyze was for enalapril, a prominent prodrug ACE inhibitor with a well-established commercial market. Given the parameters as outlined in the patent covering enalapril, an estimation of the total number of compounds included in the generic claim for enalaprilat, the active

Chemoinformatics in Drug Discovery. Edited by Tudor I. Oprea
Copyright © 2004 WILEY-VCH Verlag GmbH & Co. KGaA, Weinheim
ISBN: 3-527-30753-2

Enalapril

Enalaprilat

ingredient, was made. The following is the reference formula as described by the patent and simplified with $R_6 = OH$, and R_2 and $R_7 = H$:

$$R_2, R_7 = H$$
$$R_6 = OH$$

Thus, one can simply enumerate the members of each class of substituent and combine them combinatorially. The following details the manner in which the number of each substituent was determined with the help of Chris Ho (Marshall and Ho, unpublished).

> Substituent R: R is described as a lower alkoxy. The patent states that substituents are "otherwise represented by any of the variables including straight and branched chain hydrocarbon radicals from one to six carbon atoms, for example, methyl, ethyl, isopentyl, hexyl or vinyl, allyl, butenyl and the like." DBMAKER [8] was used to generate a database of compounds containing any combination of one to six carbon atoms, interspersed with occasional double and triple bonds, as well as all possible branching patterns. Constraints were employed to forbid the generation of chemically impossible constructs. Concord 3.01 [9] was used to generate and validate the chemical integrity of all compounds. 290 unique substituents were generated as a minimal estimate.
>
> Substituent R3: This substituent is identical to substituent R, only that it is an alkyl instead of an alkoxy. Again, 290 unique substituents of six or fewer carbon atoms were generated.

Substituent R1: R1 is described as a substituted lower alkyl wherein the substituent is a phenyl group. The patent is vague with regard to where this phenyl group should reside. If the phenyl group always resides at the carbon farthest away from the main chain, then again, 290 different substituents will result. However, if the phenyl group can reside anywhere along the 1- to 6-member chain, then approximately 1000 substituents are chemically and sterically possible.

Substituents R4 & R5: These two substituents are described by the patent as being lower alkyl groups, which may be linked to form a cyclic 4- to 6-membered ring in this position. This produces two scenarios: if these groups remain unlinked, then, as before, 290 substituents are found at *each* position.

To determine the number of possible compounds when R4 and R5 are cyclized, a different approach was used. The patent states, "R4 and R5 when joined through the carbon and nitrogen atoms to which they are attached form a 4- to 6-membered ring". Preferred ring has the formula:

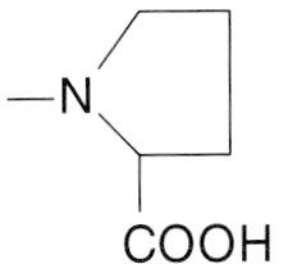

The patent is again vague in describing the generation of these cyclic systems. However, given that R4 and R5 are each 1–6 carbon alkyl groups with various branching patterns that are linked together, what results is a 4- to 6-membered ring system that may contain none, one or two side chains depending upon how R4 and R5 are connected. The overall requirement is that the total number of atoms comprising this ring system be less than or equal to 12.

To construct these ring systems, two databases were generated. The first database ("ring database") contained three compounds – a 4-, 5- and 6-membered ring as specified by the patent. The second database ("side-chain database") was constructed by cleaving each of the 290 alkyl compounds in half. One would assume that the first half of the alkyl chain would generate the ring, leaving the second half to dangle and form a side chain. A program DBCROSS (Ho, unpublished) was then used to join one compound from the ring database with up to two structures from the side-chain database at chemically appropriate substitution sites. Again, the overall requirement was that the number of atoms be less than or equal to 12. Approximately 4100 different cyclic systems were generated in this manner.

Total number of compounds

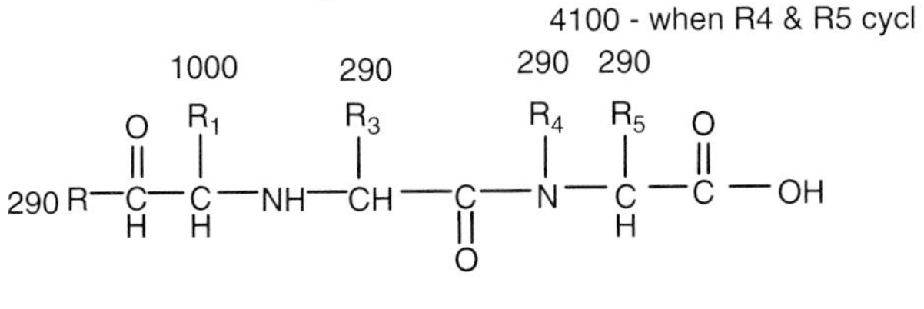

Summation $(290)(1000)(290)(290)(290)$ $= 7.07 \cdot 10^{12}$ R4/R5 noncyclic

$(290)(1000)(290)(4100)$ $= 3.44 \cdot 10^{11}$ R4/R5 cyclized

$Sum = 7.41 \cdot 10^{12} \rightarrow 3$ chiral centers (carbons where R_1, R_3 and R_5 are attached to the backbone) in this molecule: $X\ 8 = 5.93 \cdot 10^{13}$ or more than 59 *trillion* compounds included in the patent.

Note: If the phenyl group of substituent R1 is limited to the position farthest from the parent chain, then the number of compounds drops to $1.72 \cdot 10^{13}$ or more than 17 *trillion* compounds included in the patent.

Actually, the number of compounds included in the patent is severalfold larger as esters of enalaprilat such as enalapril were also included. Of the 100 trillion or so compounds included in the patent, how many could be predicted to lack druglike properties (molecular weight too large? logP too high?)? How many would be predicted to be inactive on the basis of the known structure-activity data available on angiotensin-converting enzyme (ACE) inhibitors such as captopril? How many would be predicted to be inactive now that a crystal structure of a complex of ACE with an inhibitor has been published? Given the structure-activity relationships (SAR) available on the inhibitors, what could one determine regarding the active site of ACE? What novel classes of compound could be suggested on the basis of the SAR of inhibitors? On the basis of the new crystal structure of the complex? Do the most potent compounds share a set of properties that can be identified and used to optimize a novel lead structure? Can a predictive equation relating properties and affinity for the isolated enzyme be established? Can a similar equation relating properties and *in vitro* bioassay effectiveness be established? These are representative questions facing the current drug design community and one focus of chemoinformatics.

One significant tool that is employed is molecular modeling. Because I have been involved more directly with computational chemistry and molecular modeling, there is a certain bias in my perspective. This is the reason I have used "A Personal View" as part of the title. I have also chosen a historical presentation and focused largely on those contributions that significantly impacted my thinking. This approach, of course, has its own limitation, and I apologize to my colleagues for any distortions or omissions.

1.2
Historical Evolution

With the advent of computers and the ability to store and retrieve chemical information, serious efforts to compile relevant databases and construct information retrieval systems began. One of the first efforts to have a substantial long-term impact was to collect the crystal structure information for small molecules by Olga Kennard. The Cambridge Structural Database (CSD) stores crystal structures of small molecules and provides a fertile resource for geometrical data on molecular fragments for calibration of force fields and validation of results from computational chemistry [10, 11]. As protein crystallography gained momentum, the need for a common repository of macromolecular structural data led to the Protein Data Base (PDB) originally located at Brookhaven National Laboratories [12]. These efforts focused on the accumulation and organization of experimental results on the three-dimensional structure of molecules, both large and small. Todd Wipke recognized the need for a chemical information system to handle the increasing numbers of small molecules generated in industry, and thus MDL and MACCS were born.

With the advent of computers and the availability of oscilloscopes, the idea of displaying a three-dimensional structure of the screen was obvious with rotation providing depth cueing. Cyrus Levinthal and colleagues utilized the primitive computer graphics facilities at MIT to generate rotating images of proteins and nucleic acids to provide insight into the three-dimensional aspects of these structures without having to build physical models. His paper in Scientific American in 1965 was sensational and inspired others (including myself [13]) to explore computer graphics (1966/1967) as a means of coping with the 3D nature of chemistry. Physical models (Dreiding stick figures, CPK models, etc.) were useful accepted tools for medicinal chemists, but physical overlap of two or more compounds was difficult and exploration of the potential energy surface hard to correlate with a given conformation of a physical model.

As more and more chemical data accumulated with its implicit information content, a multitude of approaches began to extract useful information. Certainly, the shape and variability in geometry of molecular fragments from CSD was mined to provide fragments of functional groups for a variety of purposes. As series of compounds were tested for biological activity in a given assay, the desire to distill the essence of the chemical requirements for such activity to guide optimization was generated. Initially, the efforts focused on congeneric series as the common scaffold presumably eliminated the molecular alignment problem with the assumption that all molecules bound with a common orientation of the scaffold. This was the intellectual basis of the Hansch approach (quantitative structure-activity relationships, QSAR), in which substituent parameters from physical chemistry were used to correlate chemical properties with biological activity for a series of compounds with the same substitution pattern on the congeneric scaffold [14, 15].

1.3
Known versus Unknown Targets

Intellectually, the application of molecular modeling has dichotomized into those methods dealing with biological systems where no structural information at the atomic level is known, the unknown receptor, and those systems that have become relatively common, where a three-dimensional structure is know from crystallography or NMR spectroscopy. The Washington University group has spent most of its efforts over the last three decades focused on the common problem encountered where one has little structural information. Others, such as Peter Goodford and Tak Kuntz, have taken the lead in developing approaches to therapeutic targets where the structure of the target was available at atomic resolution. The seminal work of Goodford and colleagues [16] on designing inhibitors of the 2,3-diphosphorylglycerate (DPG) binding site on hemoglobin

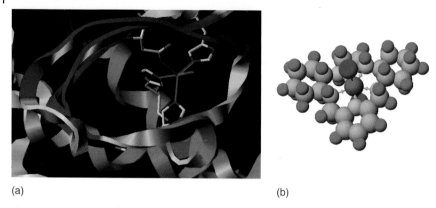

(a)   (b)

Fig. 1.1 **(a)** Active site of Mn superoxide dismutase (three histidine and one aspartic acid ligand to manganese) and **(b)** M40403, synthetic enzyme with 5 nitrogens (yellow) and two chloride (green) ligands.

for the treatment of sickle-cell disease certainly stimulated many others to obtain crystal structures of their therapeutic target. The most dramatic example of computer-aided drug design of which I am aware is the development of superoxide dismutase mimetics of below 500 molecular weight by Dennis Riley of Metaphore Pharmaceuticals. By understanding the redox chemistry of manganese superoxide reductase, Riley was able to design a totally novel pentaazacrown scaffold complexed with manganese (Figure 1.1) that catalyzes the conversion of superoxide to hydrogen peroxide at diffusion-controlled rates [17, 18]. This is the first example of a synthetic enzyme with a catalytic rate equal to or better than nature's best. The advances in molecular biology provided the means of cloning and expressing proteins in sufficient quantities to screen a variety of conditions for crystallization. Thus, it is almost expected that a crystal structure is available for any therapeutic target of interest. Unfortunately, many therapeutic targets such as G-protein-coupled receptors are still significant challenges to structural biology.

1.4
Graph Theory and Molecular Numerology

Considerable literature developed around the ability of numerical indices derived from graph theoretical considerations to correlate with SAR data. This was a source of mystery to me for some time. A colleague, Ioan Motoc, from Romania, with experience in this arena and a very strong intellect, helped me understand the ability of various indices to be useful parameters in QSAR equations [19–21]. Ioan correlated various indices with more physically relevant (at least to me) variables such as surface area and molecular volume. Since computational time was at a premium during the early days of QSAR and such indices could be calculated with minimal computations, they played a useful role and continue to be used. As a chemist, however, I am much more comfortable with parameters such as surface area or volume.

1.5
Pharmacophore

The success of QSAR led to efforts to extend the domain to noncongeneric series, where the structural similarity between molecules active in the same bioassay was not obvious. Certainly, the work of Beckett and Casey on opioids [22] to define those parts of the active molecules (pharmacophoric groups) essential for activity was seminal. Kier further developed the concept of pharmacophore and applied it to rationalize the SAR of several systems [23]. Peter Gund and Todd Wipke implemented the first *in silico* screening methodology with a program to screen a molecular database for pharmacophoric patterns in 1974 [24, 25]. Leslie Humber of Ayerst in Montreal exposed me to the wide variety of structures active in the dopaminergic system [26]. Overlaps of apomorphine, chlorpromazine and butaclamol to align the amines while maintaining the coplanarity of an aromatic ring led to a plausible hypothesis of a receptor-bound conformation. The least-squares fitting of atomic centers did not allow such an overlap, but the use of a centroid of the aromatic ring with normals to the plane for least-squares fitting accomplished the overlap. There still continues to be developments of methods to generate overlaps of hydrogen-bond donors and acceptors, aromatic rings, and so on, to generate a pharmacophore hypothesis from a set of compounds active at a given receptor/enzyme. One method developed early at Washington University was minimization of distances between groups in different molecules assigned by the

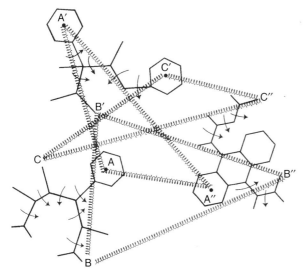

Fig. 1.2 Schematic diagram of minimization approach to the overlap of pharmacophoric groups (A with A′ with A″, B with B′ with B″, C with C′ with C″) by the introduction of constraints (springs) with intermolecular interactions ignored and only intramolecular interactions considered.

investigator with no intermolecular interactions. In effect, adding springs to cause the groups to overlap as the energy of the entire set of molecules was minimized excluding any interatomic interactions except those imposed by the springs (Figure 1.2). As a minimization procedure, the results were dependent on the starting conformation of the set of molecules being minimized, and multiple starting conformations were used to generate multiple pharmacophoric hypotheses.

1.6
Active-Analog Approach

The early work by medicinal chemists to try and rationalize their structure-activity relationships (SAR) with three-dimensional models as well as the success of Hansch and others in correlating SAR with physical properties led to exploration of molecular modeling as a means of combining the two approaches. Clearly, overall physical properties such as hydrophilicity, steric volume, charge and molar refractivity would be more meaningful in the context of a specific subsite within the receptor rather than when considered as an overall molecular average. One expected models with greater resolution and the ability to discriminate between stereoisomers, for example, as a result of the inclusion of geometrically sensitive parameters. By 1979, the group at Washington University had developed a systematic computational approach to the generation of pharmacophore hypotheses, the active-analog approach, which was disclosed at the ACS National Meeting that year [27].

The basic premise was that each compound tested against a biological assay was a three-dimensional question for the receptor. In effect, one was playing "Twenty Questions" in three dimensions. But each molecule was, in general, flexible and could present a plethora of possible three-dimensional arrays of interactive chemical groups. By computationally analyzing the sets of possible pharmacophoric patterns associated with each active molecule, one could find those pharmacophoric patterns common to a set of actives. In the simplest case, each inactive molecule would be geometrically precluded from presenting the given pharmacophoric pattern common to active molecules by steric or cyclic constraints. In practice, inactives that were capable of presenting the hypothetical pharmacophoric pattern were often found, so some other rationale for their inactivity was necessary to invoke. Aligning each active molecule to the candidate pharmacophoric pattern allowed determination of the volume requirements of the set of actives. One possible explanation for an inactive compound that could present the correct pharmacophoric pattern was a requirement for extra volume that was occupied by the receptor. When an inactive was aligned with the pharmacophore as scaffold, subtraction of the active volume space could identify such novel requirements. We had developed a Gaussian representation of molecular volume earlier [28], which readily allowed mathematical manipulation of atomic volumes. The best example of this rationalization of a data set occurred with a set of rigid bicyclic amino acids that inhibited the enzyme responsible for the synthesis of the active methyl donor in biology, S-adenosylmethionine, by Sufrin et al. [29]. In this case, the amino acid portion provided a common frame of reference that revealed that the compounds with

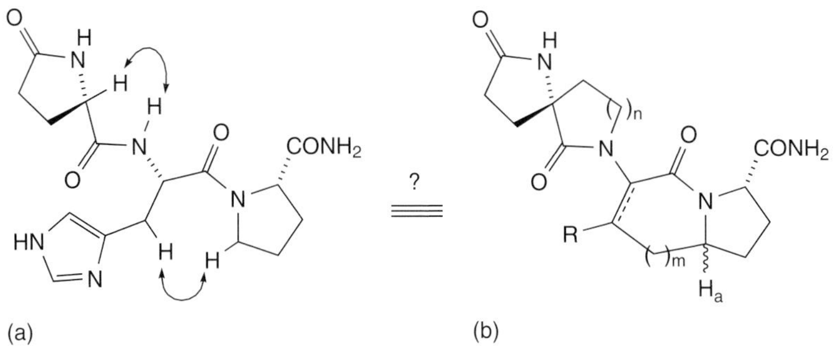

Fig. 1.3 Analysis of TRH **(a)** analogs by the active-analog approach by Font and Marshall led to a proposal for the receptor-bound conformation compatible with internal cyclization to generate polycyclic analogs **(b)**.

loss of the ability to inhibit the enzyme shared a small volume not required by active compounds, presumably required by an atom of the enzyme active site. Because the physical properties of the actives and inactives in this series were effectively identical, and the amino acid portion was clearly required for enzyme recognition, no other plausible suggestion for the data set has ever been suggested. Two other examples from analysis of SAR data were published on the glucose sensor [30] and on the GABA receptor [31].

One example of the early determination of the receptor-bound conformation of a biologically active peptide using the active-analog approach was the thesis work of Jose Font on the tripeptide TRH (thyrotropin releasing hormone), pyroglutamyl-histidyl-prolineamide. Only six torsional angles needed to be specified to determine the backbone conformation and the relative position of the imidazole ring of the bioactive conformation (Figure 1.3). Two alternative conformers were consistent with the conformational constraints required by the set of analogs analyzed. Font designed several polycyclic analogs, which were intractable for the synthetic procedures available at the time. In fact, these compounds served as a catalyst for the design of some novel electrochemical approaches by Prof. Kevin Moeller of Washington University [32–35]. Once the compounds could be prepared, their activity fully supported the receptor-bound conformation derived a decade before.

1.7
Active-Site Modeling

Any examination of crystal structures of complexes of a series of ligands binding to a protein (the set of complexes of thermolysin with a variety of inhibitors determined in the Brian Mathews lab, for example; see references in DePriest et al. [36]) shows clearly a major limitation of the pharmacophore assumption. Ligands do not optimize overlap of similar chemical functionality in complexes but find a way to maintain correct hydrogen-bonding geometry, for example, while accommodating other molecular interactions.

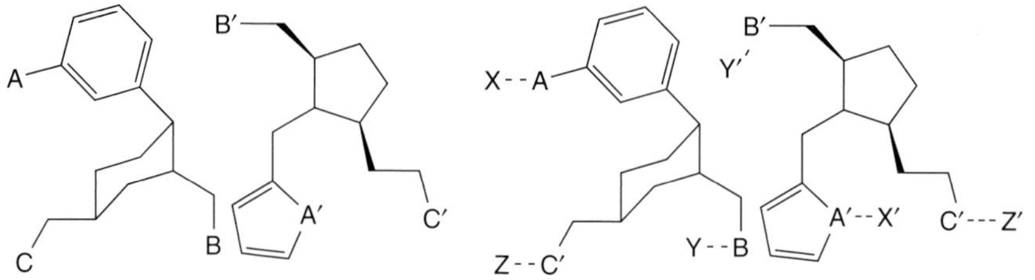

Fig. 1.4 Pharmacophore modeling with assumed ligand groups A = A′, B = B′ and C = C′. Active-site modeling with receptor groups X = X′, Y = Y′ and Z = Z′.

Thus, a transformation from overlapping chemical groups to the use of optimal sites of interaction with a stable active site of the receptor by extending hydrogen-bond donors to include an acceptor with optimal geometry, or adding a zinc to zinc-binding ligands was utilized to map the enzyme binding site (Figure 1.4). This was first emphasized for me in a study of the binding of inhibitors of angiotensin-converting enzyme (ACE) by Andrews et al. [37] in 1985. At the time, ACE was an object of intense interest in the pharmaceutical industry as captopril and enalapril, the first two approved drugs inhibiting ACE, were being extensively used to treat hypertension. Thus, each company was endeavoring to design novel chemical structures that inhibited ACE to gain a piece of the market (an activity that still occurs, witness the plethora of HMGCoA reductase and COX-2 inhibitors on the market, or in clinical studies). Analysis of the minimum energy conformations of eight ACE inhibitors revealed a common low-energy conformation of the Ala-Pro segment. Appending the pendant carboxy of enalapril or the sulfhydryl group of captopril determined a plausible site for the zinc atom involved in the enzymatic activity in the active site of ACE.

We had initiated a similar investigation of ACE inhibitors with the active-analog approach. By including additional geometrical parameters, a carboxyl group could include the zinc atom with optimal geometry from crystal structures of zinc-carboxyl complexes. Similarly, the sulfhydryl group could be expanded to include the zinc site as well with additional parameters to allow for appropriate geometrical variation. It seemed much more reasonable to assume that the groups involved in chemical catalysis and substrate recognition in the enzyme must have a relatively stable geometrical relationship, in contrast to chemical groups in a set of diverse ligands. Mayer et al.[38] analyzed 28 ACE inhibitors of diverse chemical structure available by 1987 and two inactive compounds with appropriate chemical functionality. On the basis of this data, a unique conformation for the core portion of each molecule interacting with a hypothetical ACE active site was deduced; the two inactive compounds were geometrically incapable of appropriate interaction.

1.8
Validation of the Active-Analog Approach and Active-Site Modeling

Inhibitors of the angiotensin-converting enzyme (ACE) served as a test bed for the active-analog approach in which one tries to deduce the receptor-bound conformation

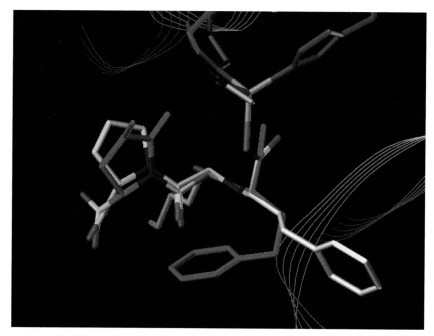

Fig. 1.5 Overlap of crystal structure of complex of the inhibitor lysinopril with angiotensin-converting enzyme and the predicted enzyme-bound conformation of ACE inhibitors by Mayer et al. [38]. Note the overlap between positions of pharmacophoric groups interacting with zinc (orange), C-terminal carboxyl and carbonyl oxygen of amide, the groups targeted by active-site modeling. The phenyl group common to enalapril analogs such as lisinopril (white ring) was not constrained (green ring) by analogs available at the time of the analysis in 1987.

of a series of active analogs based on the assumption of a common binding site. After a long delay, the crystal structure of the complex of lisinopril with ACE was finally determined [39] (not for lack of trying over two decades). The common backbone conformation of ACE inhibitors and the location of the zinc atom, hydrogen-bond donor and cationic site of the enzyme determined by Mayer et al. [38] essentially overlaps that seen in the crystal structure of the complex (Figure 1.5) arguing that, at least for this case, the assumption regarding the relative stability of groups important in catalysis or recognition is valid.

1.9
PLS/CoMFA

Dick Cramer provided insight and inspiration that led to my interest in 3D QSAR methodology [40] and was the impetus (the precursor of CoMFA was a lattice model [41] developed by Cramer and Milne at SKF) behind the development of CoMFA (Comparative Molecular Field Analysis) by Tripos [42]. The success of CoMFA in generating predictive models rested entirely on the shoulders of a new statistical approach, Partial Least Squares of Latent Variables (PLS) [43], applied to chemistry by Prof. Svante Wold of the University of Umeå, Sweden. What was revolutionary at the time was the concept that you could extract useful correlations from situations where there were more variables than observations. Traditional linear regression analysis protects the user from chance correlations when too many variables are used. PLS recognized and corrects for cross-correlation between variables, and avoids chance correlations in models by systematically determining the sensitivity of the predictability of a model to omission of training data [44].

One seminal paper [45] by Cramer examined the principal components derived from examining the physical property data of a large set of chemicals from the Handbook of Chemistry and Physics. In effect, only two principal components were responsible for a significant amount of the variance of the data in the model derived. The nature of these two properties has always intrigued me, as well as the impact/possible simplification that derivation of chemical principles in terms of these two properties might have had. My viewpoint on the potential impact of the frame of reference is analogous to the impact that transformation of variables of coordinate systems can have on simplification of mathematical equations. A good example is the simplification that arises from using internal coordinates, that is, distances between atoms rather than coordinates, in structural comparisons so that the global orientation of each molecule is eliminated from consideration.

1.10
Prediction of Affinity

In order to prioritize synthesis and testing of a compound, an accurate estimate of the binding affinity to compare with the synthetic effort is of practical utility. Unfortunately, even for systems where the crystal structure of a complex of the compound with the receptor/enzyme is already available, accurate prediction of affinity *de novo* is still challenging and relates, among other limitations of current methodology, to difficulties in estimating changes in entropy as well as lack of inclusion of multipoles and polarizability in the electrostatics used by force fields currently employed. While interpolation based on experimental data can be effective as with CoMFA and other predictive models, and simulation techniques that mutate a related compound with known activity into the compound of interest often give useful predictions of affinity, accurate *de novo* prediction of the activity is still elusive. Oprea and Marshall have recently reviewed this topic [46].

Head et al. developed a PLS-based model VALIDATE [47] to scale the relative contributions of entropy and enthalpy to binding affinity for a variety of complexes whose crystal structures had been determined. Molecular mechanics were used to calculate several parameters most correlated with enthalpy of binding, while changes in surface area, number of rotatable bonds fixed upon binding and other parameters more related to the entropy of binding were also included in the model. Of interest was that the principal components of the model were dominated by two terms (ΔH and ΔS, hopefully), several other terms had significant weight for the relative accurate model derived. Of course, doing the statistical mechanics right with a next-generation force field is an obvious solution, but a scoring function to quickly discard compounds with low affinity is still desired as witnessed by the amount of effort expended.

1.11
Protein Structure Prediction

Despite the Levinthal paradox [48, 49] that suggests the futility of attempting to predict the structure of a small protein based solely on sequence, efforts continue with increasing evidence that predictions are becoming more reliable [50]. This is based on the realization that all combinations of dihedral angles are not systematically explored by a protein in solution; rather, a funnel-like potential energy landscape guides the process. In general, the process of prediction can be dichotomized into conformer generation and conformer scoring. Obviously, one must generate a set of folds as candidates that contain the correct fold at some level of resolution to be identified and refined. Homology modeling, where one has a crystal structure of a homologous sequence, has proven a powerful approach that can generate useful models. Other approaches assemble models from homologous peptide segments [51]. The David Baker group has had considerable success in the recent CASP competitions with this approach [52], and recently designed, expressed and crystallized a small protein with a novel fold [53]. Prof. Stan Galaktionov developed a novel *ab initio* approach to fold prediction based on constraints from the contact matrix of predicted folds (Figure 1.6) that restricted possible folds to those with the correct density seen in experimental structures [54, 55]. By eliminating folds that were extended or overly compact, the fold space could be efficiently explored to generate sets of carbon alphas for further consideration. To generate a low-resolution structural model, a polyalanine chain was threaded through the carbon alphas and the orientation of the peptide planes optimized for hydrogen bonding (Welsh and Marshall, unpublished). What was needed to discriminate the candidate folds was a scoring function to evaluate the polyalanine trace utilizing the amino acid side chain information from the sequence to determine those folds worthy of full atomic representation and refinement.

We have recently published a low-resolution scoring function ProVal [56] developed with PLS that uses a multipole representation of side chains centered on the carbon alphas and betas that can distinguish the correct structure in the midst of plausible decoy folds in a large percentage of the 28 test cases studied (Figure 1.7). For 18 of the protein sets (~64%), the crystal structure scored best. In 24 sets (~86% of the cases), including the previous 18, the crystal structure ranked in the top 5%, and the crystal structure was ranked in the top 10% in all 28 cases. A second objective was to obtain a favorable correlation between the root-mean-square values for the α-carbons of the amino acids (CRMS value) of decoys and the experimental structure and the calculated score that was obtained for many of the test sets (Figure 1.7). In effect, ProVal can eliminate approximately 90% of "good" fold predictions from further consideration without specifying the coordinate position of side chains past the carbon beta. The

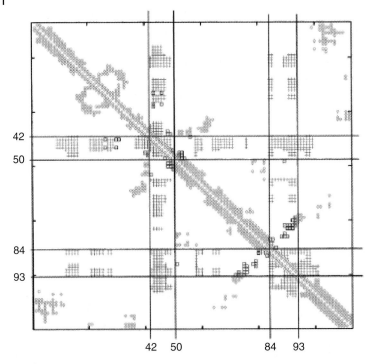

Fig. 1.6 Residue–residue contact matrix for predicted 3D structure of 3c2c (blue and green lines). The constant part Ac is shown in red, the "noncontact" matrix An is shown in green, and predicted variable contacts Ax are shown in blue. Numbers correspond to the predicted loops.

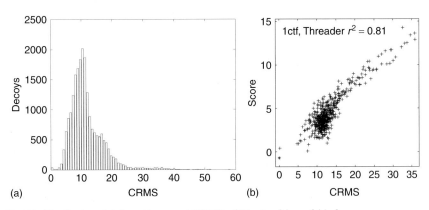

Fig. 1.7 Distribution of alpha carbon rmsd (CRMS) of 28 sets of decoy folds from crystal structures **(a)**. ProVal score versus CRMS for one decoy set generated by Threader 2.5 [57] for 1 ctf (Brookhaven PDB code) **(b)**.

details of atomic resolution are avoided because of the precision required to pack side chains efficiently with van der Waals overlap. As the quality of fold predictions increase, however, the ability to discriminate between alternatives becomes more difficult and requires a high resolution force field including multipole electrostatics and polarization (for example, the next-generation AMOEBA, Atomic Multipole Optimized Energetics for Biomolecular Applications) force field now being validated in Prof. Jay Ponder's group at Washington University, http://dasher.wustl.edu/ponder/.

1.12
Structure-Based Drug Design

As determination of crystal structures has become more commonplace, the efforts to design ligands to compliment a cavity on a molecular surface have become more sophisticated. Certainly, the pioneering effects [16] of Goodford and colleagues to design compounds binding to the DPG site on hemoglobin were dominated by chemical intuition, physical models and very primitive modeling systems. The development of DOCK by Kuntz et al. [58] was a major innovation, and sons of DOCK (AutoDock [59, 60], DREAM [61], etc.) are readily available over the web for exploring possible complex formation that include solvation approximations [62] and flexible ligands [63, 64]. Another major innovation from UCSF was the application of distance geometry as a means of generating three-dimensional coordinates from a set of distance constraints (bond lengths, sum of van der Waals radii, experimental distance constraints from NMR, etc.) [65]. Goodford has developed the use of probe atoms and chemical groups in GRID to map the binding site and identify optimal binding subsites [66]. This was certainly a prelude to experimentally determining subsite binding with subsequent assembly of fragments either by crystallography [67, 68] or by NMR [69]. The Washington University group led by Chris Ho has developed its share of structure-based design tools [8, 70–72]. Recently, a new generation software package RACHEL developed by Chris has been commercialized by Tripos, as discussed in Chapter 8 [73]. The ability to include synthetic feasibility and generate candidates with druglike properties has become a dominant theme in structure-based drug design.

1.13
Real World Pharmaceutical Issues

In reality, the approaches discussed above are all focused on the relatively simple part of developing a therapeutic, namely, lead generation. The reality of drug development is that there are many ways to interact with a given active site on a macromolecule. For example, look at the diversity of the structures capable of inhibiting HIV protease or ACE. The difficulty is in predicting adsorption, distribution, metabolism and elimination (ADME), which determine the pharmacokinetics, dosage regime and quantity of drug required. Even more problematic is prediction of toxicity, the ultimate filter that eliminates many compounds from clinical studies, and the major determinate of therapeutic ratio. At one of the first QSAR Gordon Conferences I attended, I paraphrased

Elizabeth Barrett Browning after hearing a discussion of toxicity prediction, "How can I kill thee, let me count the ways". A recent article by Stouch et al. [74] presents a thoughtful analysis of the validation effort for four such ADME/Tox models. Oprea et al. [75, 76] have compared drugs leads with compounds in development and in the marketplace and shown that compounds increase in molecular weight and logP as they progress to the bedside. *In silico* approaches certainly have their place in the pharmaceutical industry as one more tool to increase the probability of success [77].

1.14
Combinatorial Chemistry and High-throughput Screens

Development of automation and *in vitro* high-throughput biological screens has had a dramatic impact on lead discovery. Molecular biology has provided the tools for identification and validation of therapeutic targets, cloning and expression of sufficient protein to accommodate high-throughput screening, and determining the impact of elimination of the therapeutic gene by knock-out mutations.

Once the ability to screen libraries developed, the pressure on medicinal chemists increased to generate large quantities of compounds for screening. Ironically, combinatorial chemistry developed utilizing the technology of solid-phase organic chemistry. Solid-phase chemistry [78] was developed by Prof. R. Bruce Merrifield of Rockefeller University, my thesis advisor, as an automated method to assemble polypeptides, and later adapted by Prof. Marvin Caruthers of the University of Colorado for automated DNA synthesis [79]. Pioneering applications of this approach to synthetic organic chemistry in general was pioneered by a Canadian chemist, Charles Leznoff, who received little academic support, despite elegant applications including synthesis of some juvenile hormone analogs [80]. A paper on the solid-phase synthesis of benzodiazepine libraries [81] was a clarion call to medicinal chemists in industry due to the known pharmacological activity of benzodiazepines. Much of the reactions utilized in modern synthetic chemistry have been adapted to solid-phase organic chemistry for the synthesis of combinatorial libraries for high-throughput screening in the pharmaceutical industry.

Once the initial diversity fetish had run its course and management realized that it was inefficient to attempt to span the entirety of chemical space in search of drugs, more rational approaches based on chemoinformatics were developed to design combinatorial libraries and select candidates for screening on the basis of properties that have proven to be associated with successful therapeutics in the past.

1.15
Diversity and Similarity

Molecular recognition is an essential process in biological processes. One assumes that similar molecules are more likely to interact with a given receptor site than molecules that differ dramatically in size, shape or electronic distribution. This has led to the desire to compare molecules computationally prior to biological testing in order to prioritize them and test only those molecules most likely to have the desired activity. For example, in a random-screening program, a compound that generates the desired biological effect may be found. One would like to examine the company's compound library of 500 000 compounds to select those 20 compounds most likely to show the desired activity. Often, one wishes to transcend a congeneric series (not choose 20 analogs of the same basic structure) and so comparisons must be done in some three-dimensional representation. Alternatively, if one is using combinatorial chemistry in a lead discovery effort, then one may want to explore as diverse a set of potential ligands as possible with a given number of assays. This leads to the concept of chemical diversity space and, if one is not careful, to a diversity fetish.

One relevant concern has been to prioritize the order of screening, or to decide which compound libraries to purchase for screening. One approach that has been used relies on the complementary concepts of diversity and similarity. Given two compounds, how do you quantitate how divergent the two structures are? One major problem is the choice of a relevant metric, what parameters are considered, how are the parameters scaled, and so on. Similarity, like beauty, is clearly in the eye of the beholder. There is no generally relevant set of parameters to explain all observations and one should expect that a given subset of parameters will be more relevant to one problem than to another. It should be pointed out that one is focused on properties of molecules in the absence of the receptor in contrast to the detailed focus on the complex in drug design studies. Many approaches to similarity fail to even consider chirality, a common discriminator of receptors. For a recent overview of the current status of virtual screening in lead discovery, see the review by Oprea and Matter [82].

1.16
Prediction of ADME

Methods for estimating the molecular properties that correlate with ADME problems have also been a very active arena for chemoinformatics. By studying the isozymes of cytochrome P450 enzymes, for example, certain molecular signatures for metabolic stability can be discerned. In a similar way, properties such as lipohilicity, pKa, number of hydrogen-bond donors and acceptors, and so on, correlate with oral bioavailability. For any drug development effort, oral bioavailability is often a requirement to compete in the marketplace with drugs already available. For a bioactive compound to succeed as a drug, it must pass many selective filters during development (toxicity, etc.) as well as in the body including metabolism, uptake, excretion, and so on. The most potent HIV protease inhibitor prepared in my lab was also an excellent substrate for excretion by the liver in first pass; thus high levels were found in the portal circulation, but nowhere else.

1.17
Failures to Accurately Predict

Why do we still have difficulty in developing useful predictive models? Where are the sources of noise? From the point of view of molecular modeling and computational chemistry, the potential functions in common use have intrinsic significant errors in electrostatics. Estimating the entropy of binding is complex unless one is willing to sample solvent configurations sufficiently to adequately represent the partition function. Solvation models such as GB/SA [83] are certainly better than ignoring the significant impact that desolvation has on energetics. Multiple binding modes are not uncommon, but often too difficult to handle while modeling. The normal assumption of rigid receptor sites or, at the very least, limited exploration of the dynamics of the structure seen in the crystal is inherently dangerous. An excellent demonstration of the fallacy of assuming stability of receptor structure comes from the work of Don Abraham, where a compound designed to bind to an allosteric site on hemoglobin actually displaces core hydrophobic residues to optimize its interactions [84], a story which is detailed in Chapter 17 [85]. Receptors, at least GPCRs, have multiple conformations and probably different modes of activation and coupling with different G-proteins. The role of dimerization of GPCRs has only recently been shown to be important for a variety of receptors. In summary, we routinely apply Occam's razor for convenience, or as a rough approximation where we hope the results will withstand scrutiny by comparison with experimental results. The reality of biological systems is that Mother Nature never shaved with Occam's razor, and we cannot expect significant signal-to-noise from systems that have not been calibrated with the tools we are applying. If one is not using accurate *ab initio* methods, then one must remember the old dictum; to extrapolate is human, to interpolate is correct, but only within a relevant data set.

1.18
Summary

Chemoinformatics is the science of determining those important aspects of molecular structures related to desirable properties for some given function. One can contrast the atomic level concerns of drug design where interaction with another molecule is of primary importance with the set of physical attributes related to ADME, for example. In the latter case, interaction with a variety of macromolecules provides a set of molecular filters that can average out specific geometrical details and allows significant models developed by consideration of molecular properties alone.

Acknowledgments

It should be clear from the text that the author has benefited from the efforts of excellent (no, exceptional) collaborators and students. He has benefited immensely from a variety of mentors over the years, who have been both encouraging and critical regarding the development and application of computer-aided drug design. Sitting at the interface between the revolution in microelectronics, where CPUs are now a commodity, and the revolution in molecular biology and genomics, which provides a plethora of therapeutic targets and interesting conundrums to consider, has been both exciting and humbling. If the past is a preface to the future, get set for an exhilarating ride.

References

1 BROWN, F. Chemoinformatics: What is it and How does it impact drug discovery. *Annu. Rep. Med. Chem.* **1998**, *33*, 375–384.

2 HANN, M., GREEN, R. Chemoinformatics – a new name for an old problem? *Curr. Opin. Chem. Biol.* **1999**, *3*, 379–383.

3 RITCHIE, T. Chemoinformatics; manipulating chemical information to facilitate decision-making in drug discovery. *Drug Discovery Today* **2001**, *6*(16), 813–814.

4 JOHNSON, D.E., BLOWER, P.E., Jr, MYATT, G.J., WOLFGANG, G.H. Chem-tox informatics: data mining using a medicinal chemistry building block approach. *Curr. Opin. Drug Disc. Dev.* **2001**, *4*(1), 92–101.

5 OPREA, T.I., GOTTFRIES, J. Chemography: the art of navigating in chemical space. *J. Comb. Chem.* **2001**, *3*, 157–166.

6 OPREA, T.I. Chemical space navigation in lead discovery. *Curr. Opin. Chem. Biol.* **2002**, *6*(3), 384–389.

7 LEACH, A.R., GILLET, V.J. *An Introduction to Chemoinformatics.* Kluwer Academic Publishers, Dordrecht, Netherlands, **2003**, 259.

8 HO, C.M., MARSHALL, G.R. DBMAKER: a set of programs to generate three-dimensional databases based upon user-specified criteria. *J. Comput.-Aided Mol. Des.* **1995**, *9*(1), 65–86.

9 PEARLMAN, R.S. *CONCORD User's Manual.* Tripos Associates, St. Louis, MO, **1992**.

10 ALLEN, F.H., MOTHERWELL, W.D. Applications of the Cambridge structural database in organic chemistry and crystal chemistry. *Acta Crystallogr, Sect B* **2002**, *58*(Pt 3 Pt 1), 407–422.

11 ALLEN, F.H., DAVIES, J.E., GALLOY, J.J., JOHNSON, O., KENNARD, O., MACREA, C.F., MITCHELL, E.M., MITCHELL, G.F., SMITH, J.M., WATSON, D.G. The developments of versions 3 and 4 of the Cambridge database system. *J. Chem. Inf. Comput. Sci.* **1991**, *31*, 187–204.

12 ABOLA, E.E., BERNSTEIN, F.C., KOETZLE, T.F. The protein data bank. In *The Role of Data in Scientific Progress*, GLAESER, P.S. (ed.) Elsevier, New York, **1985**.

13 BARRY, C.D., ELLIS, R.A., GRAESSER, S., MARSHALL, G.R. Display and manipulation in three dimensions. In *Pertinent Concepts in Computer Graphics*, FAIMAN, M. NIEVERGELT, J. (Eds). University of Illinois Press, Chicago, IL, **1969**, 104–153.

14 FUJITA, T., IWASA, J., HANSCH, C. A new substituent constant, π, derived from partition coefficients. *J. Am. Chem. Soc.* **1964**, *86*(December 5), 5175–5180.

15 HANSCH, C., LEO, A. *Substituent Constants for Correlation Analysis in Chemistry and Biology.* Wiley & Sons, New York, **1979**.

16 GOODFORD, P.J. Drug design by the method of receptor fit. *J. Med. Chem.* **1984**, *27*(5), 557–564.

17 RILEY, D.P. Rational design of synthetic enzymes and their potential utility as human pharmaceuticals: development of manganese(II)-based superoxide dismutase mimics. *Adv. Supramol. Chem.* **2000**, *6*, 217–244.

18 RILEY, D.P., HENKE, S.L., LENNON, P.J., ASTON, K. Computer-aided design (CAD) of synzymes: use of molecular mechanics (MM) for the rational design of super-oxide dismutase mimics. *Inorg. Chem.* **1999**, *38*(8), 1908–1917.

19 MOTOC, I., MARSHALL, G.R., LABANOWSKI, J. Molecular shape descriptors. 3. Steric mapping of biological receptor. *Z. Naturforsch.* **1985**, *40a*, 1121–1127.

20 MOTOC, I., MARSHALL, G.R. Molecular shape descriptors. 2. Quantitative structure-activity relationships based upon three-dimensional molecular shape descriptor. *Z. Naturforsch.* **1985**, *40a*, 1114–1120.

21 MOTOC, I., MARSHALL, G.R., DAMMKOEHLER, R.A., LABANOWSKI, J. Molecular shape descriptors. 1. Three-dimensional molecular shape descriptor. *Z. Naturforsch.* **1985**, *40a*, 1108–1113.

22 BECKETT, A.H., CASEY, A.F. Synthetic analgesics: stereochemical considerations. *J. Pharm. Pharmacol.* **1954**, *6*, 986–999.

23 KIER, L.B., ALDRICH, H.S. A theoretical study of receptor site models for trimethylammonium group interactions. *J. Theor. Biol.* **1974**, *46*, 529–541.

24 GUND, P., WIPKE, W.T., LANGRIDGE, R. Computer searching of a molecular structure file for pharmacophoric patterns. *Comput. Chem. Res. Educ. Technol.* **1974**, *3*, 5–21.

25 GUND, P. Three-dimensional pharmacophoric pattern searching. *Prog. Mol. Subcell. Biol.* **1977**, *11*, 117–143.

26 HUMBER, L.G., BRUDERLIN, F.T., PHILIPP, A.H., GOTZ, M., VOITH, K. Mapping the dopamine receptor. 1. Features derived from modifications in ring E of the neuroleptic butaclamol. *J. Med. Chem.* **1979**, *22*, 761–767.

27 MARSHALL, G.R., BARRY, C.D., BOSSHARD, H.E., DAMMKOEHLER, R.A., DUNN, D.A. The conformational parameter in drug design: the active analog approach. In *Computer-Assisted Drug Design*, OLSON, E.C. CHRISTOFFERSEN, R.E. (eds). American Chemical Society, Washington, DC, **1979**, 205–226.

28 MARSHALL, G.R., BARRY, C.D. Functional representation of molecular volume for computer-aided drug design. *Abstr. Amer. Cryst. Assoc., Honolulu, Hawaii* **1979**.

29 SUFRIN, J.R., DUNN, D.A., MARSHALL, G.R. Steric mapping of the L-methionine binding site of ATP: L-methionine S-adenosyltransferase. *Mol. Pharmacol.* **1981**, *19*, 307–313.

30 WEAVER, D.C., BARRY, C.D., MCDANIEL, M.L., MARSHALL, G.R., LACY, P.E. Molecular requirements for recognition at glucoreceptor for insulin release. *Mol. Pharmacol.* **1979**, *16*(2), 361–368.

31 KLUNK, W.E., KALMAN, B.L., FERRENDELLI, J.A., COVEY, D.F. Computer-assisted modeling of the picrotoxinin and γ-butyrolactone receptor site. *Mol. Pharmacol.* **1982**, *23*, 511–518.

32 RUTLEDGE, L.D., PERLMAN, J.H., GERSHENGORN, M.C., MARSHALL, G.R., MOELLER, K.D. Conformationally restricted TRH analogs: a probe for the pyroglutamate region. *J. Med. Chem.* **1996**, *39*(8), 1571–1574.

33 SIMPSON, J.C., HO, C.M.C., SHANDS, E.F.B., GERSHENGORN, M.C., MARSHALL, G.R., MOELLER, K.D. Conformationally restricted TRH analogs: constraining the pyroglutamate region. *Bioorg. Med. Chem.* **2002**, *10*, 291–302.

34 SLOMCZYNSKA, U., CHALMERS, D.K., CORNILLE, F., SMYTHE, M.L., BEUSEN, D.D., MOELLER, K.D., MARSHALL, G.R. Electrochemical cyclization of dipeptides to form novel bicyclic, reverse-turn peptidomimetics. 2. Synthesis and conformational analysis of 6,5-bicyclic systems. *J. Org. Chem.* **1996**, *61*(4), 1198–1204.

35 TONG, Y.S., OLCZAK, J., ZABROCKI, J., GERSHENGORN, M.C., MARSHALL, G.R., MOELLER, K.D. Constrained peptidomimetics for TRH: cis-peptide bond analogs. *Tetrahedron* **2000**, *56*(50), 9791–9800.

36 DEPRIEST, S.A., MAYER, D., NAYLOR, C.B., MARSHALL, G.R. 3D-QSAR of angiotensin-converting enzyme and thermolysin inhibitors: a comparison of CoMFA models based on deduced and experimentally determned active site geometries. *J. Am. Chem. Soc.* **1993**, *115*, 5372–5384.

37 ANDREWS, P.R., CARSON, J.M., CASELLI, A., SPARK, M.J., WOODS, R. Conformational analysis and active site modelling of angiotensin-converting enzyme inhibitors. *J. Med. Chem.* **1985**, *28*(3), 393–399.

38 MAYER, D., NAYLOR, C.B., MOTOC, I., MARSHALL, G.R. A unique geometry of the active site of angiotensin-converting enzyme consistent with structure-activity studies. *J. Comput.-Aided Mol. Des.* **1987**, *1*(1), 3–16.

39 NATESH, R., SCHWAGER, S.L., STURROCK, E.D., ACHARYA, K.R. Crystal structure of the human angiotensin-converting enzyme-lisinopril complex. *Nature* **2003**, *421*(6922), 551–554.

40 MARSHALL, G.R., CRAMER III, R.D. Three-dimensional structure-activity relationships. *Trends Pharmacol. Sci.* **1988**, *9*, 285–289.

41 CRAMER III, R.D., MILNE, M. The lattice model: a general paradigm for shape-related structure/activity correlation. In *19th National Meeting of the American Chemical Society*. American Chemical Society, Washington, DC, **1979**, COMP 44.

42 CRAMER III, R.D., PATTERSON, D.E., BUNCE, J.D. Comparative molecular field analysis (CoMFA). 1. Effect of shape on binding of steroids to carrier proteins. *J. Am. Chem. Soc.* **1988**, *110*(18), 5959–5967.

43 WOLD, S., ABANO, C., DUNN III, W.J., ESBENSEN, K., HELLBERG, S., JOHANSSON, E., LINDBERG, W., SJOSTROM, M. Modelling data tables by principal components and PLS: class patterns and quantitative predictive relations. *Analysis,* **1984**, *12*(10), 477–485.

44 CRAMER III, R.D., BUNCE, J., PATTERSON, D., FRANK, I. Crossvalidation, boot-strapping, and partial least squares compared with multiple regression in conventional QSAR studies. *Quantum Struct. Act. Relat.* **1988**, *7*, 18–25.

45 CRAMER, I., RICHARD, D. BC(DEF) parameters. 1. The intrinsic dimensionality of intermolecular interactions in the liquid state. *J. Am. Chem. Soc.* **1980**, *10*(6), 1837–1849.

46 MARSHALL, G.R., ARIMOTO, R., RAGNO, R., HEAD, R.D. Predicting affinity: the sina qua non of activity. *Abstr. Pap. Am. Chem. Soc.* **2000**, *219*, 056–COMP.

47 HEAD, R.D., SMYTHE, M.L., OPREA, T.I., WALLER, C.L., GREEN, S.M., MARSHALL, G.R. Validate – a new method for the receptor-based prediction of binding affinities of novel ligands. *J. Am. Chem. Soc.* **1996**, *118*(16), 3959–3969.

48 ZWANZIG, R., SZABO, A., BAGCHI, B. Levinthal's paradox. *Proc. Natl. Acad. Sci. U.S.A.* **1992**, *89*, 20–22.

49 KARPLUS, M. The Levinthal paradox: yesterday and today. *Fold. Des.* **1997**, *2*(4), S69–S75.

50 SIMONS, K.T., STRAUSS, C., BAKER, D. Prospects for ab initio protein structural genomics. *J. Mol. Biol.* **2001**, 1191–1199.

51 BYSTROFF, C., BAKER, D. Prediction of local structure in proteins using a library of sequence-structure motifs. *J. Mol. Biol.* **1998**, *281*(3), 565–577.

52 SIMONS, K.T., BONNEAU, R., RUCZINSKI, I., BAKER, D. Ab initio protein structure prediction of CASP III targets using ROSETTA. *Proteins* **1999**, 171–176.

53 KUHLMAN, B., DANTAS, G., IRETON, G.C., VARANI, G., STODDARD, B.L., BAKER, D. Design of a novel globular protein fold with atomic-level accuracy. *Science* **2003**, *302*(5649), 1364–1368.

54 GALAKTIONOV, S., NIKIFOROVICH, G.V., MARSHALL, G.R. Ab initio modeling of small, medium, and large loops in proteins. *Biopolymers* **2001**, *60*(2), 153–168.

55 GALAKTIONOV, S.G., MARSHALL, G.R. Properties of intraglobular contacts in proteins: an approach to prediction of tertiary structure. *Proceedings of the 27th Hawaii International Conference on System Sciences.* IEEE Computational Society, Washington, DC, **1994**, 326–335.

56 BERGLUND, A., HEAD, R.D., WELSH, E., MARSHALL, G.R. ProVal: a protein scoring function for the selection of native and near-native folds. *Proteins: Struct., Funct., Bioinf.* **2004**, *54*, 289–302.

57 JONES, D.T., TAYLOR, W.R., THORNTON, J.M. A new approach to protein fold recognition. *Nature* **1992**, *358*, 86–89.

58 KUNTZ, I.D., BLANEY, J.M., OATLEY, S.J., LANGRIDGE, R., FERRIN, T.E. A Geometric Approach to Macromolecule-Ligand Interactions. *J. Mol. Biol.* **1982**, *161*, 269.

59 GOODSELL, D.S., LAUBLE, H., STOUT, C.D., OLSON, A.J. Automated docking in crystallography: analysis of the substrates of aconitase. *Proteins: Struct., Funct., Genet.* **1993**, *17*, 1–10.

60 MORRIS, G.M., GOODSELL, D.S., HALLIDAY, R.S., HUEY, R., HART, W.E., BELEW, R.K., OLSON, A.J. Automated docking using a Lamarckian genetic algorithm and an empirical binding free energy function. *J. Comput. Chem.* **1998**, *19*(14), 1639–1662.

61 MAKINO, S., EWING, T.J.A., KUNTZ, I.D. DREAM++: Flexible docking program for virtual combinatorial libraries. *J. Comput.-Aided Mol. Des.* **1999**, *13*(5), 513–532.

62 SHOICHET, B.K., LEACH, A.R., KUNTZ, I.D. Ligand solvation in molecular docking. *Proteins* **1999**, *34*(1), 4–16.

63 LAMB, M.L., BURDICK, K.W., TOBA, S., YOUNG, M.M., SKILLMAN, K.G., ZOU, X.Q., ARNOLD, J.R., KUNTZ, I.D. Design, docking, and evaluation of multiple libraries against multiple targets.

Proteins: Struct., Funct., Genet. **2001**, 42(3), 296–318.

64 ABAGYAN, R., TOTROV, M. High-throughput docking for lead generation. *Curr. Opin. Chem. Biol.* **2001**, 5(4), 375–382.

65 CRIPPEN, G.M. Distance geometry and conformational calculations. In *Chemometrics Research Studies*, Vol. 1, BAWDEN, D. (ed.) John Wiley, Chichester, UK, **1981**.

66 GOODFORD, P.J. A computational procedure for determining energetically favorable binding sites on biologically important macromolecules. *J. Am. Chem. Soc.* **1985**, 28(7), 849–856.

67 RINGE, D. Structure-aided drug design: crystallography and computational approaches. *J. Nucl. Med.* **1995**, 36 (6 Suppl), 28S–30S.

68 RINGE, D., MATTOS, C. Analysis of the binding surfaces of proteins. *Med. Res. Rev.* **1999**, 19(4), 321–331.

69 SHUKER, S.B., HAJDUK, P.J., MEADOWS, R.P., FESIK, S.W. Discovering high-affinity ligands for proteins: SAR by NMR. *Science* **1996**, 274(5292), 1531–1534.

70 HO, C.M., MARSHALL, G.R. Cavity search: an algorithm for the isolation and display of cavity-like binding regions. *J. Comput.-Aided Mol. Des.* **1990**, 4(4), 337–354.

71 HO, C.M.W., MARSHALL, G.R. SPLICE – a program to assemble partial query solutions from three-dimensional database searches into novel ligands. *J. Comput.-Aided Mol. Des.* **1993**, 7(6), 623–647.

72 HO, C.M., MARSHALL, G.R. FOUNDATION: a program to retrieve all possible structures containing a user-defined minimum number of matching query elements from three-dimensional databases. *J. Comput.-Aided Mol. Des.* **1993**, 7(1), 3–22.

73 HO, C.M.W. In silico lead optimization. In *Chemoinformatics in Drug Discovery*, OPREA, T.I. (ed.) Wiley-VCH, Weinheim, **2004**, 199–219.

74 STOUCH, T.R., KENYON, J.R., JOHNSON, S.R., CHEN, X.-Q., DOWYKO, A., LI, Y. In silico ADME/Tox: why models fail.

J. Comput.-Aided Drug Des. **2003**, 17, 83–92.

75 OPREA, T.I., DAVIS, A.M., TEAGUE, S.J., LEESON, P.D. Is there a difference between leads and drugs? A historical perspective. *J. Chem. Inf. Comput. Sci.* **2001**, 1308–1315.

76 OPREA, T.I. Current trends in lead discovery: Are we looking for the apporpriate properties? *J. Comput.-Aided Drug Des.* **2002**, 16, 325–334.

77 KOPPAL, T. Pharmacology's resurrection. *Drug Disc. Dev.* **2003**, 6(11), 28–32.

78 MERRIFIELD, R.B. Solid phase peptide synthesis. I. The synthesis of a tetrapeptide. *J. Am. Chem. Soc.* **1963**, 2149–2154.

79 CARUTHERS, M.H. Gene synthesis machines: DNA chemistry and its uses. *Science* **1985**, 230(4723), 281–285.

80 LEZNOFF, C.C. The use of insoluble polymer supports on general organic synthesis. *Acc. Chem. Res.* **1978**, 11, 327–333.

81 BUNIN, B.A., ELLMAN, J.A. A general and expedient method for the solid-phase synthesis of 1,4-benzodiazepine derivatives. *J. Am. Chem. Soc.* **1992**, 114, 10 997–10 998.

82 OPREA, T.I., MATTER, H. Integrating virtual screening in lead discovery. *Curr. Opin. Chem. Biol.* **2004**, 8, 349–358.

83 STILL, W.C., TEMPCZYK, A., HAWLEY, R.C., HENDRICKSON, T. Semianalytical treatment of solvation for molecular mechanics and dynamics. *J. Am. Chem. Soc.* **1990**, 112(16), 6127–6129.

84 WIREKO, F.C., KELLOGG, G.E., ABRAHAM, D.J. Allosteric modifiers of hemoglobin. 2. Crystallographically determined binding-sites and hydrophobic binding interaction analysis of novel hemoglobin oxygen effectors. *J. Med. Chem.* **1991**, 34(2), 758–767.

85 ABRAHAM, D.J. Drug discovery in academia. In *Chemoinformatics in Drug Discovery*, OPREA, T.I. (ed.) Wiley-VCH, Weinheim, **2004**, 457–484.

Part I
Virtual Screening

Chemoinformatics in Drug Discovery. Edited by Tudor I. Oprea
Copyright © 2004 WILEY-VCH Verlag GmbH & Co. KGaA, Weinheim
ISBN: 3-527-30753-2

2
Chemoinformatics in Lead Discovery

Tudor I. Oprea

2.1
Chemoinformatics in the Context of Pharmaceutical Research

The vastness of chemical space cannot be better illustrated than by using David Weininger's example [1]: Considering all the derivatives of *n*-hexane, starting from a list of 150 substituents, and enumerating all the possibilities, from mono- to 14-substituted hexanes, one can reach over 10^{29} structures. Most of them might be synthetically inaccessible, but even a very small number of building blocks, for example, twenty natural amino acids in proteins and four nucleotides in DNA, can still give an infinite number of possibilities in living systems. Chemical space is limited only at the low-molecular weight end. Our methods to probe this space in the context of pharmaceutical research, and in particular for lead discovery, are being placed under scrutiny in the context of maximizing efficiency.

There has been a strong impetus to minimize R&D costs and the time lapse between an idea (e.g., target or lead identification) and the drug, and to maximize the rate of success of preclinical R&D projects. Noting the lack of congruence between *in vitro* and *in vivo* models and the diminishing interest in good clinical research, the relevance of modern pharmaceutical research has recently been questioned [2]. The need for clinical drug developers to guide drug discovery was recognized [3], starting with Alexander Pope's observation that "the proper study of mankind is man". Simply put, the major hurdle in understanding pharmacodynamics, pharmacokinetics and toxicology is to study the effects of drugs in man, and "a rigid policy of clinical drug development that relies heavily on preclinical testing would be considered irrational" [3].

Unfortunately, given the current regulations, there is little choice but to study, with a vast array of methods, the behavior of drugs in surrogate models, ranging from ligand–receptor *in vitro* assays to humanized (transgenic) animals, before a compound is progressed from preclinical status to Phase I. The real challenge of computer-assisted lead discovery includes the evaluation of receptor–ligand binding, of accessible conformational states for both ligand and receptor, of multiple binding modes, affinity versus selectivity versus efficacy, metabolic stability (site of reactivity and turnover), absorption, distribution and excretion, as well as low toxicity and no side effects, while in the same type seeking a favorable intellectual property position (see also Figure 2.1). These difficulties have been hardly surmountable by any single software to date.

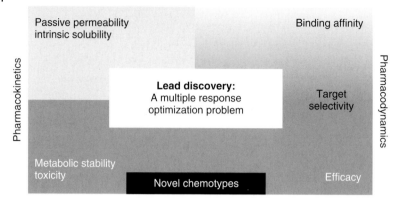

Fig. 2.1   Lead discovery requires an integrated approach to address a multiple response surface optimization problem: properties related to pharmacokinetics, toxicology and pharmacodynamics require optimization for novel chemotypes. The increased difficulty of finding such optima is suggested by darker backgrounds. Modified from [4] with permission.

The fact that lead discovery requires multiple simultaneous optimizations has been only recently recognized [5], which may explain the innovation deficit witnessed in the past decades [6]. Another reason this process is less than optimal relates to the lack of appropriate documentation with respect to decisions taken in the field of preclinical drug discovery: Why a particular chemotype was chosen, or rejected, and why certain chemical steps were taken during the selection and optimization of leads are decisions that are typically lost to posterity, since the pharmaceutical industry does not have institutional memory. Chris Lipinski has suggested a "debriefing" procedure for retiring scientists as a method to capture the invaluable information that is likely to vanish once experienced drug discovery scientists retire [7]. We have suggested chemoinformatics as the major tool to document the decision-making process [8] and perhaps to capture such information from scientists who leave a company.

Chemoinformatics integrates data via computer-assisted manipulation of chemical structures [9]. Chemical inventory and compound registration are vital to chemoinformatics, but it is their combination with other theoretical tools linked to physical (organic) chemistry, pharmacodynamics and pharmacokinetics that brings unique capabilities in the area of lead and drug discovery – thus setting the stage for multiple response optimization. While the goal of transforming data into information and information into knowledge [10] is chemoinformatics, we consider access to biological, and indeed chemical data, *but not the data themselves*, as part of chemoinformatics [8]. By the same token, several number-crunching activities traditionally associated with computational chemistry, for example, physical and chemical property calculations, are generating more numbers, rather than information. Not all computational chemistry is part of chemoinformatics; only the activities that assist decision-makers to interpret or to clarify the initial physicochemical, structural and biological data fall under this category. For example, virtual screening and quantitative structure-activity relationships

(QSARs) transform numbers (data) into models (information, even knowledge) and are, therefore, key activities in chemoinformatics.

Similar to the role envisioned for medicinal chemistry in the new millennium by Erhardt [11], chemoinformatics will play a major role in drug discovery in the next decades, as the central instrument for chemistry-related information. This information, however, is unlikely to be transformed into knowledge by medicinal chemists alone [11], as the interpretation basis for such decisions requires in-depth knowledge in several unrelated fields: physical organic chemistry (for solubility and passive transcellular permeability), medicinal and combinatorial chemistry (for synthetic prioritization and retrosynthetic analyses), process chemistry (reducing the number of synthetic steps), pharmacology and physiology (for pharmacodynamics and efficacy), and pharmacokinetics and toxicology (for Absorption, Distribution, Metabolism and Elimination, acute and chronic toxicity – collectively termed ADMET), to name disciplines illustrated in Figure 2.1. Although not enabling such competence *per se*, chemoinformatics is the computer-based tool primed to assist decision-makers in the lead discovery processes [8]. While apparently objective, its computational techniques are ultimately dependent on the (subjective) concepts and thinking that go into the models and assumptions onto which they were built. Thus, chemoinformatics provides a forum for the necessary dialogue between theoretical sciences, and experimental chemistry and biology.

2.2
Leads in the Drug Discovery Paradigm

The pharmaceutical industry is using the following (investigational) protocol for drug discovery [12]:

- Compounds that show activity during HTS are designated as *HTS hits*. These are subsequently retested to confirm activity and structure. The usual success rate for single dose is below 0.1%. Less than 1 in 1000 compounds from a randomly selected library are likely to be active in a primary HTS run on a random target. False positives [13, 14] have been shown to interfere with the biological assay (e.g., chemically reactive species, dyes or fluorescent compounds), being responsible for the low success rate. "Frequent hitters" [15], compounds that sometimes aggregate forming particles with 20 to 400 nm in diameter, can perhaps be filtered electronically [16, 17].
- Secondary screening provides the experimental means to weed out HTS hits, and provide *HTS actives* or *validated hits*, that is, compounds with known structure and activity (at this stage, dose-response curves are standard).
- If multiple *HTS actives* of the same chemical family are to some extent leadlike *(vide infra)*, they become a *lead series*, that is, structures that are amenable for further chemistry optimization and exhibit favorable patent situation [18, 19]. Lead series are further queried and derivatized in order to derive analogs with the appropriate blend of pharmacodynamic and pharmacokinetic properties.

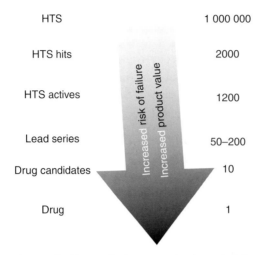

HTS	1 000 000
HTS hits	2000
HTS actives	1200
Lead series	50–200
Drug candidates	10
Drug	1

Fig. 2.2 Attrition rate in the current pharmaceutical discovery protocol. These estimates are for illustrative purposes; exact figures are difficult to average across the industry, as most companies do not disclose screening success/failure rates. Modified from [12] with permission.

When all criteria are met, the optimized lead becomes a *drug candidate*. Typically, one in 100 000 *HTS actives* reaches this stage [20]. Following approval from regulatory authorities, drug candidates proceed to clinical trials. If all three phases of clinical trials prove successful, which only one in 10 candidate drugs [21] does, the compound is approved and becomes a *launched* (or marketed) *drug*.

One in a million compounds from those initially submitted to HTS has, therefore, the probability to reach the launched drug status [12], assuming no target-bias or high-quality leads were available at the onset of the project – see also Figure 2.2. The progression HTS hits → HTS actives → lead series → drug candidate → launched drug has shifted the focus from good-quality candidate drugs, to *good-quality leads*. In fact, a comprehensive survey of the drugs launched in 2000 [22] shows that, for the vast majority of the compounds, extremely small changes were introduced to the lead structure before they reached the market.

The need to evaluate, at the lead discovery stage, permeability and solubility has been emphasized by Lipinski [23], who analyzed the trends in two sets of compounds from Merck and Pfizer. He concluded that the "rational design approach" at Merck leads to clinical candidates with poorer permeability, while the "HTS approach" at Pfizer results in clinical candidates with poorer solubility. This prompted us to survey the high-activity molecules (HAM, better than 1 nM) compounds published in the *Journal of Medicinal Chemistry* between 1997 and 1998 [12] with respect to MW (molecular weight), water/octanol partition (ELogP [24]) and intrinsic aqueous solubility (ELogSw [25]). We found that over 50% of HAMs are either large (MW $\geq$ 425), or hydrophobic (ELogP $\geq$ 4.25), or poorly soluble (ELogSw $\leq$ −4.75).

2.3
Is There a Trend for High Activity Molecules?

In an effort to understand if HAMs are, in any way, different from LAMs (low activity molecules), we extended this survey to compounds published between 1991 and 2002, as indexed in WOMBAT [26]. This database [27] contains 4927 unique structures with at least one measured activity better than 1 nM (HAMs), and 34 028 unique structures with at least one activity less than 1 µM (LAMs). Between HAMs and LAMs, 1080 molecules are common, that is, they have at least one activity above 1 nM and at least one activity below 1 µM. This is not uncommon for, for example, highly selective molecules. We did not exclude these from either set since we monitor trends, not exact figures. We studied these trends using 2-D-(two-dimensional) descriptors, that is, descriptors that do not use information related to the three-dimensional characteristics of model compounds. These descriptors can be classified as follows:

- *Size-related*: molecular weight (MW); calculated [28] molecular refractivity (CMR); the number of heavy (non-hydrogen) atoms (N_{at});
- *Complexity-related*: counts for rotatable (RTB) and rigid bonds (RGB), and the number of rings (RNG) [29];
- *Hydrophobicity-related*: the logarithm of the octanol–water partition coefficient, LogP [30], included in ClogP [31] and ELogP [24]; the logarithm of the (molar) aqueous solubility [32, 33], included in ELogSw [25]; the nonpolar surface area, both in Å^2 (NPSA) and normalized to the total surface area (%NPSA);
- *Descriptors related to electronic effects*: counts of the positive (POS) and negative (NEG) ionic centers, based on fixed substructure searches (i.e., without predicting pK_ϕ);
- *Hydrogen bonding descriptors*: the counts of hydrogen bond acceptors (HAC) or donors (HDO); the number of polar atoms, P_{at} [34]; the polar surface area in Å^2 (PSA) and normalized to the total surface area (%PSA);
- *The rule-of-five (RO5)*, a stepwise descriptor proposed by Lipinski et al. [35], based on ClogP, MW, HDO and HAC. The 90th percentile [35] for the distribution of MW ($\leq$500), LogP ($\leq$5), HDO ($\leq$5) and the sum of nitrogen and oxygen, accounting for hydrogen bond acceptors (HAC $\leq$ 10) defines the RO5 compliance, given 2245 compounds from the WDI, the World Drug Index [36], that had reached phase II clinical trials or higher. If any one of the four properties is higher than the above limits, RO5 = 1; if any two properties are higher, RO5 = 2, and so on. Any RO5 > 1 is likely to yield molecules that are not orally available [35].

The distribution of these properties (minimum, 25th and 75th percentile, median and maximum) is given in Table 2.1 for HAMs, and in Table 2.2 for LAMs. LAMs are smaller in size (MW, N_{at}), less hydrophobic (ClogP, ELogP), less complex (RNG, RTB, RGB), and better soluble (ELogSw). This trend is not statistically significant since SD (the standard deviation) is higher than the difference in each property. The trend is, however, consistent since it is supported by other descriptors: CMR for size, NPSA for hydrophobicity, PSA, %PSA and P_{at} for solubility. It is significant, however, that over 25% of HAMs, or 1500 molecules (as opposed to 10% of LAMs – data not shown) have

Tab. 2.1 The property distribution and standard deviation for HAMs ($N = 4927$)

Property	MIN	25%	MEDIAN	75%	MAX	SD
MW	111.1	364.5	456.7	574.7	1228.4	172.0
N_{at}	5	26	32	41	85	12.2
CMR	2.8	10.0	12.5	15.5	33.0	4.6
RNG	0	3	4	5	13	1.3
RTB	0	5	8	11	42	6.0
RGB	1	19	24	29	60	7.9
P_{at}	1	5	7	10	34	4.4
POZ	0	0	1	2	7	0.9
NEG	0	0	0	0	6	0.6
HDO	0	1	2	3	23	2.8
HAC	0.0	2	3	5	18	2.6
ELogP	−9.86	2.37	3.84	5.28	18.91	2.5
ELogSw	−20.49	−7.43	−5.68	−4.21	0.95	2.5
ClogP	−9.84	2.54	3.99	5.29	20.43	2.4
RO5	0	0	1	2	3	0.9
PSA	0.6	51.3	89.5	132.3	579.4	76.7
NPSA	77.4	338.4	418.9	530.4	1159.4	151.9
%PSA	0.2	11.4	17.2	23.4	68.5	9.3
%NPSA	31.5	76.6	82.8	88.6	99.8	9.3

Tab. 2.2 The property distribution and standard deviation for LAMs ($N = 34\,028$)

Property	MIN	25%	MEDIAN	75%	MAX	SD
MW	69.1	297.4	374.3	465.6	1403.2	158.0
N_{at}	5	21	26	33	85	11.2
CMR	1.7	8.2	10.3	12.8	33.7	4.2
RNG	0	2	3	4	12	1.4
RTB	0	4	6	9	43	5.6
RGB	0	15	19	24	64	7.6
P_{at}	0	4	6	8	35	4.1
POZ	0	0	1	1	8	0.9
NEG	0	0	0	0	6	0.6
HDO	0	1	2	3	23	2.6
HAC	0	1	3	4	23	2.5
ELogP	−11.40	1.68	3.25	4.79	16.94	2.6
ELogSw	−20.93	−6.36	−4.68	−3.11	1.16	2.5
ClogP	−15.51	1.67	3.25	4.72	18.50	2.6
RO5	0	0	0	1	4	0.8
PSA	0.0	47.3	75.3	114.3	623.4	72.1
NPSA	9.7	260.9	343.0	436.4	1341.5	146.6
%PSA	0.0	12.2	18.5	26.4	90.9	11.0
%NPSA	9.1	73.6	81.5	87.8	100.0	11.0

two or more RO5 violations: The active molecules in this subset are difficult to optimize for oral availability [23, 35]. Out of 1500 structures, 336 have three or more peptide units. Thus, the majority of these large, polar or hydrophobic compounds are in fact affinity-optimized, typical medicinal chemistry molecules.

To further investigate the qualities of small, but active compounds, we performed a query in WOMBAT [26] for molecules with activity better than 10 nM and MW $\leq$ 200. We found 192 unique structures, with 252 recorded activities on 46 targets (see also Figure 2.3). A vast majority of them (176 structures) are likely to be charged at pH 7.4, as they are aliphatic amines, amidines or guanidines, and sometimes carboxylic acids. This indicates that low MW compounds are likely to require salt bridge interactions

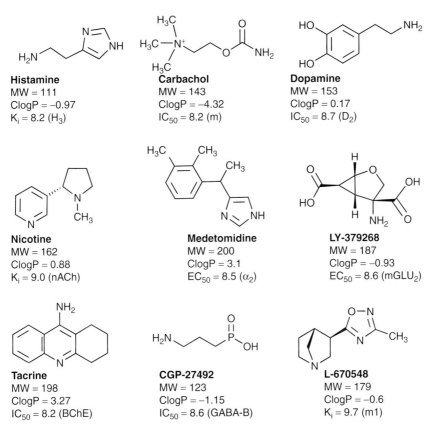

Histamine
MW = 111
ClogP = –0.97
K_i = 8.2 (H_3)

Carbachol
MW = 143
ClogP = –4.32
IC_{50} = 8.2 (m)

Dopamine
MW = 153
ClogP = 0.17
IC_{50} = 8.7 (D_2)

Nicotine
MW = 162
ClogP = 0.88
K_i = 9.0 (nACh)

Medetomidine
MW = 200
ClogP = 3.1
EC_{50} = 8.5 (α_2)

LY-379268
MW = 187
ClogP = –0.93
EC_{50} = 8.6 (mGLU$_2$)

Tacrine
MW = 198
ClogP = 3.27
IC_{50} = 8.2 (BChE)

CGP-27492
MW = 123
ClogP = –1.15
IC_{50} = 8.6 (GABA-B)

L-670548
MW = 179
ClogP = –0.6
K_i = 9.7 (m1)

Fig. 2.3 Examples of small molecules (MW $\leq$ 200) that have biological activity better than 10 nM. Under each molecule, the following information is included: molecule name, MW, ClogP, the biological activity type, value and target. Target names are as follows: H_3 – histamine receptor subtype 3; m – muscarine receptor; D_2 – dopaminergic receptor type 2; nACh – nicotine receptor; α_2 – alpha adrenergic receptor subtype 2; mGLU$_2$ – metabotropic glutamate receptor subtype 2; BChE – butyryl choline esterase; GABA-B – gamma-amino butyric acid receptor subtype B; m1 – muscarinic receptor subtype 1.

with the receptor in order to achieve high activity. Andrews et al. made a similar observation when evaluating the "average intrinsic binding energy" of drugs, and found that coefficients for charged atom types are approximately 10 times higher when compared with the coefficients for neutral atom types based on the 200 molecules that were investigated [37].

The "average intrinsic binding energy" [37], an empirical scheme better known as the Andrews binding energy (ABE), is used today to estimate the bioactivity potential of chemical libraries [38, 39]. When tested on $N = 55\,658$ activities better than $0.1\,\mu M$ from WOMBAT [26], ABE parallels the increase in MW ($R^2 = 0.72$), but not the increase in activity ($R^2 = 0.01$). However, as MW increases from 150 to, for example, 1200 Da, the affinity of ligands can increase from, for example, $1\,\mu M$ to $1\,pM$, or between 8.5 and $17\,kcal\,mol^{-1}$. Thus, while MW may increase ninefold, the free energy of binding can only double. Kuntz et al. surveyed the strongest-binding ligands for a large number of targets and concluded that binding affinity improves very little once the number of heavy atoms increases above 15 [40]. Therefore, because ABE makes no correction for higher MW, its use to evaluate chemical libraries should perhaps be restricted to small molecules (MW < 300), where the presence of charged moieties can indeed improve activity, as illustrated in Figure 2.3.

While not convincing from a statistical perspective, the results in this section are consistent with a trend: high-activity molecules published in the past decade of medicinal chemistry literature are more likely to be found in the large, hydrophobic and poor solubility corner of chemical property space. These results are not consistent with, for example, cell-based [41] and median-based [42] partitioning of biologically active compounds; however, such analyses were performed in the presence of "inactive" compounds selected from MDDR [41] or ACD [42], with quite probably unrelated chemotypes. ACD, the Available Chemicals Directory [43], and MDDR, the MDL Drug Data Report [43], are databases commonly used by the pharmaceutical industry.

The results presented above were obtained from a set of chemically related compounds, where the only discriminating factor is biological activity. The introduction of activity as the Y variable in our study might provide better insight into the particular molecular recognition features of each target. Our aim was, however, to look for features that distinguish active compounds as a group from inactive analogs, and to understand what qualities are present in HAMs from a lead discovery perspective, as discussed below.

2.4
The Concept of Leadlikeness

The concept of leadlikeness is central to drug discovery [44, 45]. As discussed earlier, the pharmaceutical industry, despite its century-old history, has done a poor job in documenting the decision process, for example, why certain chemical steps were followed to reach a particular compound. This turns out to be a particularly important question today, since the industry is under pressure to reduce costs, increase productivity and provide high-quality leads in early preclinical discovery. Chemical aspects of the history of drug discovery are discussed in the three-volume series *Chronicles in Drug Discovery* [46–48], in Walter Sneader's *Drug Prototypes and their Exploitation* [19] and in *Integration of Pharmaceutical Discovery and Development: Case Histories* [49]. These accounts, while rich in historical chemical information, do not focus explicitly on the choice of lead structures in drug discovery.

The first discussion of leads and their role in drug discovery [18] by DeStevens illustrates why, in the author's opinion, structured management does not work in the context of preclinical drug discovery – an observation still valid 15 years later [50]. DeStevens' argument was, basically, that scientists should concern themselves with science, and managers with management, and not *vice versa*. Such reasoning, however, does not stand the argument of economic pressure. One can only wonder what his reaction would have been to the financial forecast from Accenture [51], which indicated that, by 2008, pharmaceutical companies need to triple their success rate (i.e., novel drugs reaching the market annually) and reduce the preclinical drug discovery phase by 46%.

The importance of restricting small molecule synthesis in the ADMET-friendly property space [35] was introduced with the "rule-of-five", which significantly changed our perception regarding leads, lead properties and lead discovery. Many library design programs, based on combinatorial chemistry or compound acquisition, subsequently included RO5 filters. The existence of a druglike space was established [52, 53] recently thereafter (1998). This provided the ability to discriminate between "drugs" and "nondrugs" on the basis of chemical fingerprints, offering a computer-based discrimination between "drugs", as represented by WDI [36] or MDDR [43], and "nondrugs", represented by ACD [43]. Although this result was reproduced by other groups [54–56], it has yet to become accepted by the chemistry community as a decision-enabling scheme. If it was truly effective, it could assist chemists to quickly evaluate, for example, what other chemists have considered worthy of synthesis (and patenting) before them.

The problem is that good druglike scores do not make a molecule a drug; rather, they indicate that more of its features are encountered in other molecules from MDDR and WDI, and less of its features in ACD. It is often assumed that Lipinski's RO5 criteria define druglikeness, but the distinction between the "druglike property space", to which the "rule-of-five" applies [35], and the "druglike chemical space", defined by fingerprint discrimination models [52, 53], is often overlooked. More ACD compounds, that is, "nondrugs", are RO5 compliant, compared to MDDR compounds, or "drugs" [29]; thus, most ACD compounds are druglike in property space, but clearly not in chemical space. Vieth et al. [57] looked at the differences in the properties of drugs having a variety of routes of administration and confirmed that oral drugs have properties associated with lower molecular weight (MW), fewer hydrogen bond acceptors (HAC) and donors (HDO), and fewer rotatable bonds (RTB) compared to drugs that have other routes of administration. Despite this extension to RO5 criteria, there remains a gulf between these rules of thumb and true discriminating power for specific design purposes. It is therefore more appropriate to think of the RO5 type criteria as necessary but not sufficient to create an oral druglike molecule, and of the druglike scores as guidelines for filtering chemical libraries, not as tools to evaluate the quality of individual compounds.

Since they are complementary, the two filters should be combined during chemical library analysis.

Unlike the druglike scores, where large numbers of chemical structures have been submitted to statistical analyses, the leadlike concept [38] is based on significantly smaller datasets [22, 58, 59] that when combined amount to only 894 structures (*vide infra*). Despite this, the concept of leadlikeness is already having a significant impact in the design of chemical libraries [60]. This is, in part, because the concepts and methods related to leadlikeness are very intuitive and fit with the current experience of what typically happens [61] in lead optimization. On the basis of current data, it appears that, on the average, effective leads have lower molecular complexity [58], when compared to drugs, as well as a fewer number of rings (RNG) and rotatable bonds [59], have lower MW and are more polar [38], as detailed further.

The three datasets used [62] to examine leads [22, 58, 59] include 176 leads (that are not marketed drugs), 186 "leadrugs" (leads *and* marketed drugs) and 532 drugs (launched, but not recorded as leads). None of these categories is fixed in time: for example, Omeprazole and Bupivacaine, drugs in two of the analyses [58, 59], have become leads in Proudfoot's analysis (their enantiopure equivalents are now marketed as drugs). In this analysis, they are categorized as "leadrugs". Sometimes drugs are discontinued, for example, Regonol™ (Pyridostigmine) [63] and Orlaam™ (Levomethadyl) [64], but there are also drugs that are voluntarily withdrawn from the market, for example, Baycol™ (Cerivastatin) [65], Redux™ (Dexfenfluramine) [66] and Pondimin™ (Fenfluramine) [66]. This illustrates the imperative of keeping such analyses up-to-date, in particular since chemists are inclined to point out that these results are based on "historical data" that bear no relevance in today's drug discovery environment.

Drugs are chemically related to leads and "leadrugs", as seen by analyzing the median of the properties summarized in Table 2.3. From the median values, we observed [62] that leads tend to be smaller than drugs (as confirmed by MW, N_{at} and CMR), less complex (lower RNG, RTB and RGB), more polar (PSA and %PSA are higher in leads), less hydrophobic (lower ClogP, ELogP, NPSA and %PSA) and more soluble (higher ELogSw). Not surprisingly, the "leadrug" category falls in between drugs and leads. The median values for positive (1) and negative (0) ionization centers, for the number of hydrogen bond acceptors (2) and for RO5 (0) did not change across all 3 columns (numbers in brackets), and were thus excluded from Table 2.3. There was no change in RO5, since this filter was designed to cover the 90th percentile of orally available drugs; however, we did not verify how many of the 532 drugs used here are orally available.

By applying the results from Table 2.3 to the work presented in Tables 2.1 and 2.2, we can further examine the proportion of high-activity and low-activity molecules with leadlike properties, that is, MW ≤ 460, LogP −4 to 4.2, ELogSw ≥ −5, RTB ≤ 10, RNG ≤ 4, HDO ≤ 5 and HAC ≤ 9 [62]. There are 21.7% HAMs and 41% LAMs that meet the computational leadlike criteria (Table 2.4). The probability of finding HAMs in WOMBAT is rather low in the leadlike space, and is higher outside the RO5-compliant space (see Hi/Lo ratios in Table 2.4). While this may not hold true across all chemical space, it does strengthen the conclusion that HAMs are "decorated" with

Tab. 2.3 The median property for leads, leadrugs, and drugs

Property	Leads	Leadrugs	Drugs
MW	295.3	284.9	335.7
N_{at}	21	21	24
CMR	8.2	8.3	9.4
RNG	2	3	3
RTB	4	4	5
RGB	15	16	18
P_{at}	5	4	5
HDO	1.5	1	1
ClogP	1.79	2.59	2.62
ELogP	1.99	2.65	2.75
ELogSw	−2.71	−3.19	−3.83
RO5	0	0	0
PSA	68.8	51.3	64
NPSA	257.2	280.9	323.1
%PSA	21.3	14.5	16.6
%NPSA	78.7	85.5	83.4

Source: Modified from [62] with permission.

hydrophobic moieties. In this manner, the amount of nonpolar (hydrophobic) surface area contact [67] with the receptor [68], and hence the affinity, can increase.

As pointed out by Kuntz et al. [40] and confirmed in our earlier work [12], higher MW does not warrant higher activity. We also showed that an increase in the NPSA is not linked to improved activity either [62]. Here, we show plots of activity versus ELogP on two receptors, 5-HT$_{1A}$ serotoninergic and μ-opioid, and two enzymes, dihydrofolate reductase (DHFR) and phosphodiesterase 4 (PDE4). They were selected for the relatively high number of structures available in WOMBAT. The plots illustrate, quite clearly, the lack of correlation between ELogP and biological activity, even at the LogP range (0 to 5) that is typical for drugs. The relationship between LogP and bioactivity was investigated at more detail by Corwin Hansch and coworkers [69], who observed that more than 60% of the 8500 biological QSAR equations stored in the Pomona College C-QSAR database [70] contain either the ClogP term (4614 equations) or the π [71] constant (784

Tab. 2.4 The proportion of HAMs and LAMs that are leadlike and RO5 compliant. The Hi/Lo ratio is obtained by dividing HAMs to LAMs and can be seen as a rough indicator of the probability of finding high-activity molecules in the chemical property space defined by leadlike and RO5 criteria, respectively

Distribution	HAMs	%HAMs	LAMs	%LAMs	Hi/Lo ratio
Leadlike	1070	21.72	13949	40.99	0.08
RO5 compliant	3622	73.51	29813	87.61	0.12
Outside RO5	1305	26.49	4215	12.39	0.31

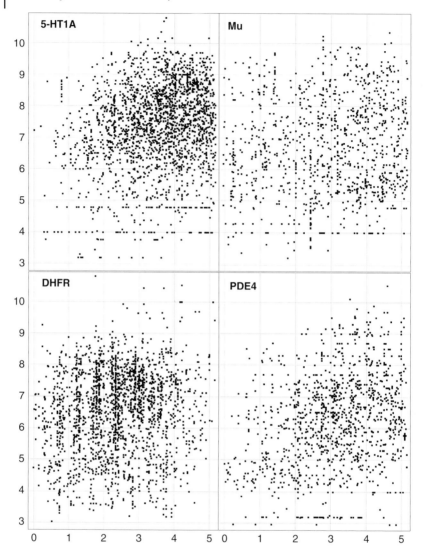

Fig. 2.4 Biological activity versus ELogP for four targets from WOMBAT: 5-HT$_{1A}$ ($N = 2306$), μ-opioid ($N = 1202$), dihydrofolate reductase (DHFR, $N = 2351$) and phosphodiesterase 4 (PDE4, $N = 1221$). The numbers indicate the amount of data points visible in each plot; the X-axis is ElogP, and the Y-axis combines K$_i$ and IC$_{50}$ values in inverse log$_{10}$ units.

equations). LogP is typically well correlated with bioactivity for congeneric series, but such correlations may not be as valid for a diverse set of compounds.

Our own evaluation [72] on the impact of hydrophobicity on biological activities examined the linear relationship between activity and the physicochemical property space defined by ELogP. From a total of 423 regressions on 14 022 activities from 162 different targets, correlating biological activity (in − log form) with ELogP, under 15% of the surveyed series had a significant relationship (i.e., $r^2 \geq 0.5$). Combined with the trends illustrated in Figure 2.4, the data indicate that transfer-related components (such as LogP) are important for biological activity, but understanding the molecular determinants responsible for biological activity may require additional descriptors [73] in the evaluation of high-quality leads.

The following computational criteria should be applied [74] for leadlike libraries: MW below 460 Da, LogP between −4 and 4.2, $LogS_w$ higher than −5,10 flexible bonds or less, 4 rings or less, 5 hydrogen bond donors or less, and 9 hydrogen bond acceptors or less. Additional experimental criteria [44], related to *in vivo* properties (e.g., in rat), should be applied individually to leadlike compounds: oral bioavailability above 30%, low clearance (e.g., below 30 mL min^{-1} kg^{-1}), $LogD_{7.4}$ (LogP at pH 7.4) between 0 and 3, poor (or no) inhibition of drug-metabolizing cytochrome P450 isozymes, plasma protein binding below 99.5%, lack of acute toxicity at the expected therapeutic window (e.g., assuming 500 mg day^{-1} P.O. regimen for 7 days), as well as poor binding to hERG. hERG is a K^+ channel implicated in sudden cardiac death and possibly responsible for the cardiac toxicity of a wide range of compounds [75], now withdrawn from the market. The experimental criteria should be applied to compounds progressed from the lead identification to the lead optimization stage.

2.5
Conclusions

Given the property distribution of existing high-activity molecules in WOMBAT [27], a significant percentage (78.28%) of them do not meet the computational criteria for leadlikeness, even though most of them are RO5 compliant. The probability of finding high-activity molecules is somewhat lower in the leadlike space, and increases fourfold outside the RO5-compliant space. Only one in five high-activity molecules matches the leadlike criteria. Perhaps not applicable in a true HTS setting [76], this points out that the intersection between leadlikeness and high activity is not negligible. The majority of high-activity molecules from WOMBAT are large, hydrophobic, and poorly soluble, even though higher MW and higher hydrophobicity are not paralleled by an increase in bioactivity. From the perspective of leads for the drugs launched in 2000 [22], it appears more and more difficult to "declare" a lead, unless one is completely confident that this molecule shows high activity, has chemical features amenable for optimization, has membership to a well-established lead series, has a favorable patent position and good pharmacokinetic properties. In addition, several experimental criteria related to, for example, bioavailability and hERG binding must be observed [44], even though the relevance of animal models to studies in man has been questioned [2, 3]. Computational models for these properties can also be applied at a much earlier stage, during virtual screening [4]. However, as outlined in Figure 2.1, we need more accurate *in silico* and experimental methods that allow for the simultaneous optimization of pharmacodynamic, pharmacokinetic and intellectual property criteria.

Acknowledgments

This work was supported by New Mexico Tobacco Settlement funds. The author thanks Prof. Hugo Kubinyi (Heidelberg, Germany) and Dr. Marius Olah (UNM) for suggestions.

References

1 WEININGER, D. Combinatorics of small molecular structures. In *Encyclopedia of Computational Chemistry*, Vol. 1, VON RAGUÉ SCHLEYER, P., ALLINGER, N.L., CLARK, T., GASTEIGER, J., KOLLMAN, P.A., SCHAEFER III, H.F. (eds.) Wiley, New York, **1998**, 425–430.

2 HORROBIN, D.F. Modern biomedical research: an internally self-consistent universe with little contact with medical reality? *Nat. Rev. Drug Disc.* **2003**, *2*, 151–154.

3 DEFELICE, S.L. *Drug Discovery – The Pending Crisis*. MEDCOM, New York, **1972**.

4 OPREA, T.I., MATTER, H. Integrating virtual screening in lead discovery. *Curr. Opin. Chem. Biol.* **2004**, *8*, 349–358.

5 OPREA, T.I. Virtual screening in lead discovery: a viewpoint. *Molecules* **2002**, *7*, 51–62.

6 DREWS, J. Innovation deficit revisited: reflections on the productivity of pharmaceutical R&D. *Drug Discovery Today* **1998**, *3*, 491–494.

7 LIPINSKI, C.A. Technical and people disconnects hindering knowledge exchange between chemistry and biology. 227th ACS National Meeting, Anaheim, CA, **2004**, CINF-009.

8 OLSSON, T., OPREA, T.I. Cheminformatics: a tool for decision-makers in drug discovery. *Curr. Opin. Drug. Disc. Dev.* **2001**, *4*, 308–313.

9 BROWN, F. Chemoinformatics: what is it and how does it impact drug discovery. *Annu. Rep. Med. Chem.* **1998**, *33*, 375–384.

10 HAHN, M.M., GREEN, R. Chem-informatics – a new name for an old problem? *Curr. Opin. Chem. Biol.* **1999**, *3*, 379–383.

11 ERHARDT, W.P. Medicinal chemistry in the new millennium. A glance into the future. *Pure Appl. Chem.* **2002**, *74*, 703–785.

12 OPREA, T.I. Lead structure searching: are we looking for the appropriate properties? *J. Comput.-Aided Mol. Des.* **2002**, *16*, 325–334.

13 RISHTON, G.M. Reactive compounds and in vitro false positives in HTS. *Drug Discovery Today* **1997**, *2*, 382–384.

14 RISHTON, G.M. Nonleadlikeness and leadlikeness in biochemical screening. *Drug Discovery Today* **2003**, *8*, 86–96.

15 MCGOVERN, S.L., CASELLI, E., GRIGORIEFF, N., SHOICHET, B.K. A common mechanism underlying promiscuous inhibitors from virtual and high-throughput screening. *J. Med. Chem.* **2002**, *45*, 1712–1722.

16 ROCHE, O., SCHNEIDER, P., ZUEGGE, J., GUBA, W., KANSY, M., ALANINE, A., BLEICHER, K., DANEL, F., GUTKNECHT, E.M., ROGERS-EVANS, M., NEIDHART, W., STALDER, H., DILLON, M., SJOEGREN, E., FOTOUHI, N., GILLESPIE, P., GOODNOW, R., HARRIS, W., JONES, P., TANIGUCHI, M., TSUJII, S., VON SAAL, W., ZIMMERMANN, G., SCHNEIDER, G. Development of a virtual screening method for identification of "frequent hitters" in compound libraries. *J. Med. Chem.* **2002**, *45*, 137–142.

17 SEIDLER, J., MCGOVERN, S.L., DOMAN, T., SHOICHET, B.K. Identification and prediction of promiscuous aggregating inhibitors among known drugs. *J. Med. Chem.* **2003**, *46*, 4477–4486.

18 DESTEVENS, G. Serendipity and structured research in drug discovery. *Prog. Drug. Res.* **1986**, *30*, 189–203.

19 SNEADER, W. *Drug Prototypes and their Exploitation.* Wiley, Chichester, UK, **1996**.

20 BOYD, D.B. Progress in rational design of therapeutically interesting compounds. In *Rational Molecular Design in Drug Research*, LILJEFORS, T., JORGENSEN, F.S., KROGSGAARD-LARSEN, P. (eds.) Munksgaard, Copenhagen, **1998**, 15–29.

21 KENNEDY, T. Managing the drug discovery/development interface. *Drug Discov. Today* **1997**, *2*, 436–444.

22 PROUDFOOT, J.R. Drugs, leads, and drug-likeness: an analysis of some recently launched drugs. *Bioorg. Med. Chem. Lett.* **2002**, *12*, 1647–1650.

23 LIPINSKI, C.A. Drug-like properties and the causes of poor solubility and poor permeability. *J. Pharmacol. Toxicol. Methods* **2000**, *44*, 235–249.

24 MEYLAN, W.M., HOWARD, P.H. Atom/fragment contribution method for estimating octanol-water partition coefficients. *J. Pharm. Sci.* **1995**, *84*, 83–92; KOWWIN is available from http://www.epa.gov/oppt/exposure/docs/episuitedl.htm.

25 MEYLAN, W.M., HOWARD, P.H., BOETHLING, R.S. Improved method for estimating water solubility from octanol/water partition coefficient. *Environ. Toxicol. Chem.* **1996**, *15*, 100–106; WSKOWIN is available from http://www.epa.gov/oppt/exposure/docs/episuitedl.htm.

26 OLAH, M., MRACEC, M., OSTOPOVICI, L., RAD, R., BORA, A., HADARUGA, N., OLAH, I., BANDA, M., SIMON, Z., MRACEC, M., OPREA, T.I. WOMBAT: World of molecular bioactivity. In *Chemoinformatics in Drug Discovery*, OPREA, T.I. (ed.) Wiley-VCH, Weinheim, **2004**, 223–239.

27 WOMBAT 2004.1 is available from Sunset Molecular Discovery, Santa Fe, New Mexico, http://www.sunsetmolecular.com and from Metaphorics LLC, Santa Fe, New Mexico, **2004**, http://www.metaphorics.com.

28 LEO, A., WEININGER, D. CMR3. Available from Daylight Chemical Information Systems, Santa Fe, New Mexico, **2004**, http://www.daylight.com/

29 OPREA, T.I. Property distribution of drug-related chemical databases. *J. Comput.-Aided Mol. Des.* **2000**, *14*, 251–264.

30 LEO, A. Estimating LogPoct from structures. *Chem. Rev.* **1993**, *5*, 1281–1306.

31 LEO, A., WEININGER, D. ClogP 4.0. Available from Daylight Chemical Information Systems, Santa Fe, New Mexico, **2004**, http://www.daylight.com/

32 RAN, Y., JAIN, N., YALKOWSKY, S.H. Prediction of aqueous solubility of organic compounds by the general solubility equation (GSE). *J. Chem. Inf. Comput. Sci.* **2001**, *41*, 1208–1217.

33 LIVINGSTONE, D.J., FORD, M.G., HUUSKONEN, J.J., SALT, D.W. Simultaneous prediction of aqueous solubility and octanol/water partition coefficient based on descriptors derived from molecular structure. *J. Comput.-Aided Mol. Des.* **2001**, *15*, 741–752.

34 OPREA, T.I. Rapid estimation of hydrophobicity for virtual combinatorial library analysis. *SAR and QSAR Environ. Res.* **2001**, *12*, 129–141.

35 LIPINSKI, C.A., LOMBARDO, F., DOMINY, B.W., FEENEY, P.J. Experimental and computational approaches to estimate solubility and permeability in drug discovery and development settings. *Adv. Drug Deliv. Rev.* **1997**, *23*, 3–25.

36 WDI, the Derwent World Drug Index, is available from Daylight Chemical Information Systems, **2004**, http://www.daylight.com/

37 ANDREWS, P.R., CRAIK, D.J., MARTIN, J.L. Functional group contributions to drug-receptor interactions. *J. Med. Chem.* **1984**, *27*, 1648–1657.

38 TEAGUE, S.J., DAVIS, A.M., LEESON, P.D., OPREA, T.I. The design of leadlike combinatorial libraries. *Angew. Chem., Int. Ed.* **1999**, *38*, 3743–3748; German version: *Angew. Chem.* **1999**, *111*, 3962–3967.

39 GOODNOW, R.A., Jr, GILLESPIE, P., BLEICHER, K. Chemoinformatic tools for library design and the hit-to-lead process: a user's perspective. In *Chemoinformatics in Drug Discovery*, OPREA, T.I. (ed.) Wiley-VCH, Weinheim, **2004**, 381–435.

40 KUNTZ, I.D., CHEN, K., SHARP, K.A., KOLLMAN, P.A. The maximal affinity of ligands. *Proc. Natl. Acad. Sci. U.S.A.* **1999**, *96*, 9997–10002.

41 PEARLMAN, R.S., SMITH, K.M. Novel software tools for chemical diversity. *Perspect. Drug Disc. Des.* **1998**, *9–11*, 339–353.

42 GODDEN, J.W., XUE, L., BAJORATH, J. Classification of biologically active compounds by median partitioning. *J. Chem. Inf. Comput. Sci.* **2002**, *42*, 1263–1269.

43 MDDR and ACD are available from MDL Information Systems, http://www.mdli.com/products/finders/database_finder/ MDDR is developed in cooperation with Prous Science Publishers, **2004**, http://www.prous.com/index.html.

44 HANN, M.M., OPREA, T.I. Pursuing the leadlikeness concept in pharmaceutical research. *Curr. Opin. Chem. Biol.* **2004**, *8*, 255–263.

45 HANN, M.M., LEACH, A., GREEN, D.V.S. Computational chemistry, molecular complexity and screening set design. In *Chemoinformatics in Drug Discovery*, OPREA, T.I. (ed.) Wiley-VCH, Weinheim, **2004**, 43–57.

46 BINDRA, J.S., LEDNICER D. (eds.) *Chronicles of Drug Discovery*, Vol. 1. Wiley-Interscience, New York, **1982**.

47 BINDRA, J.S., LEDNICER D. (eds.) *Chronicles of Drug Discovery*, Vol. 2. Wiley-Interscience, New York, **1983**.

48 LEDNICER, D. (ed.) *Chronicles of Drug Discovery*, Vol. 3. ACS Publishers, Washington, DC, **1993**.

49 BORCHARDT, R.T., FREIDINGER, R.M., SAWYER, T.K., SMITH, P.L. (eds.) *Integration of Pharmaceutical Discovery and Development: Case Histories*. Plenum Press, New York, **1998**.

50 HORROBIN, D.F. Innovation in the pharmaceutical industry. *J. R. Soc. Med.* **2000**, *93*, 341–345.

51 ANON, Speed to value. Accenture Report, **2000**, 23.

52 AJAY, WALTERS, W.P., MURCKO, M.A. Can we learn to distinguish between "drug-like" and "nondrug-like" molecules? *J. Med. Chem.* **1998**, *41*, 3314–3324.

53 SADOWSKI, J., KUBINYI, H.A. Scoring scheme for discriminating between drugs and nondrugs. *J. Med. Chem.* **1998**, *41*, 3325–3329.

54 WAGENER, M., VAN GEERENSTEIN, V.J. Potential drugs and nondrugs: prediction and identification of important structural features. *J. Chem. Inf. Comput. Sci.* **2000**, *40*, 280–292.

55 FRIMURER, T.M., BYWATER, R., NAERUM, L., LAURITSEN, L.N., BRUNAK, S. Improving the odds in discriminating "druglike" from "nondruglike" compounds. *J. Chem. Inf. Comput. Sci.* **2000**, *40*, 1315–1324.

56 BRÜSTLE, M., BECK, B., SCHINDLER, T., KING, W., MITCHELL, T., CLARK, T. Descriptors, physical properties and drug-likeness. *J. Med. Chem.* **2002**, *45*, 3345–3355.

57 VIETH, M., SIEGEL, M.G., HIGGS, R.E., WATSON, I.A., ROBERTSON, D.H., SAVIN, K.A. Characteristic physical properties and structural fragments of marketed oral drugs. *J. Med. Chem.* **2004**, *47*, 224–232.

58 HANN, M.M., LEACH, A.R., HARPER, G. Molecular complexity and its impact on the probability of finding leads for drug discovery. *J. Chem. Inf. Comput. Sci.* **2001**, *41*, 856–864.

59 OPREA, T.I., DAVIS, A.M., TEAGUE, S.J., LEESON, P.D. Is there a difference between leads and drugs? A historical perspective. *J. Chem. Inf. Comput. Sci.* **2001**, *41*, 1308–1315.

60 DAVIS, A.M., TEAGUE, S.J., KLEYWEGT, G.J. Application and limitations of X-ray crystallographic data in structure-based ligand and drug design. *Angew. Chem., Int. Ed.* **2003**, *42*, 2718–2736; German edition: *Angew. Chem.* **2003**, *115*, 2822–2841.

61 KENAKIN, T. Predicting therapeutic value in the lead optimization phase of drug discovery. *Nat. Rev. Drug Disc.* **2003**, *2*, 429–438.

62 OPREA, T.I. Cheminformatics and the quest for leads in drug discovery. In *Handbook of Cheminformatics*, Vol. 4, GASTEIGER, J., ENGEL, T. (eds.) VCH-Wiley, New York, **2003**, 1508–1531.

63 The list of drugs to be discontinued is continuously updated at the US Food and Drug Administration website, **2004**,

http://www.fda.gov/cder/drug/
shortages/#disc

64 Orlaam(tm) (Levomethadyl hydrochloride
acetate) was discontinued in the spring of
2004 because it caused severe cardiac
toxicity, including Q-T prolongation,
Torsade des pointes and cardiac arrest.
See http://www.fda.gov/cder/drug/
shortages/orlaam.htm.

65 Baycol(tm) was voluntarily withdrawn by
Bayer in August 2001 because of reports
of sometimes fatal rhabdomyolysis
(a severe muscle averse reaction). See
http://www.fda.gov/cder/drug/infopage/
baycol/default.htm.

66 Redux(tm) and Pondimin(tm) were
voluntarily withdrawn in September 1997
because 30% of the patients taking these
two drugs had abnormal echocardio-
grams, **2004**. See
http://www.fda.gov/cder/news/phen/
fenphenpr81597.htm.

67 HEAD, R.D., SMYTHE, M.L., OPREA, T.I.,
WALLER, C.L., GREEN, S.M., MARSHALL,
G.R. VALIDATE: a new method for the
receptor-based prediction of binding
affinities of novel ligands. *J. Am. Chem.
Soc.* **1996**, *118*, 3959–3969.

68 DAVIS, A.M., TEAGUE, S.J. Hydrogen
bonding, hydrophobic interactions, and
failure of the rigid receptor hypothesis.
Angew. Chem., Int. Ed. **1999**, *38*, 736–749;
German edition: *Angew. Chem.* **1999**, *111*,
778–792.

69 HANSCH, C., HOEKMAN, D., LEO, A.,
WEININGER, D., SELASSIE, C.D.
Chem-bioinformatics: comparative QSAR
at the interface between chemistry and
biology. *Chem. Rev.* **2002**, *102*, 783–812.

70 HANSCH, C., HOEKMAN, D., LEO, A.,
WEININGER, D., SELASSIE, C.D. C-QSAR
Database. Available from the BioByte
Corporation, Claremont, CA, **2004**,
http://www.biobyte.com/

71 HANSCH, C., FUJITA, T. ρ-σ-π analysis. A
method for the correlation of biological
activity and chemical structure. *J. Am.
Chem. Soc.* **1964**, *86*, 1616–1626.

72 OPREA, T.I. 3D-QSAR modeling in drug
design. In *Computational Medicinal
Chemistry and Drug Discovery*,
TOLLENAERE, J., DE WINTER, H.,
LANGENAEKER, W., BULTINCK, P. (eds.)
Marcel Dekker, New York, **2004**,
571–616.

73 OLAH, M., BOLOGA, C., OPREA, T.I. An
automated PLS search for biologically-
relevant QSAR descriptors. *Perspect.
Drug Disc. Des.*, **2004** in press.

74 OPREA, T.I., BOLOGA, C., OLAH, M.
Compound selection for virtual
screening. In *Virtual Screening in Drug
Discovery*, ALVAREZ, J.C., SHOICHET, B.
(eds.) CRC Press, Boca Raton, **2005**,
in press.

75 VANDENBERG, J.I., WALKER, B.D.,
CAMPBELL, T.J. HERG K+ channels:
friend or foe. *Trends Pharm. Sci.* **2001**, *22*,
240–246.

76 OPREA, T.I., LI, J., MURESAN, S., MATTES,
K.C. High throughput and virtual
screening: choosing the appropriate
leads. In *EuroQSAR 2002 –
Designing drugs and crop protectants:
Processes, problems and solutions*,
FORD, M., LIVINGSTONE, D., DEARDEN, J.,
VAN de WATERBEEMD, H. (eds.) Blackwell
Publishing, New York, **2003**, 40–47.

3
Computational Chemistry, Molecular Complexity and Screening Set Design

Michael M. Hann, Andrew R. Leach, and Darren V.S. Green

3.1
Introduction

A large number of new technologies have been introduced into pharmaceutical research in recent years. These include the various "omic" techniques such as genomics and proteomics that provide information about potential new drug targets and their disease association. Methods such as automated compound synthesis and biological screening are intended to facilitate the discovery of new drug molecules against such targets. A common feature of many of these technologies is that they provide us with greatly increased levels of fundamental data; a key challenge then is to effectively process this data to extract the relevant information and knowledge. Our emphasis in this chapter is on some of the ideas that we have developed for enhancing lead discovery and optimization and thereby to capitalize on these significant investments in infrastructure.

Our particular focus will be on the use of computational techniques to assist in the design and construction of collections of molecules for screening. We recognize two distinct scenarios when constructing screening sets. The first scenario corresponds to the situation where one does not have any specific target in mind at the time when the set is constructed. The most obvious example would be the construction of a set of compounds for High-Throughput Screening (HTS). Also in this first category is the construction of sets of "leadlike" compounds that are screened at higher concentrations than is typical in HTS. The second situation arises when constructing a set of compounds for a specific target (or group of related targets), using relevant target-based information in order to focus or bias the selection. We will focus on the use of virtual screening methods using both pharmacophore and protein structural information. In both situations, some common underlying principles are employed based on concepts of diversity both in molecular structure (2D) and pharmacophore and shape descriptors (3D) while also operating within the type of limits of other properties such as those espoused by Lipinski [1].

Chemoinformatics in Drug Discovery. Edited by Tudor I. Oprea
Copyright © 2004 WILEY-VCH Verlag GmbH & Co. KGaA, Weinheim
ISBN: 3-527-30753-2

3.2
Background Concepts: the Virtual, Tangible and Real Worlds of Compounds, the "Knowledge Plot" and Target Tractability

In an ideal world, we would want to reliably screen the maximum number of molecules that we can afford against every target. This is ultimately the only way to ensure that we discover all possible leads (that already physically exist) for all targets. While the number of biological targets can be considered to have some effective upper limit based on the human genome/proteome and disease association, there is no effective limit for the number of compounds that can be made or acquired. It has been estimated that there are far in excess of 10^{60} druglike molecules that could be made [2]. This is referred to as the *Virtual* world of compounds because they cannot all be made but they are essentially a "resource" that can be mined as needed. Having appropriate informatics systems to access these virtual compounds via 2D, 3D and other property spaces is a key part of our strategy. In reality, molecules only of the order of 10^8 have so far probably ever been synthesized as discrete compounds. Of these, we could buy perhaps 10^7 from suppliers. An unknown number of compounds (probably of the order 10^8) have also been made as libraries and many of these are now available for purchase. Of this large Virtual set, some compounds can be reliably made via an appropriate chemical route. We refer to these accessible virtual compounds (whether they be from external or internal sources) as *Tangibles* because we could readily make or acquire them as needed. GSK currently has a yet smaller subset of these compounds available for screening. These are the compound samples that have been accumulated over many years. To these historical samples are being added novel compounds using our new automation facilities and also compounds made in lead optimization projects. We also have an active program for acquiring samples from commercial vendors. We refer to those discrete entities that physically exist within the company and are actually available for screening as *Reals*. This Virtual/Tangible/Real (VTR) description of compounds provides a framework for considering how we build screening sets.

Another concept that has proved useful in our considerations of screening set design is the knowledge plot shown in Figure 3.1. This relates the different levels of knowledge about targets to the level of diversity (or its corollary focus) required in the molecules of interest.

Thus, the need for diversity in order to find a hit in screening is inversely proportional to the knowledge that is available on the biological target. While this is obviously a very qualitative summary of the situation, it does help in understanding the different roles that different "sets" of compounds can have, be they structure-based, pharmacophore-based, class-focused or true-diversity inspired. A good screening collection strategy incorporates all these facets as inputs because only in this way can the varying levels of knowledge and experience about targets be taken into account.

Another important issue that contributes to the ultimate success of lead generation is target tractability. The tractability of targets is clearly not uniform – for example, it is easier to find molecules for optimization against targets such as kinases and most 7TMs than it is to find inhibitors of protein–protein interactions. Understanding these probabilities and selecting appropriate (i.e. easier) targets from disease-related

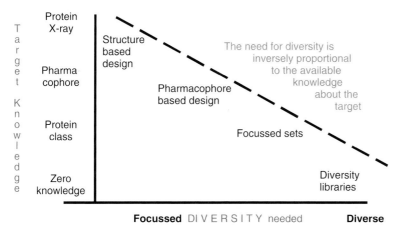

Fig. 3.1 The knowledge plot.

biological pathways is, therefore, as important as developing a screening collection. It seems likely on current evidence that some targets really are intractable to intervention by small-molecule compounds. Thus, even if there is excellent biological rationale for some targets, they remain inaccessible to small-molecule drug discovery, with protein therapeutics being the only viable way to address them.

3.3
The Construction of High Throughput Screening Sets

Compounds assembled for HTS are selected for reasons less specific than for more targeted work, but it would be incorrect to consider the design of an HTS screening set as random. In particular, the vast experience of medicinal chemists across a wide spectrum of targets and an understanding of the shortcomings of molecules from the point of view of likely developability (i.e., ADMET) issues provides important knowledge to drive designs and selections.

In the early days of building our screening collection (i.e., in the 1990s), work focused on acquiring compounds from suppliers such that the compounds bought were dissimilar to compounds that we already had. This dissimilarity was typically based on the Daylight fingerprint and as such addressed diversity using a metric related to the molecular 2-D connectivity. However, such 2-D methods do not address the issue of how proteins actually interact with ligands, which is more concerned with the properties projected by a molecule outside its internal structure. Indeed, recent studies by Martin et al. [3] have determined that a molecule with a 0.85 similarity to a lead molecule has only a 30% probability of sharing the biological activity. With this in mind we moved to the use of 3-D descriptors based on pharmacophore keys [4]. In order to generate a pharmacophore key, a conformational search is performed on each molecule. As each conformation is generated, the locations of the pharmacophore features within it are identified (comprising acid, base, H-bond acceptor, H-bond donor, lipophilic and aromatic pharmacophoric centers). All possible combinations of three pharmacophore points are enumerated (in the case of 3-center keys) and the interfeature distances calculated. Each distance is allocated to a "bin" depending on its value. The simplest binning scheme corresponds to a uniform bin width, but typically we use uneven bins to achieve some uniformity in the distribution. Each combination of three pharmacophore points with their associated distances corresponds to a particular position in the pharmacophore key, which is stored as a bitstring. Of particular relevance is that the pharmacophore key provides a descriptor space that is finite and can be partitioned into discrete bins representing pharmacophores of predefined geometry. We term this a GaP space (for Gridding and Partitioning) [5]. This approach has the further advantage of allowing us to identify the occupancy level of bins (particularly those not occupied), which is not possible with the variable clustering methods that are usually used with 2-D structural descriptors.

Having established this GaP space, we have been able to consider how well our Real compounds are distributed within it. We can then explore whether we have sufficient Real molecules with the required pharmacophores to find hits and which Tangible molecules we should convert to Real status. In Figure 3.2, part of the distribution across this GaP space (corresponding to a pharmacophore key containing approximately 220 000 elements) of ca. 420k discrete molecules from part of the collection is shown.

The figure only shows the far left-hand side of the distribution, which is very long tailed (i.e, Zipfian) in nature. In the part shown, it can be seen that, for instance, there are 2118 of the three-point pharmacophores that are represented by only one molecule and 1396 of those that have only two molecules to represent them. Because of the Zipfian nature of the distribution, the figure is not the best way to convey the contents of the entire data set. This is better represented in Table 3.1. In this representation, the curve of Figure 3.2 is in effect summed in logarithmic portions from the right-hand

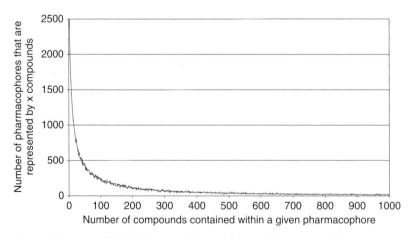

Fig. 3.2 Occupancy of 222705-element GaP space for a typical screening collection.

Tab. 3.1 Summed distribution of data from Figure 3.2 across the ca. 222k pharmacophore GaP space

Number of molecules representing pharmacophores	Number of pharmacophores covered	% of total
> or = 0	222 705	100
>1	198 311	89
>10	189 454	85
>100	162 450	73
>1000	91 120	41
>10 000	17 374	8

end of the fully extended distribution. Thus, there are 17 374 pharmacophores (i.e., 8% of the total) that have more than 10 000 compounds presenting that pharmacophore in this particular screening set. Similarly, there are 91 120 pharmacophores (41%) that reach the coverage level of over 1000 available samples. Our experience with using pharmacophore-based screening selections for tractable targets suggests that this level of coverage (i.e., 1 in 1000) is appropriate for having a realistic chance of finding an active compound in an *in vitro* bioassay requiring that pharmacophore. Thus, for this set, only 41% of possible pharmacophores have sufficient depth of coverage to find real hits at the 1:1000 *in vitro* to *in silico* success rate. If we could improve our *in silico* targeted selection processes to give a success rate of 1:100 (a tractable ambition, as there are several examples of success rates around the 1:200 mark) then we might expect that 73% of pharmacophores are now covered effectively. This naturally suggests an alternative strategy, which is to select compounds for acquisition or synthesis that increase the pharmacophore representation.

This type of analysis has, for the first time, provided us with a model that predicts what the addition of specific classes of compounds actually offers in a proactive or design sense. There are many related issues that we are also exploring using these concepts. One is *how is biological activity distributed over this GaP space?* This gives us clues as to the relative importance of some of the partitions in the GaP space. With this type of information, decisions can be made as to whether to concentrate more specifically on some partitions that may be well represented already but provide excellent returns against some targets.

3.4
Compound Filters

When building HTS screening collections, it has long been the practice not only to proactively select molecules for inclusion but also to deselect compounds that have undesirable properties. Typical deselection criteria are *in silico* derived chemical reactivity descriptors designed to remove compounds that might give an inappropriate mode of action or result in toxic events [6]. Other properties derived from the molecular structures form the basis of rules such as the rule-of-five, which helps direct us to molecules with good prospective oral absorption characteristics [1]. It is increasingly the trend to try and address these issues earlier by having more developable compounds in the screening collection and thereby reduce attrition at the later and more expensive stages of the drug discovery pipeline. A further issue is that samples for HTS are almost universally stored as DMSO solutions for extended periods of time (sometimes up to several years). Significant resource is often required to identify false positive hits following a high-throughput screen; one way to reduce this is to have a better idea of what has actually been screened. Thus, GSK has recently embarked on a large-scale project to re-analyze all the samples available for HTS. This has involved primarily the use of automated hplc/ms methods backed up by NMR. The new GSK screening collection being built from the legacy collections from the parent companies will only contain those compounds that are pure and are the correct compounds and, in addition, pass the requisite *in silico* filters.

By trying to address all of these issues in parallel, our aim is to increase the probability that any particular screening campaign will be successful. The factors that influence the outcome of each roll of the HTS dice are multifactorial, and altering the odds in our favor for each of these factors is the purpose of much of the methods-development activities currently carried out in most research-based pharmaceutical companies. Only by understanding the odds of different strategies and components can one build an approach that balances long and short odds.

3.5
"Leadlike" Screening Sets

As discussed above, *in silico* filters (e.g., reactive compounds and Lipinski counts) are useful for ruling out some unsuitable starting points and give guiding principles for the type of properties compounds should have for successful drug discovery programs. We have recently developed some additional concepts on the basis of how the complexity of molecules interacting with receptor sites affects the probability of such an event taking place [7]. These and related concepts have led us and others to develop screening strategies that are complementary to more traditional HTS methods [8–11]. The general approach is to try to find start points for lead optimization that are more "leadlike" and typically less complex than those derived solely on "druglike" criteria [12, 13].

"Druglike" criteria do however remain very valid during the lead optimization phase. Such criteria have usually been developed as statistical models, which bias sets to be "druglike". These have invariably been based on the analysis of the properties of generic sets of drugs that have reached the market and how these properties differ from molecules that have either failed or are nondruglike. Our approach [7] and that of Oprea et al. [9] was based on the observation that, when the history of past drug discovery successes are analyzed, it turns out that the starting points (i.e., the leads) for drug discovery programs have statistically different properties when compared to those of the final resultant drugs. Our own analysis used the excellent compendium of drug discovery histories compiled by Walter Sneader [14] while Oprea et al. selected

Tab. 3.2 Average property values for the Sneader lead set, average change on going to Sneader drug set and percentage change

Av # arom	Δ arom	%	Av ClogP	Δ ClogP	%	Av CMR	Δ MR	%
1.3	0.2	15	1.9	0.5	26	7.6	1.0	14.5
Av # HBA	Δ HBA	%	Av # HBD	Δ HBD	%	Av # heavy	Δ heavy	%
2.2	0.3	14	0.85	-0.05^+	(4)	19.0	3.0	16
Av MW	Δ MW	%	Av MV	Δ MV	%	Av # Rot B	Δ Rot B	%
272	42.0	15	289	38.0	13	3.5	0.9	23

Source: Redrawn from Ref. [7] with permission.

data directly from the published literature. By a number of criteria such as molecular weight (MW), Daylight fingerprint content, number of heavy atoms or hydrogen bond acceptors, it is observed that leads are historically less complex than optimized drugs. Table 3.2 illustrates these differences based on the Sneader data set.

There are several possible explanations for these differences that collectively account for the observed facts. Firstly, medicinal chemists find it easier to synthetically add mass to a lead in pursuit of potency. Secondly, many of the drugs considered by Sneader were derived from small hormones such as biogenic amines. These starting points are of such low complexity (i.e., low molecular weight) that it is only possible to add mass to them! We also explored another aspect of this problem, which examines the probability of finding matching molecular properties between ligands and target receptors as the complexity of the interactions grows.

We were able to show, using a very simple model of molecular recognition, that the probability of an effective ligand–receptor match falls dramatically as the complexity of the required interactions increases [7]. This is because, as the complexity of the number of interactions increases, there are many more ways of getting things wrong than there are of getting them right. The simple model we use is based on bitstring representations of the features of the ligand and binding sites. We require that, for a ligand to be able to fit the binding site, all the elements of the ligand and site bitstrings must match. The elements of the bitstring represent (in some undefined way) all the molecular properties of the ligand and binding site that might influence binding. Such properties are shape (global and local), electronic (local and global monopoles, dipoles, and so on, representing hydrogen bonding and electrostatic interactions) and other properties such as lipophilicity (global and local). The variation in binding probability is illustrated in Figure 3.3, where the probability of a correct match for ligands of different complexity are considered.

The requirement that all elements match may seem a harsh criterion, but even if this is relaxed, the conclusions of the resulting model are not radically different. It should also be noted that, in the published protein database structures, only one structure of a protein–ligand complex exists where a "wrong interaction" is observed and cannot be rationalized by, for example, the mediation of an unobserved water molecule [15].

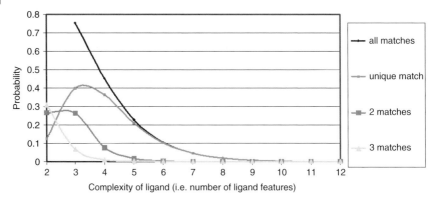

Fig. 3.3 Probabilities of ligands of varying complexity (i.e., number of features) matching a binding site of complexity 12 (redrawn from [7] with permission).

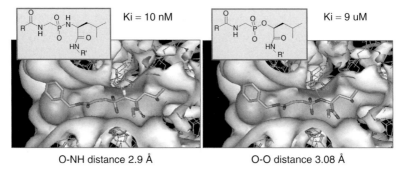

Fig. 3.4 The price of a single bad interaction.

Interestingly, this wrong interaction was actually engineered into the ligand as part of a study by Peter Kollman and Paul Barlett into the free energy changes associated with substituting an NH of a phosphoramidate inhibitor of Thermolysin with an O to give the corresponding phosphonate inhibitor (see Figure 3.4). As there are so many other strong interactions between the rest of the ligand and the protein, the phosphonate adopts a very similar binding mode but with the O of the phosphonate at only 3.08 Å from a rigid backbone carbonyl of the enzyme to which the NH of the phosphoramidate hydrogen binds. The free energy cost of this single wrong interaction is equivalent to three orders of magnitude loss of affinity as measured experimentally or calculated by FEP methods [16, 17]. The further subtleties of protein–ligand interactions revealed by protein crystallography have been reviewed recently [18].

The second part of our model incorporates the probability of being able to measure the binding of a ligand as the complexity of interactions increases. Clearly, as the number of effective pairwise interactions grows, the probability of experimentally measuring the summed interactions grows. In Figure 3.5, this probability is represented by the

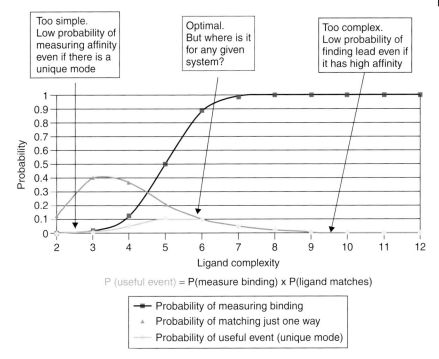

Fig. 3.5 Effect of adding potency.

curve with □; the actual shape of this measurement probability is not relevant save that it must go from zero probability when the interactions are too weak to be measured to a maximum of one. Also in this diagram is the probability of a unique binding mode (curve with △), which is taken directly from Figure 3.3. We then define the "probability of a useful event" (the curve with ◊) as the product of the probability distributions shown as curves with □ and △. This probability reflects the true balance of getting interactions correct and measuring them and thus points the way to the type of compounds to have in a screening collection.

The problem then is to determine where, for real systems comprising biological target and molecules for screening, the maximum in the curve with ◊ lies. This is a challenging problem. Nevertheless, an understanding of the reasons for the distributions can help focus synthetic and acquisition efforts toward particular types of compounds. The curves with □ and △ have competing distributions in that the first goes from high to low, while the latter goes from low to high. These distributions reflect the probabilities being considered and will, of course, differ for different model criteria and for real data. However, the fundamental shape of the resulting "probability of useful event" will always have a bell-shaped form because of the competing component distributions. An interpretation of this bell-shaped curve is shown in the annotation boxes in Figure 3.5. At low complexity, the probability of a useful event is zero because, even if its properties match, there are not enough of them to contribute to an observable binding in an assay.

At high complexity, the probability of getting a complete match is very small though, if such a match does happen, it will be easily measured. In the intermediate region, there is the highest probability of a useful event being found. Here, there is a realistic probability of both matching and measuring.

While the exact details of this model can be debated, it clearly highlights the need for caution in having over-functionalized (i.e., highly complex) molecules in a screening collection. Our complexity arguments add weight to the observations made earlier that medicinal chemists tend to add functionality (and hence mass) to molecules in lead optimization. Another aspect concerns the sampling rates that can be achieved with Real compounds of a given complexity within the vast space of Tangible or Virtual compounds. This can be explored with the aid of Figure 3.6, which shows the number of carboxylic acids (of all types) registered in the GSK registry system plotted as their

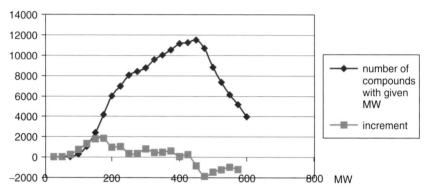

Fig. 3.6   Distribution of carboxylic acids in GSK collection.

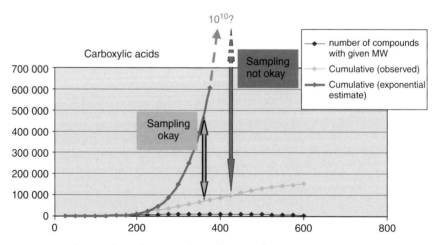

Fig. 3.7   Growth in numbers of compounds as a function of MW.

molecular weight distribution (curve with ◊). The curve with □ shows the incremental number for each 25-Da increase and is effectively the rate of increase in the number of compounds at a particular MW. There is an initial steep rise in the number of acids with molecular weight and this appears to follow an exponential curve. However, at around 150 Da, the observed MW increase of these Real compounds ceases to be exponential. Why does the initial exponential rise not continue beyond 150 Da? Our explanation is that we significantly undersample the potential carboxylic acids (i.e., the Virtual space) and that this undersampling gets worse as the MW increases. Figure 3.7 shows the same data in a different view together with the cumulative count (middle curve). The y-axis scale is also modified to provide an indication of the true cumulative number of virtual carboxylic acids that probably exist with MW ~ 400. A nominal figure of 10^{10} is suggested but this is probably an underestimate. Indeed, the precise numbers are not relevant; the key conclusion is that at lower MWs (say 350) we have more effective sampling of the Virtual world with our Real compounds than we have at a higher MW (say 425). If biological activity (albeit weak) can be found within a chemical series in the lower MW region, then it may be more effective to use this as the starting point for sampling the higher MW compounds with related chemotypes and pharmacophores (i.e., the process of lead optimization) than to try and directly sample the exponentially large number of higher MW compounds.

Such lower MW starting points are likely to have less potency and will not normally be found in HTS, where the screening concentration is typically of the order of 10 μM. The obvious solution is to screen compounds at higher concentrations but this then introduces other problems related to compound solubility, purity and interference with readout, for example, by fluorescence quenching. Nevertheless, with careful selection of compounds and robust screens, we have been able to screen several targets (mainly enzymes) at up to 1 mM concentration and still extract useful information. The so-called "reduced complexity" screening set that we have used for this purpose was assembled using a number of criteria such as 2-D property filters (e.g., mean MW of set ca. 350, rotatable bonds ≤ 6, heavy atoms ≤ 22, donors ≤ 3, acceptors ≤ 8, Clog-P ≤ 2.2) and 3-D pharmacophore patterns (again using the GaP approach). Our selection criteria also require that there is a synthetic handle present in order to facilitate the rapid

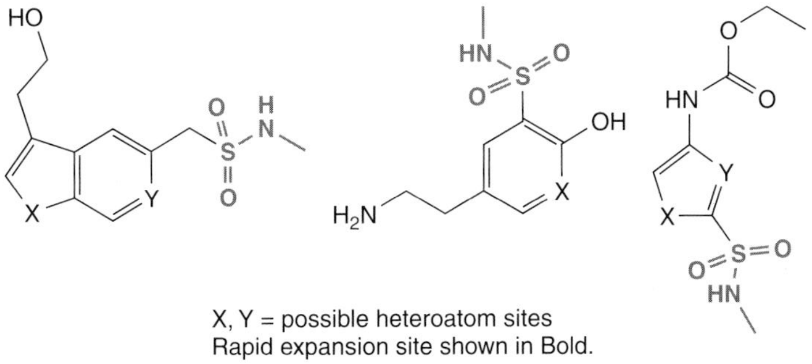

X, Y = possible heteroatom sites
Rapid expansion site shown in Bold.

Fig. 3.8 Typical generic structures selected for the reduced complexity screening set.

synthesis of further analogues. Examples of the types of generic structures selected as a result of this process are shown in Figure 3.8. In these particular examples, the *N*-methyl sulphonamide is envisaged as the diversity point through which rapid synthetic expansion could be carried out should one of these templates show activity. Similarity-searching for related compounds in our entire compound collection and external suppliers files is an alternative and, sometimes, is a faster follow-up procedure. Wherever possible, we also aim to obtain experimental data on the binding mode of the compounds to the protein by protein Xray or NMR methods.

3.6
Focused and Biased Set Design

When designing or selecting compounds for a specific target or group of targets, it is usually possible (indeed, it is highly desirable!) to make use of the available information about that target and/or the molecules known to interact with it. This contrasts with the situation described above when designing "general" HTS or other screening sets where no specific target information is taken into account.

In common with other practitioners, our typical strategy for constructing targeted screening sets is to employ relevant virtual screening and filtering methodologies to reduce the (potentially large) space of possible compounds to a smaller, more manageable number of structures that can then be selected, synthesized or purchased. Robotic technologies currently enable "cherry-picked" sets containing thousands of samples to be extracted from liquid stores and dispensed for screening, but the size of a focused screening set is typically much smaller than the usual HTS screening deck. However, it is when exploring the world of Tangible molecules (and their conversion into Real samples for screening) that virtual screening techniques are of particular importance (in part because it is increasingly possible to screen one's entire set of available samples against many different biological targets using high-throughput screening).

Critical to a strategy for the effective exploration of the Tangible world of compounds is the construction of databases containing such molecules. As outlined above, the two main sources of Tangible compounds are external compound suppliers and virtual libraries of compounds that could be made using automated synthesis methods. A large number of suppliers to provide compounds for screening now exist. Initially, many of these suppliers sourced their compounds from academic laboratories, but an increasing number of such suppliers now have their own internal synthesis capabilities. The numbers of samples available from such suppliers continues to grow rapidly and obviously requires effective database systems. To date, we have found that the chemoinformatics systems developed by Daylight CIS provide the flexibility and performance required for such an application [19].

An approach we term "Virtual Screening of Virtual Libraries" (VSVL) is intended to take advantage of our in-house automated synthesis capabilities. This involves the creation of large 3-D databases of virtual libraries using the Catalyst system [20]. Catalyst uses a prestored set of conformations for each molecule, so while construction of the database can be time-consuming, the actual searches can be performed very rapidly. The relatively high cost in terms of computational burden and disk storage means that it is important to ensure that only molecules of real interest are considered for inclusion in such databases. Our VSVL databases are therefore constructed in close collaboration with our synthetic chemistry colleagues and are restricted to contain only molecules derived from synthetic schemes that have already been enabled in-house and use building blocks that have already been demonstrated to work in those schemes (i.e., true Tangible molecules). This ensures that any "hits" from a virtual screen should have a high probability of being synthesized rapidly in order to meet the needs and timescales of the real biological assay.

VSVL libraries can be both generic and specific. Generic VSVL libraries are typically constructed using well-established chemistry and building blocks that are widely accessible. The databases of generic libraries are made available to all projects, alongside databases of in-house registered compounds and those from external vendors. More specific VSVL libraries may also be constructed for individual projects. Here, one may wish to identify so-called "privileged" building blocks that are relevant to that particular target or family of targets. The RECAP procedure [21] is one way to perform such an analysis. RECAP is a fragmentation procedure that disassembles molecules at certain retrosynthetic disconnection points. The resulting structural fragments can then be analyzed to identify the "privileged" fragments and thence potential building blocks, which provide the feedstock for the virtual library construction.

3.7
Conclusion

A variety of screening approaches are now available for the identification of hits against drug targets. Each of these screening methodologies requires some form of compound selection activity, in which computational chemistry and chemoinformatics tools play a key role. In addition, theoretical considerations and models of the biophysical processes involved in molecular recognition can provide the inspiration for new avenues to pursue and to try and ensure that no stone is left unturned in the search for new leads, particularly against challenging targets.

We have used these methods in many different research projects and have been successful on several occasions in finding good starting points for further optimization. This gives us confidence to continue with the integration of these concepts into our working practices [22].

Acknowledgements

Many present and former colleagues have contributed to the thoughts presented in this chapter and we would like to thank them all. Principal among them are Gavin Harper, Rich Green, Giampa Bravi, Andy Brewster, Dave Langley, Robin Carr, Brian

Evans, Albert Jaxa-Chamiec, Iain Mclay, Stephen Pickett, Duncan Judd, Xiao Qing Lewell, Steve Watson, Derek Reynolds, Barry Ross, Mike Cory, Malcolm Weir and John Bradshaw. We would also like to thank Drake Eggleston for careful reading of the manuscript and helpful suggestions.

References

1 LIPINSKI, C.A., LOMBARDO, F., DOMINY, B.W., FEENEY, P.J. Experimental and computational approaches to estimate solubility and permeability in drug discovery and development settings. *Adv. Drug Deliv. Rev.* **1997**, *23*, 3–25.

2 BOHACEK, R.S., MARTIN, C., GUIDA, W.C. The Art and Practice of Structure-Based Drug Design: A Molecular Modeling Perspective. *Med. Res. Rev.* **1996**, *16*, 3–50.

3 MARTIN, Y.C., KOFRON, J.L., TRAPHAGEN, L.M. Do structurally similar molecules have similar biological activity? *J. Med. Chem.* **2002**, *45*, 4350–4358.

4 MASON, J.S., MORIZE, I., MENARD, P.R., CHENEY, D.L., HULME, C., LABAUDINIERE, R.F. New 4-point pharmacophore method for molecular similarity and diversity applications: overview of the method and applications, including a novel approach to the design of combinatorial libraries containing privileged substructures. *J. Med. Chem.* **1999**, *42*(17), 3251–3264.

5 LEACH, A.R., GREEN, D.V.S., HANN, M.M., JUDD, D.B., GOOD, A.C. Where are the GaPs? A rational approach to monomer acquisition and selection. *J. Chem. Inf. Comput. Sci.* **2000**, *40*, 1262–1269.

6 See for example appendix in: HANN, M.M., HUDSON, B., LEWELL, X-Q., LIFELY, R., MILLER, L., RAMSDEN, N. Strategic pooling of compounds for high-through-put screening. *J. Chem. Inf. Comput. Sci.* **1999**, *39*, 897–902.

7 HANN, M.M., LEACH, A.R., HARPER, G. Molecular complexity and its impact on the probability of finding leads for drug discovery. *J. Chem. Inf. Comput. Sci.* **2001**, *41*, 856–864.

8 HAJDUK, P.J., BURES, M., PRAESTGAARD, J., FESIK, S.W. Privileged molecules for protein binding identified from NMR-based screening. *J. Med. Chem.* **2000**, *43*, 3443–3447.

9 OPREA, T.I., DAVIS, A.M., TEAGUE, S.J., LEESON, P.D. Is there a difference between leads and drugs? A historical perspective. *J. Chem. Inf. Comput. Sci.* **2001**, *41*, 1308–1315.

10 ERLANSON, D.A., BRAISTED, A.C., RAPHAEL, D.R., RANDAL, M., STROUD, R.M., GORDON, E.M., WELLS, J.A. Site-directed ligand discovery. *Proc. Nat. Acad. Sci. U.S.A.* **2000**, *97*, 9367–9372.

11 CARR, R. JHOTI, H. Structure-based screening of low-affinity compounds. *Drug Disc. Today* **2002**, *7*, 522–527.

12 GILLET, V.J., WILLETT, P., BRADSHAW, J. Identification of biological activity profiles using substructural analysis and genetic algorithms. *J. Chem. Inf. Comput. Sci.* **1998**, *38*, 165–179.

13 SADOWSKI, J. KUBINYI, H. A scoring scheme for discriminating between drugs and nondrugs. *J. Med. Chem.* **1998**, *41*, 3325–3329.

14 SNEADER, W. *Drug Prototypes and their Exploitation.* John Wiley & Sons, Chichester and New York, **1996**.

15 ALEXANDER, A. Unpublished observations. PDB files are: 6TMN and 5TMN.

16 BARTLETT, P.A. MARLOWE, C.K. Evaluation of intrinsic binding energy from a hydrogen bonding group in an enzyme inhibitor. *Science* **1987**, *235*, 569–571.

17 MERZ, K.M., Jr., KOLLMAN, P.A. Free energy perturbation simulations of the inhibition of thermolysin: prediction of the free energy of binding of a new inhibitor. *J. Am. Chem. Soc.* **1989**, *111*, 5649–5658.

18 DAVIS, A.M., TEAGUE, S.J., KLEYWEGT, G.J. Application and limitations of X-ray crystallographic data in structure-based

ligand and drug design. *Angew. Chemie. Int. Ed.* **2003**, *42*, 2718–2736.

19 BRADSHAW, J. D. LANGLEY. More fun with chemical catalogues. Talk at *EuroMUG Meeting,* **1999**. See http://www. daylight.com/meetings/mug99/ Bradshaw/Presentation.html

20 *Catalyst.* Accelrys Inc. 9685 Scranton Road, San Diego, CA 92121-3752, USA.

21 LEWELL, X-Q., JUDD, D.B., WATSON, S.P., HANN, M.M. RECAP-retrosynthetic combinatorial analysis procedure: a powerful new technique for identifying privileged molecular fragments with useful applications in combinatorial chemistry. *J. Chem. Inf. Comput. Sci.* **1998**, *38*, 511–522.

22 HANN, M.M. OPREA, T.I. Pursuing the leadlikeness concept in pharmaceutical research. *Curr. Opin. Chem. Biol.* **2004**, *8*, 255–263.

4
Algorithmic Engines in Virtual Screening

Matthias Rarey, Christian Lemmen, and Hans Matter

4.1
Introduction

Several technological advances [1] complementing high-throughput screening have recently gained attention for identifying novel lead compounds in the pharmaceutical industry [2]. The commonly applied strategies range from knowledge-driven approaches starting from literature, patents, related research programs and biostructural information to random approaches, exemplified by technologically sophisticated combinatorial chemistry and high-throughput screening strategies. The combination of both extremes into knowledge-driven lead finding covering a wider area of the available, leadlike chemistry space is anticipated to deliver more lead series fulfilling the quality criteria for continuation of a research program [2]. As the identification of promising leads has significant implications for a drug discovery program, a clear process driven by relevant metrics for compound progression is required [3].

Virtual screening is one that is complementary to high-throughput screening *in silico* lead finding approach to identify novel entry points for drug discovery programs [4, 5]. It is ultimately related to a selection of individual compounds (cherry picking) or a focused sublibrary (biased library design) having a significantly increased chance to be active at a particular molecular target. This assumption is based on data analysis and computational models to convert this information into predictive filters. The amount and quality of input data determines the level of abstraction to describe molecules to set up a filter. It is also the basic information for selecting appropriate software tools and a sequential virtual screening strategy, which often leads to significant enrichment rates.

Virtual screening might be often used to identify novel chemotypes to enrich the basis for a drug discovery program (scaffold hopping). It could also result in new starting points based on a single active molecule, X-ray structure or homology model. Alternatively, one might want to retrieve all related compounds on an identified chemotype that have a high chance to display significant biological activity (similarity searching). The goal of a virtual screening run consequently influences the selection of algorithms, methods, physicochemical models, and descriptors.

Till now, it is unfortunately the case that none of the available screening tools solve any of the screening problems exactly, and it is unlikely that this situation will change soon. The reasons for this are manifold, the two most important ones being our lack of understanding allowing an exact and efficient model of the real microscopic world and the complexity of the optimization problems we face. Even worse, in most cases it is impossible to provide a reliable estimation for the error that might be produced by a software tool: no similarity measure guarantees to identify all compounds within a certain activity range and no molecular docking tool guarantees to find a docking solution within a certain maximal root-mean square derivation (RMSD) when compared to the experimental structure. Even if an almost "bioactive" structure is generated during the search procedure, it is not guaranteed that current scoring functions would be able to pick this solution. From the user's perspective, the heuristic nature of software for virtual screening raises at least two very pragmatic questions:

- If several software tools for the same problem are available, which one is the best for my specific problem?
- With what probability is the predicted result correct? Or: what success rates can be expected when applying the software to my problem?

Estimates for these questions can only be provided after retrospective analysis. Since many possible errors are mixed, the experimental data are usually not complete and the input given to the software is more or less biased; retrospective analysis can only give hints rather than answers to these questions. In summary, it should be emphasized that virtual screening software tools cannot be applied as black boxes. A good understanding of the underlying models and algorithms turns out to be extremely helpful in judging the appropriateness, usefulness, and quality of a virtual screening tool. The ultimate answer of its usefulness, however, will only be available after a prospective application to a novel target. If the identified structures are active and useful for medicinal chemists to start a new lead optimization program, the first milestone is reached.

The major goal of this chapter is to give a general introduction to the most frequent physicochemical models and algorithmic techniques applied in virtual screening software tools. We will focus on what is frequently used and important in our point of view, rather than providing a complete survey, which would be impossible anyway. While the physicochemical models are associated with a particular application in virtual screening, most of the algorithms are applied in several tools addressing different variations of virtual screening problems. For each of them, we will describe the algorithmic technique itself and give concrete examples of applications. In addition, we try to identify potential problems and drawbacks.

Before we can start the explorative tour into the world of algorithms for virtual screening, we will span the space of screening applications in the next section. Sections 4.3 and 4.4 then cover the models and algorithms for virtual screening. After this excursion, we will reenter the real world of molecular design giving several examples of practical virtual screening applications using a variety of different approaches.

4.2
Software Tools for Virtual Screening

Software tools for virtual screening can be best classified by the input data available for screening. On the one side, there is always a collection of compounds to be screened, which differs in size (from a few tens to several millions) and in structure (from structurally unrelated compounds via combinatorial libraries to chemistry spaces). On the other side, there is the data that is used to create the screening query, which can be a protein structure, a known active compound or a pharmacophore created from several known actives (see Figure 4.1). In summary, we are ending up with four classes of screening tools:

– Molecular docking (structure-based virtual screening): on the basis of a target protein structure, ligand molecules that bind to the active site of the target are searched for.
– Structural alignments (ligand-based virtual screening): on the basis of a known active compound, molecules that show a good superimposition in shape and physicochemical features are searched for.

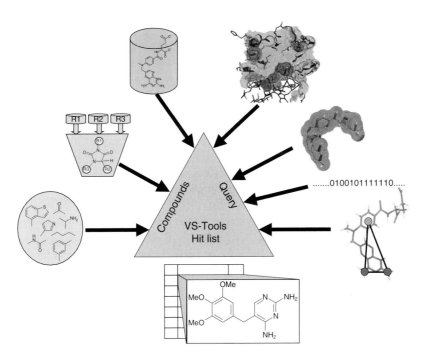

Fig. 4.1 Virtual screening tools can be categorized by the compound data to be screened (compound collection, combinatorial library, chemistry space) and the query type (structure-based, ligand-based, descriptor-based, pharmacophore-based). The output is always a list of compounds together with a score quantifying the fit to the query.

– Molecular similarity (similarity-based virtual screening): on the basis of one or a small set of known actives, molecules showing a high similarity concerning specific features stored in a molecular descriptor are searched for.

– Pharmacophore mapping (pharmacophore-based virtual screening): given the relative orientation of a few functional groups in space (a pharmacophore), ligand molecules that are able to fulfill the pharmacophore are searched for.

For most of the classes, recent review articles can easily be found: Molecular docking: [6–10], Structural alignments: [11, 12], Molecular similarity: [13–15], Pharmacophore mapping: [16–18].

For the development of software tools in any of the four classes, two ingredients are needed: a model for describing the relevant physicochemical properties of the problem and some algorithms for tackling the upcoming optimization tasks. The physicochemical model usually covers the description of chemical properties, the handling of conformational space, or the scoring issues, which is either the estimation of a binding affinity based on the protein–ligand complex or the quantification of the molecular similarity between compounds. An overview of models frequently used in virtual screening is given in Section 4.3.

Concerning the algorithms, we can usually distinguish between the overall optimization method of a software tool and algorithms that are used to solve specific key problems within the tool efficiently. Owing to the universal nature of algorithms, most of them are used in multiples of the four different classes of screening tools, often even beyond the world of computer-aided molecular design. After introducing frequently used models in virtual screening, an overview of such algorithms is given in Section 4.4.

4.3
Physicochemical Models in Virtual Screening

This section discusses physicochemical models for virtual screening applications. A comprehensive review is beyond the scope of this chapter, while trends, useful models, and practical considerations are provided. These models describe to a certain level of abstraction physicochemical properties of molecules and their interaction with receptors. They also capture conformational preferences, which allows the identification of binding modes and reliable molecular superpositions. Any model depends on the amount and quality of input data, while there is a trade-off between the level of description and computational efficiency. As virtual screening is aiming to select compounds from large or very large virtual and existing compound libraries, the complexity and generality of the underlying molecular representation is important. Quantum chemistry approaches for ligand conformations or atomic charges and free-energy calculations for an accurate description of energy differences associated with protein–ligand interactions are often not suitable here, while they impact later stages in the discovery process.

4.3.1
Intermolecular Forces in Protein–Ligand Interactions

4.3.1.1 General Considerations

The exponentially increasing number of three-dimensional protein structures [19] allows to use them for structure-based virtual screening. The interaction between a ligand and its macromolecular target is a significant event driven by steric and electrostatic surface complementarity for both partners [20]. However, even experimental binding modes still turn affinity prediction into a major challenge. Any structure-based virtual screening protocol involves models for prediction of binding affinity based on molecular recognition considerations. Models of increasing complexity exist – from scoring functions to estimate structural and conformations contributions to protein–ligand recognition to free-energy calculation from first principles. However, only relatively fast empirical force-field energies and scoring functions find practical application in virtual screening to evaluate the protein–ligand interactions.

Even the knowledge of a molecular target at atomic resolution does not guarantee a correct solution, as there are other critical factors, namely, the requirement for a relevant experimental crystal structure, the knowledge of the detailed binding mode, and the problems with scoring functions [21–23]. Several limitations do not allow for accurate *in silico* affinity prediction, as the effect of entropy, protein flexibility, multiple binding modes [24, 25], conserved water in binding sites [26], pH-related protonation and binding modes [27], tautomerization and others could hardly be accurately described with current tools.

The knowledge on protein–ligand interactions has extensively been reviewed [20, 28, 29] and will only be summarized here. The physicochemical basis of intermolecular interactions is the free-energy change ΔG upon binding of a ligand to its receptor, containing an enthalpic and a temperature-dependent entropic contribution. The negative logarithm of the binding affinity (Ki) from biological assays correlates with ΔG. This free-energy change is typically expressed as the sum of terms for electrostatic, inductive and nonpolar intermolecular interactions, and conformational energy and entropy.

4.3.1.2 Electrostatic Interactions

Electrostatic interactions are long-range and result in attractive noncovalent bonds for partners with oppositely charged groups. Many druglike molecules contain charged groups at physiological pH, like carboxylates or aliphatic amines, which strongly interact with complementary partners in a binding site. Charge–dipole and dipole–dipole interactions are weaker than ionic bonds, but also occur in druglike molecules. The strength of this interaction is related to the distance of both charges and the dielectricity constant, although values for the latter are still debated. While the dielectricity constant is ~ 78 for water, it adopts values between 2 and 5 for protein cores [30]. Common approximations are to use either a value between solute and solvent constants or a distance-dependent dielectricity constant [31]. A more accurate calculation of the electrostatic energy can be done using the Poisson–Boltzmann equation [32, 33].

The accuracy of electrostatic interaction energies depends on charges of ligands and proteins, respectively. The most common approach is to use atomic point charges, often derived from *ab initio* calculations [34, 35]. Simpler approaches are based on electro negativity equalization. All charges in the studied system must be assigned using one consistent approach.

The formation of protein–ligand complexes involves charge redistribution, either within one molecule (polarization) or both partners (charge-transfer). Inductive polarization occurs when a temporary dipole is induced in a nonpolar group by the field of a permanently charged dipole or ion [20]. Charge-transfer interactions take place when electron donors and acceptors are sufficiently close in space to allow electron transfer. Although both effects occur frequently upon ligand recognition, they are often omitted from interaction energy and binding affinity approximations.

4.3.1.3 Nonpolar Interactions

Significant interactions are also observed between nonpolar parts of molecules, preferably at short intermolecular distances. These dispersions or London forces are present in the liquid phase, as even nonpolar molecules show at a given moment a temporary dipole moment due to electron movements. These fluctuating dipoles induce opposite dipoles in neighboring molecules, which result in weak attractive forces. A close fit between nonpolar ligand and protein parts leads to significant forces by summing up many individual short-range contributions. The balance between opposite forces is often approximated by a computationally convenient Lennard–Jones 6–12 potential. Simpler potential forms are implemented to speed up docking and scoring [36]. The simplest form is used in the hard-sphere approximation [37], where the energy is assigned to large values for short distances and to zero for larger ones.

Interactions between hydrophobic partners are favorable in an aqueous environment, as the association reduces the net hydrophobic surface in contact with water, thus reducing structured water around the solutes and providing a positive association entropy. This hydrophobic effect is estimated from solubility measurements of alkanes in water. Although the effect per atomic surface segment is relatively small, its contribution to protein–ligand interactions is significant due to the large number of hydrophobic atomic compensations in a complex.

4.3.1.4 Hydrogen Bonding Interactions

Hydrogen bonds are essential motifs in biomolecules, they are best described as electrostatic interaction between two electronegative atoms and one proton covalently bound to one partner. This allows to estimate the strength of hydrogen bonds by electrostatic and van der Waals forces without explicit additional term for their calculation [38]. Their directionality arises from steric effects and an anisotropic, mutually polarized charge distribution. The strongest hydrogen bonds are formed between groups with the greatest electrostatic character. For strong bonds, the distance between donor and acceptor is less than the sum of van der Waals radii. Statistical analyses of hydrogen bonds in small-molecule crystals from CSD (Cambridge Crystallographic Database) reveal directional preferences in agreement with theory [39].

While some force fields do not contain a term to model hydrogen bonds, others (cf [40]) utilize such an additional term. The treatment of hydrogen bonds using experimentally determined protein–ligand complexes is difficult due to limitations arising out of lack of hydrogen and lone-pair coordinates, ambiguous assignment of protonation and tautomerization and indirect deduction of the hydrogen bond network at the interface between protein and ligand, especially, if water or terminal amide groups are involved.

Weak hydrogen bonding networks [39] might also contribute to stabilize protein tertiary structures or protein–ligand recognition. Here, the CH–O hydrogen bond is the most prominent example. Although these interactions are quite commonly observed in protein 3-D structures [41], they are expected, from theoretical studies, to be only up to one-half the strength of a traditional hydrogen bond [42, 43]. Recently, a combined approach including X-ray crystallography and thermodynamical stability measurements on mutants of the membrane protein bacteriorhodopsin did not reveal any significant contribution of this CH–O interaction to membrane protein stabilization [44]. At best, this interaction was found to be energetically neutral or overwhelmed by other contributions, while they still might be important to allow for closer protein packing due to electron orbital overlap.

A comparative analysis of 184 3-D structures of kinase protein–ligand complexes and 358 other complexes from the protein databank (PDB) provides additional evidence for this nonstandard interaction in protein–ligand recognition involving ligands with aromatic and heterocyclic CH fragments interacting with protein carbonyl oxygens [45]. From *ab initio* calculations complementing the statistical analysis of this study, a considerable range of CH–O hydrogen bonding potentials in various aromatic ring systems was observed, with those substructures from kinase inhibitors exhibiting this nonstandard interactions being among the most favorable ones.

4.3.1.5 $\pi-\pi$ and Cation–π Interactions

Sixty percent of protein aromatic side chains are involved in aromatic–aromatic pairs with particular geometrical requirements ($\pi-\pi$ interactions), 80% of which form networks with more than three interacting partners [46]. Those networks are important for tertiary structure stabilization, but they might also influence ligand recognition. Nonbonded energy calculations indicate that such an interaction has an energy of between -4 and -8 kJ mol^{-1}, while free-energy contributions depend on the environment of the aromatic pair.

Cation–π interactions are another important motif, dominated by electrostatic interactions [47] within mainly hydrophobic subsites. Such a stabilizing interaction takes place between cations like ammonium ions and electron-rich π-systems of neighboring aromatic rings. Both interactions are often not explicitly contained in common force fields or scoring functions.

4.3.1.6 **Penalty Terms**

In addition to attractive forces, there are penalty terms upon formation of the protein–ligand complex, which are enthalpic or entropic in nature. Enthalpic contributions result from ligand binding not in its global minimum, but its "bioactive" conformation, which might be significantly different and higher in energy. Furthermore, polar ligand groups have to break bonds to solvent molecules before binding to the protein. The entropic contributions are related to losses of rotational, translational, and conformational freedom during the ligand binding process. The extent of entropy loss per internal rotation is estimated to be 5 to 6 kJ mol^{-1}, while this clearly depends also on the tightness of the protein–ligand complex. In the case of a rigid ligand binding, this factor provides an entropic advantage to more flexible analogs.

In order to achieve extremely high ligand affinities, both enthalpic and entropic optimizations have to be considered. One of the main obstacles in the lead optimization process originates from the phenomenon of enthalpy–entropy compensation, that is, favorable changes in binding enthalpy are compensated by opposite changes in entropy and vice versa, resulting only in small changes in binding affinity [48].

4.3.2
Scoring Functions for Protein–Ligand Recognition

These theoretical considerations and requirements of processing larger databases led to the development of simple schemes for estimating protein–ligand interactions. The resulting scoring functions allow to guide flexible docking algorithms and rank order series of docked molecules. Although substantial progress was reported in both fields docking and scoring, these tools are still not accurate enough to guarantee success in most of the cases. A variety of scoring functions belonging to three main categories exist today: force-field methods, empirical functions, and knowledge-based approaches, as discussed in many recent reviews [23, 29, 49–52]. The earlier studies concentrated on the statistical analysis of essential terms for protein–ligand interactions with attempts to establish their relative importance by correlations to structurally and biologically diverse sets of experimental protein–ligand complexes. The success of this approach, however, is limited by the amount and consistency of experimental protein–ligand complexes as well as by the consistency and quality of the biological data used for regression. This problem points to a major source of uncertainty in most derived functions. Moreover, experimental protein–ligand complexes are mostly available for high affinity ligands, while no structural information for a chemically related analog with low (or no) binding affinity is available. The statistical analysis of a larger number of complexes in the PDB database led to the generation of knowledge-based scoring functions like PMF [53], DrugScore [54], Bleep [55]. These functions still rely on the availability and consistency of protein–ligand complexes, while they systematically mine and extract the inherent geometrical knowledge at the protein–ligand interface.

Recent advances in computational power have made it possible to use efficient approximations of the Poisson–Boltzmann equation also to estimate the electrostatic component of protein–ligand interactions [56].

Difficulties in these approaches include that, according to literature [57–59] and the authors' own experience, there is no general preference to one superior docking tool and scoring function that adequately works for different problems. Furthermore, it is an open question whether the detection of the best docking mode and the ranking of molecules require different scoring functions.

These limitations prompted for alternative approaches in structure-based virtual screening. In consensus scoring [60], predictions from different functions are combined to rank protein–ligand geometries. This approach consistently led to an increased performance in terms of hit rates. Another approach combines experimental X-ray information with binding poses obtained from docking calculations. Here, one recently described approach uses a genetic algorithm to optimize functions based on binding property data from multiple docking runs [61].

Targeted scoring functions are a promising alternative for proteins with more structural information about ligands, that is, structurally related molecules with known or hypothetical binding modes in the active site of this protein. Gohlke et al. developed AFMoC (adaption of fields for molecular comparison) [29] to derive tailored scoring functions on the basis of knowledge-based pair-potentials [54], which are adapted to a single protein binding site by incorporating ligand information. Others derive tailor-made empirical functions by adding more descriptors and determine weights using a training set for one protein cavity [62, 63].

4.3.3
Covering Conformational Space

The use of 3-D descriptors, structural alignments, and docking for virtual screening requires reliable methods for conformational analysis to closely reproduce the unknown "bioactive" conformation. State-of-the-art conformational analysis for pharmaceutial applications covers a wide area, which will not be reviewed here, while relevant searching and optimization algorithms will be discussed in the algorithmic section below. However, the quality and speed of a conformational analysis critically determines the results of 3-D searching and virtual screening. In general, there are two fundamentally different approaches to incorporate conformational flexibility into docking and alignment calculation. The first approach relies on an *a priori* conformational analysis, resulting in a series of acceptable conformers per molecule. Those conformers are stored in a database and are subsequently used for rigid-body docking and alignment. In contrast, the second approach starts from a single molecule and builds acceptable conformations on the fly during the docking or alignment calculation. Current tools for conformational analysis are often based on a combination of force-field approaches with empirical rules on conformational preferences for particular substructures. For high-quality geometries as a basis for structure-activity relationships and virtual screening, detailed force fields including implicit solvation effects should be considered. Some tools for conformational analysis and flexible superpositions work in torsion space on internal coordinates with knowledge-based conformational preferences from databases of experimental structures to limit the search space [64, 65]. Recent comparisons of conformational analysis tools [66, 67] revealed that although conformers closely resembling the "bioactive" conformation can be produced, there are still significant deviations from known X-ray structures in complexes with relevant protein binding sites. Those deviations clearly influence superposition and docking results.

Considering only a single 3-D conformation generated by standard 2-D/3-D converters often is inadequate for 3-D-based comparisons and most classes of pharmaceutically relevant molecules. Averaging over the accessible conformational space as an alternative gives only a rough view on the possible arrangement of key functional groups in 3-D space. A rough approximation for virtual screening could include a single, preferably rigid molecule as search query or an approach to build and validate a reliable pharmacophore hypothesis based on flexible superposition of a series of active molecules or information about the protein binding site prior to employing this hypothesis for virtual screening query by 3-D searching. The use of rigid analogs maintaining some activity at the target of interest significantly limits the searching space for deriving the pharmacophore, while there is no guarantee that superimposed flexible molecules show similar activity due to enthalpic and entropic penalties to the free energy of binding.

4.3.4
Scoring Structural Alignments

As molecules are known to interact with protein binding sites through their respective molecular fields, comparisons could also employ potentials around ligands, such as the molecular electrostatic potential (MEP), hydrogen-bonding potential, van der Waals potential, hydrophobic potential and others, which additionally offers the advantage to compare structurally less-related molecules. Lennard–Jones-type steric and electrostatic potentials cover the major part of intermolecular interaction energies. Additional hydrophobic fields have been incorporated into the program HINT [68]. Thus, derived molecular interaction fields even in the absence of a binding site are relevant descriptors to probe putative interactions, identify similarities among interaction motifs between two ligands, and drive molecular alignments. A quantitative comparison of field and potential differences between molecules supports structural alignments and thus quantitative structure-activity relationships. There are several possibilities in comparing fields. The most common way is to evaluate interaction energies between each molecule in a data set and an appropriate probe atom at regularly spaced grid points in a common reference frame. This also forms the conceptual basis for field-based 3-D QSAR approaches like comparative molecular field analysis (CoMFA) [69] and comparative molecular similarity index analysis (CoMSIA) [70]. Electrostatic and steric molecular interaction fields are computed between each ligand and a probe atom located on predefined grid points for CoMFA, while for CoMSIA those interaction fields are replaced by fields based on similarity indices to probe atoms. These molecular similarity fields in CoMSIA are approximated by Gaussian functions, which results in significantly smoother distance dependence compared to the Lennard-Jones 6–12 potential without singularities at the atomic positions. The program GRID [71–74] allows to sample potential energies of putative interactions between a large variety of relevant probe atoms and functional groups to molecules, which also could guide comparisons. Those approaches lead to statistical relationships between molecular property fields and biological activities.

For any comparison of molecules based on fields, a quantitative measure for similarity has to be employed. The most important point to consider in comparisons based on molecular subgraphs, atomic coordinates or interaction fields is whether a global unweighted similarity comparison or a local similarity based only on necessary features for activity is desirable. For any description of local similarity, some weighting scheme describing the importance of particular features has to be derived, preferably from chemometrical analysis of structurally related analogs. If there is only one lead structure without data on the importance of particular functional groups, similarity has to be globally expressed, treating all groups equally.

4.4
Algorithmic Engines in Virtual Screening

In the following section, several algorithms that have been proved to be quite useful in virtual screening applications are summarized. The algorithms are divided into four classes: mathematical concepts, algorithmic concepts, descriptor technology and global search methods. Only the third class is somehow bound to the class of one screening application, namely, descriptor-based virtual screening. The importance of this area and the elegance of the algorithms justify a separate section from our point of view. For each class, a set of important problems and related algorithms are presented. Each section has the same structure starting with a problem description, followed by outlines of algorithms to solve it, and the range of applications in virtual screening. Most importantly, the sections are finalized highlighting advantages, potential drawbacks or limitations relevant for applications built around the algorithms.

4.4.1
Mathematical Concepts

4.4.1.1 Numerical Optimization
Numerical optimization is quite a broad general concept for finding optima, given an *objective function* of your problem at hand. In general, an objective function has a number of input variables (*multivariate*) and a real-valued output. For the sake of simplicity, we will assume that the objective function should be minimized.

In general, linear functions and correspondingly linear optimization methods can be distinguished from nonlinear optimization problems. The former, being in itself the wide field of linear programming with the predominant Simplex algorithm for routine solution [75] shall be excluded here.

The area of nonlinear optimization can be subdivided further into various classes of problems. Firstly, there is constrained versus unconstrained optimization. Normally, the inclusion of constraints generates only a special case for the sibling, unconstrained optimization problem. Therefore, for the sake of generality, here we will focus on the unconstrained optimization case. Secondly, discrete versus continuous optimization can be distinguished. There are numerical techniques for discrete (or combinatorial) optimization. However, this is predominantly the area where specific algorithms are applied to solve the problem at hand most efficiently (some of which will be discussed in the other sections) [76]. Therefore, we will focus on continuous optimization. Third, the characteristics of the objective function determine the applicability of general versus specific, and frequently more effective optimization techniques. We identify two general types of functions.

First, and most general, is the case of an objective function that may or may not be smooth and may or may not allow for the computation of a gradient at every point. The nonlinear Simplex method [77] (not to be confused with the Simplex algorithm for linear programming) performs a pattern search on the basis of only function values, not derivatives. Because it makes little use of the objective function characteristics, it typically requires a great many iterations to find a solution that is even close to an optimum.

Second are the sufficiently smooth functions that possess continuous derivatives at least up to the second order. This again applies to quite a broad class of functions and numerous methods for their optimization. Newton methods are probably the most widely known ones [78]. They are the prototype of second derivative methods, operating on the gradient vector $\nabla f(x)$ and the Hessian matrix $\nabla^2 f(x)$. The gradient contains the first derivative of each component $(\delta f(x)/\delta x_i)$ and the Hessian contains the second partial derivatives $(\delta^2 f(x)/\delta x_i \delta x_j)$. The basic Newton method proceeds iteratively by forming a quadratic model of the objective function around the current iterate and finding its minimum approximately. An update of gradient and Hessian at that point provides the basis for the next iteration. In the multivariate case, a search direction has to be defined (e.g., steepest descent) and in general precautions have to be taken in order to guarantee convergence.

Some of the most important variations are the so-called Quasi-Newton Methods, which update the Hessian progressively and therefore economize compute requirements considerably. The most successful scheme for that purpose is the so-called BFGS update. For a detailed overview of the mathematical concepts, see [78, 79]; an excellent account of optimization methods in chemistry can be found in [80].

So far, only techniques, starting from some initial point and searching locally for an optimum, have been discussed. However, most optimization problems of interest will have the complication of multiple local optima. Stochastic search procedures (cf. Section 4.4.4.1) attempt to overcome this problem. Deterministic approaches have to rely on rigorous sampling techniques for the initial configuration and repeated application of the local search method to reliably provide solutions that are reasonably close to globally optimal solutions.

Practically all virtual screening procedures rely at least in part on some numerical optimization, be it an optimization of overlap (as in many alignment programs) [81–90], the generation of energetically favorable conformations of a molecule (for example CONCORD [91] and CORINA [92]), or the relaxation of a compound in complex with the protein (for example [93–97]). The particular virtual screening problem as a whole may be solved this way. Once a decent scoring function is defined, numerical methods can be used to find optimal solutions. However, usually these approaches are quite time consuming and should be applied only as a last resort, as the numerical analyst Forman Acton may be quoted, "...minimum seeking methods are often used when a modicum of thought would disclose more appropriate techniques" [98].

FlexX [95] and FlexS [99] contain both an implementation of the Quasi-Newton optimizer with BFGS update to optionally either readjust the rigid-body fit of the placement during or after the incremental construction process, or to flexibly postoptimize, that is, relax the resulting solutions. Both also contain an implementation of the Tripos force field [100] and allow for energy minimization of the ligand as a preprocessing step.

4.4.1.2 RMS fitting

If the correspondences of at least three reference points are available for two rigid-body objects, RMS fitting minimizes the sum of the squared distances of corresponding points.

Since this is such an important basic problem with applications (not only in virtual screening) everywhere, it received much attention. Since with the optimal superposition of two point sets the centers of mass always align, invariably the first step in the optimization is to determine the centers of mass and translate one of them by the difference vector. Then the optimal rotation remains to be determined.

Basically, two types of approaches are developed here: iterative (optimization-based) approaches like the one by Sippl et al. [101] and direct approaches like the one by Kabsch [102, 103], based on Lagrange multipliers. Unfortunately, the much expedient direct methods may fail to produce a sufficiently accurate solution on some degenerate cases. Redington [104] suggested a hybrid method with an improved version of the iterative approach, which requires the computation of only two 3×3 matrix multiplications in the inner loop of the optimization.

Superimposing three points (triangles) only, is a special case that further simplifies the problem. The normal vectors to the planes defined by the triangles are aligned and then the rotation angle about the normal vector originating from the centers of mass needs to be determined. Both the steps are geometric manipulations that have simple analytic solutions.

An important extension to rigid-body fitting is the so-called *directed tweak* technique [105]. Directed tweak allows for an RMS fit, simultaneously considering the molecular flexibility. By the use of local coordinates for the handling of rotatable bonds, it is possible to formulate analytical derivatives of the objective function. With a gradient-based local optimizer flexible RMS fits are obtained extremely fast. However, no torsional preferences may be introduced. Therefore, directed tweak may result in energetically unfavorable conformations.

FlexX (and FlexS respectively) carries out on the order of 10 000 (25 000) superpositions in the course of docking or aligning a molecule respectively. This requires around 60 (140) ms on average on a common-day PC (AMD 2.0 GHz). A triangle-matching step is one of the most frequent ones carried out while docking (resp. aligning) so-called base fragments with FlexX (FlexS). The corresponding algorithm is called on the order of 480 000 (270 000) times with a runtime of 0.011 ms per triangle fit. An extension of the docking procedure FlexX, called FlexX-Pharm, allows for constraining the docking solutions to those matching a number of pharmacophoric points (in the protein). FlexX-Pharm utilizes an internal implementation of the directed tweak algorithm and exhibits a runtime of 40 to 400 ms, depending on the length of the chain to be "tweaked".

Numerous virtual screening procedures have variants of this basic functionality somewhere in their inner machinery. Most obvious of course is the application for placing molecules onto a set of pharmacophoric points with predefined correspondence, the latter in turn being the biggest problem if no correspondence is defined or obvious from the data. Examples of placement procedures based on such principles are DOCK [106] (placing atoms on sphere-centers), GOLD [107] (placing atoms on interaction centers) and FlexX/S [108, 109] (placing atom-triangles onto triangles of interaction points).

Further on, the measure RMS distance that is to be optimized is a valuable point of information in itself. It is used, for example, to compare predictions with crystal structures and invaluable for clustering similar placements. However, caution must be taken to avoid problems with symmetry in the molecules. Again, the problem of correspondence must be treated carefully, since, for example, a rotation of 180° of a phenyl ring should not affect the result of such a quality assessment.

4.4.1.3 Distance Geometry

Distance geometry is a fairly old mathematical theory dealing with the description of objects by distances and transforming them into Euclidean space [110, 111]. Since structure determination methods like nuclear magnetic resonance (NMR) spectroscopy produce distance information, there are natural applications of distance geometry in computer-aided molecular design. Distance geometry is frequently applied in generation of conformers, molecular similarity, and in early-days docking tools [112, 113].

The geometry of a molecule is mostly described by atom coordinates (in Euclidean space) or by internal coordinates (bond lengths and angles, torsions). An alternative way to describe the geometry is to specify all atom-to-atom distances in a molecule. This description allows to deal with conformational flexibility by defining interatom upper and lower distance bounds rather than fixed distances. An additional advantage of distance matrices is their invariance against translation and rotation. Several concepts in computer-aided molecular design can be much easier expressed via distances, for example, close contacts between atoms, molecular interactions like hydrogen bonds or hydrophobic contacts, or low-energy conformational constraints resulting from bond lengths and angles or ring closures.

Obviously, working with distance matrices also has some potential pitfalls. A distance matrix is invariant against mirror imaging so that enantiomers cannot be distinguished directly. Describing the geometry of a molecule in Euclidean (three-dimensional) or internal coordinate space requires $O(N)$ parameters, where N is the number of atoms. The distance matrix, however, contains $N(N-1)/2$ interatom distance ranges, the matrix is therefore over determined. As a consequence, not every matrix represents a valid geometry in 3-D space. The process of creating a 3-D structure given a distance matrix is called *embedding*. Distance geometry provides tools in order to do the embedding of distance matrices. The most important step here is the proper manipulation of distance bounds. If we reduce a distance range within an embeddable matrix, there might be values in other ranges that cannot be achieved by any 3-D structure fulfilling all distance constraints. The process of reducing the distance ranges to embeddable distances is called tightening. For tightening, the triangle [114] and tetrangle inequality [115] which are obeyed by distances in space can be used to iteratively reduce distance ranges.

Operating in distances rather than in Euclidean space is of importance for several virtual screening techniques. Pharmacophores are usually described in distance space (see [116] for a distance geometry based matching algorithm). Well-known programs for protein–ligand docking like DOCK [93, 117], ADAM [118] or SLIDE [119] perform matching in distance space. Recently, the docking program DOCKIT [120] was presented following the lines of distance geometry and embedding. Moreover, several known conformation generators are based on distance geometry [121–123]. Finally, similarity calculations and structural alignments [124, 125] between molecules can also be performed in distance space. A recent approach following this idea can be found in [126].

The problems arising when working in distance space are sometimes reflected in the final software tools. Care has to be taken concerning stereochemistry due to the lack of distinguishing enantiomers with distance matrices. When embedding is used in a docking procedure, the conformation of the ligand molecule is fixed in a very late phase of the calculation, which may result in distortions. The quality of conformations should therefore be checked carefully when applying these tools.

4.4.1.4 FFT-based Methods

The *Fourier transform*, in essence, decomposes a function into sinus functions of different frequency that sum to the original function. It is often useful to think of functions and their transforms as occupying two domains. These domains are often referred to as real and Fourier space, which are in most physics applications time and frequency. Operations performed in one domain have corresponding operations in the other. Moving between domains allows for operations to be performed where they are easiest or most advantageous [127].

One important example is the convolution operation. For two functions f and g, the convolution $c(t)$ is defined as the integral over all x of $f(x)g(t-x)$. The Fourier-transformed C is simply the product of the Fourier-transformed F and G.

$$\int f(x)g(t-x) \bullet\!\!-\!\!\circ F(t)G(t) \tag{1}$$

By applying a variant of the extremely powerful convolution theorem stated above, computing the overlap integral of one scalar field (e.g., an electron density), translated by t relative to another scalar field for all possible translations t, simplifies to computing the product of the two Fourier-transformed scalar fields. Furthermore, if periodic boundary conditions can be imposed (artificially), the computation simplifies further to the evaluation of these products at only a discrete set of integral points (Laue vectors) in Fourier space.

Function approximation comes naturally with the Fourier transition. Since tiny details of a function in real space relate to high-frequency components in Fourier space, restricting to low-order components when transforming back to real space (low-pass filtering) effectively smoothes the function to any desirable degree. There are special function decomposition schemes, like spherical harmonics, which especially build on this ability [128].

For obvious reasons, Fourier transformations are widely used to solve problems in X-ray crystallography [129]. With innumerable replications of a molecule in a crystal, all being oriented the same way, approximate periodic boundary conditions are given. Periodic functions become discrete when Fourier transformed. In fact, the diffraction pattern of an X-ray shot on a crystal amounts to the Fourier transform of the square of the absolute values of the real space function [130]. The measurements of intensities and different reflection angles from the crystal relate to the Fourier transform of the electron densities in the crystal.

One major problem crystallographers have to deal with is the so-called phase problem, which states that of the two components of an irrational Figure (magnitude and phase) only the magnitude can be measured. A technique called molecular replacement is an approach to deal with this problem [131].

The most obvious drawback of Fourier space approaches is the computational cost of the Fourier transformation itself. However, this can be circumvented in some virtual screening applications. Gaussian functions are frequently used to approximate electron densities. Interestingly, the Fourier transform of a Gaussian function is again a Gaussian function and hence amenable to analytic transformation.

A special case is the application domain of discrete functions (e.g., measurements on some spatial grid). The Fourier transform of a discrete function can be computed quite efficiently by a special algorithm (Fast Fourier Transform) at discrete points in Fourier space [132].

The rigid-body superposition method RigFit [133] in FlexS is based on some of the above Fourier transform principles. Other examples of the usefulness of Fourier transformation in virtual screening and modeling can be found in [134–138].

4.4.1.5 Machine Learning

Machine Learning as a general technique is quite broad topic and its application in virtual screening could easily fill a chapter on its own [139]. Therefore, similar to the topic on numerical optimization, only the tip of the iceberg can be covered here.

The input of a Machine Learning method is a *descriptor* (like a set of measurable properties of a molecule), which characterizes a single data instance at hand. Most generally, the descriptor is a vector of real values with a *descriptor component* being a single number representing a molecular property (like a log P value). However, integer or binary descriptors (often called fingerprints) are quite common too (cf. Section 4.4.3).

Two generally different scenarios can be found for applications of machine learning technology: so-called supervised and unsupervised learning. The difference is the presence or absence of observation of the desired output on a training data set.

Methods for unsupervised learning invariably aim at compression or the extraction of information present in the data. Most prominent in this field are clustering methods [140], self-organizing networks [141], any type of dimension reduction (e.g., principal component analysis [142]), or the task of data compression itself. All of the above may be useful to interpret and potentially to visualize the data.

In supervised learning, feedback is given to the learner by labels on a subset of the data called training set. Again, two generally different tasks can be identified: classification and regression. Regression aims at predicting a real-valued label (e.g., an activity) as closely as possible correlated to the true labels. The whole field of quantitative structure-activity relationships (QSAR) is about finding regression models for the data [143]. The range is from simple linear equations to polynomial or other nonlinear models as can be produced, for example, by Neural Networks. Linear models may be derived, for example, by least-squares methods, with its most prominent representative being partial least square (PLS) [144, 145] which is used for 3-D QSAR applications like CoMFA [146, 147].

Classification, on the other hand, aims at predicting a binary label (two-class classification) or at most an integer label (multiclass classification) with a ratio of true versus false positives as a typical measure of success and failure (separation rate). A data set is called separable if a classifier can do the classification without error. Among a broad range of different methods for classification, there are a few frequently used ones:

- Neural Nets (NNs) relate a set of input neurons with an output neuron (providing the prediction label of a data point) by a network of layers of neurons in the interior. They are certainly among the most frequently used Machine Learning methods in the field [148] and allow for a high degree of customization since the architecture of the network itself is part of the parameters the user may define.

- Support Vector Machines (SVMs) generate either linear or nonlinear classifiers depending on the so-called *kernel* [149]. The kernel is a matrix that performs a transformation of the data into an arbitrarily high-dimensional feature-space, where linear classification relates to nonlinear classifiers in the original space the input data lives in. SVMs are quite a recent Machine Learning method that received a lot of attention because of their superiority on a number of hard problems [150].

- Decision Trees are also a well-known technique in the field [151]. They arrange a subset of the descriptor components in a hierarchical fashion (a binary tree) such that on a particular node in the tree a classification on a single descriptor component decides whether the left or the right branch underneath is followed. The leaves of the tree determine the overall classification label. Decision trees have been found useful, especially on large-scale descriptors like binary pharmacophore descriptors [152].

- Boosting methods, similar to Decision trees, builds upon multiple decisions of many so-called weak learners (like a decision on a single descriptor component). The weak learners are assembled in a sequence, where the best separating weak learner (on the training data) is taken first. Then the training data is re-weighted, giving the mislabeled data points higher weight on the basis of which the next weak learner is selected. The final result is a linear combination of weak learners [153].

One problem that occurs in virtually all Machine Learning applications is that of *overfitting*. The more complex a classifier or regression model gets, the more likely it is that it would perform well on the training data; however, it does not generalize well to unseen test data. Think, for example, of the memorizing of the active compounds in the training set as an extreme case.

Controlling the complexity of a model is called regularization. To this end, hold-out data is important. In order to benefit from a training set that is as large as possible and still to be able to measure the performance on unseen data, *cross validation* is used. It does multiple iterations of training and testing on different partitionings of the data. *Leave-one-out* is certainly the most prominent concept here [154]; however, other ways to partition are in use as well.

A recent new trend called *Active Learning* substitutes the often assumed static setting of training and test set in which a learning machine is applied by the probably more realistic scenario of a continuous flow of data. The outcome of experiments influences the choice and generation of subsequent data points [155]. Active Learning provides tools that help select the most promising next subset of data to be subjected to experimentation [156].

The common measure of success is usually finding the highest number of active molecules with the least possible number of experiments. Note that while this may well be the ultimate goal, it is not necessarily the best choice for the intermediate learning stage. Here, the goal must be to improve the model to the best level. In the classification case, this means that failures (i.e., picking inactive compounds) have to be admitted. Think of defining a boundary between active and inactive while seeing active compounds only.

Machine Learning applications in virtual screening are numerous. All of the above-mentioned techniques like clustering, classification and regression are more suitable for the task of analyzing experimental data in order to extract its most valuable ingredient, namely, information. This may be the structure-activity relation, some essential functional group, or a suggested set of high-potential candidates for further testing. On the basis of the Feature Trees approach (cf. Section 4.4.3.2), we developed an analysis tool for high-throughput screening data including a multitude of Machine Learning tools [157].

A general remark on cautious use of Machine Learning methods shall be made here. Like any other sophisticated software, it should not be used as a black box. While these algorithms are general purpose by design, some thought should go into the most appropriate design of a machine learning experiment. The encoding of a problem has to be compatible to the specific powers of the learning machine. Note that what is obvious to the human observer may, by definition, be not at all accessible to the algorithm.

4.4.2
Algorithmic Concepts

4.4.2.1 Graph Matching
The standard way of describing the topology or 2-D structure of a molecule is by a labeled graph. Not surprisingly, graph algorithms play an important role in chemistry and are

also useful for virtual screening purposes: most descriptors for similarity searching rely on the molecular topology only. The algorithmic key problem in basically all topology-based similarity measures is a variant of graph matching. In *subgraph matching*, we would like to know whether a small graph F representing, for example, a functional group is contained in a molecule graph G. Given a set of functional groups, the similarity between two molecules can be expressed by the number of common functional groups from this set. In order to identify the functional groups, subgraph matching is used. Similarity between molecules can also be expressed by the largest subgraph the two molecules have in common. Finding this maximum common subgraph is called the *MCS problem*. Both problems, subgraph matching and maximum common subgraph (MCS), belong to the class of non-deterministic polynomial (NP-) hard problems [158]; in other words, it is likely that no asymptotically efficient algorithm exists for them. Fortunately, in most applications in chemistry, the graphs are small so that the existing algorithms are sufficient in practice although they have exponential run-time behavior in theory. A recent review on MCS algorithms can be found in [159].

A fast and easy-to-implement algorithm for subgraph matching is the Ullmann Algorithm [160] (see also Figure 4.2). The algorithm is based on a 0/1 matrix representation of the matching between the subgraph and the graph, having a row for each subgraph node and a column for each graph node. Initially, all matrix entries representing assignments of compatible nodes are set to 1 and all others are set to 0. In order to describe an assignment of the subgraph to the graph, each row should contain exactly one entry set to 1 and each column should at most contain one entry set to 1. The algorithm now converts the matrix row by row into an assignment matrix by setting exactly one entry to 1 and all others to 0. During the conversion, all possibilities resulting in valid matchings of the nodes belonging to the rows processed so far are tried in turn. With this strategy, all possible matches of the subgraph to the graph are found.

In order to find the maximum common subgraph between two graphs G and H, the following approach is frequently used. Firstly, a so-called product graph is created. The graph contains a node for all possible assignments of nodes from one graph to the other. Two nodes in the product graph are connected via an edge if the two assignments are compatible to each other, that is, the corresponding nodes in graph G and H are either both connected by an edge or not. Each common subgraph in G and H corresponds to a fully connected subgraph, a so-called clique, in the product graph P.

The detection of cliques is a well-studied problem in computer science. The most frequently used algorithm was developed by Bron and Kerbosch [161]. The algorithm belongs to the class of branch & bound algorithms. Similar to Ullmann's algorithm, cliques are iteratively constructed, carefully enumerating all possibilities of extending a clique by a single node. Also in analogy to the Ullmann algorithm, the Bron–Kerbosch algorithm lists all maximum cliques within a graph.

Clique detection algorithms have other applications besides that of finding maximum common subgraphs. Whenever a matching between two objects can be expressed by pairwise compatibility of object parts, the method discussed above can be applied. The initial DOCK algorithm [106] formulates the docking problem as a matching problem between spheres in the active site and atoms in the ligand molecule. Two matches

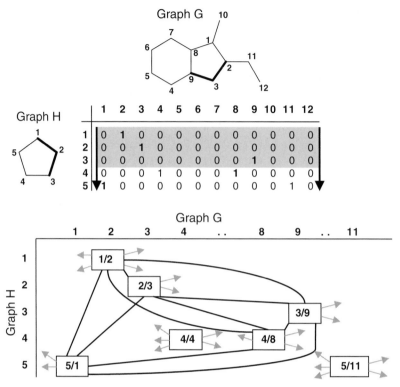

Fig. 4.2 Graph matching algorithms: For finding subgraphs, Ullmann's algorithm creates a matrix with all possible assignments. When the algorithm proceeds, rows of the matrix will be converted from top to bottom such that only one entry per row remains. The matrix is shown after matching the first three nodes of H to G. When the largest common subgraph is searched, the product graph having a node for each potential match is created (see below). Edges represent compatible matches such that a clique (a fully connected subgraph) represents a valid matching. Here, only a few nodes together with the edges forming a clique are shown for simplicity.

(between spheres and ligand atoms) are compatible if the distance between the spheres is approximately the same as the distance between the atoms. The DOCK algorithm searches for cliques in the product graph of receptor spheres and ligand atoms to find initial orientations of the ligand within the active site of the protein. The same principle has been used to address the structural alignment problem as well as the pharmacophore identification and matching problem [162].

The subgraph matching, the MCS, and the clique search algorithm mentioned above are correct, that is, they always terminate and always return the correct answer. Concerning software tools using these algorithms, care has to be taken with respect to the models that are used in order to end up with the graph problem. For geometrical applications like in DOCK, the matching tolerance is of importance. If set too low, only close to exact matches will be found which is inappropriate due to the conformational flexibility of the molecules. If set too high, basically everything can be matched on everything.

4.4.2.2 Geometric Hashing and Pose Clustering

In the mid-90s, two techniques from pattern recognition, geometric hashing [163] and pose clustering [164], were adapted to geometric problems in computer-aided molecular design (see Figure 4.3). Both techniques are now applied not only in molecular docking tools [108, 165, 166] but also in tools for calculating small-molecule structural alignments [109, 167].

A well-known problem in pattern recognition is the detection of certain objects in a camera scene. The problem can be tackled by firstly identifying characteristic points in the scene (like the corners of objects) and then comparing distances between 3 to 4 of these points to the distances in the database of known objects. Hashing techniques can then be used to quickly retrieve the objects matching the scene. Obviously, there is some parallelism to structure-based virtual screening applications: If we understand the protein as our camera scene, the potential sites of interaction as the characteristic points, the potential ligands as the database of known objects, the same hashing techniques based on comparing distance patterns between points might be applicable. Since parts of an object might be hidden in the scene, even the aspect of partial matching (some part of the ligand molecule might interact with surrounding water) is reflected. Here, we will explain the algorithms in terms of molecular docking.

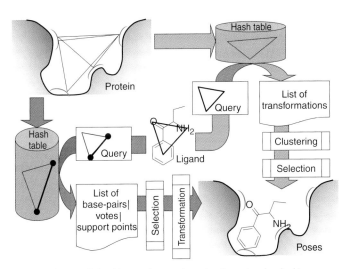

Fig. 4.3 Geometric hashing and pose clustering in molecular docking: interaction triangles from the protein are stored in hash tables, in the case of geometric hashing in the form of bases and support points. Triangle queries are created from the ligand and processed until poses result. The sequence of steps is illustrated by the red (pose clustering) and blue (geometric hashing) arrow.

In geometric hashing [163], the basic idea is to find a transformation such that as many points of an object as possible fall onto points in the scene. In a first step, a hash table is created storing all atom triplets within a set of ligand molecules. Each entry is addressed by the relative distances between the three atoms. Two of the three atoms are considered as the basis of the triangle. The goal during the initial phase of the matching procedure is to find a match of a ligand atom basis to a pair of spheres representing empty interaction sites in the protein. For this task, for each potential pair of spheres, triangles with a third sphere are created and used as queries to find matching ligand atom triangles. Each matching triangle is considered as a vote for the match of the corresponding ligand basis to the pair of protein atoms. Matches with high votes indicate a high degree of complementarity between the protein and the ligand and will be further analyzed.

A problematic point in geometric hashing is that, in principle, a basis consisting of three atoms is necessary to define an unambiguous orientation of the ligand into the active site. As a consequence, quadruplets instead of triplets have to be stored in the hash table resulting in extremely large hash tables. Using only two atoms as a basis leaves one degree of freedom open (namely, the rotation around the common axis between the two atoms in the ligand and the two spheres in the protein) such that only a small portion of the found matches of basis with a high vote can be transformed into a ligand pose with a good fit.

Similar to geometric hashing, in pose clustering [164], a hash table with triangles is created, however, this time with triplets of interaction points from the protein active site. Every triplet of ligand atoms or interaction centers is used as a query to find matching triplets of interaction points from the protein. Again, the distances between the interaction centers are used to address the hash table. For every match, the two triplets can be superimposed uniquely defining the orientation of the ligand within the active site. Instead of actually transforming the ligand, only the transformation data for each match is calculated. On the basis of the transformation, a clustering can be performed providing an initial set of orientations for the ligand, which can then be further processed.

In pose clustering, the hash table also grows asymptotically with the third power, as it is the case in geometric hashing. However, all three points are used for calculating a unique transformation instead of using one point for voting. Using list merging techniques, the asymptotic growth of the hash table can be reduced to be quadratic only [108]. In summary, a more detailed view containing more interaction points in the protein can be processed with pose clustering. It is therefore more appropriate if chemical interactions like hydrogen bonds or hydrophobic interactions are modeled instead of pure steric fit.

Geometric hashing and pose clustering are efficient algorithms calculating geometric matchings between point sets in 3-D space. Applied to molecular docking, steric fit and/or molecular interactions have to be modeled by matching points [95, 108]. This is only possible by approximation: usually, a pocket has to be filled or an interaction counter group has to be placed in a specific region. Neither the pocket nor the interaction region is geometrically pointlike. In order to achieve good results, it is extremely important that the spacing between points and the tolerance in the matching procedure are coordinated. Therefore, parametrization plays an important role in these docking approaches.

4.4.3
Descriptor Technology

Whenever no structure or model of the target protein is available in a drug design project, the only remaining concept is molecular similarity [168, 169]. On the basis of known actives and/or inactives, similarity searches [15], structural superpositions [170], and QSAR studies [171] can be performed. The retrieval of similar molecules therefore plays an important role and belongs to the oldest computer applications in molecular design. Similarity between molecules can be expressed in various ways. One possibility is to look for the largest common substructure. An algorithm for this problem was explained in Section 4.3.2. Since, for target protein binding, aspects like the molecular shape or the relative arrangement of functional groups are much more meaningful than the molecular topology, alternative approaches are of interest. Typically, large sets of molecules should be analyzed; therefore, the computational efficiency of the comparison algorithm has also to be kept in mind.

Here, we introduce frequently used descriptors classified by the structure of the descriptor itself: linear, treelike, and field descriptors.

4.4.3.1 **Linear Descriptors**
A simple way of describing similarity between two objects is by counting the number of common features. In molecular similarity, there are many ways to realize this concept. What they all have in common is that they result in a linear descriptor, a bitstring or a vector for the molecule; each position in the string represents a single feature. The bitstring for a specific molecule contains a 1 at this position if the feature is there and a 0 otherwise. Linear descriptors have three very useful properties: they can be easily stored and processed with computers, they can easily be combined just by concatenating the corresponding strings to a larger string, and a similarity value can easily be calculated by comparing two strings position by position. Mathematically, several functions for calculating a similarity value from two bitstrings are available. Frequently used functions are the Tanimoto [172] and Cosine [173] coefficient (see [15] for an overview).

The most frequently used linear molecular descriptors are purely topology-based. In the structural keys approach [174], a list of small molecular fragments is assigned to bit positions. If the molecule contains the fragment, the corresponding bit is set to 1, otherwise to 0. Originally designed to speed up substructure search [175], structural keys give quite a good performance on similarity searching. The most significant drawback is the fact that only a small portion of functional groups is directly covered in the bitstrings. Molecules containing more rare fragments cannot be handled properly, simply because the relevant fragments which would be necessary to describe them are not present.

The problem can be addressed by taking a more flexible set of fragments resulting in the hashed fingerprints approach [176, 177]. Instead of checking the occurrence of individual fragments, all atom paths up to a certain length (five or seven in practice) are enumerated. Since there are too many different paths to reserve a bit for each, a hash function that maps the path to a certain range of bit positions is applied. Hashed fingerprints, therefore, do not rely on a fixed fragment set; this is, however, paid for by loosing the one-to-one correspondence between fragments and bit positions. An alternative way of reducing the fingerprint length was recently presented [178]. Here, a statistical analysis was employed to filter out the most informative fragments for bitstrings resulting in so-called minifingerprints.

A frequent criticism about structural keys and hashed fingerprints is that they only represent topological features neglecting the geometric arrangement of functional groups. As a consequence, 3-D linear descriptors were developed. Atom-pair descriptors, originally developed as a topological descriptor measuring the topological distance between atoms (see for example [179], or the CATS descriptor [180]) can be converted into 3D [181, 182]. Each bit position represents two atom types occurring in a specific distance range, for example a nitrogen and an oxygen being between 5 and 6 Å away from each other. This concept can be extended to triplets and instead of atom types, pharmacophore points can be used [183]. The most problematic issue in 3-D linear descriptors is their conformation dependency. There are two approaches tackling the problem: The first one is to enrich the database by conformer sets for each contained molecule. The second one is to merge the bitstrings of individual conformers into a single string using the logical OR operation [183]. Both solutions are very rough approximations of the true problem: In the first case, the conformational space has to be dramatically restricted in order to keep the similarity search computational feasible. In the second case, the dependencies between the occurrences of individual atom-pair distances get lost. The lack of an exact and efficient representation of conformational flexibility is likely to be the main reason for the inferior performance of 3-D descriptors compared to topological ones with respect to biological activity [184].

Linear representations are by far the most frequently used descriptor type. Apart from the already mentioned structural keys and hashed fingerprints, other types of information are stored. For example, the topological distance between pharmacophoric points can be stored [179, 180], auto- and cross-correlation vectors over 2-D or 3-D information can be created [185, 186], or so-called BCUT [187] values can be extracted from an eigenvalue analysis of the molecular adjacency matrix.

4.4.3.2 Similarity Based on Tree Matching Algorithms

Linear descriptors are clearly the most frequently used descriptor types showing a good performance in similarity searching. Nevertheless, there are good reasons for studying nonlinear data types in the hope of finding a representation that might give better results in similarity searching applications. One possibility to describe molecules in a nonlinear fashion is to use a graph. For example, the molecular graph can be used in combination with a maximum common subgraph algorithm to describe similarity [159] (see also Section 4.4.2.1). An exact subgraph match, however, is not a very good similarity measure since the relative orientation of interaction patterns, and not the exact molecule topology, is relevant for binding. Instead of the molecular graph, one is interested in a descriptor that allows a fuzzy matching of functional groups. The first approach in this direction was the reduced graph representation by Dethlefsen and Gillet [188, 189].

Here, groups of atoms are considered as single entities, each entity falling in one out of a few categories. On the basis of a reduced graph representation, similarity can again be expressed via common substructures. The Feature Tree [190] approach summarized here extends the idea of reduced graphs in several ways. Especially, more detailed shape and chemical descriptors are used for the nodes and specially designed matching algorithms are applied allowing for more flexibility during matching. Compared to linear descriptors, the relative arrangement of functional groups is preserved in the graph structure.

In a feature tree descriptor, each node represents a small building block of the molecule. The nodes are connected according to the topology of the molecule. For the comparison algorithms described later, it is important that a Feature Tree is indeed a tree, that is, that it has no cycles. In order to achieve this, rings have to be represented by single nodes (see also Figure 4.4). For each node, the steric and physicochemical properties of the building block are stored by shape and chemistry descriptors. As a default, the volume and the number of ring closures are used as shape descriptors and an interaction profile accounting for hydrogen bonding capabilities and hydrophobic contact surface is used as a chemistry descriptor. Since the Feature Tree technology is not dependent on a certain shape or chemistry descriptor, other descriptors can alternatively be used. For example, partial charges can be used as an additional chemistry descriptor. Note that no three-dimensional information is used for generating the feature tree; the descriptor is therefore conformation independent.

A linear descriptor naturally contains an ordering of its elements, which makes it very easy to compare them. In a tree structure however, it is not clear which parts should be mapped onto each other. Different mappings will result in different similarity values;

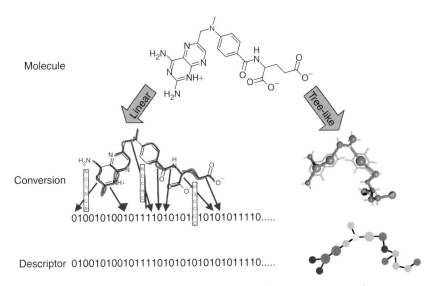

Fig. 4.4 Linear versus treelike molecular descriptors: Linear descriptors represent the occurrence of certain features like fragments or paths, the relative arrangement gets lost. Treelike descriptors represent the building blocks of the molecule as well as their relative arrangement.

therefore, an algorithm that calculates a mapping under optimization of the similarity of the two molecules is needed. The strength of the Feature Tree approach is that it contains mapping algorithms that allow some fuzziness. The grouping of nodes and even the introduction of nonmatched intermediate nodes (gaps) facilitate this instead of simply performing a one-on-one assignment of feature tree nodes. At the same time, the algorithms perform a topology-maintaining matching such that the overall structure of the molecule is considered in the matching process. For a detailed description of the algorithms, we refer to [190].

A disadvantage of treelike descriptors is their comparison time. A Feature Tree matching – although being calculated at a rate of more than 50 000 comparisons per minute – is slower than the comparison of a linear descriptor based on bitstrings. So, why should one consider these descriptors at all? Treelike descriptors combine the advantages of 2-D and 3-D descriptors: applied to practical screening exercises, they show enrichment rates that are in the same order of or even superior to 2-D descriptors. They are less dependent on the detailed topology of the molecule, consider the overall distribution of functional groups within the molecule while at the same time not being conformation dependent as 3-D descriptors are. A further advantage is that since treelike descriptors represent fragments of the molecules by independent subunits of the descriptor, they are more suitable for similarity searches in combinatorial libraries or chemistry spaces [191].

4.4.3.3 Field-based Descriptors

"Molecules recognize each other by their surfaces . . . not through their underlying bonding skeletons", as remarked by Klebe in [192]. Now a surface can be defined in various ways. Quite common besides variants of the van der Waals surface are iso-contour surfaces of the different fields modeling the different properties of a molecule.

Most commonly used is certainly the molecular electrostatic potential. It can be derived from any kind of charge distribution. Usually, the MEP is first calculated on a grid and subsequently transformed to the sphere or Gaussian representation. Quite important is the electron density distribution, which closely models the steric occupancy by a molecule. Other approaches utilize artificial fields for physicochemical properties commonly associated with binding, like a field for the hydrophobicity [193] or H-bonding potential [133, 194].

Fields as property values in 3-D space can either be evaluated and encoded on a regular (often cubic) grid [71], or approximated by certain distribution functions. Most often, Gaussians [83] have been employed here, as they have some desirable mathematical properties and can usually approximate the original field reasonably well with not too many parameters.

Fields can be utilized in virtual screening applications for assessing the similarity (alignment) or complementarity (docking) of molecules. Two similarity measures have achieved the most attention. These are the so-called Carbo- [195] and Hodgkin indexes [196] respectively. Others are Pearson's product moment correlation coefficient [169] and Spearman's rank correlation coefficient [169].

The successful application of any of these measures requires a reasonable orientation of the two fields relative to each other as a prerequisite (see the introductory remarks on docking and alignment).

The fields that are derived on a grid can be encoded into lengthy binary [167] or real-valued descriptors [147]. Also, an abstract description as the so-called field graphs has been attempted [197]. Another approach to convert fields back into linear descriptors is to extract the characteristic features [198].

4.4.4
Global Search Algorithms

Virtual screening applications based on superposition or docking usually contain difficult-to-solve optimization problems with a mixed combinatorial and numerical flavor. The combinatorial aspect results from discrete models of conformational flexibility and molecular interactions. The numerical aspect results from describing the relative orientation of two objects, either two superimposed molecules or a ligand with respect to a protein in docking calculations. Problems of this kind are in most cases hard to solve optimally with reasonable compute resources. Sometimes, the combinatorial and the numerical part of such a problem can be separated and independently solved. For example, several virtual screening tools enumerate the conformational space of a molecule in order to address a major combinatorial part of the problem independently (see for example [199]). Alternatively, heuristic search techniques are used to tackle the problem as a whole. Some of them will be covered in this section.

4.4.4.1 Stochastic Search Methods
The landscape of an energy or scoring function is often not at all smooth but is instead rugged and loaded with multiple local optima. In such cases, local search methods like the optimization methods described in Section 4.3.1.1 depend heavily on a good starting point in order to arrive at a reasonable optimum, or rather multiple starting points to increase chances of finding the global optimum. Such mechanisms are usually built in global search methods.

Monte Carlo methods sample many data points on the energy surface by randomly choosing new parameters to generate them. This way, information of the system under consideration can be gathered [200]. However, in real applications parameter space is easily too vast and not all parameter combinations really make sense. *Metropolis Monte Carlo* methods [201] sample the space by tweaking the parameters a little bit and accepting a configuration based on its energy [202–204]. If the energy is lower (in the case of minimization), the configuration is accepted. If it is higher, the configuration is accepted only with a certain probability depending on the magnitude of the rise.

Simulated Annealing uses this technique in its inner loop. Derived from statistical mechanics [205], it relates to a procedure to cool liquid glass to the most stable solid state. The so-called cooling schedule changes the criterion of acceptance of configurations with higher energy over time by permitting large changes in the beginning and only smaller changes later, as the system cools off. Simulated Annealing can be proven to find the global optimum if the cooling schedule permits an infinite number of iterations. It was found useful in solving complex optimization problems with large search spaces [205] and found its application in virtual screening methods too [206, 207].

Genetic Algorithms are a general purpose, global optimization technique [208] that provides promising results in the entire area of computational chemistry [209] and structural biology [210]. Genetic Algorithms mimic the process of evolution. A generation within this process comprises a set of configurations that are coded via chromosomes. Chromosomes are subjected to manipulation by some genetic operators such as crossover and mutation. The information content of the chromosomes varies depending on the application. Typically, it comprises the intramolecular matches or a coding of the orientational degrees of freedom and a coding of the torsional degrees of freedom in the case of considered molecular flexibility. The fitness function used to enable the process of selection typically comprises an efficiently "evaluable" energy function. Examples of Genetic Algorithms in virtual screening can be found in [94, 107, 211]. Genetic Algorithms can also be applied in a partially computational – partially experimental setup for identification of new lead structures. For example, Weber et al. [212] applied a genetic algorithm to successively optimize a focused combinatorial library.

Similar to the general-purpose numerical optimization procedures (cf. Section 4.4.1.1) these stochastic global optimization techniques also have the advantage that they are easy to use, extensible, and comparably robust about model changes. However, this has its price in redundant computation and generally high-computational cost due to the generally little exploitation of the problem structure. There is no magic in these methods to find a global optimum and in several cases a great number of optimization cycles may be necessary to find a promising solution.

4.4.4.2 Fragment-based Search Methods

All optimization problems have the property that they become significantly easier to solve when the input complexity decreases. Strategies based on solving small subproblems and then combining or extending the achieved results to a solution for the original problem are therefore quite common in computer science. Divide & Conquer, Dynamic Programming, Branch & Bound, and Greedy Algorithms [76] are examples for such strategies. It is therefore not surprising that they are also developed for virtual screening applications. Matching problems are frequently addressed by these strategies and we have mentioned some algorithms above, for example Bron–Kerbosch for Clique Searching [161] or Dynamic Programming for Tree Matching [190]. Here, we want to focus on structural superposition and docking algorithms implementing the idea of problem subdivision by molecule fragmentation.

A very small molecule is significantly simpler to align or to dock because of its limited conformational space. The molecule is either rigid or has a low number of conformations, which can be enumerated. This is not the case anymore for molecules with more than, say, five rotatable bonds. Here, the number of low-energy conformations is typically too high to justify an enumeration. Since several algorithmic engines for dealing with rigid molecules are already in place, a reasonable approach is to cut the molecule into smaller pieces or fragments and consider them somehow individually. The remaining open issue is then how to combine or extend placements for fragments to placements for the whole molecule.

One possible option is to place fragments individually and then try to combine solutions for fragments to placements for the whole molecule. In order to avoid the testing of every combination of fragment placements for geometric compatibility, that is, whether the length and the angles at the connecting bond are within reasonable geometric tolerances, hashing techniques can be used. Sandak et al. [166, 213] give an extension of geometric hashing for this task. The strategy, we call it *place & join* [214], is applied in early approaches to molecular docking[215]. Most frequently, however, it is used in *de novo* design algorithms (for example LUDI [216, 217], SPROUT [218], PRO_LIGAND [219], see [220–222] for a review) in which a large collection of fragments is used to construct molecules instead of looking at a single molecule at a time.

An alternative strategy called *incremental construction* takes the placements for a single fragment and extends the placements by further fragments until the whole molecule is constructed. In each round, a partial ligand for which placements are calculated is extended by one fragment exploring the conformational space of the fragment and combined with the placements found for the partial ligand. Several *de novo* design tools, for example [223], [216] and most molecular docking tools based on fragmentation like DOCK 4.0 [93], FlexX [95, 224] and Hammerhead [225] as well as some structural alignment tools [226] are following the incremental construction strategy. The overall incremental construction strategy looks simple and in fact is so. The success, however, depends on important details, for example, how many partial placements can be covered within each round, how does the selection of placements for the next round work, and how are the partial placements scored and optimized? A good implementation should consider as many partial placements as possible. The selection process should keep a certain degree of diversity within the solution set rather than selecting by score alone. Finally, an optimization of each partial ligand orientation is necessary in order to reduce the influence of the initial base placement [95].

Incremental construction compared over place & join has the advantage that it considers the ligand conformation in the first place. In other words, bond lengths and angles are correct. Place & join considers the protein–ligand interactions of the fragment in the first place, which might result in distorted conformations. Of course, the structures can be minimized in order to improve the ligand conformation; however, it is possible that the hydrogen bonds formed and matching hydrophobic surfaces between the fragment and the protein are broken up. This is probably the most important reason why incremental construction is preferred to place & join in molecular docking applications.

4.4.4.3 Dealing with Combinatorial Input Spaces

Combinatorial chemistry, developed in the mid-90s [227–229], allows the efficient synthesis of large sets of compounds with diverse features. It is therefore a widely used technology for creating screening libraries. For experimental screening, the most important facts that count are the number, the size, and the diversity of molecules that can be created. For virtual screening, the question of dealing with a historically grown set of compounds with little relationships between each other or with a combinatorial library is completely different. The combinatorial library is not just larger – it is a well-structured search space compared to the historic compound collection. Obviously, every screening tool – structure- or ligand-based – can be applied to an enumerated combinatorial library. The interesting point, however, is whether and how the structure of the combinatorial library can be exploited to speed up the virtual screening process.

Depending on the computational resources necessary to process a single molecule within a screening process, the need for exploiting the library structure is more or less pressing. Linear descriptors, for example, can be searched with such a high speed that combinatorial approaches are of less importance [187]. Since the generation of linear descriptors is time consuming, Barnard et al. [230] developed a method for creating bitstring representations based on the combinatorial library in its closed form. Roughly spoken, this algorithm works in three phases. First, information is created on a per fragment basis, that is, for the core and for each R-group independently. Second, this information is combined for each molecule within the combinatorial library and third, corrections are made for the interface region between the connected fragments.

For more complex descriptors, the question arises whether the explicit enumeration of all molecules of the library can be avoided during screening. We can consider the process of searching for a similar molecule in a combinatorial library as an optimization process, namely, the optimization of similarity while navigating through the space of molecules defined by the combinatorial library. In contrast to a screening collection, a combinatorial library naturally defines a neighborhood relationship among its molecules by the number of R-group instances two molecules have in common. From this viewpoint, every general-purpose optimization algorithm like simulated annealing or a genetic algorithm can be applied to the problem. The TOPAS [231] system allows the creation of similar molecules on the basis of an evolutionary algorithm. Starting with a random set of molecules, the molecule's similarity to the target structure is evaluated. Then, parent molecules are selected and variants are created, which then are evaluated again. TOPAS was developed for searching in chemistry spaces, which is a much broader concept than a combinatorial library. An advantage of applying a generic optimization algorithm like the genetic algorithm in TOPAS is the flexibility concerning the similarity measure: everything that returns a similarity measure for two molecules as input will be sufficient (see also Section 4.5.2.4 for applications of TOPAS).

For treelike descriptors, special purpose algorithms can be designed for combinatorial searching [191]. In a treelike descriptor, a fragment of a molecule belongs to a subtree within the descriptor. Therefore, an exchange of an R-group is reflected by a change of a subtree within the descriptor. On the basis of a dynamic programming approach, it is possible to directly calculate the most similar molecule within a combinatorial library or chemistry space. The advantage is that the algorithm is deterministic, resulting in the optimal solution in extremely short computing times. This is paid, however, by not being flexible concerning the choice of the descriptor anymore. A third example is the ChemSpace approach [232, 233], which combines field-based descriptors for virtual combinatorial libraries during the search.

Owing to the much longer per-molecule computing time, exploiting the structure of combinatorial libraries becomes much more important for molecular docking algorithms.

Docking of combinatorial libraries is highly related to structure-based *de novo* design [220–222]. If the algorithm allows a clear definition of the rules describing how fragments can be combined, it can be directly used as a combinatorial docking algorithm just by selecting the appropriate fragment sets. This is done, for example, with LUDI in order to create new thrombin inhibitors by a single-step reaction [234]; see also Section 4.5.

For molecular docking software suites like DOCK and FlexX, combinatorial library extensions were developed. Since both tools follow the idea of incremental construction, they start with docking an initial fragment and exploring the combinatorial library space based on the resulting placements for it. The CombiDOCK [235] approach starts with the core fragment and then considers the R-groups individually. High-scoring R-group placements are then combined and the resulting molecule is tested for intraligand overlap. Owing to the independent consideration of R-groups in the initial docking phase, CombiDOCK is very fast. The disadvantages are that the starting point has to be the core fragment and that intraligand clash is detected in a very late stage of the docking algorithm. FlexX-C [236] initiates its combinatorial docking procedure by placing all instances of a user-selected R-group (or the core). From there on, all molecules of the combinatorial library are subsequently created and build-up in the active site. Although every molecule of the library is constructed in the end, there is a significant speed-up since placements for partially build-up molecules are recycled. Ideas similar to those for dealing with combinatorial libraries in docking applications can obviously also be used for structural alignment algorithms based on incremental construction like FlexS.

Whenever large combinatorial libraries or chemistry spaces should be screened, a good exploitation of the combinatorial structure of the input is crucial. For most kinds of screening problems, approaches are available. Linear descriptors seem to be less suited for dealing with combinatorial libraries. However, since the similarity calculation is very fast, they can typically be applied even on enumerated libraries. Some representations like field-based descriptors (topomeric shape descriptors) or tree-based descriptors (Feature Trees) keep the fragment structure within the descriptor, allowing the processing of much larger libraries. Concerning structure-based screening, the algorithmic choice depends on the structure of the library. If the core is likely to have a well-defined orientation within the active site, the CombiDOCK approach of screening R-groups individually can be applied. If one of the R-groups seems to dominate the binding, the FlexX-C approach seems to be favorable.

4.5
Entering the Real World: Virtual Screening Applications

4.5.1
Practical Considerations on Virtual Screening

There is no standard procedure for structure- or ligand-based virtual screening. Every prospective application requires understanding and tuning of key parameters. All available information on the problem should first be analyzed to generate or select validated models and define the work flow accordingly. Filters are useful at each stage. These are based on tools of increasing complexity and thus reduce computational speed. An important point to consider is the appropriate treatment of ionization and tautomerization states in the input data. This is currently a challenging task, as there is no software tool that automatically performs the required corrections in an automatic fashion. The correct tautomer for a particular ligand also depends on protein binding site requirements. However, any correct treatment significantly influences the quality of results for structure- and ligand-based virtual screening.

An interesting discussion on practical aspects of virtual screening including a work flow for structure-based virtual screening was given by Lyne [237]. He pointed out that any successful virtual screening campaign encompasses a number of sequential computational phases from database filtering and preparation to target preparation, predocking filters, docking, postdocking analysis and prioritization of compounds for testing.

Individual steps in a virtual screening work flow could be subdivided into at least four different tasks on the basis of the level of complexity of input [2]. Very simple one-dimensional filters should be used early to significantly reduce the number of candidates. This is a step that tries to eliminate unwanted chemical structures with features that make them less likely to be convertible into drug candidates. Such criteria describe physicochemical properties of "druglike" molecules; these criteria include "rule-of-five" [238] or related ones [239], summarizing simple, intuitive parameters as alert for compounds that are unlikely to be orally absorbed. Other simple criteria for filtering include reactive, toxic or otherwise undesirable fragments, potential metabolic liabilities, leadlike or druglike property filters [240, 241] and others, which recently have been summarized [242]. Those filters have to be applied only once before using a database for multiple virtual screening applications. Strategies for virtual screening by early and efficient combination of informative descriptors for scoring of protein–ligand binding modes and ADME optimization in parallel have also been described [243, 244]. This early integration of descriptors toward simultaneously addressing two tasks in virtual screening protocols appears to be very promising. In the past years, several virtual screening filters based on straightforward 2-D descriptors have also been established to weed out compounds with a potentially undesirable side effect at an early stage before evaluating their capability to fit to the target. Those filters for liabilities include the prediction of frequent hitters [245], hERG channel blocking [246], CYP3A4 and CYP2D6 inhibition [247, 248], and others.

Experience in medicinal chemistry suggests that successful drug discovery programs are also limited by nature of the molecular target. It has often been suggested that organic compounds can module more readily some privileged protein target classes, which led many groups [249] to the development of target family related projects and a preselection of tractable targets in addition to careful selection following therapeutic requirements. Similar considerations about privileged substructures, motifs or properties could be incorporated at this step.

The second step comprises similarity searches from known ligands using 2-D descriptors like fingerprints, topological descriptors like atom-pair fingerprints, topological pharmacophores (CATS) or feature trees. As this step might introduce a significant 2-D bias into the resulting hit list, the balance between generality and previous knowledge has to be adjusted on a case-by-case basis. While 3-D pharmacophore searching now can be classified as third step, it requires more information on active molecules for a particular target. The fourth and most complex step requires even more structural knowledge, when molecules are docked into protein binding sites and compounds are selected using scoring functions. Each virtual screening run requires a postanalysis phase toward the minimization of false positives in the hit lists and propagating the true hits to the top of these lists. Techniques like consensus scoring, postscoring using an additional scoring function or using more rigorous approaches to compute some important protein–ligand interaction terms like electrostatics or desolvation contributions can be applied subsequently. Furthermore, final geometrical filtering by necessary or undesirable, unrealistic interactions, shape complementarity, exposed hydrophobic parts of a ligand, and unrealistic conformations might be useful, while some tasks are still most efficiently accomplished by visual inspection of top-ranking compounds in the result list.

   These applications of virtual screening filters and methods with increased complexity are dependent on the stage at which they are used in the virtual screening work flow. A sequential process should always be set up to focus the compound set, as much as possible, for time-consuming investigations like flexible docking.

4.5.2
Successful Applications of Virtual Screening

The primary focus of virtual screening is to identify starting points for high-quality lead structures having the potential to be convertible into drug candidates. This motivation limits also the number of success stories in the recent literature, as many projects in early phases have to invest a significant amount of resources to convert a promising hit from an *in silico* hit finding approach into a promising lead series prior to clinical development, while results are still unpredictable and the attrition rate is significant.

   Virtual screening is often seen as a modern technology complementing or augmenting traditional high-throughput screening approaches in pharmaceutical settings. An increasing number of prospective applications of virtual screening are described. However, most studies are retrospective in nature, as their primary focus is to demonstrate the utility of a software tool or virtual screening strategy. It is useful for assessing its real value to separate retrospective validation studies from prospective applications in a drug discovery setting. Consequently the discussion in this section will focus on a selection of different prospective applications.

4.5.2.1   Structure-based Virtual Screening
Furet et al. described the discovery of a novel, potent and selective kinase CK2 inhibitor by a high-throughput docking protocol [250]. A large subset of the Novartis corporate database with ~400 000 compounds was flexibly docked using DOCK 4.01 [93, 117]

into the ATP binding site of a human CK2α homology model, derived using the X-ray structure of *Zea mays* CK2α (PDB entry 1DAW). The work flow encompassed first a filtering step on molecular weight, number of rotatable bonds, and undesirable substructures prior to 3-D conversion and adjustment of protonation states using a rule-based method. The authors conclude that structure-based virtual screening is useful for lead identification, but only if postdocking filtering and reranking procedures are applied to the primary hit lists directly after docking. To this end, the compounds with best DOCK scores were further pruned using a three-step postprocessing treatment. A filtering of the hit list by additionally available pharmacophore information, that is, hydrogen bonds to the kinase hinge region, was done, followed by reranking of acceptable compounds using the scoring function from Wang et al. [251], before visual inspection of the best compounds, resulting in a total of 12 compounds for biological testing. Visual inspection is seen as a crucial step to discard compounds with unfavorable interactions in the binding site or unrealistic conformations. The best rigid inhibitor showed an IC$_{50}$ value of 80 nM (see Figure 4.5a).

Peng et al. describe the identification of novel BCR-ABL tyrosine kinase inhibitors using a related docking protocol [252] from a database of 200 000 commercially available compounds using DOCK 4.01. From 1000 candidates with best DOCK energy scores, 15 compounds were tested after grouping them in structurally diverse classes and applying druglikeness filters. Eight of them showed significant activity with IC$_{50}$ values between 10 and 200 μM. Other successful applications of DOCK in drug discovery have been reported for targets like HIV-1 protease [253], thymidylate synthase [254, 255], influenza hemagglutinin [256], parasitic proteases [257], PTP1b [258], Kinesin [259], and others.

Grüneberg et al. successfully applied a sequential virtual screening protocol in search for novel inhibitors of human carbonic anhydrase II [260, 261]. Their innovative virtual screening protocol is based on several consecutive hierarchical filters applied to a database of $\sim$ 100 000 compounds involving a preselection based on privileged functional groups from an analysis of available ligand and protein structural data, followed by pharmacophore searching using a protein binding site–derived pharmacophore. The similarity of candidates with known ligands is used to rerank the hit list from this step. The acceptable 100 compounds after this step were subjected to flexible docking using FlexX [95, 224]. After docking and the analysis of affinity predictions, 13 compounds were tested, showing remarkable affinities with 3 candidates in the subnanomolar range and structural novelty. Moreover, the binding modes from docking poses could be experimentally validated by X-ray structure analysis of two inhibitors (cf. Figure 4.5b).

In another application from this group by Brenk et al., a similar sequential virtual screening protocol was used to identify submicromolar lead structures for tRNA-guanine transglycosylase (TGT) on the basis of an X-ray structure analysis of a previous micromolar inhibitor [262]. On the basis of its rather unexpected binding mode, three differing protein-based pharmacophore hypotheses were generated and employed for virtual screening, resulting in a diverse set of candidates. The most acceptable compounds were docked into one of the two alternating TGT binding site conformers using FlexX, resulting in a final set of 9 candidates with significant activity in the micromolar and submicromolar range (cf. Figure 4.6).

The use of structure-based virtual screening toward novel nuclear hormone receptor antagonists was described by Schapira et al. [263]. As only X-ray structures of ligand binding domains of agonist-bound nuclear receptors with exception of the antagonist-bound estrogen receptor-α are available, the authors constructed an "inactive"

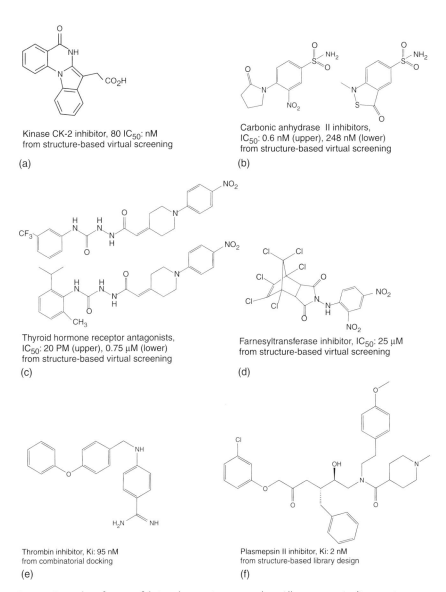

Kinase CK-2 inhibitor, 80 IC$_{50}$: nM
from structure-based virtual screening

(a)

Carbonic anhydrase II inhibitors,
IC$_{50}$: 0.6 nM (upper), 248 nM (lower)
from structure-based virtual screening

(b)

Thyroid hormone receptor antagonists,
IC$_{50}$: 20 PM (upper), 0.75 µM (lower)
from structure-based virtual screening

(c)

Farnesyltransferase inhibitor, IC$_{50}$: 25 µM
from structure-based virtual screening

(d)

Thrombin inhibitor, Ki: 95 nM
from combinatorial docking

(e)

Plasmepsin II inhibitor, Ki: 2 nM
from structure-based library design

(f)

Fig. 4.5 Examples of successful virtual screening approaches. All structures (a–l) are active against different target proteins or receptors, they were discovered in prospective applications using a variety of virtual screening approaches and work flows. See text for details.

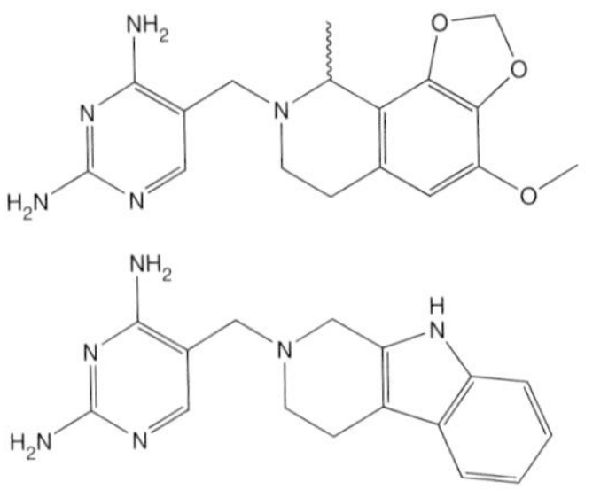

Dihydrofolate reductase inhibitors from structure-based library design

(g)

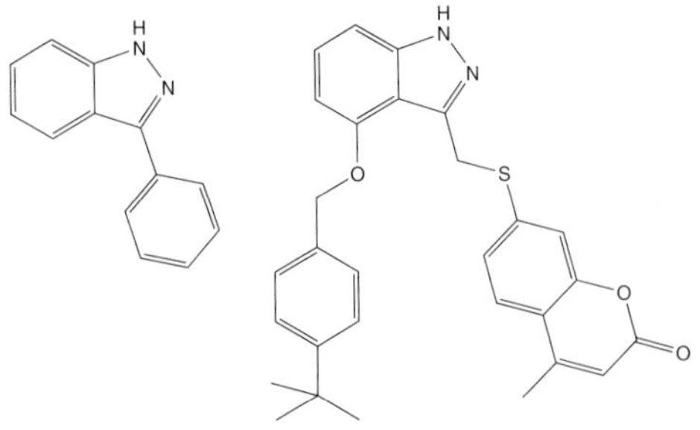

DNA gyrase inhibitors from structure-based fragment screening (left, 8 µg/mL) and subsequent lead optimization (right, 0.03 µg/mL)

(h)

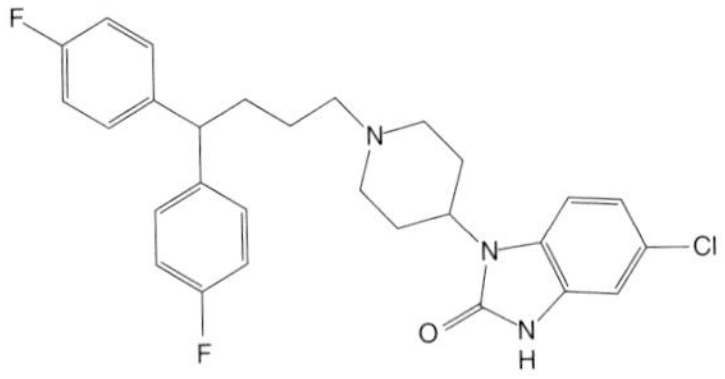

Calcium antagonist Clopimozid, IC$_{50}$ < 1µM from topological pharmacophore searching

(i)

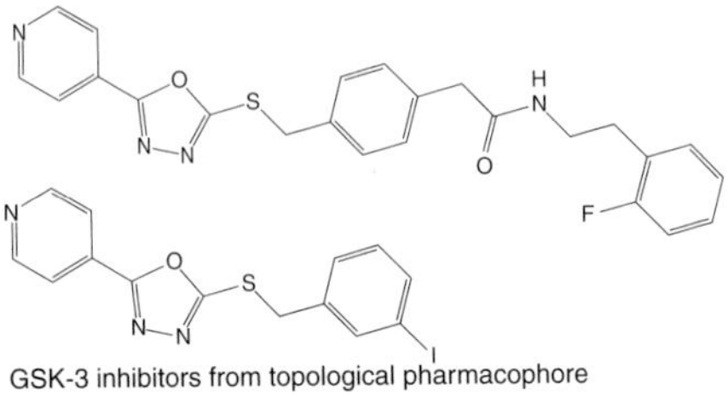

GSK-3 inhibitors from topological pharmacophore searching (IC$_{50}$:1.2 µM,upper) and subsequent lead optimization (0.39 µM, lower)

(j)

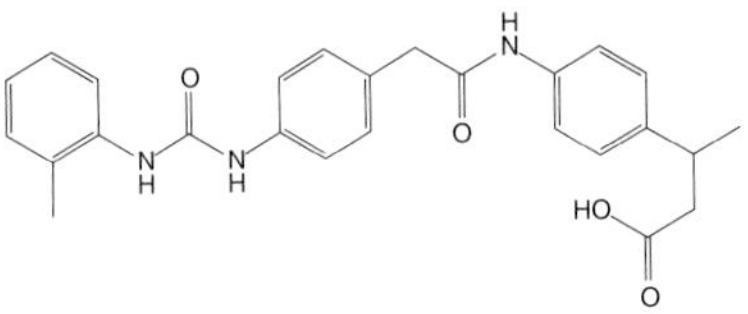

D4 E1 (VLA-4) antagonist, IC$_{50}$: 1 nM from 3D pharmacophore searching

(k)

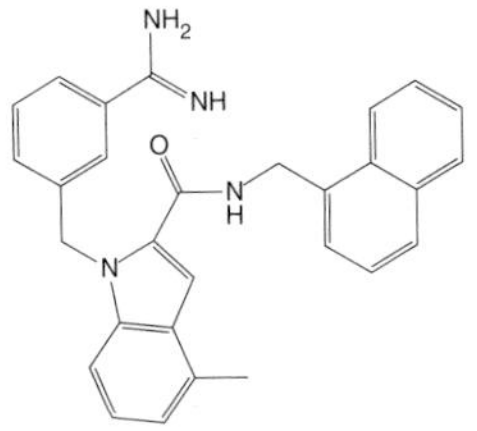

Urotensin II antagonist, IC$_{50}$:400 nM from 3D pharmacophore searching

(l)

Fig. 4.5 *(Continued)*

conformational model for the human retinoic acid receptor-α from this estrogen receptor structure and successfully applied it for structure-based virtual screening using ICM [264]. The best scoring compounds were attributed to conformational postfiltering and visual inspection. A total of 32 compounds were tested and this led to two structurally novel antagonists. The authors emphasize their approach to exclude any *a priori* privileged substructure and pharmacophore knowledge from their protocol in order to retrieve an unbiased set of candidates with potential interactions with the receptor. The authors attribute the success of this approach to the quality of the docking

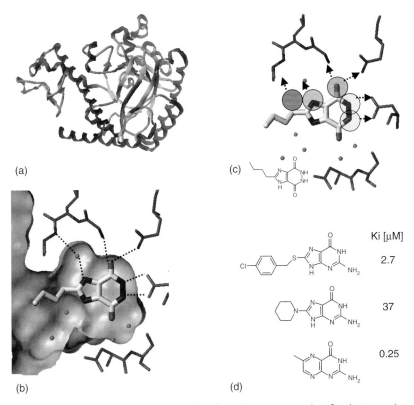

	Ki [μM]
	2.7
	37
	0.25

Fig. 4.6 Example for successful structure-based virtual screening to identify submicromolar tRNA-guanine transglycosylase (TGT) inhibitors based on X-ray structures of weaker ligands. **(a)** X-ray structure of *Zymomonas mobilis* TGT at 2.1 A resolution (pdb entry 1N2V). **(b)** Experimental binding mode for pyridazinedione scaffold (Ki: 83 μM) including structure water from 1N2V. **(c)** Definition of structure-based pharmacophore with ligand interactions as spheres (red – acceptor, light blue – donor, magenta – water) and protein-side points indicated by arrows. One alternative ligand donor (dark blue) has been identified from an X-ray structure of a different ligand (not shown), leading to slightly differing pharmacophore hypotheses. **(d)** New TGT inhibitors discovered by virtual screening protocol.

program in combination with its scoring function. This could partially also be due to binding site requirements. Protein binding sites with different hydrogen bonding and hydrophobicity characteristics favor different docking and scoring tools, as unveiled in a comparative study of different binding pockets [59]. The results on the estrogen receptor binding site suggest that steric complementarity and hydrophobic interactions are in fact determining the docking into some nuclear receptors.

A similar approach was used toward the identification of antagonists of the thyroid hormone receptor [265]. Again, a model of this receptor in its predicted antagonist-bound conformation served to select 100 antagonist candidates out of a library of ∼ 250 000 compounds. Fourteen of 75 tested molecules were found to antagonize this receptor with IC_{50} values ranging from 1.5 to 30 μM. Furthermore, one of the best ligands from this run was used to design a follow-up virtual library. After synthesis, some of these second-generation antagonists showed activities in the submicromolar range (cf Figure 4.5c). The structural diversity of these novel antagonist demonstrates that structure-based virtual screening is able identify diverse chemistries that comply with the structural rules of nuclear receptor antagonism.

These authors also showed by docking 5000 structurally diverse compounds including 78 known nuclear receptor ligands against a total of 18 nuclear receptor X-ray structures and one computational model significant enrichment rates for all but one target [266]. They point to the problem that receptor flexibility still remains to be fully included and that consequently the choice of the appropriate binding site geometry critically determines virtual screening results. This statement again should prompt for a careful validation of structural and pharmacophore model prior to using them for virtual screening.

Perola et al. successfully applied structure-based virtual screening on a database with ~220 000 commercially available compounds to identify novel inhibitors of farnesyltransferase (FT) [267]. After a number of consecutive filters to remove undesirable structures from the database, the authors applied their program EUDOC for rigid docking. They again highlighted the importance of protonation states and relevant predocking filters, which depend on the nature of the investigated target structure. The final postdocking filtering included the use of AMSOL solvation energies and visual inspection to reject compounds with unrealistic binding modes. Among 21 final candidates, 4 inhibited FT with IC_{50} values between 25 to 100 μM, while none of 21 randomly selected compounds had an IC_{50} lower than 100 μM (cf Figure 4.5d).

4.5.2.2 Structure-based Library Design

The integration of combinatorial chemistry, structure-based library design and virtual screening [268, 269] also resulted in successful applications [270, 271]. It ultimately should result in broader SAR information about directionality and physicochemical requirements of acceptable building blocks. This concept is based on feasible scaffolds for exploring protein subsites using parallel or combinatorial synthesis.

Böhm described combinatorial docking [272] implemented in LUDI [216] for reductive amination in search for thrombin inhibitors. The binding mode of the most active one with a Ki value of 95 nM was even confirmed by X-ray structure analysis [234] (see Figure 4.5e).

Murray et al. [273] described an implementation of the preoriented template approach, which also forms the basis for recent extensions of QXP [204] and GOLD [274].

The DOCK combinatorial docking implementation [275] was also applied for the design of novel enzyme inhibitors [276, 277]. In one of these prospective examples, Haque et al. reported potent low-nanomolar plasmepsin II aspartyl-protease inhibitors [278] from a set of several focused libraries with a best K_i value of 2 nM (cf Figure 4.5f).

The discovery of novel dihydrofolate reductase inhibitors by structure-based library design based on a 5-(dialkylamino)-2,4-diaminopyrimidine scaffold was reported by Wyss et al. [279] (cf Figure 4.5g). On the basis of a diaminopyrimidine core, a virtual library of ~9000 compounds was constructed. One sublibrary with 252 compounds for synthesis was selected from the virtual library by docking using FlexX, resulting in 54 hits (21%). In contrast, a second, low-scoring library did only produce 4 hits (1%), while another diversity-based selection resulted in 17 hits (3%). The authors concluded that compounds selected using structure-based virtual screening and library design methods resulted in both a higher fraction of active molecules and lead compounds with higher activities than did compounds selected using diversity considerations.

Liebeschuetz et al. described the identification of potent and selective factor Xa inhibitors by combining structure-based virtual screening and array-based chemistry [280] as follow-up to a previous virtual screening study for factor Xa, thrombin, and estrogen receptor ligands [281]. Their in-house program PRO_SELECT is applied for evaluating virtual libraries to fit the protein binding site. This led the authors to explore a small seed fragment in three iterations of library design, virtual screening, rapid synthesis and testing, resulting in highly potent and selective factor Xa inhibitors.

Those examples show that the integration of structure-based design with combinatorial chemistry was accompanied with novel algorithms toward focused library design and subset selection using steric, electrostatic and hydrophobic binding site constraints. These emerging techniques allow to computationally build and screen virtual combinatorial libraries for binding site complementarity, while selection criteria might involve empirical scoring functions, diversity assessment and synthetic feasibility.

4.5.2.3 Fragment-based Virtual Screening

In search for new, active motifs to start a drug discovery program, it has been recently discussed in literature that leads with a relatively low structural complexity can be better detected and show a more favorable optimization potential. Hence, some groups tried to identify those small molecular starting points also by structure-based virtual screening. Such an approach is not directed to search for final molecules, but for fragments aiming to bind only to one particular subpocket of the protein binding site. This approach could serve to identify anchor points for a subsequent structure-based library design by extending the fragment into other binding site areas. Alternatively, one could link different fragments to gain additional binding affinity. Recent examples include the discovery of DNA gyrase inhibitors by means of 3-D structure-based fragment screening combined with biophysical assays and protein structure-based optimization at Roche [282] (cf. Figure 4.5h) and the discovery of IMPDH (inosine 5'-monophosphate dehydrogenase) inhibitors [283] from a virtual screening protocol tailored to discover small, alternative fragments to phenyloxazole anilines.

4.5.2.4 Ligand-based Virtual Screening

Ligand-based design approaches have been used for some time to extract information from databases using hypotheses on motifs relevant for biological activity. The section is not aiming at reviewing all successful applications of 3-D searching, pharmacophore searching, similarity searching or substructure searching; while using recent prospective examples, the scope of ligand-based work flows could be demonstrated.

Schneider et al. [180] reported the successful application of a novel topological pharmacophore-like descriptor named CATS for "scaffold hopping", which refers to the identification of isofunctional structures with significantly different chemotypes in the virtual screening context. This approach was applied to the prediction of novel cardiac T-type Ca^{2+} channel blocking agents using mibefradil as query. From the 12 highest-ranking molecules using the CATS descriptors, one compound was found to have an IC_{50} value <1 μM (see Figure 4.5i).

This descriptor was also reported to be successful in ligand-based combinatorial design of selective purinergic receptor (A2A) antagonists [284]. A virtual screening procedure based on a topological pharmacophore similarity metric was applied to optimize combinatorial products functioning as P1 purinergic receptor antagonists. The target was the human A2A receptor. Using preliminary structure-activity relationship models with self-organizing neural networks, a combinatorial library design was performed by projecting virtual molecules onto this network. Selection, synthesis and testing of 17 combinatorial products resulted in a threefold lower binding affinities and 3.5-fold higher selectivity than the initial library. The most selective antagonist revealed a 121-fold relative selectivity for A2A with Ki (A2A) = 2.4 nM, and Ki (A1) = 292 nM, which demonstrates again the use of this approach.

The same descriptor was used by Naerum et al. for identification of structurally novel glycogen synthase kinase-3 inhibitors [285]. Using a virtual screening strategy based on CATS descriptor, a novel chemotype with activity against the GSK-3 enzyme was identified through scaffold hopping (cf Figure 4.5j). After focused parallel synthesis around the identified motifs, interesting kinase inhibitors with GSK-3 activity below 1 μM were identified.

A prospective study using 3-D topomeric fields [232] was done by Tripos and BMS via synthesis of a focused library resulting from searching a huge virtual library for angiotensin-II antagonists [233] using the ChemSpace technology. Interestingly, the most similar compounds to the previously known antagonists, used as search query, were the most active ones, suggesting that modified steric fields are useful to describe bioisosterism.

Milne et al. described the discovery of HIV-1 protease inhibitors based on a pharmacophore derived from X-ray crystal structures. After searching the NCI database with 206 000 entries and postfiltering the resulting hits, a total of 50 molecules were tested, leading to submicromolar activity for the best two compounds [286]. In their search for HIV-1 integrase inhibitors, another pharmacophore produced up to 340 hits, resulting in 10 structurally different classes and four compounds with affinities below 30 μM [287].

The discovery of farnesyltransferase inhibitors via 3-D database searching was reported by Kaminski et al. [288]. A pharmacophore was constructed using Catalyst [289] and was used to search the Schering–Plough corporate database. From a total of 718 hits after searching, the five best compounds showed IC_{50} values below 5 μM after biological testing, resulting in a novel series on a dihydrobenzothiophene scaffold. The successful use of a protein-derived 3-D pharmacophore and its refinement by adding ~100 excluded volumes was described by Greenidge et al. in a search for thyroid hormone receptor ligands [290].

Marriott et al. reported the discovery of muscarinic (M3) antagonists by pharmacophore searching [291]. Three out of 177 hits were potent antagonists, although structurally diverse from known M3 antagonist series. In a search for novel angiotensin II receptor antagonists, Kiyama et al. applied a three-feature pharmacophore to search the MDDR database [292]. The resulting 139 hits led to the identification of a novel lead series with activities in the nanomolar range. A combined sequential database searching strategy encompassing 2-D substructure searching and bioisosterism considerations followed by pharmacophore searching was described by Leysen et al. to discover highly selective 5-HT2c ligands [293].

Catalyst was also used by Singh et al. for the identification of potent and novel integrin $\alpha 4\beta 1$ antagonists [294]. The authors describe a series of potent inhibitors of $\alpha 4\beta 1$ that were discovered using pharmacophore-based virtual screening for replacements of the peptide region of an existing tetrapeptide-based inhibitor derived from fibronectin. The search query was constructed using a model of this peptide that was based upon the X-ray conformation of the related integrin binding region of vascular cell adhesion molecule-1 (VCAM-1). The computational screen identified 12 from a virtual library of 8624 molecules as satisfying the model and additional synthetic filters. All synthesized molecules were found to inhibit the interaction between $\alpha 4\beta 1$ and VCAM-1, with the most potent candidate showing an IC_{50} value of 1 nM, comparable to the starting compound (cf Figure 4.5k). Hence, the authors were able to demonstrate that pharmacophore searches led to the identification of nonpeptidic structures using conformational information about peptides as a starting point.

A related pharmacophore-based virtual screening strategy was applied by Flohr et al. at Aventis to identify nonpeptidic Urotensin II receptor antagonists [295]. In addition to focused screening of the internal GPCR-biased library, a rational approach by gathering structure-activity information through an alanine scan and by NMR studies on Urotensin II was employed. The identified message sequence Trp7-Lys8-Tyr9 and an NMR-derived solution structure of the starting peptide Urotensin II served to construct a hypothetical receptor-bound conformation. By assuming this as "bioactive" conformation, a three-point pharmacophore was built, containing two hydrophobic features from the aromatic moieties of tryptophan and tyrosine and a positive ionizable center from the terminal amino group of lysine. In addition to these three features, the shape of the message sequence was included in the pharmacophore. Subsequent virtual screening returned six different scaffold classes, with the most active Urotensin II receptor antagonist showing an IC_{50} value of 400 nM (cf. Figure 4.5l). Again, relevant information about the original peptide could be converted into nonpeptide structures as a starting point for a drug discovery program.

4.6
Practical Virtual Screening: Some Final Remarks

Every currently available virtual screening approach in structure- or ligand-based design is heuristic and thus will neither provide any quality guarantee nor an error bound on the obtained results. In every case, the fundamental model describing similarity or affinity is only a rough approximation of the underlying biological reality. In most cases, the optimization strategy is heuristic in addition. The biological nature of data makes it very difficult to extract those molecular features that are responsible for high affinity of certain ligands. From the user's perspective, the nontrivial question arises how to select the best-suited method or sequential combination of methods for a specific application.

In an optimal scenario, a test set of ligands for the same or a related target should be compiled to compare the performance of individual methods and to serve as positive control during the prospective virtual screening run. It is essential to extract as much information as possible on ligands and protein binding site requirements in order to choose the appropriate virtual screening tool and set parameters accordingly. There is no *a priori* guideline on which approach performs best – often a combination of techniques could lead to substantial improvements in the results. Relevant approaches have to be validated in the context of the problem before applying them for virtual screening in a prospective way. If this is not the case, it is worthwhile to study published applications with a similar input scenario. Sometimes comparison studies or the method itself gives hints on the applicability or preference of a certain method. Concerning individual algorithms, the advantages and disadvantages have already been mentioned in the corresponding sections above.

Current virtual screening applications can be grouped into at least four different categories of increasing complexity, while they all depend on physicochemical models based on different approximation levels for the biological reality. Recent years have witnessed not only rapid developments but also limitations of scoring functions. Here, some novel alternative approaches do not attempt to solve this problem for all binding sites but generate local, tailored scoring functions based on existing knowledge of a problem. This also underscores the importance of incorporating as much knowledge as possible into a sequential virtual screening work flow.

Most virtual screening campaigns in literature encompassed a number of consecutive computational phases from database filtering to target preparation, predocking filters, docking, postdocking analysis, and prioritization of compounds for testing. Docking under pharmacophore constraints here could be one option to integrate different steps into a single task during virtual screening (see for example [296, 297]). Filters are useful at each stage of this process, while the preparation of the input database needs to consider realistic ionization and tautomerization states. Every virtual screening run requires a postanalysis phase to remove false positives in the final hit lists and focus on the true hits. This could again be done only by applying additional knowledge to the problem – and visual inspection of the hits. The latter is a crucial step to discard compounds with unfavorable interactions or unrealistic conformations. A high-throughput approach without reflecting methods and parameter settings and a subsequent validation for a particular target is not very likely to succeed.

Virtual screening is not limited to available compounds. Its search space could be expanded by leadlike fragments to fill characteristic protein subpockets, potentially followed by structure-based library design. Both of the different approaches have different requirements on the performance of virtual screening engine. Fragment-based searches, for example, are more directed toward the identification of polar, rigid molecules, while the combinatorial nature of the problem in library design clearly prompts for algorithms capturing on this efficiency. Here the search space is expanded to potentially available compounds by combinatorial or parallel synthesis using robust, validated and efficient chemistry steps.

The pharmacophore concept has been shown to be of high relevance to virtual screening as a filter in structure-based applications or a stand-alone approach. Input for a pharmacophore can come from different sources: a ligand, a ligand series, NMR-derived structure of a related macromolecule, or protein X-ray structure. This information has to be generated, if needed, and productively used for virtual screening.

The user should also be pragmatic and not expect to retrieve all hits, but rather to detect meaningful starting points for hit exploration, which put additional requirements on validated hits. This clearly is an iterative process; more knowledge on a series might require another virtual screening run incorporating this novel data to focus on a different aspect of this project. Hence, virtual screening is not a disconnected science before a project starts; it should be tightly integrated within the work flow of a project team, optimally interfaced and resourced for rapid turnaround times, which means that novel biological results might lead to enhanced models for that target as basis for subsequent virtual screening exercises.

In this chapter, we gave a broad overview of virtual screening methods with a focus on the basic algorithmic techniques and models. More detailed information on physicochemical models describing molecular interactions, conformational space, or binding affinity can be found in other chapters of this book. Although not in the focus here, the models are at least as important in judging the individual performance as the software tools applied. When taking the advantages and drawback of algorithms and models together, a more realistic picture of virtual screening, its applicability, and limitations emerges.

Note added in proof

A recent review evaluating virtual screening lists in terms of predicted ADME and toxicology properties demonstrates some advantages of the integration of in silico approaches in search for viable lead structures [298].

References

1 HILLISCH, A. and HILGENFELD, R. (Eds.). *Modern Methods of Drug Discovery.* Birkhäuser, Basel, **2003**.

2 BLEICHER, K.H., BÖHM, H.-J., MÜLLER, K., and ALANINE, A.I. Hit and lead generation: beyond high throughput screening. *Nat. Rev. Drug Disc.* **2003**, *2*, 369–378.

3 ALANINE, A., NETTEKOVEN, M., ROBERTS, E., and THOMA, A.W. Lead generation – enhancing the success of drug discovery by investing in the hit to lead process. *Comb. Chem. High Throughput Screen.* **2003**, *6*, 51–66.

4 BÖHM, H.-J. and SCHNEIDER, G. (Eds.). *Virtual Screening for Bioactive Molecules.* Wiley-VCH, Weinheim, **2000**.

5 BAJORATH, J. Integration of virtual and high throughput screening. *Nat. Rev. Drug Disc.* **2002**, *1*, 882–894.

6 MUEGGE, I. and RAREY, M. Small molecule docking and scoring. In

Reviews in Computational Chemistry, LIPKOWITZ, K.B. and BOYD, D.B. (Eds.). Wiley, New York, **2001**, 1–60.

7 SCHNEIDER, G. and BÖHM, H.-J. Virtual screening and fast automated docking methods. *Drug Disc. Today.* **2002**, *7*, 64–70.

8 HALPERIN, I., MA, B., WOLFSON, H., NUSSINOV, R. Principles of docking: An overview of search algorithms and a guide to scoring functions. *PROTEINS* **2002**, *47*, 409–443.

9 BROOIJMANS, N. and KUNTZ, I.D. Molecular recognition and docking algorithms. *Annu. Rev. Biophys. Biomol. Struct.* **2003**, *32*, 335–373.

10 STAHL, M. and SCHULZ-GASCH, T. Practical database screening with docking tools. In *Small Molecule-Protein Interaction*, WALDMANN, H. and KOPPITZ, M. (Eds.). Springer-Verlag, Berlin and Heidelberg, **2003**, 127–150.

11 BURES, M.G. Recent techniques and applications in pharmacophore mapping. In *Practical Application of Computer-Aided Drug Design*, CHARIFSON, P.S. (Ed.). Marcel Dekker, New York, **1997**, 39–72.

12 LEMMEN, C. and LENGAUER, T. Computational methods for the structural alignment of molecules. *J. Comput.-Aided Mol. Des.* **2000**, *14*, 215–232.

13 DOWNS, G.M. and WILLETT, P. Similarity searching in databases of chemical structures. *Reviews in Computational Chemistry.* Wiley, New York, **1996**, 1–57.

14 GOOD, A.C. and MASON, J.S. Three-dimensional structure database searches. *Reviews in Computational Chemistry.* Wiley, New York, **1996**, 67–117.

15 WILLETT, P. Chemical similarity searching. *J. Chem. Inf. Comput. Sci.* **1998**, *38*, 983–996.

16 MILNE, G.W., NICKLAUS, M.C., and WANG, S. Pharmacophores in drug design and discovery. *SAR QSAR Environ. Res.* **1998**, *9*, 23–38.

17 KUROGI, Y. and GUNER, O.F. Pharmacophore modeling and three-dimensional database searching for drug design using catalyst. *Curr. Med. Chem.* **2001**, *8*, 1035–1055.

18 PATEL, Y., GILLET, V.J., BRAVI, G., and LEACH, A.R. A comparison of the pharmacophore identification programs: Catalyst, DISCO and GASP. *J. Comput Aided Mol. Des.* **2002**, *16*, 653–681.

19 BERMAN, H.M., WESTBROOK, J., FENG, Z., GILLILAND, G., and BHAT, T.N. The protein data bank. *Nucleic Acids Res.* **2000**, *8*, 235–242.

20 ANDREWS, P.R. Drug-receptor interactions. In *3D-QSAR in Drug Design. Theory, Methods and Applications*, KUBINYI, H. (Ed.). ESCOM, Leiden, NL, **1993**, 13–40.

21 KUBINYI, H. Structure-based design of enzyme inhibitors and receptor ligands. *Curr. Opin. Drug Disc. Dev.* **1998**, *1*, 4–15.

22 MURCKO, M.A., CARON, P.R., and CHARIFSON, P.S. Structure-based drug design. *Annu. Rep. Med. Chem.* **1999**, *34*, 297–306.

23 BÖHM, H.-J. and STAHL, M. Structure-based library design: molecular modeling merges with combinatorial chemistry. *Curr. Opin. Chem. Biol.* **2000**, *4*, 283–286.

24 MATTOS, C. and RINGE, D. Multiple binding modes. In *3D QSAR in Drug Design. Theory, Methods and Applications*, KUBINYI, H. (Ed.). ESCOM, Leiden, **1993**, 226–254.

25 MAIGNAN, S., GUILLOTEAU, J.-P., CHOI-SLEDESKI, Y.M., BECKER, M.R., and EWING, W.R. Molecular structures of human factor xa complexed with ketopiperazine inhibitors: preference for a neutral group in the S1 pocket. *J. Med. Chem.* **2003**, *46*, 685–690.

26 LADBURY, J.E. Just add water! The effect of water on the specificity of protein-ligand binding sites and its potential application to drug design. *Chem. Biol.* **1996**, *3*, 973–980.

27 STUBBS, M.T., REYDA, S., DULLWEBER, F., MOLLER, M., and KLEBE, G. pH-dependent binding modes observed in trypsin crystals: lessons for structure-based drug design. *Chembiochem.* **2002**, *46*, 685–690.

28 BABINE, R.E. and BENDER, S.L. Molecular recognition of protein-ligand complexes: applications to drug design. *Chem. Rev.* **1997**, *97*, 1359–1472.

29 GOHLKE, H. and KLEBE, G. Approaches to the description and prediction of the binding affinity of small-molecule ligands to macromolecular receptors. *Angew. Chem. Int. Ed. Engl.* **2002**, *41*, 2644–2676.

30 GILSON, M.K. and HONIG, B.H. The dielectric constant of a folded protein. *Biopolymers.* **1986**, *25*, 2097–2119.

31 MCCAMMON, J.A., WOLYNES, P.G., and KARPLUS, M. Picosecond dynamics of tyrosine side chains in proteins. *Biochemistry.* **1979**, *18*, 924–942.

32 WARWICKER, J. and WATSON, H.C. Calculation of the electric potential in the active site cleft due to ff-helix dipoles. *J. Mol. Biol.* **1982**, *157*, 671–679.

33 DAVIS, M.E. and MCCAMMON, J.A. Electrostatics in biomolecular structure and dynamics. *Chem. Rev.* **1990**, *90*, 509–521.

34 COX, S.R. and WILLIAMS, D.E. Representation of the molecular electrostatic potential by a net atomic charge model. *J. Comput. Chem.* **1981**, *3*, 304–323.

35 BRENEMAN, C.M. and WIBERG, K.B. Determining atom-centered monopoles from molecular electrostatic potentials. The need for high sampling density in formamide conformational analysis. *J. Comput. Chem.* **1990**, *11*, 361–373.

36 GELHAAR, D.K., VERKHIVKER, G.M., REJTO, P.A., SHERMAN, C.J., and FOGEL, D.B. Molecular recognition of the inhibitor AG-1343 by HIV-1 protease: conformationally flexible docking by evolutionary programming. *Chem. Biol.* **1995**, *2*, 317–324.

37 KITAGORODSKY, A.I. Non-bonded inter-actions of atoms in organic crystals and molecules. *Chem. Soc. Rev.* **1978**, *7*, 133–163.

38 DAUBER, P. and HAGLER, A.T. Crystal packing, hydrogen bonding, and the effect of crystal forces on molecular conformation. *Acc. Chem. Res.* **1980**, *13*, 105–112.

39 TAYLOR, R., KENNARD, O., and VERSICHEL, W. Geometry of the imino-carbonyl (N–H···O:C) hydrogen bond. 1. Lone-pair directionality. *J. Am. Chem. Soc.* **1983**, *105*, 5761–5766.

40 WADE, R.C. Molecular interaction fields. In *3D-QSAR in Drug Design. Theory, Methods and Applications*, KUBINYI, H. (Ed.). ESCOM, Leiden, NL, **1993**, 486–505.

41 DEREWENDA, Z.S., LEE, L., and DEREWENDA, U. The occurrence of C–H···O hydrogen bonds in proteins. *J. Mol. Biol.* **1995**, *252*, 248–262.

42 SCHEINER, S., KAR, T., and GU, Y. Strength of the CaH···O hydrogen bond of amino acid residues. *J. Biol. Chem.* **2001**, *276*, 9832–9837.

43 VARGAS, R., GARZA, J., DIXON, D.A., and HAY, B.P. How strong is the C–H···O=C hydrogen bond?. *J. Am. Chem. Soc.* **2000**, *122*, 4750–4755.

44 YOHANNAN, S., FAHAM, S., YANG, D., GROSFELD, D., CHAMBERLAIN, A.K., and BOWIE, J.U. A Ca–H···O hydrogen bond in a membrane protein is not stabilizing. *J. Am. Chem. Soc.* **2004**, *126* 2284–2285.

45 PIERCE, A.C., SANDRETTO, K.L., and BEMIS, G.W. Kinase inhibitors and the case for CH...O hydrogen bonds in protein-ligand binding. *Proteins: Struc., Func., Genet.* **2002**, *49*, 567–576.

46 BURLEY, S.K. and PETSKO, G.A. Aromatic–aromatic interaction: a mechanism of protein structure stabilization. *Science.* **1985**, *229*, 23–28.

47 DOUGHERTY, D.A. and STAUFFER, D.A. Acetylcholine binding by a synthetic receptor: implications for biological recognition. *Science.* **1990**, *250*, 1558–1560.

48 LUQUE, I. and FREIRE, E. Structural parameterization of the binding enthalpy of small ligands. *Proteins: Struc., Func., Genet.* **2002**, *49*, 181–190.

49 TAME, J.R.H. Scoring functions: a view from the bench. *J. Comput.-Aided Mol. Des.* **1999**, *13*, 99–108.

50 OPREA, T.I. and MARSHALL, G.R. Receptor-based prediction of binding affinities. In *3D-QSAR in Drug Design*, KUBINYI, H., FOLKERS, G., and MARTIN, Y.C. (Eds.). Kluwer/ESCOM, Dordrecht, NL, **1998**, 35–61.

51 BÖHM, H.-J. and STAHL, M. The use of scoring functions in drug discovery applications. *Rev. Comput. Chem.* **2002**, *18*, 41–87.

52 GOHLKE, H. and KLEBE, G. Statistical potentials and scoring functions applied to protein-ligand binding. *Curr. Opin. Struct. Biol.* **2001**, *11*, 231–235.

53 MUEGGE, I. and MARTIN, Y. A general and fast scoring function for protein-ligand interactions: a simplified potential approach. *J. Med. Chem.* **1999**, *42*, 791–804.

54 GOHLKE, H., HENDLICH, M., and KLEBE, G. Knowledge-based scoring function to predict protein-ligand interactions. *J. Mol. Biol.* **2000**, *295*, 337–356.

55 MITCHELL, J.B.O., LASKOWSKI, R.A., ALEX, A., and THORNTON, J.M. BLEEP – A potential of mean force describing protein-ligand interactions: I. Generating the potential. *J. Comput. Chem.* **1999**, *20*, 1165–1177.

56 GRANT, J.A., PICKUP, B.T., and NICHOLLS, A. A smooth permittivity function for poisson-boltzmann solvation methods. *J. Comput. Chem.* **2001**, *22*, 608–640.

57 BISSANTZ,C., FOLKERS, G., and ROGNAN, D. Protein-based virtual screening of chemical databases. 1. Evaluation of different docking/scoring combinations. *J. Med. Chem.* **2000**, *43*, 4759–4767.

58 STAHL, M. and RAREY, M. Detailed analysis of scoring functions for virtual screening. *J. Med. Chem.* **2001**, *44*, 1035–1042.

59 SCHULZ-GASCH, T. and STAHL, M. Binding site characteristics in structure-based virtual screening: evaluation of current docking tools. *J. Mol. Model.* **2003**, *9*, 47–57.

60 CHARIFSON, P.S., CORKERY, J.J., MURCKO, M.A., and WALTERS, W.P. Consensus scoring: a method for obtaining improved hit rates from docking databases of three-dimensional structures into proteins. *J. Med. Chem.* **1999**, *42*, 5100–5109.

61 SMITH, R., HUBBARD, R.E., GSCHWEND, D.A., LEACH, A.R., and GOOD, A.C. Analysis and optimization of structure-based virtual screening protocols (3). New methods and old problems in scoring function design. *J. Mol. Graph. Model.* **2003**, *22*, 41–53.

62 ROGNAN, D., LAUEMOLLER, S.L., HOLM, A., BUUS, S., and TSCHINKE, V. Predicting binding affinities of protein ligands from three-dimensional models: application to peptide binding to class i major histocompatibility proteins. *J. Med. Chem.* **1999**, *42*, 4650–4658.

63 MURRAY, C.W., AUTON, T.R., and ELDRIDGE, M.D. Empirical scoring functions. II. The testing of an empirical scoring function for the prediction of ligand-receptor binding affinities and the use of Bayesian regression to improve the quality of the model. *J. Comput.-Aided Mol. Des.* **1998**, *12*, 503–519.

64 KLEBE, G., MIETZNER, T., and WEBER, F. Different approaches toward an automatic structural alignment of drug molecules: applications to sterol mimics, thrombin and thermolysin inhibitors. *J. Comput.-Aided Mol. Des.* **1994**, *8*, 751–778.

65 LEMMEN, C., LENGAUER, T., and KLEBE, G. FLEXS: a method for fast flexible ligand superposition. *J. Med. Chem.* **1998**, *41*, 4502–4520.

66 BOSTROM, J. Reproducing the conformations of protein-bound ligands: a critical evaluation of several popular conformational searching tools. *J. Comput.-Aided Mol. Des.* **2001**, *15*, 1137–1152.

67 GOOD, A.C. and CHENEY, D.L. Analysis and optimization of structure-based virtual screening protocols (1): exploration of ligand conformational sampling techniques. *J. Mol. Graph. Model.* **2003**, *22*, 23–30.

68 KELLOGG, G.E., JOSHI, G.S., and ABRAHAM, D.J. New tools for modeling and understanding hydrophobicity and hydrophobic interactions. *Med. Chem. Res.* **1992**, *1*, 444–453.

69 CRAMER, R.D., PATTERSON, D.E., and BUNCE, J.E. Comparative molecular field analysis (CoMFA). 1. Effect of shape on binding of steroids to carrier proteins. *J. Am. Chem. Soc.* **1988**, *110*, 5959–5967.

70 KLEBE, G., ABRAHAM, U., and MIETZNER, T. Molecular similarity indices in a comparative analysis (CoMSIA) of drug molecules to correlate and predict their biological activity. *J. Med. Chem.* **1994**, *37*, 4130–4146.

71 GOODFORD, P.J. A computational procedure for determining energetically favorable binding sites on biologically important macromolecules. *J. Med. Chem.* **1985**, *28*, 849–857.

72 BOOBBYER, D.N.A., GOODFORD, P.J., MCWHINNIE, P.M., and WADE, R.C. New hydrogen-bond potentials for use in determining energetically favorable binding sites on molecules of known structure. *J. Med. Chem.* **1989**, *32*, 1083–1094.

73 WADE, R.C., CLERK, K.J., and GOODFORD, P.J. Further development of hydrogen bond functions for use in determining energetically favorable binding sites on molecules of known structure. 1. Ligand probe groups with the ability to form two hydrogen bonds. *J. Med. Chem.* **1993**, *36*, 140–147.

74 WADE, R.C. and GOODFORD, P.J. Further development of hydrogen bond functions for use in determining energetically favorable binding sites on molecules of known structure. 2. Ligand probe groups with the ability to form more than two hydrogen bonds. *J. Med. Chem.* **1993**, *36*, 148–156.

75 CHVATAL, V. *Linear Programming.* Freeman Publisher, New York, **1983**.

76 CORMAN, T.H., LEISERSON, C.E., RIVEST, R.L., and STEIN, C. *Introduction to Algorithms*, 2nd edition. MIT Press, Cambridge, **2001**.

77 NELDER, J.A. and MEAD, R. A simplex method for function minimization. *Comput. J.* **1965**, *7*, 308–313.

78 FLETCHER, R. *Practical Methods of Optimization*, 2nd edition. John Wiley & Sons, Tiptree, Essex, UK, **1987**.

79 DENNIS, J.E. and SCHNABEL, R.B. *Numerical Methods for Unconstrained Optimization and Nonlinear Equations.* Prentice Hall, Englewood Cliffs, **1983**.

80 SCHLICK, T. Optimization methods in computational chemistry. In *Reviews in Computational Chemistry*, LIPKOWITZ, K.B. and BOYD, D.B. (Eds.). VCH Publishers, New York, **1992**, 1–69.

81 KEARSLEY, S.K. and SMITH, G.M. An alternative method for the alignment of molecular structures: maximizing electrostatic and steric overlap. *Tetrahedron Comput. Methodol.* **1990**, *3*, 615–633.

82 MOOCK, T.E., HENRY, D.R., OZKABAK, A.G., and ALAMGIR, M. Conformational searching in {ISIS/3D} databases. *J. Chem. Inf. Comput. Sci.* **1994**, *34*, 184–189.

83 GOOD, A.C., HODGKIN, E.E., and RICHARDS, W.G. Utilization of Gaussian functions for the rapid evaluation of molecular similarity. *J. Chem. Inf. Comput. Sci.* **1992**, *32*, 188–191.

84 GRANT, J.A., GALLARDO, M.A., and PICKUP, B.T. A fast method of molecular shape comparison: a simple application of a Gaussian description of molecular shape. *J. Comb. Chem.* **1996**, *17*, 1653–1666.

85 KLEBE, G., MIETZNER, T., and WEBER, F. Different approaches toward an automatic structural alignment of drug molecules: application to sterol mimics, thrombin and thermolysin inhibitors. *J. Comp. Aided Mol. Des.* **1994**, *8*, 751–778.

86 MASEK, B.B., MERCHANT, A., and MATTHEW, J.B. Molecular shape comparison of angiotensin {II} receptor antagonists. *J. Med. Chem.* **1993**, *36*, 123–1238.

87 MCMAHON, A.J. and KING, P.M. Optimization of carbo molecular similarity index using gradient methods. *J. Comput. Chem.* **1997**, *18*, 151–158.

88 MESTRES, J., ROHRER, D.C., and MAGGIORA, G.M. MIMIC: a. molecular-field matching program. Exploiting applicability of molecular similarity approaches. *J. Comput. Chem.* **1997**, *18*, 934–954.

89 PETITJEAN, M. Geometric molecular similarity from volume-based distance minimization: application to saxitoxin and tetrodotoxin. *J. Comb. Chem.* **1995**, *16*, 80–90.

90 SANZ, F., MANAUT, F., SANCHEZ, J.A., and LOZOYA, E. Maximum electrostatic similarity between biomolecules optimizing both relative positions and conformations. *J. Mol. Struc.* **1991**, *230*, 437–446.

91 PEARLMAN, D.A. CONCORD: rapid generation of high quality approximate 3D molecular structures. *Chem. Des. Automat. News.* **1987**, *2*, 1–7.

92 GASTEIGER, J., RUDOLPH, C., and SADOWSKI, J. Automatic generation of

3D-atomic coordinates for organic molecules. *Tetrahedron Comput. Methodol.* **1990**, *3*, 537–547.

93 EWING, T.J.A., MAKINO, S., SKILLMAN, A.G., and KUNTZ, I.D. DOCK 4.0: search strategies for automated molecular docking of flexible molecule databases. *J. Comput.-Aided Mol. Des.* **2001**, *15*, 411–428.

94 JONES, G., WILLETT, P., GLEN, R.C., LEACH, A.R., and TAYLOR, R. Development and validation of a genetic algorithm for flexible docking. *J. Mol. Biol.* **1997**, *267*, 727–748.

95 RAREY, M., KRAMER, B., LENGAUER, T., and KLEBE, G.A Fast flexible docking method using an incremental construction algorithm. *J. Mol. Biol.* **1996**, *261*, 470–489.

96 ABAGYAN, R., TOTROV, M., and KUZNETSOV, D. ICM – a new method for protein modeling and design: applications to docking and structure prediction from the distorted native conformation. *J. Comput. Chem.* **1994**, *15*, 488–506.

97 GOODSELL, D.S., MORRIS, G.M., and OLSON, A.J. Automated docking of flexible ligands: applications of autodock. *J. Mol. Recognit.* **1996**, *9*, 1–5.

98 ACTON, F.S. *Numerical Methods that Usually Work.* Mathematical Association of America, Washington, **1990**, Chapter 17.

99 LEMMEN, C., LENGAUER, T., and KLEBE, G. FLEXS: a method for fast flexible ligand superposition. *J. Med. Chem.* **1998**, *41*, 4502–4520.

100 CLARK, M., CRAMER III, R.D., and VAN OPDENBOSCH, N. Validation of the general purpose tripos 5.2 force field. *J. Comput. Chem.* **1989**, *10*, 982–1012.

101 SIPPL, M.J. and STEGEBUCHNER, H. Superposition of three dimensional objects: a fast and numerically stable algorithm for the calculation of the matrix of optimal rotation. *Comput. Chem.* **1991**, *15*, 73–78.

102 KABSCH, W. A solution for the best rotation to relate two sets of vectors. *Acta Crystallogr.* **1976**, *A32*, 922–923.

103 KABSCH, W. A discussion of the solution for the best rotation to relate two sets of vectors. *Acta Crystallogr.* **1978**, *A34*, 827–828.

104 REDINGTON, P.K. MOLFIT: a computer program for molecular superposition. *Comput. Chem.* **1992**, *16*, 217–222.

105 HURST, T. Flexible 3D searching: the directed tweak technique. *J. Chem. Inf. Comput. Sci.* **1994**, *34*, 190–196.

106 KUNTZ, I.D., BLANEY, J.M., OATLEY, S. J., LANGRIDGE, R. L., and FERRIN, T.E. A geometric approach to macromolecule–ligand interactions. *J. Mol. Biol.* **1982**, *161*, 269–288.

107 JONES, G., WILLETT, P., and GLEN, R.C. Molecular recognition of receptor sites using a genetic algorithm with a description of desolvation. *J. Mol. Biol.* **1995**, *245*, 43–53.

108 RAREY, M., WEFING, S., and LENGAUER, T. Placement of medium-sized molecular fragments into active sites of proteins. *J. Comput.-Aided Mol. Des.* **1996**, *10*, 41–54.

109 LEMMEN, C. and LENGAUER, T. Time-efficient flexible superposition of medium-sized molecules. *J. Comput Aided Mol. Des.* **1997**, *11*, 357–368.

110 MENGER, K. Untersuchungen "uber allgemeine metrik. *Math. Ann.* **1928**, *100*, 75–163.

111 BLUMENTHAL, L.M. *Theory and Applications of Distance Geometry.* Clarendon Press, Oxford, **1953**.

112 KUHL, F.S., CRIPPEN, G.M., and FRIESEN, D.K. A combinatorial algorithm for calculating ligand binding. *J. Comput. Chem.* **1984**, *5*, 24–34.

113 GHOSE, A.K. and CRIPPEN, G.M. Geometrically feasible binding modes of a flexible ligand molecule at the receptor site. *J. Comput. Chem.* **1985**, *6*, 350–359.

114 DRESS, A.W.M. and HAVEL, T.F. Shortest path problems and molecular conformation. *Discrete Appl. Math.* **1988**, *19*, 129–144.

115 EASTHOPE, P.L. and HAVEL, T.F. Computational experience with an algorithm for tetrangle inequality bound smoothing. *Bull. Math. Biol.* **1989**, *51*, 173–194.

116 CLARK, D.E., WILLETT, P., and KENNY, P.W. Pharmacophoric pattern matching in files of three-dimensional chemical structures: Implementation of flexible searching. *J Mol. Graph.* **1993**, *11*, 146–156.

117 EWING, T.J.A. and KUNTZ, I.D. Critical evaluation of search algorithms for automated molecular docking and database screening. *J. Comput. Chem.* **1997**, *18*, 1175–1189.

118 MIZUTANI, M.Y., TOMIOKA, N., and ITAI, A. Rational automatic search method for stable docking models of protein and ligand. *J. Mol. Biol.* **1994**, *243*, 310–326.

119 SCHNECKE, V., SWANSON, C.A., GETZOFF, E.D., TAINER, J.A., and KUHN, L.A. Screening a peptidyl database for potential ligands to proteins with side-chain flexibility. *Proteins: Struct., Func., Genet.* **1998**, *33*, 74–87.

120 DIXON, J.S. *DockIt.* **2000**.

121 GLUNT, W., HAYDEN, T.L., and RAYDAN, M., Molecular Conformations from Distance Matrices. *J. Comput. Chem.* **1993**, *14*, 114–120.

122 SPELLMEYER, D.C., WONG, A.K., BOWER, M.J., and BLANEY, J.M. Conformational analysis using distance geometry methods. *J. Mol. Graph. Model.* **1997**, *15*, 18–36.

123 FEUSTON, B.P., MILLER, M.D., CULBERSON, J.C., NACHBAR, R.B., and KEARSLEY, S.K. Comparison of knowledge-based and distance geometry approaches for generation of molecular conformations. *J. Chem. Inf. Comput. Sci.* **2001**, *41*, 754–763.

124 SHERIDAN, R.P., NILAKANTAN, R., DIXON, J.S. and VENKATARAGHAVAN, R. The ensemble approach to distance geometry: application to the nicotinic pharmacophore. *J. Med. Chem.* **1986**, *29*, 899–906.

125 STREICH, W.J. DIGFIT: a superposition generator for flexible molecules. In *Trends in QSAR and Molecular Modelling 92*, WERMUTH, C.G. (Ed.). Kluwer, Leiden, **1992**, 416–417.

126 RAYMOND, J.W. and WILLETT, P. Similarity searching in databases of flexible 3D structures using smoothed bounded distance matrices. *J. Chem. Inf. Comput. Sci.* **2003**, *43*, 908–916.

127 BRACEWELL, R.N. *The Fourier Transform and its Applications*, Electrical and Electronic Engineering Series, 2nd edition. McGraw-Hill, New York, **1978**.

128 MAX, N.L. and GETZOFF, E.D. Spherical harmonic molecular surfaces. *IEEE Comput. Graph. Appl.* **1988**, *8*, 42–50.

129 RAMACHANDRAN, G. and SRINIVASAN, R. *Fourier Methods in Crystallography.* Wiley-Interscience, New York, **1970**.

130 ROSSMANN, M.G. and BLOW, D.M. The detection of sub-units within the crystallographic asymmetric unit. *Acta Cryst.* **1962**, *15*, 24–31.

131 ROSSMANN, M.G. *The Molecular Replacement Method*. Gordon & Breach, New York, **1972**.

132 COOLEY, J.W. and TURKEY, J.W. An algorithm for the machine calculation of complex Fourier series. *Math. Comput.* **1965**, *19*, 297–301.

133 LEMMEN, C., HILLER, C., and LENGAUER, T. RigFit: a new approach to superimposing ligand molecules. *J. Comput.-Aided Mol. Des.* **1998**, *12*, 491–502.

134 RITCHIE, W. and KEMP, G.J.L. Protein docking using spherical polar Fourier correlations. *Proteins: Struc., Func., Genet.* **2000**, *39*, 178–194.

135 DIEDERICHS, K. Structural superposition of proteins with unknown alignment and detection of topological similarity using a six-dimensional search algorithm. *Proteins: Struc., Func., Genet.* **1995**, *23*, 187–195.

136 COOPER, D.L. and ALLAN, N.L. A novel approach to molecular similarity. *J. Comp. Aided Mol. Des.* **1989**, *3*, 253–259.

137 NISSINK, J.W.M., VERDONK, M.L., KROON, J., MIETZNER, T., and KLEBE, G. Superposition of molecules: electron density fitting using fourier transforms. *J. Comb. Chem.* **1997**, *18*, 638–645.

138 KATCHALSKI-KATZIR, E., SHARIV, I., EISENSTEIN, M., FRIESEM, A.A., and AFLALO, C. Molecular surface recognition: determination of geometric fit between proteins and their ligands by correlation techniques. *Proc. Nat. Acad. Sci. U. S. A.* **1992**, *89*, 2195–2199.

139 MITCHELL, T. *Machine Learning*. McGraw Hill, New York, **1997**.

140 BUHMANN, J. Data clustering and learning. In *Handbook of Brain Theory and Neural Networks*, ARBIB, M. (Ed.). Bradfort Books/MIT Press, Cambridge, **1995**.

141 KOHONEN, T. Information sciences. *Self-Organizing Maps*. Springer-Verlag, Berlin, **1997**.

142 JOLLIFFE, I.T. *Principal Component Analysis*. Springer-Verlag, New York, **1986**.

143 KUBINYI, H. Methods and principles in medicinal chemistry. In *QSAR: Hansch Analysis and Related Approaches*, MANNHOLD, R., KROGSGAARD-LARSEN, P., and TIMMERMAN, H. (Eds.). VCH, Weinheim, New York, **1993**.

144 DUNN, W.J., WOLD, S., EDLUND, U., HELLBERG, S., and GASTEIGER, J. Multivariate structure-activity relationships between data from a battery of biological tests and an ensemble of structural descriptors: The PLS method. *Quant. Struc-Act. Relat.* **1984**, *3*, 31–137.

145 WOLD, H. Partial least squares. *Encyclopedia of Statistical Sciences*, KOTZ, S. and JOHNSON, N.L. (Eds.). Wiley, New York, **1985**, 581–591.

146 WOLD, S., RUHE, A., WOLD, H. and DUNN,W.J. III. The collinearity problem in linear regression: the partial least squares approach to generalized inverses. *SIAM J. Sci. Stat. Comput.* **1984**, *5*, 735–743.

147 CRAMER, R.D., PATTERSON, D.E., and BUNCE, J.D. Comparative molecular field analysis (CoMFA). 1. Effect of shape on binding of steroids to carrier proteins. *J. Am. Chem. Soc.* **1988**, *110*, 5959–5967.

148 ZUPAN, J. and GASTEIGER, J. *Neural Networks in Chemistry and Drug Design*. Wiley-VCH, Weinheim, Germany, **1999**.

149 CRISTIANINI, N. and SHAWE-TAYLOR, J. *An Introduction to Support Vector Machines*. Cambridge University Press, Cambridge, **2000**.

150 SCHÖLKOPF, B., BURGES, C.J.C., and SMOLA, A.J. *Advances in Kernel Methods – Support Vector Learning*. MIT Press, Cambridge, MA, **1999**.

151 QUINLAN, J.R. *C4.5: Programs for Machine Learning*. Morgan Kaufmann, San Francisco, CA, **1993**.

152 CHEN, X., RUSINKO, A.R. III., TROPSHA, A., and YOUNG, S.S. Automated pharmacophore identification for large chemical data sets. *J. Chem. Inf. Comput. Sci.* **1999**, *39*, 887–896.

153 FREUND, Y. and SCHAPIRE, R.E. A decision-theoretic generalization of on-line learning and an application to boosting. *J. Comput. Sys. Sci.* **1997**, *55*, 119–139.

154 STONE, M. An asymptotic equivalence of choice of model by cross-validation and akaike's criterion. *J. R. Stat. Soc., Ser. B.* **1977**, *38*, 44–47.

155 CAMPBELL, C., CRISTIANINI, N., and SMOLA, A. Query learning with large margin classifiers. *International Conference on Machine Learning*. Morgan Kaufmann, Stanford, CA, **2000**, 8.

156 WARMUTH, M.K., LIAO, J., RÄTSCH, G., MATHIESON, M., and PUTTA, S. Active learning with support vector machines in the drug discovery process. *J. Chem. Inf. Comput. Sci.* **2003**, *43*, 667–673.

157 ZIMMERMANN, M., RAREY, M., NAUMANN, T., MATTER, H., HESSLER, G., and LENGAUER, T. Extracting knowledge from high-throughput screening data: towards the generation of biophore models. In: Euro QSAR 2002. FORD, M., LIVINGSTONE, D., DEARDEN, J., VAN DE WATERBEEMD, H. (Eds.), Blackwell, Oxford, **2003**.

158 JOHNSON, D.S. The NP-completeness column: an ongoing guide. *J. Algorithms*. **1984**, *5*, 147–160.

159 RAYMOND, J.W. and WILLETT, P. Maximum common subgraph isomorphism algorithms for the matching of chemical structures. *J. Comput.-Aided Mol. Des.* **2002**, *16*, 521–533.

160 ULLMANN, J.R. An algorithm for subgraph isomorphism. *J. Assoc. Comput. Machinery*. **1976**, *23*, 31–42.

161 BRON, C. and KERBOSCH, J. Finding all cliques of an undirected graph [H]. *Commun. Assoc. Comput. Machinery*. **1973**, *16*, 575–577.

162 MARTIN, Y.C., BURES, M.G., DANAHER, E.A., DELAZZER, J., and LICO, I. A fast new approach to pharmacophore mapping and its application to dopaminergic and benzodiazepine agonists. *J. Comput.-Aided Mol. Des.* **1993**, *7*, 83–102.

163 LAMDAN, Y., WOLFSON, H. J. Geometric Hashing: a general and efficient model-based recognition scheme.

Proceedings of the IEEE International Conference on Computer Vision. **1988**, 238–249.

164 LINNAINMAA, S., HARWOOD, D., and DAVIS, L.S. Pose determination of a three-dimensional object using triangle pairs. *IEEE Trans. Pattern Anal. Machine Intell.* **1988**, *10*, 634–646.

165 FISCHER, D., LIN, S.L., WOLFSON, H.L., and NUSSINOV, R. A geometry-based suite of molecular docking processes. *J. Mol. Biol.* **1995**, *248*, 459–477.

166 SANDAK, B., NUSSINOV, R., and WOLFSON, H.J. A method for biomolecular structural recognition and docking allowing conformational flexibility. *J. Comput. Biol.* **1998**, *5*, 631–654.

167 PUTTA, S., LEMMEN, C., BEROZA, P., and GREENE, J. A novel shape-feature based approach to virtual library screening. *J. Chem. Inf. Comput. Sci.* **2002**, *42*, 1230–1240.

168 MAGGIORA, G.M. and JOHNSON, M.A. *Concepts and Applications of Molecular Similarity.* Wiley, New York, **1990**, 99–117s.

169 DEAN, P.M. *Molecular Similarity in Drug Design.* Chapman & Hall, London, UK, **1995**.

170 LEMMEN, C. and LENGAUER, T. Computational methods for the structural alignment of molecules. *J. Comput.-Aided Mol. Des.* **2000**, *14*, 215–232.

171 HANSCH, C. and LEO, A. *Exploring QSAR: Fundamentals and Applications in Chemistry and Biology.* American Chemical Society, Washington, DC, **1995**.

172 WILLETT, P. and WINTERMAN, V. A comparison of some measures for the determination of inter-molecular structural similarity: measures of inter-molecular structural similarity. *Quant. Struc.-Act. Relat.* **1986**, *5*, 18–25.

173 HOLLIDAY, J.D., RANADE, S.S., and WILLETT, P. A fast algorithm for selecting sets of dissimilar molecules from large chemical databases. *Quant. Struc.-Act. Relat.* **1995**, *14*, 501–506.

174 *MACCS II.* MDL Information Systems Inc., CA.

175 BARNARD, J.M. Substructure searching methods: old and new. *J. Chem. Inf. Comput. Sci.* **1993**, *33*, 572–584.

176 *UNITY Chemical Information Software Version 4.0.* Tripos Inc, St. Louis, MO, **1994**.

177 JAMES, C.A., WEININGER, D. and DELANEY, J., *DAYLIGHT Software Manual.* DAYLIGHT Inc., Mission Viejo, CA, **1994**.

178 XUE, L., GODDEN, J.W., and BAJORATH, J. Mini-fingerprints for virtual screening: design principles and generation of novel prototypes based on information theory. *SAR QSAR Environ. Res.* **2003**, *14*, 27–40.

179 CARHART, R.E., SMITH, D.H., and VENKATARAGHAVAN, R. Atom pairs as molecular features in structure-activity studies: definition and applications. *J. Chem. Inf. Comput. Sci.* **1985**, *25*, 64–73.

180 SCHNEIDER, G., NEIDHART, W., GILLER, T., and SCHMIDT, G. "Scaffold-Hopping" by topological pharmacophore search: a contribution to virtual screening. *Angew. Chemie.* **1999**, *111*, 3068–3070.

181 SHERIDAN, R.P. and MILLER, M.D. A method for visualizing recurrent topological substructures in sets of active molecules. *J. Chem. Inf. Comput. Sci.* **1998**, *38*, 915–924.

182 SHERIDAN, R.P., MILLER, M.D., UNDERWOOD, D.J., and KEARSLEY, S.K. Chemical similarity using geometric atom pair descriptors. *J. Chem. Inf. Comput. Sci.* **1996**, *36*, 128–136.

183 MASON, J.S., MORIZE, I., MENARD, P.R., CHENEY, D.L., and HULME, C. New 4-point pharmacophore method for molecular similarity and diversity applications: overview of the method and applications, including a novel approach to the design of combinatorial libraries containing privileged sub-structures. *J. Med. Chem.* **1999**, *42*, 3251–3264.

184 BROWN, R.D. and MARTIN, Y.C. Use of structure-activity data to compare structure-based clustering methods and descriptors for use in compound selection. *J. Chem. Inf. Comput. Sci.* **1996**, *36*, 572–584.

185 MOREAU, G. and BROTO, P. The auto-correlation of a topological structure:

a new molecular descriptor. *Nouv. J. Chim.* **1980**, *4*, 359–360.

186 PASTOR, M., CRUCIANI, G., MCLAY, I., PICKETT, S., and CLEMENTI, S. GRid-INdependent Descriptors (GRIND): a novel class of alignment-independent three-dimensional molecular descriptors. *J. Med. Chem.* **2000**, *43*, 3233–3243.

187 PEARLMAN, D.A. and SMITH, D.H. Novel software tools for chemical diversity. *Perspect. Drug Disc. Des.* **1998**, *9/10/11* 339–353.

188 DETHLEFSEN, W., LYNCH, M.F., GILLET, V.J., DOWNS, G.M. and HOLLIDAY, J.D. Computer storage and retrieval of generic chemical structures in patents. 12. Principles of search operations involving parameter lists: matching-relations, user-defined match levels, and transition from the reduced graph search to the refined search. *J. Chem. Inf. Comput. Sci.* **1991**, *31*, 253–260.

189 GILLET, V.J., WILLETT, P., and BRADSHAW, J. Similarity searching using reduced graphs. *J. Chem. Inf. Comput. Sci.* **2003**, *43*, 338–345.

190 RAREY, M. and DIXON, J.S. Feature trees: a new molecular similarity measure based on tree matching. *J. Comput.-Aided Mol. Des.* **1998**, *12*, 471–490.

191 RAREY, M. and STAHL, M. Similarity searching in large combinatorial chemistry spaces. *J. Comput.-Aided Mol. Des.* **2001**, *15*, 497–520.

192 KLEBE, G. Structural alignment of molecules.*3D QSAR in Drug Design.* ESCOM Science Publishers, Leiden, Netherlands, **1993**, 173–199.

193 KELLOGG, G.E., SEMUS, S.F., and ABRAHAM, D.J. HINT – a new method of empirical hydrophobic field calculation for CoMFA. *J. Comput.-Aided Mol. Des.* **1991**, *5*, 545–552.

194 CRUCIANI, G., PASTOR, M., and GUBA, W. VolSurf: a new tool for the pharmacokinetic optimization of lead compounds. *Eur. J. Pharm. Sci.* **2000**, *11*, 29–39.

195 CARBO, R., LEYDA, L., and ARNAU, M. How similar is a molecule to another? An electron density measure of similarity between two molecular

structures. *Int. J. Quantum Chem.* **1980**, *17*, 1185–1189.

196 HODGKIN, E.E. and RICHARDS, W.G. Molecular similarity based on electro-static potential and electric field. *Int. J. Quantum Chem.* **1987**, *14*, 105–110.

197 THORNER, D.A., WILLETT, P., WRIGHT, P.M., and TAYLOR, R. Similarity searching in files of three-dimensional chemical structures: Representation and searching of molecular electrostatic potentials using field-graphs. *J. Comput.-Aided Mol. Des.* **1997**, *11*, 163–174.

198 CRUCIANI, G., PASTOR, M., and MANNHOLD, R. Suitability of molecular descriptors for database mining. A comparative analysis. *J. Med. Chem.* **2002**, *45*, 2685–2694.

199 KEARSLEY, S.K., UNDERWOOD, D.J., SHERIDAN, R.P., and MILLER, M.D. Flexibases: a way to enhance the use of molecular docking methods. *J. Comput.-Aided Mol. Des.* **1994**, *8*, 565–582.

200 DOLL, J. and FREEMAN, D.L. (M)onte (C)arlo methods in chemistry. *IEEE Comput. Sci. Eng.* **1994**, 22–32.

201 METROPOLIS, N., ROSENBLUTH, M.N., TELLER, A. H., and TELLER, E. Equation of state calculation by fast computer machines. *J. Chem. Phys.* **1953**, *21*, 1087–1092.

202 TOTROV, M. and ABAGYAN, R. Detailed ab initio prediction of lysozym-antibody complex with 1.6 Å accuracy. *Struct. Biol.* **1994**, *1*, 259–263.

203 PARRETTI, M.F., KROEMER, R.T., ROTHMANN, J.H., and RICHARDS, W.G. Alignment of molecules by the {M}onte {C}arlo optimization of molecular similarity indices. *J. Comput. Chem.* **1997**, *18*, 1344–1353.

204 MCMARTIN, C. and BOHACEK, R.S. QXP: powerful, rapid computer algorithms for structure-based drug design. *J. Comput.-Aided Mol. Des.* **1997**, *11*, 333–344.

205 KIRKPATRIK, S., GELATT, C.D.J., and VECCHI, M.P. Optimization by simulated annealing. *Science* **1983**, *220*, 671–680.

206 GOODSELL, D.S. and OLSON, A.J. Automated docking of substrates to proteins by simulated annealing. *Proteins: Struc., Func. Genet.* **1990**, *8*, 195–202.

207 PERKINS, T.D.J., MILLS, J.E.J., and DEAN, P.M. Molecular surface-volume and property matching to superpose flexible dissimilar molecules. *J. Comput.-Aided Mol. Des.* **1995**, *9*, 479–490.

208 GOLDBERG, D.E. *Genetic Algorithms in Search Optimization and Machine Learning.* **1989**.

209 CLARK, D.E. *Evolutionary Algorithms in Molecular Design.* Wiley-VCH, Weinheim, **2000**.

210 DEVILLERS, J. *Genetic Algorithms in Molecular Modelling.* Academic Press, London, **1996**.

211 POIRETTE, A.R., ARTYMIUK, P.J., RICE, D.W., and WILLETT, P. Comparison of protein surfaces using a genetic algorithm. *J. Comput.-Aided Mol. Des.* **1997**, *11*, 557–569.

212 WEBER, L., WALLBAUM, S., BROGER, C., and GUBERNATOR, K. Optimization of the biological activity of combinatorial compound libraries by a genetic algorithm. *Angew. Chem. Int. Ed.* **1995**, *34*, 2280–2282.

213 SANDAK, B., NUSSINOV, R., and WOLFSON, H.J. An automated computer vision and robotics-based technique for 3D flexible biomolecular docking and matching. *Comput. Appl. Biol. Sci.* **1995**, *11*, 87–99.

214 RAREY, M. Protein-ligand docking in drug design. *Bioinformatics.* Wiley-VCH, Weinheim, **2001**, 315–360.

215 DESJARLAIS, R.L., SHERIDAN, R.P., DIXON, J.S., KUNTZ, I.D., and VENKATARAGHAVAN, R. Docking flexible ligands to macromolecular receptors by molecular shape. *J. Med. Chem.* **1986**, *29*, 2149–2153.

216 BÖHM, H.-J. The computer program LUDI: a new method for the de novo design of enzyme inhibitors. *J. Comput.-Aided Mol. Des.* **1992**, *6*, 61–78.

217 BÖHM, H.-J. LUDI: rule-based automatic design of new substituents for enzyme inhibitor leads. *J. Comput.-Aided Mol. Des.* **1992**, *6*, 593–606.

218 GILLET, V.J., MYATT, G., ZSOLDOS, Z., and JOHNSON, A.P. SPROUT, HIPPO, and CAESA: tools for de novo structure generation and estimation of synthetic accessibility. *Perspect. Drug Disc. Des.* **1995**, *3*, 34–50.

219 CLARK, D.E., FRENKEL, D., LEVY, S.A., LI, J., and MURRAY, C.W. PRO_LIGAND: an approach to de novo molecular design. 1. Application to the design of organic molecules. *J. Comput.-Aided Mol. Des.* **1995**, *9*, 13–32.

220 HONMA, T. Recent advances in de novo design strategy for practical lead identification. *Med. Res. Rev.* **2003**, *23*, 606–632.

221 BOHACEK, R.S. and MCMARTIN, C. Modern computational chemistry and drug discovery: structure generating programs. *Curr. Opin. Chem. Biol.* **1997**, *1*, 157–161.

222 BÖHM, H.-J. Current computational tools for de novo ligand design. *Curr. Opin. Biotechnol.* **1996**, *7*, 433–436.

223 MOON, J.B. and HOWE, W.J. Computer design of bioactive molecules: a method for receptor-based de novo ligand design. *Proteins: Struc., Func., Genet.* **1991**, *11*, 314–328.

224 KRAMER, B., RAREY, M., and LENGAUER, T. Evaluation of the FlexX incremental construction algorithm for protein-ligand docking. *Proteins: Struc., Func., Genet.* **1999**, *37*, 228–241.

225 WELCH, W., RUPPERT, J., and JAIN, A.N. Hammerhead: fast, fully automated docking of flexible ligands to protein binding sites. *Chem. Biol.* **1996**, *3*, 449–462.

226 LEMMEN, C., LENGAUER, T., and KLEBE, G. FlexS: a method for fast flexible ligand superposition. *J. Med. Chem.* **1998**, *41*, 4502–4520.

227 GALLOP, M.A., BARRETT, R.W., DOWER, W.J., FODOR, P.A., and GORDON, E.M. Applications of combinatorial technologies to drug discovery. 1. Background and peptide combinatorial libraries. *J. Med. Chem.* **1994**, *37*, 1233–1251.

228 GORDON, E.M., BARRETT, R.W., DOWER, W.J., FODOR, P.A., and GALLOP, M.A. Applications of combinatorial technologies to drug discovery. 2. Combinatorial organic synthesis, library screening strategies, and future directions. *J. Med. Chem.* **1994**, *37*, 1386–1401.

229 GORDON, E.M. Strategies in the design and synthesis of chemical libraries. In

Combinatorial Chemistry and Molecular Diversity in Drug Discovery, Gordon, E.M. and Kerwin, J.F. Jr. (Eds.). Wiley, New York, **1998**, 17–38.

230 Barnard, J.M., Downs, G.M., Scholley-pfab, A.V., and Brown, R. D. Use of Markush structure analysis techniques for descriptor generation and clustering of large combinatorial libraries. *J. Mol. Graph.* **2000**, *18*, 452–463.

231 Schneider, G., Lee, M.L., Stahl, M., and Schneider, G. De novo design of molecular architectures by evolutionary assembly of drug-derived building blocks. *J. Comput.-Aided Mol. Des.* **2000**, *14*, 487–494.

232 Cramer, R.D., Clark, R.D., Patterson, D.E., and Ferguson, A.M. Bioisosterism as a molecular diversity descriptor: steric fields of single "topomeric" conformers. *J. Med. Chem.* **1996**, *39*, 3060–3069.

233 Cramer, R.D., Poss, M.A., Hermsmeier, M.A., Caulfield, T.J., and Kowala, M.C. Prospective identification of biologically active structures by topomer shape similarity searching. *J. Med. Chem.* **1999**, *42*, 3919–3933.

234 Böhm, H.J., Banner, D.W., and Weber, L. Combinatorial docking and combinatorial chemistry: design of potent non-peptide thrombin inhibitors. *J. Comput.-Aided Mol. Des.* **1999**, *13*, 51–56.

235 Sun, Y., Ewing, T.J.A., Skillman, A.G., and Kuntz, I.D. CombiDOCK: structure-based combinatorial docking and library design. *J. Comput.-Aided Mol. Des.* **1999**, *12*, 597–604.

236 Rarey, M. and Lengauer, T. A recursive algorithm for efficient combinatorial library docking. *Perspect. Drug Disc. Des.* **2000**, *20*, 63–81.

237 Lyne, P.D. Structure-based virtual screening: an overview. *Drug Disc. Today.* **2002**, *7*, 1047–1055.

238 Lipinski, C.A., Lombardo, F., Dominy, B.W., and Feeney, P.J. Experimental and computational approaches to estimate solubility and permeability in drug discovery and development settings. *Adv. Drug Deliv. Rev.* **1997**, *23*, 3–25.

239 Veber, D.F., Johnson, S.R., Cheng, H.Y., Smith, B.R., and Ward, K.W.

Molecular properties that influence the oral bioavailability of drug candidates. *J. Med. Chem.* **2002**, *12*, 2615–2623.

240 Oprea, T.I., Davis, A.M., Teague, S.J., and Leeson, P.D. Is there a difference between leads and drugs? A historical perspective. *J. Chem. Inf. Comput. Sci.* **2001**, *41*, 1308–1315.

241 Teague, S.J., Davis, A.M., and Leeson, P.D. The design of leadlike combinatorial libraries. *Angew. Chem. Int. Ed.* **1999**, *38*, 3743–3748.

242 Matter, H., Baringhaus, K.H., Naumann, T., Klabunde, T., and Pirard, B. Computational approaches towards the rational design of drug-like compound libraries. *Comb. Chem. High Throughput Screen.* **2001**, *4*, 453–475.

243 Oprea, T.I. Virtual screening in lead discovery: a viewpoint. *Molecules.* **2002**, *7*, 51–62.

244 Zamora, I., Oprea, T., Cruciani, G., Pastor, M., and Ungell, A.-L. Surface descriptors for protein-ligand affinity prediction. *J. Med. Chem.* **2003**, *46*, 25–33.

245 Roche, O., Schneider, P., Zuegge, J., Guba, W., and Kansy, M. Development of a virtual screening method for identification of "Frequent Hitters" in compound libraries. *J. Med. Chem.* **2002**, *45*, 137–142.

246 Roche, O., Trube, G., Zuegge, J., Pflimlin, P., and Alanine, A. A virtual screening method for prediction of the herg potassium channel liability of compound libraries. *Chembiochem.* **2002**, *3*, 455–459.

247 Zuegge, J., Fechner, U., Roche, O., Parrott, N., Engkvist, O. A fast virtual screening filter for cytochrome P450 3A4 inhibition liability of compound libraries. *Quant. Struc.-Act. Relat.* **2002**, *21*, 249–256.

248 Ekins, S., Berbaum, J., and Harrison, R.K. Generation and validation of rapid computational filters for CYP2D6 and CYP3A4. *Drug Metab. Dispos.* **2003**, *31*, 1077–1080.

249 Wess, G., Urmann, M., and Sickenberger, B. Medicinal chemistry: challenges and opportunities. *Angew. Chemie Int. Ed.* **2001**, *40*, 3341–3350.

250 VANGREVELINGHE, E., ZIMMERMANN, K., SCHOEPFER, J., PORTMANN, R., and FABBRO, D. Discovery of a potent and selective protein kinase CK2 inhibitor by high-throughput docking. *J. Med. Chem.* **2003**, *46*, 2656–2662.

251 WANG, R., LIU, L., LAI, L., and TANG, Y. SCORE: a new empirical method for estimating the binding affinity of a protein-ligand complex. *J. Mol. Model.* **1998**, *4*, 379–394.

252 PENG, H., HUANG, N., QI, J., XIE, P., and XU, C. Identification of novel inhibitors of BCR-ABL tyrosine kinase via virtual screening. *Bioorg. Med. Chem. Lett.* **2003**, *13*, 3693–3699.

253 FILIKOV, A.V. and JAMES, T.L. Structure-based design of ligands for protein basic domains: application to the HIV-1 Tat protein. *J.Comput.-Aided Mol. Des.* **1998**, *12*, 229–240.

254 SHOICHET, B.K., STROUD, R.M., SANTI, D.V., KUNTZ, I.D., and PERRY, K.M. Structure-based discovery of inhibitors of thymidylate synthase. *Science.* **1993**, *259*, 1445–1450.

255 TONDI, D., SLOMCZYNSKA, U., COSTI, M. P., WATTERSON, D. M., and GHELLI, S. Structure-based discovery and in-parallel optimization of novel competitive inhibitors of thymidylate synthase. *Chem. Biol.* **1999**, *6*, 319–331.

256 BODIAN, D.L., YAMASAKI, R.B., BUSWELL, R.L., STEARNS, J.F., and WHITE, J.M. Inhibition of the fusion-inducing conformational change of influenza hemagglutinin by benzoquinones and hydroquinones. *Biochemistry.* **1993**, *32*, 2967–2978.

257 RING, C.S., SUN, E., MCKERROW, J.H., LEE, G.K., and ROSENTHAL, P.J. Structure-based inhibitor design by using protein models for the development of antiparasitic agents. *Proc. Natl. Acad. Sci. U. S. A.* **1993**, *90*, 3583–3587.

258 DOMAN, T.N., MCGOVERN, S.L., WITHERBEE, B.J., KASTEN, T.P., and KURUMBAIL, R. Molecular docking and high-throughput screening for novel inhibitors of protein tyrosine phosphatase-1B. *J. Med. Chem.* **2002**, *45*, 2213–2221.

259 HOPKINS, S.C., VALE, R.D., and KUNTZ, I.D. Inhibitors of kinesin activity from structure-based computer screening. *Biochemistry.* **2000**, *39*, 2805–2814.

260 GRÜNEBERG, S., STUBBS, M.T., and KLEBE, G. Successful virtual screening for novel inhibitors of human carbonic anhydrase: strategy and experimental confirmation. *J. Med. Chem.* **2002**, *45*, 3588–3602.

261 GRÜNEBERG, S., WENDT, B., and KLEBE, G. Subnanomolar inhibitors from computer screening: a model study using human carbonic anhydrase II. *Angew. Chem. Int. Ed.* **2001**, *40*, 389–393.

262 BRENK, R., NAERUM, L., GRAEDLER, U., GERBER, H.-D., and GARCIA, G.A. Virtual screening for submicromolar lead inhibitors of TGT based on a new binding mode detected by crystal structure analysis. *J. Med. Chem.* **2003**, *46*, 1133–1143.

263 SCHAPIRA, M., RAAKA, B.M., SAMUELS, H.H., and ABAGYAN, R. Rational discovery of novel nuclear hormone receptor antagonists. *Proc. Natl. Acad.Sci. U. S. A.* **2000**, *97*, 1008–1013.

264 TOTROV, M. and ABAGYAN, R. Flexible protein-ligand docking by global energy optimization in internal coordinates. *Proteins Suppl.* **1997**, *1*, 215–220.

265 SCHAPIRA, M., RAAKA, B.M., DAS, S., FAN, L., and TOTROV, M. Discovery of diverse thyroid hormone receptor antagonists by high-throughput docking. *Proc. Natl. Acad. Sci. U. S. A.* **2003**, *100*, 7354–7359.

266 SCHAPIRA, M., ABAGYAN, R., and TOTROV, M. Nuclear hormone receptor targeted virtual screening. *J. Med. Chem.* **2003**, *46*, 3045–3059.

267 PEROLA, E., XU, K., KOLLMEYER, T.M., KAUFMANN, S.H., and PRENDERGAST, F.G. Successful virtual screening of a chemical database for farnesyl-transferase inhibitor leads. *J. Med. Chem.* **2000**, *43*, 401–408.

268 ANTEL, J. Integration of combinatorial chemistry and structure-based drug design. *Curr. Opin. Drug Disc. Dev.* **1999**, *2*, 224–233.

269 SALEMME, F.R., SPURLINO, J., and BONE, R. Serendipity meets precision: the integration of structure-based drug design and combinatorial chemistry for efficient drug discovery. *Structure*. **1997**, 5, 319–324.

270 ILLIG, C., EISENNAGEL, S., BONE, R., RADZICKA, A., MURPHY, L., RANDLE, T., SPURLINO,J., JAEGER, E., and SALEMME, F.R. Expanding the envelope of structure-based drug design using chemical libraries: application to small-molecule inhibitors of thrombin. *Med. Chem. Res.* **1998**, 8, 244–260.

271 BRADY, S.F., STAUFFER, K.J., LUMMA, W.C., SMITH, G.M., RAMJIT, H.G., LEWIS, S.D., LUCAS, B.J., GARDELL, S.J., LYLE, E.A., APPLEBY, S.D., COOK, J.J., HOLAHAN, M.A., STRANIERI, M.T., LYNCH,Y J., LIN, J.H., CHEN, I.W., VASTAG, K., NAYLOR-OLSEN, A.M., and VACCA, J.P. Discovery and development of the novel potent orally active thrombin inhibitor N-(9-Hydroxy-9-fluorenecarboxy)prolyl trans-4-aminocyclohexylmethyl amide (L-372,460): coapplication of structure-based design and rapid multiple analog synthesis on solid support. *J. Med. Chem.* **1998**, 41, 401–406.

272 BÖHM, H.-J. Combinatorial docking. In *Computer-Assisted Lead Finding and Optimization*, VAN DE WATERBEEMED, H., TESTA, B., and FOLKERS, G. (Eds.). Verlag Helvetica Chimica Acta, Basel, CH, **1997**, 125–133.

273 MURRAY, C.W., CLARK, D.E., AUTON, T.R., FIRTH, M.A., and LI, J. PRO_SELECT: combining structure-based drug design and combinatorial chemistry for rapid lead discovery. 1. Technology. *J. Comput.-Aided Mol. Des.* **1997**, 11, 193–207.

274 JONES, G., WILLETT, P., GLEN, R.C., LEACH, A.R., and TAYLOR, R. Further development of a genetic algorithm for ligand docking and its application to screening combinatorial libraries. *ACS Symp. Ser.; Am. Chem. Soc.* **1999**, 719, 271–291.

275 SUN, Y., EWING, T.J.A., SKILLMAN, A.G., and KUNTZ, I.D. CombiDOCK: structure-based combinatorial docking and library design. *J. Comput.-Aided Mol. Des.* **1998**, 12, 597–604.

276 ARONOV, A.M., MUNAGALA, N.R., MONTELLANO, P.R.O.D., KUNTZ, I.D., and WANG, C.C. Rational design of selective submicromolar inhibitors of tritricho-monas foetus hypoxanthine-guanine-xanthine phosphoribosyl-transferase. *Biochemistry*. **2000**, 39, 4684–4691.

277 BI, X., HAQUE, T.S., ZHOU, J., SKILLMAN, A.G., and LIN, B. Novel cathepsin D inhibitors block the formation of hyperphosphorylated tau fragments in hippocampus. *J. Neurochem.* **2000**, 74, 1469–1477.

278 HAQUE, T.S., SKILLMAN, A.G., LEE, C.E., HABASHITA, H., and GLUZMAN, I.Y. Potent, low-molecular-weight non-peptide inhibitors of malarial aspartyl protease plasmepsin II. *J. Med. Chem.* **1999**, 42, 1428–1440.

279 WYSS, P.C., GERBER, P., HARTMAN, P.G., HUBSCHWERLEN, C., and LOCHER, H. Novel dihydrofolate reductase inhibitors. structure-based versus diversity-based library design and high-throughput synthesis and screening. *J. Med. Chem.* **2003**, 46, 2304–2312.

280 LIEBESCHUETZ, J.W., JONES, S.D., MORGAN, P.J., MURRAY, C.W., and RIMMER, A.D. PROSELECT: combining structure-based drug design and array-based chemistry for rapid lead discovery. 2. The development of a series of highly potent and selective factor Xa inhibitors. *J. Med. Chem.* **2002**, 45, 1221–1232.

281 BAXTER, C.A., MURRAY, C.W., WASZKOWYCZ, B., LI, J., and SYKES, R.A. New approach to molecular docking and its application to virtual screening of chemical databases. *J. Chem. Inf. Comp. Sci.* **2000**, 40, 254–262.

282 BÖHM, H.-J., BOEHRINGER, M., BUR, D., GMUENDER, H., HUBER, W.Novel inhibitors of DNA gyrase: 3D structure based biased needle screening, hit validation by biophysical methods, and 3D guided optimization. A promising alternative to random screening. *J. Med. Chem.* **2000**, 43, 2664–2674.

283 PICKETT, S.D., SHERBORNE, B.S., WILKINSON, T., BENNETT, J., and BORKAKOTI, N. Discovery of low molecular weight inhibitors of IMPDH Via virtual needle screening. *Bioorg. Med. Chem. Lett.* **2003**, 13, 1691–1694.

284 SCHNEIDER, G. and NETTEKOVEN, M. Ligand-based combinatorial design of selective purinergic receptor (A2A) antagonists using self-organizing maps. *J. Combinat. Chem.* **2003**, *5*, 233–237.

285 NAERUM, L., NORSKOV-LAURITSEN, L., and OLESEN, P.H. Scaffold hopping and optimization towards libraries of glycogen synthase kinase-3 inhibitors. *Bioorg. Med. Chem. Lett.* **2002**, *12*, 1525–1528.

286 WANG, S., MILNE, G.W., YAN, X., POSEY, I., and NICKLAUS, M.C. Discovery of novel, non-peptide HIV-1 protease inhibitors by pharmacophore searching. *J. Med. Chem.* **1996**, *39*, 2047–2054.

287 HONG, H., NEAMATI, N., WANG, S., NICKLAUS, M.C., and MAZUMDER, A. Discovery of HIV-1 integrase inhibitors by pharmacophore searching. *J. Med. Chem.* **1997**, *40*, 930–936.

288 KAMINSKI, J.J., RANE, D.F., SNOW, M.E., WEBER, L., and ROTHOFSKY, M.L. Identification of novel farnesyl protein transferase inhibitors using three-dimensional database searching methods. *J. Med. Chem.* **1997**, *40*, 4103–4112.

289 GREENE, J., KAHN, S., SAVOJ, H., SPRAGUE, S., and TEIG, S. Chemical function queries for 3D database search. *J. Chem. Inf. Comp. Sci.* **1994**, *34*, 1297–1308.

290 GREENIDGE, P.A., CARLSSON, B., BLADH, L.G., and GILLNER, M. Pharmacophores incorporating numerous excluded volumes defined by x-ray crystallographic structure in three-dimensional database searching: application to the thyroid hormone receptor. *J. Med. Chem.* **1998**, *41*, 2503–2512.

291 MARRIOTT, D.P., DOUGALL, I.G., MEGHANI, P., LIU, Y.J., and FLOWER, D.R. Lead generation using pharmacophore mapping and three-dimensional database searching: application to muscarinic M3 receptor antagonists. *J. Med. Chem.* **1999**, *42*, 3210–3216.

292 KIYAMA, R., HONMA, T., HAYASHI, K., OGAWA, M., HARA, M., FUJIMOTO, M., and FUJISHITA, T. Angiotensin II receptor antagonists. Design, synthesis, and in vitro evaluation of dibenzo[a,d]cycloheptene and dibenzo[b,f]oxepin derivatives. Searching for bioisosteres of biphenyltetrazole using a three-dimensional search technique. *J. Med. Chem.* **1995**, *38*, 2728–2741.

293 LEYSEN, D. and KELDER, J. Ligands for the 5-HT2C receptor as potential antidepressants and anxiolytics. *Pharmacochem. Libr.* **1998**, *29*, 49–61.

294 SINGH, J., VLIJMEN, H.V., LIAO, Y., LEE, W.-C., and CORNEBISE, M. Identification of potent and novel a4b1 antagonists using in silico screening. *J. Med. Chem.* **2002**, *45*, 2988–2993.

295 FLOHR, S., KURZ, M., KOSTENIS, E., BRKOVICH, A., and FOURNIER, A. Identification of nonpeptidic urotensin II receptor antagonists by virtual screening based on a pharmacophore model derived from structure-activity relationships and nuclear magnetic resonance studies on urotensin II. *J. Med. Chem.* **2002**, *45*, 1799–1805.

296 HINDLE, S.A., RAREY, M., BUNING, C., and LENGAUER, T. Flexible docking under pharmacophore constraints. *J. Comput.-Aided Mol. Des.* **2002**, *16*, 129–149.

297 JOSEPH-MCCARTHY, D., THOMAS IV, B.E., BELMARSH, M., MOUSTAKAS, D., and ALVAREZ, J.C. Pharmacophore-based molecular docking to account for ligand flexibility. *Proteins: Struc., Func. Genet.* **2003**, *51*, 172–188.

298 OPREA, T.I. and MATTER, H. Integrating virtual screening in lead discovery. *Curr. Opin. Chem. Biol.* **2004**, *8*, 349–358.

5

Strengths and Limitations of Pharmacophore-Based Virtι Screening

Dragos Horvath, Boryeu Mao, Rafael Gozalbes, Frédérique Barbosa, and Sherry L. Rogalski

5.1
Introduction

The increasing cost of finding new molecular entities for therapeutic use has led to the development of high-throughput screening (HTS) technologies. Combinatorial chemistry and parallel synthesis have held promise in generating lead compounds that can be developed into drug molecules. Computational sciences have been important contributors, allowing for the *in silico* identification [1–3] of chemical leads for target enzymes or receptors ("lead finding" or "virtual screening"), and enhancing the development of these leads into fully marketable entities for human health ("lead optimization" via property and activity modeling) [4, 5]. The new opportunities and challenges for computational tools arise from the need for investigating a large number of compounds in an efficient manner. The size of these compound libraries precludes a complete screening by even the high-throughput methods, and suitable *in silico* technologies would provide an initial virtual filtering. A successful virtual screening must be sufficiently accurate with regard to desired screening criteria (such that any missed opportunities due to false-negatives would be kept at a minimum), though the *in silico* scoring may not be entirely relied upon in its precision for identifying desirable biological activities. Thus the challenge is great, yet with a potentially very high pay-off.

5.2
The "Pharmacophore" Concept: Pharmacophore Features

The understanding of three-dimensional molecular structure and the explanation of ligand-site affinity on hand of shape and functional group complementarity ("lock and key" hypothesis) naturally lead to the introduction of the "pharmacophore" concept in medicinal chemistry and implicitly in computational chemistry; see [6] and references therein. The specific physicochemical mechanisms controlling the macromolecule–ligand interactions could be, in principle, understood on a purely

Chemoinformatics in Drug Discovery. Edited by Tudor I. Oprea
Copyright © 2004 WILEY-VCH Verlag GmbH & Co. KGaA, Weinheim
ISBN: 3-527-30753-2

thermodynamic basis. However, for practical purposes, medicinal chemists have adopted a simplified view assuming that the total binding free energy could be broken down into pairwise contributions stemming from interactions of some functional groups of ligand and site respectively. "Interchangeable" functional groups of similar physicochemical behavior, accordingly expected to induce similar effects on binding affinities, were therefore loosely classified into "pharmacophore types". The fundamental types are "hydrophobic", "polar positive" and "polar negative" and may be further split into "hydrophobic-alkyl" and "aromatic", "hydrogen bond (HB) donors" and "cations", "HB acceptors" and "anions" respectively. This classification is easy to perform computationally (e.g., using SMARTS [7] to describe the functional groups fitting each category). Such an algorithm would easily allow for more detailed categorizing [8, 9] (splitting, e.g., cations into "protonated groups", "quaternary ions" and "delocalized" charges), but this is rarely done in practice. Pharmacophore modeling inherently favors conceptual simplicity instead of physicochemical rigor, replacing the need for detailed modeling of subtle enthalpic and entropic effects – such as the "hydrophobic effect" [10, 11] – by the straightforward assumption of "interchangeability" of functional groups of a same pharmacophore type. In spite of this – or, rather, *because* of this – the computational implementations of the pharmacophore concept, typically used for a quick filtering of compound libraries in order to selectively discard obviously inappropriate ligand candidates, are currently an important drug discovery tool.

5.3
Pharmacophore Models: Managing Pharmacophore-related Information

Various computational approaches can be used to capture and exploit the information concerning the relative arrangement of the pharmacophore "features" of molecules in order to derive predictive models. Pharmacophore models can be roughly classified into "similarity-based" and "hypothesis-based". The former is the expression of the molecular similarity principle [12, 13] in a structure space defined on hand of molecular descriptors that capture information about the *overall* pharmacophore pattern. Certainly, if a similarity-driven search manages to retrieve analogs that match *every* feature present in the active reference compound, the probability of finding actives among these analogs will be high. As we would expect, in real life, some pharmacophore parts will be better matched than others, and the knowledge of the moieties that are relevant for binding would improve the predictive quality of models, as failure to match irrelevant reference features needs no longer be computationally penalized. This insight must however be first extracted [14] under the form of a pharmacophore hypothesis matching the observed trends in activity within a learning set of both active and inactive compounds. Hypothesis-based models are therefore linear or nonlinear QSAR [15, 16] approaches, involving some "machine learning"; for example, selecting and weighing a minimal set of pharmacophore-related structural elements seen to occur in all actives and only in the active compounds.

5.4
The Main Topic of This Paper

According to the previous discussion, pharmacophore-based virtual screening techniques could be ranked in terms of increasing complexity, information content and expected success rate, from the simplest similarity-based methods (recognizing an overall pharmacophore pattern), to hypothesis-based methods (singling out the key elements that are responsible for activity), to docking [17] (exploiting the knowledge of the receptor structure). This scheme is intellectually appealing and was therefore never seriously challenged to our knowledge. It is the goal of this paper to have a closer look at this well-established paradigm, and notably to investigate whether hypothesis-based approaches truly represent a fundamental improvement over similarity-based models – do they really teach us anything about how the ligand–site interaction occurs? We will first present a brief and nonexhaustive review of both similarity-based and hypothesis-based pharmacophore models. Activity models representing each of these categories will be built with either our proprietary or commercially available modeling tools, using a Cyclooxygenase-II (Cox2) [18] activity data set. In this way, the relative merits and pitfalls can be easily and quantitatively compared, and also explained in the light of the experimental knowledge about the Cox2 active site and its interactions with cocrystallized ligands [19].

5.5
The Cox2 Data Set

This data set was obtained by the merging of a set [20] of 326 Cox2 inhibitors with 1914 BioPrint [5] reference drugs and bioactive compounds, for which the Cox2 affinities were measured at Cerep. The comprehensive set of 2240 molecules regroups, to our knowledge, most of the important selective and nonselective Cox2 inhibitors, as well as a wealth of very diverse inactives. The Cox2 inhibitory potencies were expressed, for both subsets, as pIC_{50} values ($-\log IC_{50}$ in mol l^{-1}). The close match of the activity values for all of the several compounds that were common to both subsets proved that the pIC_{50} activity data of either origin are equivalent.

A cell-based diversity analysis has been performed on the merged sets, in both a topological and a pharmacophore descriptor space, in order to monitor the scattering of potent Cox2 inhibitors with respect to the "Universe" of reference drugs. The topological descriptor space featured the first three Principal Components (PC) [21] extracted from a set of topological descriptors [22, 23] and E-state keys [24], whereas the pharmacophore space was defined on hand of the first three PCs derived from ComPharm overlay descriptors ([25], see Figure 5.1 for a brief explanation) with respect to the reference compound in Figure 5.2. These spaces (Figure 5.3) were spanned by regular 3-D grids, where the number of divisions per axis was chosen such as to generate, in each space, a roughly equal number of grid cells populated by molecules. $N_T = 28$ divisions per axis were needed to split the topological space into 28^3 cells, out of which 867 were

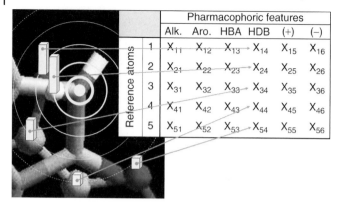

		Pharmacophoric features					
		Alk.	Aro.	HBA	HDB	(+)	(−)
	1	X_{11}	X_{12}	X_{13}	X_{14}	X_{15}	X_{16}
	2	X_{21}	X_{22}	X_{23}	X_{24}	X_{25}	X_{26}
Reference atoms	3	X_{31}	X_{32}	X_{33}	X_{34}	X_{35}	X_{36}
	4	X_{41}	X_{42}	X_{43}	X_{44}	X_{45}	X_{46}
	5	X_{51}	X_{52}	X_{53}	X_{54}	X_{55}	X_{56}

Fig. 5.1 Principle of ComPharm field descriptors. The hydrogen bond donor (HBD) in the candidate molecule (cylinder rendering) generates a "HBD field" in the surrounding space. Its intensity is probed at the centers of the atoms of the reference compound (in ball and stick). Local intensities produced by each donor of the candidate at every atom *i* of the reference are summed up to yield the components $X_{i,4}$ of the field descriptor matrix $X_{N,6}$ monitoring the intensities of every field type – hydrophobic "Alk(yl)", aromatic "Aro", acceptor "HBA", donor "HBD", cation "(+)", anion "(−)" – produced by the candidate at each of the *N* reference atoms (including hydrogens). Note that the >NH donor group has been positioned such as to match a −OH donor group in the reference, for example, the high HBD field level at the reference OH group signals that this candidate successfully replaces that hydrogen bond donor in the reference. If the −OH group contributed a lot to the binding affinity of the reference, this candidate might also be active.

populated, whereas only $N_P = 20$ divisions are required to obtain 860 populated cells out of the 20^3 defined in the pharmacophore space. The distribution of actives (defined by a $pIC_{50} \geq 7$) within these cells was assessed in order to evaluate the relative diversity of the pool of actives with respect to the entire data set.

5.6
Pharmacophore Fingerprints and Similarity Searches

The concept of pharmacophore fingerprint is rooted in the early practice of substructure searching, where chemical structures are encoded in the natural form of connection table (atomic identities and bonding connectivity) and the chemical information so encoded is processed in the search for substructures of chemical and biological interest [26, 27]. In addition to substructures based on 2-D topological connectivity, 3-D molecular conformation gave rise to the concept of 3-D substructures (or pharmacophores) within the context of ligand–receptor recognition, as discussed in the introductory chapter, and methods of searching for such pharmacophore have also been developed [28–30].

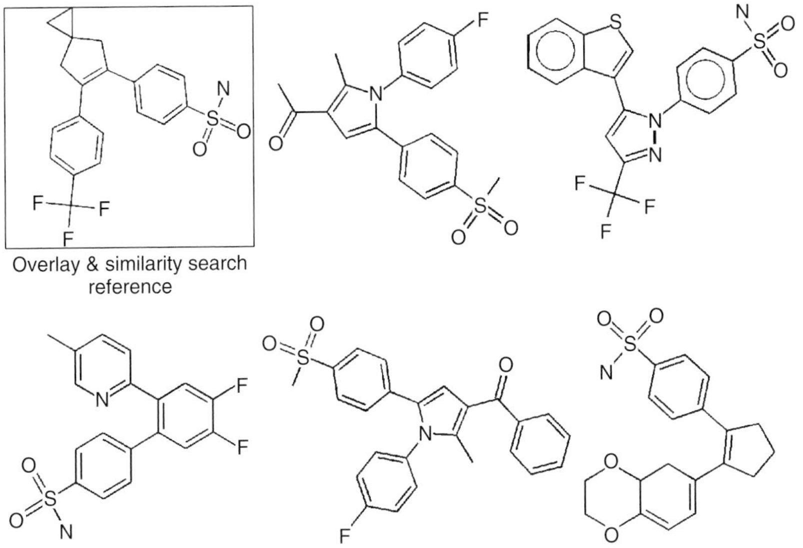

Fig. 5.2 Typical chemotypes represented in the Cox2 data set [20].

Once sufficient progress was made in the development of substructure searching techniques, it became clear that more general schemes of molecular representation beyond connection table might be useful. For example, all possible pharmacophore feature arrangements can be enumerated to form the basis of a bitstring in which each bit records the presence or absence of a given pharmacophore arrangement in a molecule [31]. In almost all practical applications, the pharmacophore construction was based on geometrical distances, although topological distances may be used if desired [32, 33]. The calculation of 3-D pharmacophore descriptors requires a reliable method for generating stable conformations, but by the same token, the 3-D structural information inherent in these fingerprints makes them more attractive than 2-D fingerprints. For a bitstring fingerprint, possible pharmacophores have been enumerated on the basis of simple centers [34], feature pairs [35, 36], three points [31, 37] and four points [8, 38]. The move toward multiple center pharmacophore increases the length of the fingerprint, but can potentially encode more useful information [38]. Extensions of the basic pharmacophore fingerprint technology include the incorporation of BCUT descriptors [39], the inclusion of pharmacophore fingerprints from inactive compounds and the generation of ensemble hypotheses [40], an efficient and flexible way for including multiplet pharmacophore representation [41].

Pharmacophore fingerprints other than binary bitstrings have also been studied. Makara [42] designed a "total pharmacophore fingerprint". Horvath [43] introduced Fuzzy Bipolar PharmacophoreFuzzy Bipolar Pharmacophore Autocorrelograms (FBPA) based on fuzzy binning. Both of these are real number vectors, shorter than binary fingerprints. In the structural spaces defined by these fingerprint representations, molecular dissimilarity can now be quantitatively defined as the "distance" between

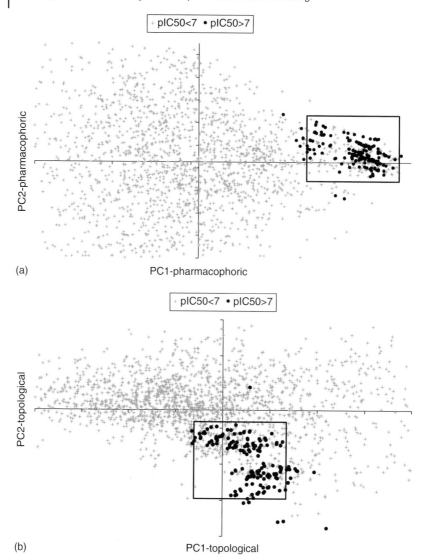

Fig. 5.3 Principal component plots in pharmacophoric and topological spaces. Rectangles schematically indicate the "activity zones".

the two "points" standing for a pair of compounds [44]. Many similarity measures have been developed [27]. Given such distances, neighbors of a compound of interest can be identified on the basis of their distances with respect to the target compound, or a group of compounds can be divided into subgroups that share more similarity within each subgroup than between subgroups [45]. In either approach, the similarity measure in the structure must be related to the chemical and/or biological behaviors with some significance. In other words, the validity of the "similarity principle" stating that similar compounds are expected to display similar properties may now be quantitatively expressed and will be a function of both the used descriptors and similarity score. The term "neighborhood behavior" was proposed to describe such relationships between structural neighborhood and similarity of a biological response [46], a concept recently extended to encompass multiple biological responses – an activity profile [33]. That work also presented an optimized fuzzy bipolar pharmacophore-based similarity score. As in other *in silico* techniques that are improved by some degree of fuzziness [47, 48], fuzzy pharmacophore fingerprint-based metrics display good neighborhood behavior [49].

In the similarity screening procedures exemplified in this work, "virtual hit" prioritization is done by ranking with respect to increasing calculated dissimilarity scores between the screened compounds and the nanomolar Cox2 reference ligand from Figure 5.2. Two different pharmacophore-based similarity searches were performed – using (1) the optimal FBPA [43] dissimilarity score [50] and (2) the ComPharm [25] dissimilarity coefficient.

5.7
Molecular Field Analysis (MFA)-Based Pharmacophore Information

Comparative Molecular Field Analysis (CoMFA) [51], one of the most widely used tools in 3-D QSARs, attempts to relate the observed molecular property values to the intensities of molecular fields monitored at well-defined points of space surrounding the molecule. A meaningful comparison of the field intensities produced by several molecules in a key point of space – which hypothetically harbors a protein residue interacting with these ligand fields – requires a prior alignment [52] of the putative ligands (ideally) representing an overlay of the bioactive conformations adopted in the common binding site. Since CoMFA models are preferentially used to build QSAR models for targets of unknown 3-D active site geometries, the molecular overlay rules *per se* – widely differing in sophistication, from rigid atom-to-atom mappings on hand of a well-conserved skeleton, to fully flexible searches aimed at maximizing the spatial covariance of molecular fields – cannot offer any guarantees that the actual "bioactive" configurations will be necessarily found. Therefore, in spite of attempts to render the interpretation of CoMFA fields both statistically more robust and more intuitive [53, 54], it is not straightforward to "translate" CoMFA QSAR models into "pseudoreceptor" [55] models mapping hypothetic features of the active site.

The classical CoMFA procedure relies on sterical and electrostatic fields or on spatial interaction energy maps of the potential ligands with standardized hydrophobic or polar "probe" molecules [56]. In this work, the modified "ComPharm" approach described in detail elsewhere [25] will be used. Its key differences with respect to classical CoMFA are the following:

- ComPharm explicitly monitors pharmacophore patterns in calculating the intensities of six empirical "pharmacophore fields" described by Gaussian functions of the distance to their sources, which are the functional groups of the corresponding pharmacophore type.

- Molecular alignment is achieved with respect to a single "reference" species, atop of which molecules are overlaid so as to realize the best compromise, simultaneously maximizing the weighed spatial covariance of all six pharmacophore fields of reference and overlaid compound. If the key features responsible for affinity are known from previous studies, the algorithm can be directed to specifically assign more weight to them. Otherwise – as is the case here and in all the studies aimed at building a pharmacophore interaction map "from scratch" – the pharmacophore features in the reference compounds were weighed by *type* only; for example, all aromatic carried the same weight, different from the one shared by all the cations (see [25] for details on the weighing scheme, as well). The Genetic Algorithm–driven procedure chooses, out of predefined conformer pools of reference and overlaid compound, the conformer pair leading to an optimal overlay when *rigidly* superimposed.

- The same empirical pharmacophore field intensities used to guide the overlay procedure are recalculated at the points occupied by the reference atoms at optimal overlay in order to be used as molecular field descriptors. They render an overlaid molecule through the "filter" of the reference compound, basically discriminating the reference features that are successfully matched from the ones that have no equivalent in the overlaid compound. Pharmacophore field intensities are probed *only* at the space points occupied by the atoms of the reference molecule (Figure 5.1) and therefore convey little information about the overall molecular size. Contributions from overlaid compound fragments that extend far beyond the zone occupied by reference molecule atoms are not captured, as their intensities at the closest "grid points" (reference atoms) fade away. In order to extend the space zone covered by the reference grid, the H atoms of the reference compound are also used as grid points. In this way, a fragment "protruding" from the zone mapped by the reference compound will be "sensed" in terms of increased field contributions at the level of the outermost reference hydrogens in that area. ComPharm is therefore able to detect the effect of substitution by a small functional group at the given point of the reference molecule. However, increasing the length of the protruding chain will eventually cease to affect the pharmacophore field values as newly added groups will be too far away to contribute any significant Gaussian intensities at the level of the reference atoms. While in "classical" CoMFA this problem can be, in principle, fixed by a larger definition of the grid map (at, however, high computational cost), the ComPharm procedure needs to rely on the choice of the reference structure in order to achieve a meaningful mapping of the alleged "active site zone".

- As the ComPharm overlay procedure is stochastic, it is typically repeated five times and the resulting average pharmacophore field intensities at the reference atom centers are used as "ComPharm Field Descriptors".

A QSAR approach allowed to select relevant ComPharm field descriptors is therefore expected to probe which of the features represented in the reference compound (typically chosen to be a potent inhibitor) must be conserved and which can be altered without an activity loss. This kind of intuitive [15] information is highly valued by medicinal chemists as it can easily be "translated" into an analog synthesis plan.

5.8
QSAR Models

For QSAR model building purposes, the Cox2 inhibitor set was split into a learning set (LS, 80% of compounds) and a Validation set (VS, 20%). Splitting was done so as to ensure an equivalent relative distribution of actives and inactives throughout both sets, and ignoring the original provenience of the compounds. These sets were used to train and validate two types of linear QSAR approaches:

- an "overlay-independent" QSAR based on two-point pharmacophore fingerprints (FBPA [43]) and related descriptors, and
- an "overlay-based" QSAR that additionally includes average ComPharm Field descriptors derived from a five-times-repeated overlay of the compounds with respect to the reference in Figure 5.2.

Each of these two QSAR model searches led to pools of several thousands of statistically valid linear equations, expressing the estimate of the Cox2 pIC_{50} value as linear combinations of molecular descriptors selected by a Genetic Algorithm (GA) [57, 58]. The linear models within each pool were subjected to a diversity analysis procedure in order to discard redundant equations based on roughly similar descriptor choices. This procedure assessed whether some of the different descriptors used by different equations were intercorrelated and, therefore, interchangeable [59]. The remaining diverse QSAR equations were further classified by "size" (number of descriptors they include). The best equations of each encountered size were kept for final validation with the VS molecules and for further analysis. *Consensus* models featuring average predictions over these equations were also generated and validated. We focus here on the discussion of the "minimalist" overlay-independent and overlay-based QSAR models, each including only six descriptors, and refer to the optimal consensus model of the overlay-based QSAR approach families for comparative purposes.

5.9
Hypothesis Models

The GA-driven selection of ComPharm field descriptors optimally related to activity may be ultimately viewed as a "pharmacophore hypothesis" – a listing of space zones where groups of specified pharmacophore type are required (positive weighing coefficient) or forbidden (negative weighing). Other well-known hypotheses-building tools, based on different statistical approaches, are HipHop, HypoGen (both subsystems of Catalyst [60]), DISCO [61] and GASP [62].

Unfortunately, HipHop and HypoGen cannot process the large training sets of the size used for QSAR model building. The set of 29 most potent Cox2 inhibitors has been submitted to HipHop and the best of the resulting hypotheses has been qualitatively compared to the overlay-based QSAR model hypothesis.

5.10
The Minimalist Overlay-Independent QSAR Model

The information regarding pharmacophore feature pair distribution in the molecules appears to be sufficient in order to explain an important fraction of the observed Cox2 activity variance. Only five FBPA descriptors and an "amphiphilic distance" term are needed to generate a statistically robust QSAR model: training RMS = 0.799 pIC_{50} units, training $R^2 = 0.637$, validation RMS = 0.794, validation $Q^2 = 0.643$ – all this with a (number of observables)/(number of variables) ratio of almost 300. Furthermore, this overlay-independent, and therefore computationally effective, model has the nonnegligible merit to be easily translated into chemical language: according to Table 5.1, a molecule may be a potent Cox2 inhibitor if

- it contains *fewer* atom pairs of both (Aromatic-Hydrogen Bond Acceptor) types at 11 Å and of (Hydrophobic-Hydrogen Bond Acceptor) types at 13 Å than the typical, current drugs on the market. This renders both zsig3(Ar-HA11) and zsig3(Hp-HA13) maximal (= 1), and increments the calculated pIC_{50} by 3.1 and 2.4 units respectively; see table footer for further details;
- atom pairs of (aromatic-hydrogen bond acceptor) types at 10 Å are, on the contrary, *more* numerous than in typical drugs, thus forcing the negatively weighed zsig3-(Ar-HA10) term to its minimum, 0;

Tab. 5.1 The six-variable linear overlay-independent Cox2 QSAR model: $pIC_{50} = \Sigma$descriptors × coefficients + intercept. Except for DistH-, a descriptor representing the distance in angstroms between the geometric centers of the hydrophobic groups and the one of polar electron-enriched groups (hydrogen bond acceptors and anions), all the other terms are FBPA descriptors [43], sometimes subjected to nonlinear transformations (see footer explanations)

Descriptor	Coeff	Average in current drugs	Variance in current drugs
zsig3[b] (Ar-HA11[a])	3.1	1.4	2.4
zsig3(Ar-HA10[a])	−2.0	1.0	2.1
Hp-HA10[a]	0.2	1.6	2.5
DistH-	0.3	2.5	1.5
zsig3(Hp-HA13[a])	2.4	0.5	1.3
Ar-Ar6[a]	0.1	3.5	4.3
Intercept	0.0	–	–

a) Each FBPA descriptor, denoted as **T1-T2d** reports the average over all conformers of the number of atom pairs of given pharmacophore types T1 and T2 respectively (**Hp** – hydrophobic, **Ar** – aromatic, **HA** – hydrogen bond acceptor) and having an interatomic distance of about d Å. These pharmacophore pair counts are truncated to a maximum value of 10, for example, Ar-Ar6 will equal 10 for all molecules with 10 or more aromatic atom pairs at about 6 Å. Reported averages and variances take this truncation into account.

b) The sigmoid function zsig3 of a descriptor D is $\mathbf{zsig3(D)} = [1+e^{3*z(D)}]^{-1}$ where $z(D) = [D − AVG(D)]/VAR(D)$ is the average-variance normalization function. The average and variance values used for normalization are the ones taken over the BioPrint [5] set and are listed in this table. zsig3(D) tends to 0 whenever the D value of the current molecule is much larger that its average over the known drugs, and to 1 when D is significantly smaller than expected in an "average" drug molecule.

- the compound includes a maximum of (aromatic–aromatic) pairs at 6 Å and (hydrophobic-hydrogen bond acceptor) pairs at 10 Å;
- the geometric center of the hydrophobic groups is as distant as possible from the one of polar electron-enriched groups (hydrogen bond acceptors and anions).

It is tempting to interpret the coefficients of the QSAR model in Table 5.1 as the absolute "propensities" to cause Cox2 activity associated with each pharmacophore feature pair. Theoretically, such "propensities" or "potentials of mean force" associated with a descriptor D could be obtained by taking, over the whole chemical space, the average activity $\langle Act \rangle|_{D \approx d}$ over the possible chemical structures for which D adopts a current value d, and monitoring its variation with respect to d. It is however unwise to assume that real-life QSAR coefficients ever approach their "ideal" values c_{ideal} = $\partial \langle Act \rangle|_{D \approx d} / \partial d$, as the sampling of the whole "druglike" chemical subspace is practically unfeasible. The presence of a pharmacophore feature pair in a molecule is weakly, or not at all, correlated with overall molecular shape or size. Therefore, out of the quasi-infinite Universe of $10^{18} \ldots 10^{56}$ potentially "drug-able" molecules (according to various and "slightly" diverging estimations), the subset of compounds matching a given pharmacophore feature pair is certain to be most likely composed of molecules of either wrong size and shape or including other features that prevent them for binding. Assuming a (probably grossly overestimated) rate of 1% of nanomolar compounds among the ones displaying the feature D (assumed, for simplicity, to be coded by a binary descriptor $D = 1$) in an "ocean" of millimolar "inactives", then $\langle pIC_{50} \rangle|_{D=1} = 0.99 \times 3 + 0.01 \times 9 = 3.06$, whereas $\langle pIC_{50} \rangle|_{D=0} = 3.00$ (assuming that absence of the feature triggers absence of activity). In other words, each individual "ideal" coefficient associated to a local molecular feature is expected to be of an order of magnitude translating to a maximal activity variation ($\Delta Act = c_{ideal}|D_{max} - D_{min}|$, where the associated descriptor spans its entire value range) of no more than a few percent of the activity scale. Even a model based on a training set that is by almost two orders of magnitude larger than the typical collection of 20 to 30 compounds currently encountered in QSAR literature, and yet using no more descriptors than would have been acceptable in one of those situations, is seen to be very far from this theoretical ideal. The variations of the average activity with respect to the presence/absence of a required pharmacophore pattern may be quite dramatic *within* the subset of molecules *of "right" size, "right" shape and void of interfering groups*, but no single candidate descriptor of the initial list subjected to this GA sampling simulation has, *per se*, the ability to account alone for either of these global constraints. So why did the GA fail to obtain a multilinear model with many – potentially up to 100, until reaching the limit of say 20 variables/explained value – of little contributing descriptors that "cooperate" in order to generate an overall, coherent picture of these shape and size constraints?

Why should Cox2 activity be at all related to the presence or absence of certain pharmacophore feature pairs in a molecule? Certainly, it can be argued – using the typical interpretation given to QSAR models based on autocorrelogram-type descriptors [63] – that the selected pharmacophore feature pairs stand for pairs of functional groups that are involved in direct interactions with the site, and these functional groups must be found at a fixed relative distance. However, this model provides an extremely sketchy description of this hypothesized Cox2 pharmacophore. While it notes what type of pharmacophore feature pairs a molecule should contain, it provides no information about the relative arrangement of these required features. It does not mention, for example, whether the aromatic-acceptor pairs must stem from a single acceptor group surrounded by aromatic atoms at 10 or 11 Å. It may be that molecules having no such pairs at all will always be inactive, but it is difficult to argue that the set of all possible compounds featuring such pairs, regardless of the locations and contexts, will be significantly enriched in actives.

This notwithstanding, within the current set, molecules matching the conditions of this minimalist QSAR do represent a subset that is significantly enriched in actives. A natural explanation thereof would be that the potent Cox2 ligands in this study form a *homogeneous* structural subfamily ("homogeneous", at least, with respect to the whole structural space occupied by druglike compounds.) The QSAR model did not learn *what are the features required to render a compound active*, but has rather learned *the features required to define a compound as a member of this structural family*. Only members of this subfamily match the outlined pharmacophore pair criteria. Therefore, whenever these criteria are fulfilled, all of the structural constraints required for binding, but not explicitly mentioned by the model, are "automatically" fulfilled in virtue of "subfamily membership". This model may act like a pattern recognition filter: on the basis of a minimal number of subfamily-specific criteria, it is able to differentiate between subfamily members and nonmembers and further *assumes* that *family membership* implies *activity*. This assumption is statistically valid within the given compound set. In other words, the selected feature pairs were picked not because of their mechanistic involvement in binding, but because they are "overexpressed" in the subset of actives. The chance to encounter molecules with an aromatic-acceptor pair at 10 Å among the BioPrint drugs and reference compounds of this study should ideally match the "natural" probability to find such a molecule within the Universe of existing drugs. The current training set has been however artificially enriched in actives from the Cox2 learning set and implicitly enriched in compounds with aromatic-acceptor pairs at 10 Å, as this feature is much more often represented in these actives than in randomly picked compounds. However, the subset of "actives" is hardly a random selection of molecules: on the contrary, they also fulfill the shape, size and other constraints required for binding. Thus, the subset of molecules displaying the Ar-HA10 element shows an artifactual enrichment of compounds of COX2-compatible shape and size, so that the pharmacophore element appears as an (undue!) "marker" of the COX2-compatible scaffold.

5.11
Minimalist and Consensus Overlay-Based QSAR Models

Upon inclusion of ComPharm field descriptors in the initial pool of molecular indices, four of these are selected by the GA-driven QSAR building procedure to enter the minimalist six-variable overlay-based QSAR model. These ComPharm "key points" are

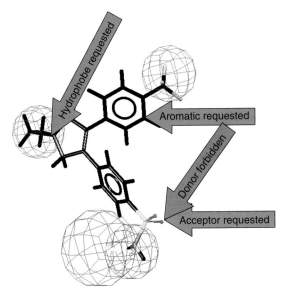

Fig. 5.4 Comparative display of a catalyst-HipHop hypothesis and the pharmacophore field descriptors selected by the minimalist ComPharm overlay-based model. ComPharm key features are pinpointed by arrows, while HipHop feature spheres stand for hydrophobes (light blue) and hydrogen bond acceptors (green).

located on the reference compound structure in Figure 5.4, together with the features of the top-ranked HipHop hypothesis.

At identical variable number, the overlay-based model outperforms its overlay-independent equivalent: training RMS = 0.712 (pIC$_{50}$ units), training R^2 = 0.712, validation RMS = 0.698, training Q^2 = 0.724. ComPharm field descriptors therefore appear to have a better predictive power than pairwise pharmacophore feature counts alone.

Yet more powerful than the minimalist approach, the optimal consensus Cox2 model displays an excellent training RMS of 0.56 pIC$_{50}$ units, for example, R^2 = 0.825, and behaves even better with respect to the external validation set (RMS = 0.53 pIC$_{50}$ units, Q^2 = 0.841).

According to the minimalist model in Table 5.2, a potent Cox2 ligand is a molecule that
- contains fewer positively charged groups than the "average drug molecule", e.g., contains *no* positively charged groups;
- features a number of aromatic-acceptor pairs at 5 Å, which closely matches the average value (5.3) of this descriptor over the set of reference drugs;
- when overlaid on top of the reference compound, it generates strong "aromatic", "hydrophobic" and "acceptor"-type pharmacophore fields at the reference atoms highlighted by green arrows in Figure 5.4, but does *not* produce any significant "donor"-type field at the location pointed to by the red arrow.

Tab. 5.2 The six-variable linear overlay-based Cox2 QSAR model

Descriptor	Coeff.	Comment
Ar@Atom#2	0.191	Aromatic pharmacophore field intensity at atom Nr.2 of reference (in upper aromatic ring of Figure 5.4)
Hp@Atom#11	0.179	Hydrophobic pharmacophore field intensity at atom Nr.11 of reference (in central ring of Figure 5.4)
HA@Atom#20	0.430	Acceptor pharmacophore field intensity at atom Nr.20 of reference (sulfonamide oxygen in Figure 5.4)
HD@Atom#20	−0.428	Donor pharmacophore field intensity at atom Nr.20 of reference (sulfonamide oxygen in Figure 5.4)
zexp[a] (Ar-HA5)	1.414	Gaussian function of the number of aromatic-acceptor atom pairs at 5 Å
zsig3(PI)	1.414	Sigmoid function of the number of positive charges of the molecule (see footer of Table 5.1 for the definition of zsig3)
Intercept	0.000	

a) **zexp(D)** = $\exp(-z(D)^2)$ with z(D) being the average/variance normalization function: also see footnote of Table 5.1.

Again, it is tempting to believe that these key points of the reference molecule match privileged zones of the active site where site–ligand interactions are particularly strong. Furthermore, the ComPharm key points are practically identical to the features of the HipHop hypothesis. Both models have designed exactly the same hydrophobic zone at the central ring system, whereas the second HipHop "hydrophobe" can be assimilated to the ComPharm aromatic feature on the nearby ring, as HipHop does not distinguish aromatics from hydrophobes. Also, the importance of a hydrogen bond acceptor is clearly outlined by both approaches. Unfortunately, HipHop designs the sulfonamide nitrogen in this role, although it could have, by a simple rotation around the C–S bond, aligned the better acceptor – the sulfonamide oxygen singled out by ComPharm – to the HBA feature.

However, the ComPharm hypothesis specifically asks for strong hydrophobic, aromatic and acceptor contributions in zones where the reference molecule itself features hydrophobic, aromatic and hydrogen bond acceptor groups, respectively. It is though legitimate to ask whether these constraints in terms of local pharmacophore field intensities are not just an alternative way to assure the selection of compounds that are *globally* similar to the reference, rather than to pinpoint the features that must be conserved. Clearly, overall pharmacophore pattern similarity between a candidate ligand and the reference is a sufficient condition to achieve the requested pharmacophore field levels. It is, though, not a necessary one – dissimilar compounds might nevertheless be appropriately aligned with respect to the reference. If such examples are however not represented in the learning set, then the selected, local pharmacophore field constraints and the overall pharmacophore feature match required to score low dissimilarity values would be equivalent. Thus, molecules that generate the needed pharmacophore fields at the expected locations are *members of the privileged structural family* and might be assumed to be active *in virtue of this membership*.

5.12
Diversity Analysis of the Cox2 Compound Set

The above-mentioned hypothesis is validated by the finding that all the potent Cox2 ligands do indeed appear to concentrate in a well-defined zone of the Topological Structure Space (the first two principal components of which are shown in the lower plot of Figure 5.3.). The vast majority of potent Cox2 ligands have a common chemotype based on a central unsaturated or aromatic 1,2-disubstituted ring with both substituents being aromatic, one of which carries a sulfonamide or sulfone group. Nonselective Cox1/Cox2 inhibitors, featuring a carboxylate group, such as diclofenac or flurbiprofen, are included in the study, but do not figure among the most potent Cox2 inhibitors as their IC_{50} values are above the 100-nM threshold. Out of the 860 populated cells of the topological structure space, 58 (6.7%) contain active molecules. The strong relatedness of the active compounds is even more evident in the Pharmacophore Field Space (first two PCs shown in the upper plot of Figure 5.3) where the actives are regrouped in 46 (5.3%) out of the 867 populated cells. The first principal component of this space significantly discriminates between actives and inactives ($R^2 = 0.42$), being a "privileged" axis with respect to Cox2 activity in the "ChemGPS" [64] sense. The topological plot reveals some "outlier" actives with unusual chemotypes, but the pharmacophore plot proves that these molecules largely "hide" the same pharmacophore pattern under their different connectivity. We will, further on, refer to the subsets of cells each containing at least two actives as "Activity Zones". In the topological space, the activity zone covers 35 cells containing 152 of the 175 actives and 142 inactives, with the local density of actives above 50%, for example, a 6.6-fold enrichment with respect to the entire library. This enrichment is more significant in the 28 cells of the pharmacophore activity zone, regrouping 157 actives and 115 inactives; for example, 7.4 times more actives than expected from a random library sample of this size.

5.13
Do Hypothesis Models Actually Tell Us More Than Similarity Models About the Structural Reasons of Activity?

It is not the goal of this paper to argue for a general answer to this question (expected to be YES according to common sense). We now, however, have all the elements in hand in order to answer the more specific question on whether the Cox2 QSAR hypotheses are actually able to recognize key activity-modulating points in the structure, and hence outperform similarity-based approaches based on general recognition of "active-enriched" structural classes. We may look at a QSAR model from two different perspectives:

- As a tool that delimits, out of the entire structural space, the zones most likely to harbor a higher density in active compounds ("Activity Zones"), all based on variations around a common pharmacophore pattern.

- As a tool that discriminates between the actives and inactives *within* the Activity zones, for example, to predict which of the local pharmacophore pattern variations will enhance and which will decrease activity.

While both similarity-based and hypothesis-based models should be successfully completing the first task, the latter are expected to have a significantly better discriminating power *within* an "Activity Zone". Indeed, the Euclidean dissimilarity metrics employed for the cell-based definition of Activity Zones are unable to provide any further way to distinguish between actives and inactives sharing a same cell. More refined pharmacophore-based dissimilarity scores might provide some discrimination between these "false similar" compound pairs, by proving that the pharmacophore patterns within the cell-based activity zone are actually not "as similar as the PC-based diversity function suggested". However, only models based on feature *selection* may account for the different activities of compounds with an *overall* pharmacophore similarity, by pinpointing the responsible *local* structural differences.

Table 5.3 monitors the degree of correlation between actual Cox2 activities and various activity predictors representing computational models of different classes:

- Pharmacophore similarity scoring schemes are represented by the FBPA and ComPharm overlay dissimilarity scores, the expectation being to see low dissimilarity scores matching high activity levels.
- QSAR models (the minimalist overlay-independent and both the minimalist and the optimal consensus overlay-based) are represented by the actual predicted pIC_{50} values – these should ideally equal the experimental activities.

These correlation coefficients are calculated both with respect to the entire set and also specifically within the pharmacophore and topological activity zones. According to the previous discussion, it would be expected to see that

- both QSAR predictions and pharmacophore dissimilarity scores strongly correlated to activity over the entire set and over the Topological Activity Zone, since both these (sub)spaces feature a significant amount of pharmacophore pattern diversity;
- pharmacophore dissimilarity scores are outperformed by hypothesis models in their attempt to make the difference between the actives and inactives within the Pharmacophore Activity Zone.

Tab. 5.3 Correlation coefficients between the similarity-based and hypothesis-based activity prediction scores and experimental activity values, taken over the whole Cox2 inhibitor set and with respect to the selected "activity zones" respectively

Activity predictor	Whole set	Pharmacophore activity zone	Topological activity zone
ComPharm similarity score	0.6223	0.6434	0.7876
FBPA similarity score	0.5128	0.1142	0.3288
Minimalist overlay-independent model	0.6884	0.4658	0.7158
Minimalist overlay-based model	0.7573	0.5552	0.7972
Consensus overlay-based model	0.8246	0.6575	0.8239

However, only the first of these expectations is met: with respect to the entire data set, variable selection-based models (overlay-based consensus > overlay-based minimalist > overlay-independent minimalist model) outperform the similarity-based methods. With one notable exception – the FBPA fuzzy similarity score [43] – the correlations obtained within the Topological Activity Zone are better than or equal to the ones within the full compound set. This makes sense, as the Topological Activity Zone subset no longer includes the wealth of "exotic" chemotypes seen within the set of reference drugs (antibiotics, steroids, etc.), hence offering fewer chances to encounter species that fortuitously match the QSAR criteria or exploit artifacts in the ComPharm similarity metric in order to score "undeservedly" low dissimilarity scores. The FBPA dissimilarity scheme [43] is based on atom pair counts scaling as the square of the size of the molecule. This score is therefore more sensitive to important structural variations: its best performance is therefore achieved with respect to the entire data set rather than within focused subfamilies.

As we would anticipate, the correlation between FBPA scores and activity breaks down within the Pharmacophore Activity Zone. However, the ComPharm similarity score fully maintained its ability to discriminate between the actives and inactives of this focused subset. This is a proof of its much improved neighborhood behavior with respect to the PC-based metric used to define the subset. Although both metrics originally rely on the same pharmacophore field descriptors, the ComPharm score (a) does not suffer from the information loss because of discarding all but the three first principal components, (b) includes tuned feature weighing in the score calculation – even though these weights were not specifically trained with respect to the Cox2 problem – and (c) is based on a Hodgkin [65]-type score, a scheme that outperforms the Euclidean distance in neighborhood behavior studies [33].

Therefore, this shows that, even within a focused subset with much less pharmacophore pattern variance than encountered within the set of all drugs, a powerful pharmacophore similarity metric may still explain 64% of the observed activity variance. This level is not even matched by any of the minimalist QSAR models, and not significantly improved by the optimal consensus approach. These models have therefore no advantage over similarity models: predicting an activity in terms of the selected, specific features is not a better strategy than prioritizing potential hits in terms of their overall relatedness with respect to a potent reference ligand. Thus, there is no evidence to support that the selected features match the key groups involved in strong ligand–substrate interactions.

However, hypothesis models did clearly outperform similarity-based scoring with respect to the entire set. This advantage does not, as previously shown, stem from a better prediction of the subtle differences between the actives and inactives within the pharmacophoric "Activity Zone". It may be ascribed to the ability of hypothesis models to correctly recognize actives that "hide" the "key elements" asked for by the model within a globally different pharmacophore pattern, which prevents them from being top ranked by the similarity search.

5.14
Why Did Hypothesis Models Fail to Unveil the Key Cox2 Site–Ligand Interactions?

The analysis of the crystal structures of Cox2 with a selective ligand SC-558 (PDB [66] code 1CX2) and the nonselective Cox1/Cox2 inhibitor flurbiprofen (PDB code 3PGH) fully confirms the expectedly weak match between the pharmacophore models built by variable selection and the actual Cox2 interaction map.

The best agreement between hypotheses and binding mode (Figure 5.5) is reached with respect to the bromophenyl moiety, which is indeed occupying a hydrophobic pocket rich in aromatic residues. In this respect, and if aromatic stacking plays a key role there, the ComPharm models, discriminating between aromatic and hydrophobic features and specifically asking for an aromatic pharmacophore field in that zone, might be closer to the truth than the HipHop demand for an unspecified "hydrophobe".

There are however no strong hydrogen bridges involving any of the sulfonamide oxygen atoms. A first look (Figure 5.5) at the interaction map would suggest that the sulfonamide – NH$_2$ acts as a hydrogen bond donor (HBD) to Histidine 90, apparently receiving a hydrogen bond from the arginine NH to its "free" electron pair, as suggested by HipHop (question mark in Figure 5.5). This interaction makes little sense, however,

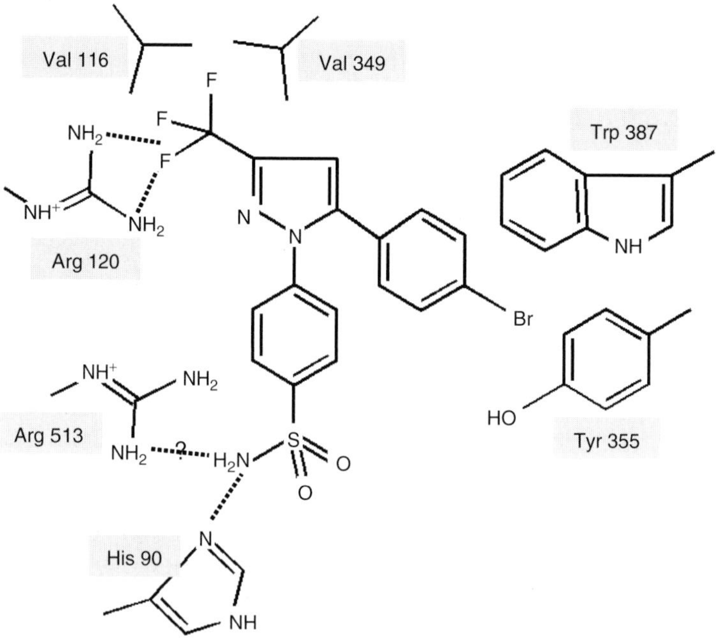

Fig. 5.5 Putative interactions of the cocrystallized selective Cox2 ligand SC-558 with its active site, assuming a neutral sulfonamide group – the state used to assign pharmacophore feature flags by the used software. Dotted lines stand for hydrogen bonds, the other residues being responsible for hydrophobic contacts. From a physicochemical point of view, an ionized SO$_2$ NH$^-$ involved in a salt bridge with Arg 513 and hydrogen bonding to the other tautomer of His 90 would make more sense.

as sulfonamide – NH_2 is a very weak, if any, hydrogen bond acceptor. This group has most likely lost a proton under the influence of the interaction with the guanidinium system of Arg 513, to form a salt bridge $-SO_2 NH^-$ ...Arg$^+$. However, pharmacophore feature assignment tools working on compounds *prior* to the elucidation of their binding mode could *not* have predicted the presence of an active-site induced anion feature at that level (for example, the feature assignment tool used with ComPharm assigns anion flags to acetyl or phenylsulfonamides only, accounting for the acidifying effect of a second electron-withdrawing group). The pharmacophore hypotheses retained the need of a hydrogen bond acceptor instead of an anion – which seems, at first sight, fair enough. However, they reached a correct conclusion on the basis of a *wrong* premise, as, indeed, acidity is caused by the electron-withdrawing effect of several neighboring heteroatoms, out of which at least some happen to also be acceptors, as not all of their free electron pairs are taken by protons. Thus, the flurbiprofen carboxylate and the sulfonamide share both the properties of being "anions" and "acceptors". The former seems to be physically relevant, but has not been recognized as a common feature by pharmacophore assignment routines assuming the ligand "default" protonation status shown in Figure 5.5. Therefore, the *second* has been (successfully) built into the QSAR models. Of course, the activity of candidate compounds featuring *strict* acceptor groups instead of the anion might be overestimated by the models – if the "pure" hydrogen bond turns out to be energetically less favorable than the salt bridge. This is not at all obvious, as salt bridge energetics is more heavily penalized by desolvation, not to mention the fact that, in this particular case, part of the achieved stabilization is counterbalanced by the somewhat "forced" deprotonation of $-SO_2 NH_2$, which does not spontaneously occur at pH $= 7$.

The $-CF_3$ substituent matches the hydrophobic feature next to the central 5-membered ring hypothesized by both ComPharm and HipHop – both approaches agreeing that the trifluoromethyl group actually is a hydrophobe. The interaction map, however, shows a more subtle picture: while the $-CF_3$ group is partly surrounded by hydrophobes, it also forms a close contact to Arg 120, a cation and hydrogen bond donor. This is surprising, as fluorine atoms in organic fluorides are not customarily known to undergo hydrogen bond interactions, as their significant partially negative charge acquired because of their electronegativity is counterbalanced by the low polarizability of their "free" electron pairs. Furthermore, with flurbiprofen (3PGH), the position of the $-CF_3$ group is taken by a *carboxylate*, forming a salt bridge with the arginine. Therefore, only a pharmacophore model that would have evolved to stipulate the equivalence of a hydrophobe with an anion could have correctly predicted the biologically relevant alignment of flurbiprofen (Figure 5.6) and SC-558 – or, such an assumption is in fundamental contradiction with the very essence of the pharmacophore "philosophy". ComPharm aligns flurbiprofen to the reference in Figure 5.2 by placing the carboxylate atop of the sulfonamide oxygens and allowing for significant overlap between the aromatic parts. Flurbiprofen, topologically completely unrelated to the family of selective Cox2 inhibitors like SC-558, actually displays a "hidden" pharmacophore pattern similarity of 67% to the reference compound representing this ligand family. ComPharm correctly detected the underlying similarity – *though for the wrong reason, based on an overlay making perfect sense from all points of view but the one of the Cox2 site*.

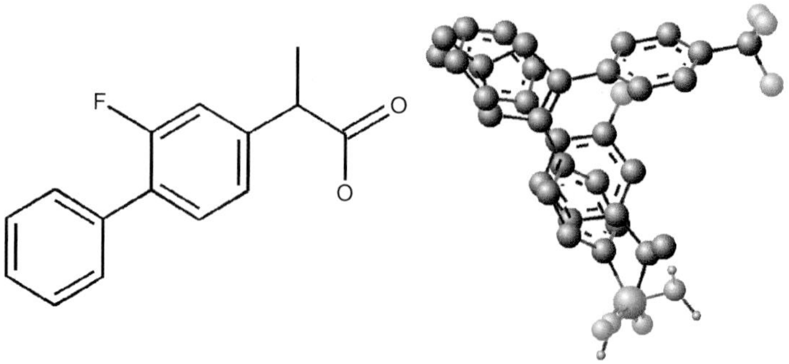

Fig. 5.6 Flurbiprofen (a) and its overlay (b) on the ComPharm reference structure shown in Figure 5.2.

5.15
Conclusions

The direct comparison of similarity-based and hypothesis-based pharmacophore approaches, at least in the present case study, suggest that similarity-based methods may be suited as a first-line virtual screening tool with respect to considerations discussed in the introduction. From this comparison, it thus becomes clear why, in this case, similarity-based pharmacophore approaches work so well and cannot be outperformed by feature extraction techniques: the reasons are intrinsic to the failure of pharmacophore reasoning to account for subtle site–ligand interaction mechanisms. As the borderline between hydrophobic and polar groups is never clearly defined, and as the flexible site residues may adapt to render a pocket more or less hydrophobic in order to accommodate the ligand, a rigid assignment of "pharmacophore features" will inherently lead to inconsistencies. This notwithstanding, the high-quality pharmacophore-based models derived here can be effectively used for the virtual search of Cox2 inhibitors. Although these models would miss out actives that are pharmacophorically dissimilar to the reference structure used here, the pharmacophore similarity detection tools are powerful enough to encompass a large spectrum of possible chemotypes, therefore being able to come up with "unexpected" new leads.

Acknowledgments

The authors wish to thank Dr Nicolas Baurin (Ribotargets, Inc.) for providing an electronic copy of the Cox2 ligand set. We are grateful to many Cerep scientists for the helpful discussions and encouragements.

References

1 BLAKE, J.F. Chemoinformatics – predicting the physicochemical properties of 'drug-like' molecules. *Curr. Opin. Biotechnol.* **2000**, *11*, 104–107.

2 OPREA, T. and MARSHALL, G.R. Receptor-based prediction of binding affinities. *Perspect. Drug Disc. Des.* **1998**, *9*(11), 35–65.

3 HOPFINGER, A.J. and DUCA, J.S. Extraction of pharmacophore information from high-throughput screens. *Curr. Opin. Biotechnol.* **2000**, *11*, 97–103.

4 ZAMORA, I., OPREA, T., CRUCIANI, G., PASTOR, M., and UNGELL, A.-L. Surface descriptors for protein-ligand affinity prediction. *J. Med. Chem.* **2003**, *46*, 25–33.

5 KRESJA, C.M., HORVATH, D., ROGALSKI, S.L., PENZOTTI, J.E., MAO, B., BARBOSA, F., and MIGEON, J. Predicting ADME properties and side effects: the BioPrint approach. *Curr. Opin. Drug Disc. Dev.* **2003**, *6*, 470–480.

6 GÜNER, O.F. *Pharmacophore Perception, Use and Development in Drug Design*, IUL Biotechnologies Series, USA, **2000**.

7 http://www.daylight.com/meetings/ summerschool02/course/basics/ smarts.html, **2003**.

8 MASON, J.S., MORIZE, I., MENARD, P.R., CHENEY, D.L., HULME, C., and LABAUDINIERE, R.F. New 4-point pharmacophore method for molecular similarity and diversity applications: overview of the method and applications, including a novel approach to the design of combinatorial libraries containing privileged substructures. *J. Med. Chem.* **1999**, *42*, 3251–3264.

9 MCGREGOR, M.J. and MUSKAL, S.M. Pharmacophore fingerprinting. 1. Application to QSAR and focused library design. *J. Chem. Inf. Comput. Sci.* **1999**, *39*, 569–574.

10 KAUZMANN, W. Some factors in the interpretation of protein denaturation. *Adv. Protein Chem.* **1959**, *14*, 1–63.

11 GROSS, E.L. and PEARSON, D.C. Jr. Brownian dynamics simulations of the interaction of chlamydomonas cytochrome f with plastocyanin and cytochrome c6. *Biophys. J.* **2003**, *85*, 2055–2068.

12 JOHNSON, M.A. and MAGGIORA, G.M. *Concepts and Applications of Molecular Similarity*. Wiley, New York, **1990**.

13 DEAN, P.M. *Molecular Similarity in Drug Design*. Chapman & Hall, New York, **1995**.

14 PATEL, Y., GILLET, V.J., BRAVI, G., and LEACH, A.R. A comparison of the pharmacophore identification programs: catalyst, DISCO and GASP. *J. Comput.-Aided Mol. Des.* **2002**, *16*, 653–681.

15 KING, R.D., HIRST, J., and STERNBERG, M.J.E. Comparison of artificial intelligence methods for modeling pharmaceutical QSARS. *Appl. Artif. Intell.* **1995**, *9*, 213–233.

16 KUBINYI, H. *Quantum Struct.-Act. Relat.* **1994**, *13*, 285–294.

17 HALPERIN, I., MA, B., WOLFSON, H., and NUSSINOV, R. Principles of docking: an overview of search algorithms and a guide to scoring functions. *Proteins: Struct., Funct., Genet.* **2002**, *47*, 409–443.

18 GILROY, D.W. and COLVILLE-NASH, P.R. New insights into the role of COX 2 in inflammation. *J. Mol. Med.* **2000**, *78*, 121–129.

19 KURUMBAIL, R.G., STEVENS, A.M., GIERSE, J.K., MCDONALD, J.J., STEGEMAN, R.A., PAK, J.Y. GILDEHAUS, D., MIYASHIRO, J.M., PENNING, T.D., SEIBERT, K., ISAKSON, P.C., and STALLINGS, W.C. Structural basis for selective inhibition of cyclooxygenase-2 by anti-inflammatory agents. *Nature* **1996**, *384*, 644–648.

20 BAURIN, N. Etude et développement de techniques QSAR pour la recherche de molécules d'intérnêt thérapeutique. Ph.D. Thesis, Université d'Orleans, **2002**.

21 EVERITT, B.S. and DUNN, G. *Applied Multivariate Data Analysis*. Oxford University Press, New York, **1992**.

22 BOLLOBAS, B. *Modern Graph Theory*. Springer, New York, **1998**.

23 The topological descriptors used here were those implemented in the QSAR+ module of Cerius2, Accelrys, http://www.accelrys.com/ **2003**.

24 HALL, L.H. and KIER, L.B. The E-state as the basis for molecular structure space definition and structure similarity. *J. Chem. Inf. Comput. Sci.* **2000**, *40*, 784–791.

25 HORVATH, D. In *QSPR/QSAR Studies by Molecular Descriptors*, DIUDEA, M. (Ed.). Nova Science Publishers, New York, **2001**, 395–439.

26 HAGADONE, T.R. Molecular substructure similarity searching: efficient retrieval in two-dimensional structure databases. *J. Chem. Inf. Comput. Sci.* **1992**, *32*, 515–521.

27 WILLETT, P., BARNARD, J.M., and DOWNS, G.M. Chemical similarity searching. *J. Chem. Inf. Comput. Sci.* **1998**, *38*, 983–996.

28 MARTIN, Y.C., DANAHER, E.B., MAY, C.S., and WEININGER, D. MENTHOR, a database system for the storage and retrieval of three-dimensional molecular structures and associated data searchable by substructural, biologic, physical, or geometric properties. *J. Comput.-Aided Mol. Des.* **1998**, *1*, 15–29.

29 VAN DRIE, J.H., WEININGER, D., and MARTIN, Y.C. ALADDIN: an integrated tool for computer-assisted molecular design and pharmacophore recognition from geometric, steric, and substructure searching of three-dimensional molecular structures. *J. Comput.-Aided Mol. Des.* **1998**, *3*, 225–251.

30 SHERIDAN, R.P., NILAKANTAN, R., RUSINKO, A., BAUMAN, N., HARAKI, K.S., and VENKATARAGHAVAN, R. 3DSEARCH: a system for three-dimensional sub-structure searching. *J. Chem. Inf. Comput. Sci.* **1989**, *29*, 255–260.

31 PICKETT, S.D., LUTTMANN, C., GUERIN, V., LAOUI, A., and JAMES, E. DIVSEL and COMPLIB – strategies for the design and comparison of combinatorial libraries using pharmacophoric descriptors. *J. Chem. Inf. Comput. Sci.* **1998**, *38*, 144–150.

32 GOOD, A.C., KRYSTEK, S.R., and MASON, J.S. High-throughput and virtual screening: core lead discovery technologies move towards integration. *Drug Discovery Today* **2000**, *5*, S61–S69.

33 HORVATH, D. and JEANDENANS, C. Neighborhood behavior of in silico structural spaces with respect to in vitro activity spaces – a benchmark for neighborhood behavior assessment of different in silico similarity metrics. *J. Chem. Inf. Comput. Sci.* **2003**, *43*, 691–698.

34 MUEGGE, I. Pharmacophore features of potential drugs. *Chem. Eur. J.* **2002**, *8*, 1977–1981.

35 BROWN, R.D. and MARTIN, Y.C. Use of structure-activity data to compare structure-based clustering methods and descriptors for use in compound selection. *J. Chem. Inf. Comput. Sci.* **1996**, *36*, 572–584.

36 GOOD, A.C. and KUNTZ, I.D. Investigating the extension of pairwise distance pharmacophore measures to triplet-based descriptors. *J. Comput.-Aided Mol. Des.* **1995**, *9*, 373–379.

37 PICKETT, S.D., MASON, J.S., and MCLAY, I.M. Diversity profiling and design using 3D pharmacophores: pharmacophore-derived queries (PDQ). *J. Chem. Inf. Comput. Sci* **1996**, *36*, 1214–1223.

38 MASON, J.S. and CHENEY, D.L. Library design and virtual screening using multiple 4-point pharmacophore fingerprints. *Pac. Symp. Biocomput.* **1999**, *4*, 456–467.

39 MASON, J.S. and BENO, B.R. Library design using BCUT chemistry-space descriptors and multiple four-point pharmacophore fingerprints: simultaneous optimization and structure-based diversity. *J. Mol. Graph. Model.* **2000**, *18*, 438–451.

40 BRADLEY, E.K., BEROZA, P., PENZOTTI, J.E., GROOTENHUIS, P.D.J., and SPELLMEYER, D.C. Miller, a rapid computational method for lead evolution: description and application to alpha(1)-adrenergic antagonists. *J. Med. Chem.* **2000**, *43*, 2770–2774.

41 ABRAHAMIAN, E., FOX, P.C., NAERUM, L., CHRISTENSEN, I.T., THØGERSEN, H., and CLARK, R.D. Efficient generation, storage, and manipulation of fully flexible pharmacophore multiplets and their use in 3D similarity searching. *J. Chem. Inf. Comput. Sci.* **2003**, *43*, 458–468.

42 MAKARA, G.M. Measuring molecular similarity and diversity: total pharma-

cophore diversity. *J. Med. Chem.* **2001**, *44*, 3563–3571.

43 HORVATH, D. In *Combinatorial Library Design and Evaluation. Principles, Software Tools, and Applications in Drug Discovery*, GHOSE, A.K. and VISWANADHAN, V.N. (Eds). Marcel Dekker, New York, **2001**, 429–472.

44 SHERIDAN, R.P. and KEARSLEY, S.K. Why do we need so many chemical similarity search methods? *Drug Discovery Today* **2002**, *7*, 903–911.

45 STANTON, D.T., MORRIS, T.W., ROYCHOUDHURY, S., and PARKER, C.N. Application of nearest-neighbor and cluster analyses in pharmaceutical lead discovery. *J. Chem. Inf. Comput. Sci.* **1999**, *39*, 21–27.

46 PATTERSON, D.E., CRAMER, R.D., FERGUSON, A.M., CLARK, R.D., and WEINBERGER, L.E. Neighborhood behavior: a useful concept for validation of "molecular diversity" descriptors. *J. Med. Chem.* **1996**, *39*, 3049–3059.

47 KEARSLEY, S.K., SALLAMACK, S., FLUDER, E.M., ANDOSE, J.D., MOSLEY, R.T., and SHERIDAN, R.P. Chemical similarity using physiochemical property descriptors. *J. Chem. Inf. Comput. Sci.* **1996**, *36*, 118–127.

48 DEBSKA, B. and GUZOWSKA-SWIDER, B. Fuzzy definition of molecular fragments in chemical structures. *J. Chem. Inf. Comput. Sci.* **2000**, *40*, 325–329.

49 HORVATH, D. and MAO, B. Neighborhood behavior – fuzzy molecular descriptors and their influence on the relationship between structural similarity and property similarity. *QSAR Comb. Sci.* **2003**, *25*, 498–509.

50 HORVATH, D. and JEANDENANS, C. Neighborhood behavior of in silico structural spaces with respect to in vitro activity spaces – a novel understanding of the molecular similarity principle in the context of multiple receptor binding profiles. *J. Chem. Inf. Comput. Sci.* **2003**, *43*, 680–690.

51 CRAMER, R.D., PATTERSON, D.E., and BUNCE, J.D. Comparative molecular field analysis (CoMFA). 1. Effect of shape on binding of steroids to carrier proteins. *J. Am. Chem. Soc.* **1988**, *110*, 5959–5967.

52 KLEBE, G. In *3D-QSAR in Drug Design*, KUBINYI, H. (Ed.). ESCOM, Leiden, **1993**, 173–199.

53 PASTOR, M., CRUCIANI, G., and CLEMENTI, S. Smart region definition: a new way to improve the predictive ability and interpretability of three-dimensional quantitative structure-activity relationships. *J. Med. Chem.* **1997**, *40*, 1455–1464.

54 SULEA, T., OPREA, T.I., MURESAN, S., and CHAN, S.L. Different method for steric field evaluation in CoMFA improves model robustness. *J. Chem. Inf. Comput. Sci.* **1997**, *37*, 1162–1170.

55 WALTERS, D.E. and HINDS, R.M. Genetically evolved receptor models: a computational approach to construction of receptor models. *J. Med. Chem.* **1994**, *37*, 2527–2536.

56 CRUCIANI, G. and WATSON, K.A. Comparative molecular field analysis using GRID force-field and GOLPE variable selection methods in a study of inhibitors of glycogen phosphorylase b. *J. Med. Chem.* **1994**, *37*, 2589–2601.

57 DAVIS, L. *Handbook of Genetic Algorithms*. Van Nostrand Rheinhold, New York, **1991**.

58 FOGEL, D.B. *Evolutionary Computation – Toward a New Philosophy of Machine Intelligence*. IEEE Press, Piscataway, **1995**.

59 TODESCHINI, R. Reality and models. Concepts, strategies and tools for QSAR. Lecture at the EuroQSAR Meeting 2002. Bournemouth, UK, **2002**.

60 Catalyst is a product of Accelrys, Inc., http://www.accelrys.com/ **2003**.

61 MARTIN, Y.C., BURES, M.G., DANAHER, E.A., DELAZZER, J., LICO, I., and PAVLIK, P.A. A fast new approach to pharma-cophore mapping and its application to dopaminergic and benzodiazepine agonists. *J. Comput.-Aided Mol. Des.* **1993**, *7*, 83–102.

62 JONES, G., WILLETT, P., and GLEN, R.C. A genetic algorithm for flexible molecular overlay and pharmacophore elucidation. *J. Comput.-Aided Mol. Des.* **1995**, *9*, 532–549.

63 ZAMORA, I., AFZELIUS, L., and CRUCIANI, G. Predicting drug metabolism: a site of metabolism prediction tool applied to the

cytochrome P450 2C9. *J. Med. Chem.*
2003, *46*, 2313–2324.

64 OPREA, T.I. Chemical space navigation in
lead discovery. *Curr. Opin. Chem. Biol.*
2002, *6*, 384–389.

65 HODGKIN, E.E. and RICHARDS, W.G.
Molecular similarity based on
electrostatic potential and electric field.
Int. J. Quantum Biol. Symp. **1980**, *14*,
105–110.

66 Protein Data Bank,
http://www.rcsb.org/pdb/ **2003**.

Part II
Hit and Lead Discovery

6
Enhancing Hit Quality and Diversity Within Assay Throughput Constraints

Iain McFadyen, Gary Walker, and Juan Alvarez

6.1
Introduction

The identification of suitable lead molecules as a starting point for optimization is a critical determinant of success in the drug discovery process. High-Throughput Screening (HTS) is currently the single most important source of leads industrywide [1]. The dramatic rise in the importance of HTS over the last 10 to 15 years is in large part due to an explosive growth in the available technology (assay automation, miniaturization, throughput, and analytical sensitivity [1, 2]). However, with corporate screening collections typically comprising hundreds of thousands of compounds, in some cases over a million entities [3], the HTS process is also extremely resource intensive. The cost of a single HTS run has been estimated at $75 000 [4], but it should be noted that this excludes all assay-development and follow-up efforts. Indeed, it has recently been suggested that biological screening and preclinical pharmacological testing account for ~14% of the total research and development expenditure of the pharmaceutical industry [5]. Despite the escalating costs associated with acquiring and managing huge compound inventories and performing HTS assays, early results in terms of hit rates and yields of promising lead candidates have been somewhat disappointing [6]. Thus, although ever-increasing numbers of compounds have been screened in HTS assays over the last decade or so, there has been no corresponding increase in the number of successfully launched drugs [3, 6]. A study of drugs launched in 2000 showed that only 1 out of 29 compounds could be traced back to a screening hit, the remainder coming from modifications to existing drugs, clinical candidates or natural products [7]. A number of factors may contribute to this, some of which are discussed below. Given the investment made in and the downstream impact of HTS, it is absolutely imperative that maximal return is obtained from every screen. One key factor in the HTS process that can be easily controlled is the selection of active compounds from the primary assay. We suggest that uncertainties in the quality of the data, operational throughput limits of secondary assays, and general naivety have contributed to the suboptimal selection of lead candidates arising from HTS.

6.1.1
What Makes a Good Lead Molecule?

Activity in a given bioactivity assay is not sufficient to qualify a molecule as a promising lead; a suitable candidate must also represent an opportunity to advance through the drug discovery process. The optimization of a lead molecule into a clinical candidate generally results in increased molecular complexity, as measured by various physiochemical properties [8]. For example, molecular weight (MW), number of rotatable bonds, number of hydrogen bonds and cLogP are all shown to increase significantly, on average. This is not unexpected; a lead arising from HTS on a novel target is likely to require a significant increase in bioassay potency, which is typically the initial focus of optimization. The desired improvement in potency is generally achieved through the creation of additional interactions, often van der Waals interactions and the hydrophobic effect [9] (hence the increase in cLogP), and typically require an increase in the number of ligand atoms (hence the increase in molecular weight). These physiochemical properties are particularly significant due to their downstream impact on ADME properties such as solubility, permeability, and bioavailability. Indeed, molecular weight and cLogP form part of the "Lipinski Rule of 5", which is widely used throughout the industry [10] to identify compounds that are likely to suffer from poor absorption or permeation. Number of rotatable bonds and polar surface area were found to be key indicators in a study correlating physiochemical properties with oral bioavailability [11]. In a theoretical analysis using a simplified model of protein–ligand interactions, Hann et al. concluded that the probability of observing a useful interaction for a randomly chosen ligand decreases dramatically as the complexity of the ligand increases [12]. Taken together, these observations suggest that leads should ideally possess physiochemical property profiles that allow for the increase in complexity to be anticipated during the optimization process, that is, they should be "leadlike" (less complex than known drugs).

Finally, Kuntz et al. made the observation that the strongest binding ligands for a variety of targets show on average a $-1.5 \, \text{kcal mol}^{-1}\text{atom}^{-1}$ contribution to binding, with a nonthermodynamic nonlinearity observed for ligands with >15 heavy atoms [9]. Proximity to the line of correlation between the number of heavy atoms and free energy of binding can provide an additional means for evaluating the efficiency (on a per-atom basis) of the lead molecule and can contribute to the selection criteria.

6.1.2
Compound Collections – Suitability as Leads

Many compounds screened in early HTS assays were simply not amenable to optimization into clinical candidates because of the stringent requirements of solubility, stability, ADME, and toxicity parameters. Put simply, they were not good leads [3]. Although many pharmaceutical companies are moving to routine ultraHigh-Throughput Screening (uHTS) of a million or more compounds, there is a growing awareness that compound quality is as important as collection size [3, 4, 6]. Thus, automated filtering is typically applied to newly synthesized or purchased compounds before addition to the screening collection [6, 13]. Removing undesirable compounds from the screening process offers significant savings in resources, both immediately in the reduced cost of HTS runs, and, more importantly, in the long term by delivering only chemically interesting, leadlike compounds that are more readily converted into promising clinical candidates. Although the filters used vary widely, they typically exclude compounds with undesirable physiochemical properties and/or those containing substructure features that are known to be reactive/unstable [14]. As discussed in Section 6.1.1, a variety of easily calculated physiochemical properties have been shown to correlate well with solubility, permeability, and/or bioavailability and are commonly used as filters. These include MW, clogP, donors, acceptors (collectively forming the Lipinski Rule of 5) [10, 15], rotatable bonds, and polar surface area [11]. The definition of what constitutes reactive or unstable functionality naturally varies somewhat, but there is also a good deal of consensus (aldehydes, alkyl halides, vinyl ketones, and so forth) [14, 16–18].

As a result of the huge overhead involved in reformatting a screening library, an established screening collection may contain many legacy compounds that are known to be undesirable. Even in the case of a carefully managed library, a project team may not be interested in all of the actives identified by a screen. For example, a well-established target may be the subject of extensive academic and patent literatures detailing many classes of well-known and well-characterized inhibitors. Obviously, an HTS is not required to identify such compounds, and yet, if any of them exist within the screening collection, one would expect them to exhibit high potency in an HTS, and following up such actives would waste valuable resources. Thus, it is often desirable to perform a filtering step on the primary HTS results, although such filters must be carefully implemented so as not to eliminate interesting compounds. Filtering schemes are already in place in a number of companies, although the details differ significantly, and can provide a significant improvement in the quality of actives being pursued [13].

6.1.3
Compound Collections – Diversity

Corporate compound collections naturally contain large families of closely related structures resulting from combinatorial chemistry and traditional medicinal chemistry "analoging" efforts. On the other hand, there are inevitably vast areas of chemical space that remain unexplored in even the largest compound collection. This uneven distribution means that screening collections suffer from both redundancy and underrepresentation simultaneously [19, 20]. The former can be reduced through diversity analysis; see reviews by Willett, Mason and Brown for more on the various descriptors, measures, and clustering techniques that are commonly used [21–23]. At the same time, many companies are pursuing aggressive purchasing programs to supplement the diversity of their in-house collections with compounds bought from outside vendors. The libraries of interest are typically compared with the existing corporate collection using diversity analysis tools to assess whether they offer significant added coverage of chemical space. Again, see reviews by Willet and Menard for more details [24, 25].

Many corporate screening collections may inadvertently contain two or more samples of the same compound, for example, from different synthetic batches. Examining the results from a large number of such replicate samples can provide a useful way to assess the upper limit of variability in the primary assay. This represents an upper limit because other factors such as compound degradation can account for an unknown amount of the observed variability. Such replicate samples can also be filtered out if desired.

One potential unifying strategy is that of "clustered diversity", a balance between coverage of chemical space (diversity) and retention of multiple representatives within each chemical series (clusters). Increased diversity increases the probability that for a novel target (that is, one for which no compound within the collection would be predicted to have a greater chance of being active than any other), a hit, albeit likely to be weak, will be found. On the other hand, the existence of multiple representatives of a chemical series reduces the likelihood of the series being completely missed through the occurrence of false negatives. In a retrospective analysis of an HTS where 1200 known actives were included in the screen, Spencer [26] determined the false negative rate to be 9%. In this case, a false negative was defined as the failure to identify a known 2 µM or better inhibitor using a 50% inhibition threshold. Thus, the presence of a second "active" member of the same series would reduce the chances of the series being missed to 0.81%. The 9% false negative rate quoted purely reflects the information collected on a single HTS run and can be higher or lower, given the quality of the assay and instrumentation, the properties of the molecules in the library, the hit definition criteria, and other parameters. Clearly, screening multiple members of a series does not ensure that all of them will be active, but a fundamental relationship remains that structurally related molecules are more likely to possess similar properties (including bioactivity) than structurally unrelated ones [27, 28]. Thus, the hit rate within an "active" cluster should be enriched relative to the background hit rate, that is, there should be a higher-than-expected number of actives, given the size of the cluster. It follows that the false positive rate within highly enriched clusters should be lower than that in the background for the assay. Here, the minimum size of a cluster is that which enables statistical analysis of the hit rate relative to background, and is, in practice, around 5. The presence of clusters of structurally related compounds and the expected enrichment of activity within certain clusters is a critical component of the HTS analysis scheme that we describe below (Section 6.1.6).

6.1.4
Data Reliability

HTS groups are encountering an ever-increasing demand for throughput due to the growing size of corporate screening collections and/or the drive to higher numbers of screens per year. This in turn leads to a real need to trim costs. These conflicting demands are being met primarily through miniaturization of assay technology and protocols. Modern HTS assays are typically performed on 384-well microplates, using reaction volumes on the low μL scale. Technologies enabling even higher plate densities (1536 or greater) and smaller reaction volumes (nL) have been under development [1, 2]. There is a certain amount of unavoidable variability in the data associated with any assay protocol and its associated instrumentation. These operational errors (assay related, liquid handling, sample evaporation, compound deterioration, human-associated, and others), which generally occur at a negligible rate, are exacerbated by the scale of HTS assays [29]. Thus, misidentification of just 0.1% of the screening library as false positives represents a 20% false positive rate when the overall hit rate is 0.5%. Since the operational hit rate for HTS is often determined by the throughput of the next assay (confirmation or secondary), these false positives represent missed opportunities for identifying additional truly active molecules. Furthermore, primary HTS assays are typically run as single replicates at a single-compound concentration, which prevents the detection of excessive variability through examination of multiple replicate data points.

Therefore, HTS data is inherently "noisy" [29]. Sills et al. [30] highlighted the variability in HTS data in a recent study. They compared the results gained from screening a 30 000 compound library against the same tyrosine kinase using three different assay formats, namely, scintillation proximity assay (SPA), homogeneous time-resolved fluorescence resonance energy transfer (HTR-FRET), and fluorescence polarization (FP). They concluded that the observed lack of correlation between the results from the three assay formats would have resulted in different compounds being identified at the end of the screening process in each case. Indeed, only four or seven compounds (depending upon the activity cutoffs used) would have been identified as validated actives from all three of the assay systems. The differences were attributed to the inherent lack of precision in the primary screening process.

In addition to false positives created by random or systematic error in the assay, as described above, the set of most active compounds in a given HTS is likely to be enriched in false positives relative to sets containing weaker, but still active, compounds [31]. There are at least three classes of false positives that contribute to this enrichment, all compound-specific. The first is promiscuous inhibitors that act through a nonspecific mechanism as aggregates. In two recent studies, McGovern and colleagues [32, 33] showed that a range of nonspecific inhibitors exhibited enzyme inhibitory activity dependent upon time, enzyme concentration, and ionic strength, and attenuated by albumin, guanidinium, or urea. These compounds also formed particles of 30 to 400 nm diameter by both light scattering and electron microscopy techniques; the size of these particles were also regulated by ionic strength. Only a small percentage of the compounds in a given screening library may exhibit this behavior, but because they are likely to exhibit nonspecific activity in most HTS screens, they may nevertheless account for a significant fraction of the most potent actives.

Compounds that interfere with the detection mechanism of the HTS assay will, in many cases, be detected as highly potent actives [31]. One example would be compounds that intrinsically emit or absorb light at the wavelengths used in a fluorescence-based assay such as fluorescence resonance energy transfer (FRET). In an HTS screen using a fluorescence-based assay at Wyeth, 1.2% of the samples tested showed not just high fluorescence but the maximum possible initial (time 0) reading on the fluorimeter.

One-third of these appeared in the actives for the assay, which was defined using a 30% inhibition threshold. The overall hit rate for the assay was 2.0%; in other words, 20% of the actives were false positives due to extreme optical interference with the assay readout. This number rises depending on the level of initial fluorescence used as a cutoff; in the selection scheme used by the project team at the time, almost half of the actives were rejected as false positives on the basis of the initial fluorescence readings. The corporate screening collection has been reformatted since then, and many of these problem compounds have been removed.

Compounds with low purity are a common source of false positives. Degradation of compounds over time or as a function of exposure to light, oxygen, moisture, or repeated freeze/thawing cycles can result in reactive, unstable, or polymerized (high molecular weight) degradation products that are detected as potent actives. Again, this can account for a sizeable fraction of the most potent actives – purity/integrity checks performed on random samples from the Wyeth screening library reveal a <10% failure rate (that is <85% purity by NMR or inconsistent MS), and yet the failure rate for "Top X" actives (see Section 6.1.5 for definition) selected from historical HTS assays varies from 20 to 65%. Thus, there is an enrichment of impure samples among the most active compounds. Further, even compounds passing purity and integrity checks can contain relatively low levels of impurities (up to 15% in the case of Wyeth compounds), and sometimes these are solely responsible for the observed activity. Needless to say, this can only be discovered after time- and resource-intensive follow-up work.

Taken together, these classes of compounds can cause a significant enrichment of false positives among the most active compounds. This is illustrated by the sudden spike in the number of actives at the highest levels of inhibition (the most active ~0.2%) in a recent primary HTS assay run here at Wyeth; see Figure 6.1. In the study by Sills

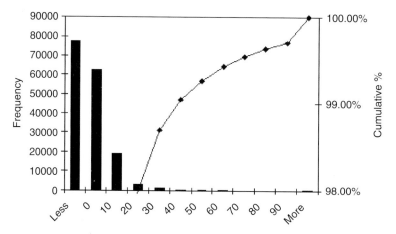

Fig. 6.1 Activity histogram (left *Y* axis) and cumulative frequency curve (right *Y* axis) versus binned percent inhibition (*X* axis) for a recent primary HTS run at Wyeth. Note the sudden spike in the cumulative frequency curve for compounds with >100% inhibition, corresponding to the ~0.2% most active compounds.

et al. mentioned above, false positive rates between 19 and 28% in the primary assay and between 17 and 43% from deconvolution (generating dose–response curves) were reported [30]. Historical data from Wyeth HTS assays suggests confirmation rates of 40 to 75%, giving false positive rates of 25 to 60%. Although the quality of the HTS assay and the expected rate of false positives and false negatives can be evaluated using statistical measures such as signal to noise ratio (S/N), coefficient of variance (CV), standard deviation (SD), and Z-factor [29], this does not help in revealing the misidentified compounds. False positives can only be identified through follow-up work, which can represent a significant drain on resources. On the other hand, potential leads among the false negatives are likely to be simply "lost", with potentially disastrous consequences for a drug discovery program. Needless to say, it is imperative that both of these outcomes are avoided if possible. This is best attempted through the use of an appropriate selection scheme.

6.1.5
Selection Methods

The selection or prioritization of compounds at the HTS stage of the discovery process obviously has far-reaching implications for later efforts. Owing to the immense costs associated with late-stage failures, especially in clinical trials, there is an increasing focus on improved decision making at the earliest stages of the process such that only suitable leads are carried forward [3]. One of the earliest decision points encountered by project teams is the choice of actives to pursue from the primary HTS. In this context, an active is defined as a compound with reliably detectable biological activity. The ideal, and simplest, approach would be to select all such actives and then apply appropriate filtering. However, confirmation assay throughput limits are relatively fixed (4000–10 000), whereas the hit rate from the primary HTS can vary widely depending on assay conditions. Furthermore, the hit rate is difficult to predict *a priori*, even if diligent assay validation and pilot studies have been performed. As mentioned in Section 6.1.4, there are heavy demands on HTS groups to deliver increasing number of assays each year. Thus, the selection of actives from primary HTS assays is heavily influenced by the twin practical considerations of confirmation assay throughput and time constraints.

Historically, a "Top X" paradigm has typically been used, whereby the most active compounds are selected solely on the basis of potency (percent inhibition) in the primary assay. The activity cutoff used to select the actives is unrelated to the limit of detection of the assay that separates true actives from inactives [29]. Instead, preselected values for potency (50% inhibition is commonly used), hit rate (e.g., the most potent 1% of the actives), or the number of actives (e.g., the most potent 5000 actives) are used. This approach has the advantages of meeting confirmation throughput limits while being simple and fast to apply. However, we suggest that it suffers from a number of significant drawbacks, all related to the inadequacy of potency alone as a sole criterion for selecting actives.

The natural tendency on first reflection is to select the most potent actives to get a "head start" on lead optimization. However, the practical reality is that the

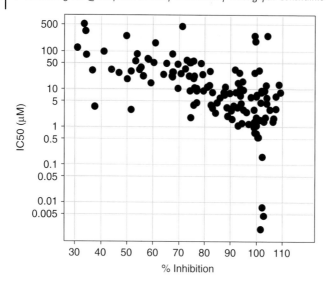

Fig. 6.2   Plot of percent inhibition in the primary HTS assay versus IC_{50} (μM) in a secondary *in vitro* assay for 118 compounds of interest identified from a recent Wyeth HTS. Note the relatively flat correlation such that compounds with low percent inhibition (e.g., 50%) have IC_{50}s, on average, only one order of magnitude less than compounds with maximal (~100%) inhibition.

potency of interesting HTS actives is bounded at the lower (weaker) end by the detection capabilities of secondary biological assays and at the upper end by probability considerations. Further, the correlation between percent inhibition in the primary HTS assay and IC_{50} as determined in follow-up *in vitro* assay is relatively flat. This is illustrated for 118 compounds of interest that were identified from a recent Wyeth HTS and taken into secondary assays (Figure 6.2) and more dramatically by Spencer for 1200 known actives included in a Pfizer screen (Figure 2 in [26]). Given the relatively narrow range of potency of interesting actives, approximately high nM to mid μM, and the lack of predictive power of primary activity, there seems to be little drawback and significant advantages to using other criteria to select weaker, but still active, compounds in place of more-potent ones.

So, if other factors are to be considered in hit selection, what should they be? We suggest that chemical structure is at least as important as potency in the primary HTS. In the following paragraphs, we describe ways in which the limitations of the "Top X" approach can be addressed, often involving the use of chemical structure. In the remainder of the chapter, we describe a method that combines these techniques and offers substantial improvements in several regards.

In many projects, the throughput limit for the confirmation assay is significantly lower than the number of actives from the primary HTS, and in these cases, a "Top X" approach based on a preselected threshold has the obvious disadvantage of failing to consider a potentially large number of true actives, leading to an artificially high false negative rate. This is highlighted by Sill et al., who attribute the high number of false negatives found in their study of three different assay systems to precisely this kind of "Top X" selection [30], although they consider this to be a standard approach. In contrast, we suggest that all active compounds are worthy of consideration, not just those appearing most potent. This can be achieved using clustering, based on the assumption that structurally similar molecules are more likely to exhibit similar biological molecules than unrelated ones [27, 28]. Where families of closely related structures contain many actives relative to the background hit rate, a representative sample is taken, helping to reduce the redundancy among the selected actives.

Any "Top X" selection includes a substantial fraction of compounds that could be safely eliminated from consideration. These include any undesirable compounds present in the screening collection, structures of no interest in the context of the specific project, any duplicate samples of individual compounds, and false positives. If some or all of these were to be removed by filtering, then weaker, but still active, compounds that would otherwise not be considered could be selected instead. Alternatively, if the number of actives is approximately equal to or lower than the throughput limit of the confirmation assay, then significant resources can be saved. We use filters based on chemical structure to remove undesirable and uninteresting compounds (physiochemical or substructure based, see Section 6.1.2), while duplicate samples can be easily filtered using name matching (Section 6.1.3). Clustering helps to pinpoint false positives arising from random or systematic error in the screen by identifying clusters that are not substantially enriched in biological activity; this is achieved using ring scaffold hashcodes as described in Section 6.2.4.1. Other types of false positives, for example, those that act nonspecifically as aggregates or by interference with the detection mechanism of the assay (Section 6.1.4), can be filtered using exclude lists.

Lastly, every research project presents a unique combination of parameters (target, assay methodology, hit rate, throughput limit, and so forth) and project team requirements for interesting actives (physiochemical properties, novelty, and so forth). This suggests that an optimal selection scheme should be easily tailored to meet these unique needs, which is certainly not true of a simple "Top X" approach.

Why is it so important to select the best actives from a HTS? Let us assume that a hypothetical screen of 500 000 samples yields a 3% hit rate, but that the throughput limit of the confirmation assay is 5000 compounds. A "Top X" selection will capture only the most active 1% of the samples. Even conservative estimates of the problems discussed above would lead us to expect that 20% of these actives would be lost upon confirmation, a further 20% would be lost during deconvolution (failure to get an appropriate dose-response), yet another 20% would be lost because of purity/integrity failures, and so on, for "promiscuous inhibitors", and so forth. Thus, the available pool of HTS hits can dwindle rapidly to less than 10% of the original number. Even after all the false positives are eliminated through resource-draining follow-up work, it is commonly the case that the majority of the remaining actives are rejected by a medicinal chemistry assessment as uninteresting, that is, non-leadlike. This is typically due to undesirably high levels of molecular complexity; for example, the molecules may be too large, have too many rotatable bonds, are too lipophilic, and so forth. That a "Top X" approach lacking any filtering is likely to yield molecules falling outside the "leadlike" criteria (depending on the potency cutoff used) is to be expected; larger, more lipophilic molecules are more likely to show the potency required for being selected, but are less "efficient" in terms of potency on a per-atom basis [9]. As described above, as the HTS follow-up process moves through each of its many steps, the attrition rates multiply together alarmingly and the number of surviving actives dwindles rapidly. Careful selection of actives resulting in a significant improvement in one or more of these survival rates, for example, an increase in confirmation rates because of a decrease in the number of false positives selected or a decrease in medicinal chemistry rejection rates because of improved physiochemical properties, can translate into a substantial increase in the yield and quality of actives remaining at the end of the process.

Another aspect of HTS analysis is the selection of actives from the confirmation assay. Historically, confirmation assays at Wyeth have been run as triplicates at a single concentration. The reproducibility of these assays is generally excellent; in one recent example, 93.7% of the samples had <5% standard deviation from four replicates. In the confirmation assay, one hopes to detect some or all of the false positives from the primary assay. This should certainly be true for compounds misidentified because of relatively low levels of random error in the assay, which is due to the low probability of encountering a false positive result for the same compound in both the primary and confirmation assays. False positives from systematic error in the assay are not likely to be detected unless the variable responsible for the error is changed, for example, the liquid-handling robot in the case of pipetting errors. Purity/integrity problems associated with false positive activity may be picked up if different samples are used for primary and confirmation assays, for example, by maintaining two separate sets of mother plates. This is the case at Wyeth, and although the two sets were derived from the same ultimate source, they have separate histories. Of course, any such changes in assay methodology or samples can introduce new false positives, but these should be limited by the same probability argument invoked above. Although the confirmation rate (i.e., the percentage of molecules selected from the primary assay that were also hit in the confirmation assay) for a carefully optimized and well-validated assay should be very high, the reality is often quite different despite the high quality of the confirmation assay. Historical confirmation rates for assays using single point and triplicates for the primary and confirmation assay, respectively (both single concentration), are between 40 and 75%. One factor that contributes to reduced confirmation rates relative to expectation is certainly "edge effects" of the kind described by Zhang [29]. Briefly, if the same threshold is used to define confirmed actives as was used to select actives from the primary assay, then compounds just above the threshold in the first instance may fail to confirm simply because of the variability associated with the assay. This suggests that the threshold for a confirmation assay should be set at a lower level than for the primary, and, furthermore, that the "window" should be determined using rigorous statistical analysis. Even if confirmations are run in a multiple-dose assay, there still remains the problem of determining what level of activity (and at what dose) qualifies a compound as a confirmed hit, as the number of concentrations and/or the experimental data are unlikely to be sufficient to yield IC_{50}s at this stage.

There is a growing interest in HTS data analysis and hit selection, as indicated by the plentiful commercial offerings from companies such as BioReason (ClassPharmer

Suite, [34]), Leadscope Inc. (Leadscope), GoldenHelix Inc. (ChemTree) and Tripos Inc. (SARNavigator). However, there are relatively few papers published in the literature dealing specifically with hit selection schemes, perhaps not surprisingly for a topic that is of immediate interest to pharmaceutical and biotechnology companies rather than academic groups [4, 35, 36].

6.1.6
Enhancing Quality and Diversity of Actives

The approach presented here addresses many of the limitations outlined above; it gives due consideration to all significant actives, is customizable to the individual needs of the project team, and offers improvements in quality of actives, diversity of actives, and

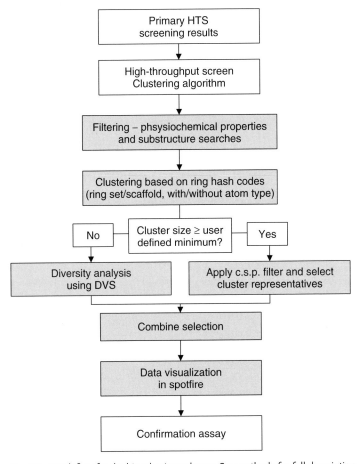

Fig. 6.3 Work flow for the hit selection scheme. See methods for full description.

confirmation rates. It involves filtering, clustering, diversity analysis, and visualization techniques. The general work flow is depicted in Figure 6.3, and details are provided in Section 6.2.

The threshold used to define an "active" is determined by rigorous statistical analysis and represents the limit of reliable detection of biological activity. Filtering based on chemical substructures and physiochemical properties is used to remove undesirable actives and thereby improve the quality of actives being pursued. Substructure searches commonly used include reactive/unstable functionality and scaffolds that represent well-known inhibitors – see Section 6.1.2. Physiochemical property filters used include MW, clogP, donors, acceptors [10], rotors, and topological polar surface area (TPSA) [11]. Filtering of reactive/unstable functionality is modeled after Rishton [17], while well-known inhibitor scaffolds are chosen on a project-by-project basis. As mentioned above, this type of filtering alone offers significant enhancement in the quality of molecules pursued.

Clustering relies on the fundamental assumption that compounds with similar chemical structures tend to exhibit similar biological activity [27, 28]. Families of highly related structures are examined for biological activity as clusters. The redundancy found in screening collections is used to add reliability to the HTS data, as clusters highly enriched in activity are more likely to contain true actives than a cluster of similar size containing fewer actives [26]. This can lead to improved confirmation rates. At the same time, redundancy in the confirmation set is reduced by selecting only representative subsets from each large cluster. Specifically, we identify structural families through clustering techniques based on ring hashcodes [37], and the level of biological activity present in each cluster relative to the expected background level is assessed using a cumulative significant probability measure based on the binomial distribution [38].

A significant percentage of any compound library will inevitably fall into small clusters unsuitable to rigorous statistical evaluation. These must be considered separately – in our case, using diversity analysis with BCUT descriptors [39] to supplement the list derived from clustering. Throughout this process, we use visualization to assess data quality, identify potential problems such as edge effects, and check trends and patterns.

This approach can reduce a very large number of actives to meet the practical limits of a confirmation assay while considering all actives and provides a rational method for hit selection. It offers improvements in quality of actives, diversity of actives, and confirmation rates. Lastly, every detail of this process is flexible by design. We present details from two diverse projects that highlight the application of the described method and the results achieved thereby.

6.2
Methods

As noted above, the selection scheme can be tailored to the individual project, but representative details are provided here.

6.2.1
Screening Library

The corporate screening library used for HTS screens comprises ~370 k individual samples, prescreened using in-house tools to exclude egregious compounds by physiochemical properties (clogP > 10, 100 > MW > 750) and a variety of substructure searches representing reactive, unstable, and greasy functionalities. The samples are stored as single solutions at 10 mg mL^{-1} in 100% dimethylsulfoxide (DMSO) on 384-well plates containing several blank wells and are typically screened in single-point fashion at a final concentration of 10 μg mL^{-1} (0 to 10% DMSO) on 384-well plates.

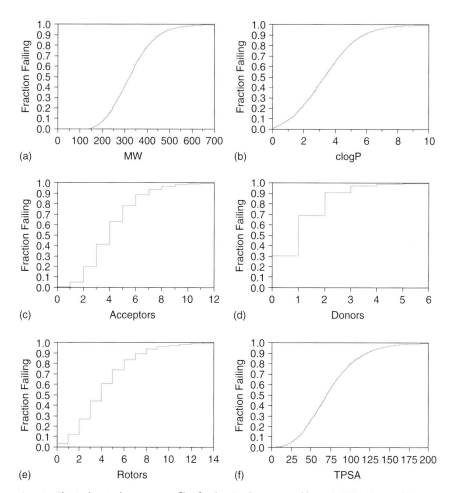

Fig. 6.4 Physiochemical property profiles for the Wyeth screening library (~370 000 samples). (a) MW; (b) clogP; (c) number of hydrogen bond acceptors; (d) number of hydrogen bond donors; (e) number of rotatable bonds; (f) TPSA. Properties were calculated in MOE [41].

Physiochemical property profiles of the screening library for MW, clogP, number of hydrogen bond acceptors, number of hydrogen bond donors, number of rotatable bonds, and TPSA are shown in Figure 6.4. This screening library is fairly typical of those found among pharmaceutical companies [26, 40].

6.2.2
Determination of Activity Threshold

There are several methods available to rigorously determine a threshold above which a compound demonstrates reliably detectable biological activity and hence can be considered an active, primarily based on analysis of the distribution of potency among either the entire population or the controls alone [29]. The distribution of activity of well-characterized negative controls known to have no activity in the assay can be used to set the activity threshold on the basis of the mean plus some multiple of the standard deviation. However, the most common method [29] is to determine the mean and standard deviation of the entire primary data set and set the activity threshold equal to the mean plus some multiple of the standard deviation (often three). This is based on the assumption that the great majority of compounds tested are not biologically active in the assay and hence compound activity is distributed in a Gaussian normal manner. This method will give a 0.13% hit rate for a perfectly normally distributed population by definition. However, in screens with a large number of actives, this assumption fails because the observed distribution becomes the sum of two overlapping distributions and hence deviates significantly from normal. In this case, the common measure of standard deviation becomes less useful and more robust measures of standard deviation such as Median Absolute Deviation (MAD) or Inter-Quartile Range (IQR) are applicable [42, 43]. Here, we have used statistical evaluation of the controls and the entire population in the two different projects.

6.2.3
Filtering

Filtering is applied to the primary assay data set in the High-Throughput Screen Clustering Algorithm (HTSCA) program (see Section 6.2.4) using exclude lists generated in RS3 Discovery for Excel [44, 45]. Exclude lists are generated dynamically to meet the needs of the individual project team and are typically based on substructure searches and physiochemical properties. The former are designed to remove well-known inhibitor classes, while the latter are based on MW, clogP, donors, acceptors, [10] rotors, and TPSA [11]. Typical starting values for the property filters are based on the literature studies, specifically $150 > MW > 500$, $clogP > 5$, donors > 5, acceptors > 10, rotatable bonds > 10 and TPSA $> 140 \text{ Å}^2$, but are often adjusted to be more or less stringent on the basis of examination of the dataset and/or feedback from the project team and, specifically, the chemists. Filtering for reactive compounds (beyond the filtering already applied to the entire screening library, see above) occurs after confirmation assays using a function in Molecular Operating Environment (MOE) (CCG Inc., Montreal, Canada [41]) based on Rishton [17].

6.2.4
High-Throughput Screen Clustering Algorithm (HTSCA)

HTSCA is a clustering program based on ring hashcodes written using Microsoft Access 97 [46], which presents the user with a graphical interface, including prompts for a variety of parameters (see Figure 6.5). It connects to the corporate data warehouse via Oracle to retrieve primary HTS data (compound ID, selected activity field, chosen ring hashcode), which is then manipulated locally. Clusters are created by simple matching of ring hashcodes (see Section 6.2.4.1 for a description of hashcodes). The final result is a table populated with clusters and their associated calculated properties, including ring hashcode, number of compounds in cluster, number of actives in the cluster, percentage of active compounds in the cluster, cumulative significant probability (c.s.p) (see Section 6.2.4.2), and highest activity value within the cluster. Options exist to sort or filter the list by any of these properties, to apply include or exclude lists, to view the structures and/or data for a given cluster or for all clusters, and to copy or export the data.

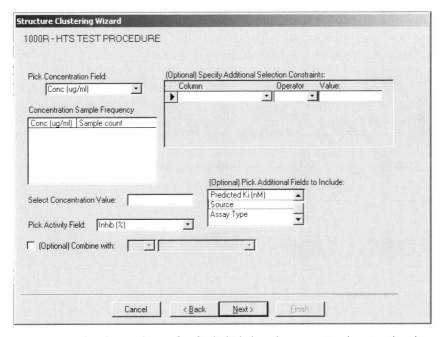

Fig. 6.5 Screenshot showing the interface for the high-throughput screening clustering algorithm. This is the second page of the wizard, where a number of options can be set for analysis of a specified HTS dataset.

User-definable parameters include project ID, limit and exclude lists, duplicate data grouping, activity threshold, hashcode type, and minimum cluster size. Limit and exclude lists apply Boolean AND or NOT operations respectively to the dataset. In other words, the compounds considered by the clustering algorithm are those present in all specified Include lists and simultaneously not present in any of the specified Exclude lists. In the present method, Exclude lists are used for filtering purposes (see Section 6.2.3). The data grouping options for the treatment of duplicate samples include best, worst, and mean result for a given compound ID. In the present method, worst result per compound ID was used. The determination of activity threshold is discussed in section 6.2.2, while the definition of ring hashcodes is described in Section 6.2.4.1. The minimum cluster size is typically set between five and eight, and compounds in clusters falling below this threshold were treated by diversity analysis (see Section 6.2.5).

6.2.4.1 Ring Hashcodes

The ring hashcodes used to create the clusters have been described previously [37]. Briefly, each hashcode consists of a seven-letter string and is derived from atom pair

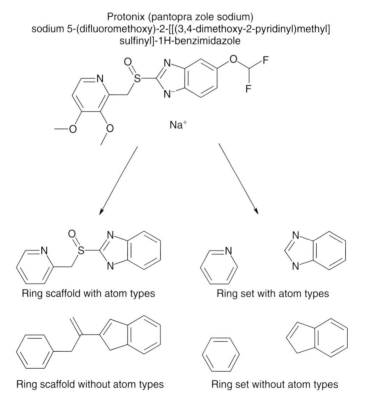

Protonix (pantopra zole sodium)
sodium 5-(difluoromethoxy)-2-[[(3,4-dimethoxy-2-pyridinyl)methyl] sulfinyl]-1H-benzimidazole

Ring scaffold with atom types   Ring set with atom types

Ring scaffold without atom types   Ring set without atom types

Fig. 6.6   Illustration of the four different types of ring hashcodes for a sample molecule.

and topological torsion descriptors. Each compound has four ring hashcodes based on the combination of two factors. The first is the choice of ring set (ring features ignoring connecting atoms) or ring scaffold (ring features including all connecting atoms) and the second is the choice of atom type perception (atom types treated distinctly) or neglect (all atom types treated equally). The four types of ring hashcode are illustrated for a sample molecule in Figure 6.6. In the present method, ring scaffolds with atom types were used. Acyclic molecules have the value "AAAAAAA" for all four ring hashcodes. Clusters are formed by simple matching of eight-letter codes.

Ring hashcodes are computed at entry for all new compounds and are stored along with other physiochemical descriptors in the corporate data collection. As a descriptor for clustering purposes, ring hashcodes have several known disadvantages. Hashcodes based on ring scaffolds and/or inclusion/exclusion of atom types suffer from over-definition in the case of reasonably complex molecules. For example, the molecule shown in Figure 6.6 would fall into a different cluster than close analogs with either a carbonyl in place of the sulfoxide or a phenyl in place of the pyridine, on the basis of the atom type. In contrast, hashcodes based on ring sets alone are often underdefined, placing all compounds with, for example, two phenyl rings in the same cluster regardless of their substitution or connectivity. All of the available ring hashcode definitions are under-defined with respect to simple molecules, placing all compounds with a single phenyl ring into one cluster and (by definition) all acyclic molecules into another. However, they offer several advantages: they are easy to calculate on compound entry; they are computationally efficient during clustering, allowing for arbitrarily large clusters (although see Section 6.2.4.2 for limitations on c.s.p. calculation); and for many molecules they present a classification that closely matches the perception of a medicinal chemist, which is typically based on ring scaffolds. We are currently working on an improved descriptor for clustering of HTS data.

6.2.4.2 Cumulative Significant Probability (c.s.p)

A compound is flagged as either active or inactive on the basis of the activity threshold determined as in Section 6.2.2. The probability that the observed number of actives x occurs in a cluster of size n given a background hit rate p (expressed as a fraction) is determined by the binomial probability mass function [38], shown as follows:

$$b(x, n, p) = \binom{n}{x} p^x (1 - p)^{n-x} \tag{1}$$

By extension, the probability that a given number of actives *or fewer* should occur in a given distribution is described by the cumulative binomial distribution function, calculated as

$$B(x, n, p) = \sum_{y=0}^{n} b(y, n, p) \tag{2}$$

This value ranges from 0 to 1. The probability that a given number of actives *or greater*, that is, the cumulative significant probability, is then simply the complement:

$$c.s.p = 1 - B(x, n, p) \tag{3}$$

These statistics are trivially calculated using the BINOMDIST function in Microsoft Access (with a built-in limitation to $n \leq 1000$). As a rough guide to expected c.s.p. values, when $x = n \times p$, that is, the number of actives is exactly that expected from the background hit rate, the c.s.p. ≈ 0.54. Similarly, for clusters significantly enriched in biological activity, that is, $x \gg n \times p$, the c.s.p. value quickly drops off and effectively becomes zero at a number of actives roughly double the expected background. The choice of c.s.p. threshold is user definable, but typically 0.05 or 0.1 is used, representing a 5 or 10% chance respectively that the observed number of actives or greater could have occurred in the cluster because of random chance and thus does not represent privileged biological activity.

6.2.5
Diversity Analysis

Diversity analysis provides a rational method for selecting compounds while simultaneously meeting confirmation throughput limits. The filtered list of compounds falling into clusters smaller than the minimum cluster size specified in HTSCA (see Section 6.2.4) is exported into RS3 [44] for Excel [45]. The structures are exported as an SD file that is read into DiverseSolutions (DVS) [47]. A chemistry space was defined in a study of 1.59 million compounds from internal and external compound collections using the auto-choose facility in DVS and effectively covers the chemical space represented in the corporate screening collection (G. Walker, unpublished results). The chemspace definition specifies the following 6 BCUT descriptors [39], all created using nonlinear scaling:

bcut_gastchrg_burden_0.20_R_H.bmf – the highest eigenvector of a bond-order matrix formed after removing hydrogens, with Gasteiger-Huckel charges on the diagonal and 0.2 times the Burden values as the off-diagonal elements;

bcut_gastchrg_invtopd_0.80_R_L.bmf – the lowest eigenvector of a matrix formed after removing hydrogens, with Gasteiger-Huckel charges on the diagonal and 0.8 times the inverse topological distance values as the off-diagonal elements;

bcut_haccept_invtopd2_1.60_R_H.bmf – the highest eigenvector of a matrix formed after removing hydrogens, with hydrogen bond accepting potentials on the diagonal and 0.2 times the square of the inverse topological distances as the off-diagonal elements;

bcut_hdonor_invtopd2_1.00_R_H.bmf – the highest eigenvector of a matrix formed after removing hydrogens, with hydrogen bond donor potentials on the diagonal and the square of the inverse topological distances as the off-diagonal elements;

bcut_tabpolar_distDtopd_0.06_R_H.bmf – the highest eigenvector of a matrix formed after removing hydrogens, with tabulated polarizabilities on the diagonal and 0.06 times the distances divided by the topological distances as the off-diagonal elements;

bcut_tabpolar_distDtopd_1.15_R_L.bmf – the lowest eigenvector of a matrix formed after removing hydrogens, with tabulated polarizabilities on the diagonal and 1.15 times the distances divided by the topological distances as the off-diagonal elements.

These descriptors were calculated for all compounds in the set of interest. Finally, cell-based subset selection was performed using uniform sampling of one compound per cell, with the choice of compound within the cell weighted by activity in the primary assay. The number of bins per axis was varied to achieve the closest possible match to the desired selection size.

6.2.6
Data Visualization

Compound IDs and biological data from HTSCA, structures from RS3, and calculated properties from RS3 and MOE are compiled in a MOE database and exported in tab delimited format. Spotfire DecisionSite with Lead Discovery and Analytics [48]

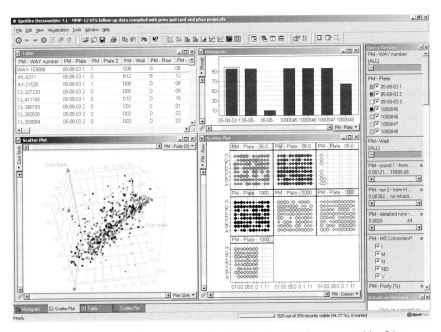

Fig. 6.7 Visualization of HTS data using Spotfire decision site [48]. Shown are a table of data (top left), a bar chart (top right, number of compounds screened from each plate), 3-D scatterplot (bottom left, percent inhibition in the primary assay versus percent inhibition in the confirmation assay versus percent purity), and a plate map (bottom right, well versus Column arranged by plate number). In the 3-D scatterplot and the plate map, the points are colored by plate number and sized by percent inhibition in the primary assay.

was used to visualize the large amounts of data involved in each HTS [35] (see Figure 6.7). Specifically, the following trends and relationships were examined: edge/row effects, distribution of primary activity, activity versus cluster code, distribution of compounds in clusters, and activity of duplicate samples. Various statistics were also calculated, including mean and standard deviation. During selection from confirmed actives, filtering was applied in Spotfire using dynamic real-time selection based on physiochemical property data calculated in MOE (CCG Inc., Montreal, Canada). Finally, structures and data for compounds/series of interest were visualized using built-in Spotfire tools.

6.3
Results

Results for three projects with diverse targets, namely, a peptide hydrolase, a protein–protein interaction, and a protein kinase are presented here. The full screening library was screened against each target at concentrations from 10 to 50 μg mL^{-1} in three different assay systems. The resulting hit rates ranged from ~1.5 to ~15% at the chosen activity thresholds, yielding between ~5 and ~55 k actives. The confirmation assay throughput limits were between 4000 and 8000 compounds. Each of the project teams had different requirements for actives in terms of acceptable physiochemical properties and chemical features. Despite these widely varying backgrounds, selection schemes following the outline described above were tailored to meet the needs of the project teams.

6.3.1
Peptide Hydrolase

Analysis of the negative control no-sample wells gave a standard deviation of ~10%, and so the activity threshold was set at 30%. On the basis of this cutoff, the primary HTS assay yielded an extremely high hit rate of 14.6%, or ~54 k actives, an order of magnitude greater than the 4 k throughput limit of the confirmation assay. Selection of the most active 4000 compounds would have disregarded ~50 k compounds with less than 89.0% inhibition, and only 1899 unique ring hashcodes would be represented compared to the 20 856 in the entire set of hits. Furthermore, at this level of inhibition, a small number of nonnovel structural families uninteresting to the project team were disproportionately represented – just two relatively specific substructure searches yielding a total of 4584 compounds from the entire corporate database (<1%) flagged 544 or 13.6% of the 4000 most active samples. A strategy was devised to use filtering, clustering, and diversity analysis to select 4000 compounds from this massive number of hits to meet the confirmation throughput limit while simultaneously improving the diversity and quality of the selected set relative to the Top 4000.

Filters were applied to the entire primary data set (372 k samples) to exclude compounds with MW > 500 or clogP > 5, duplicate samples, and the results from two substructure queries based on previously known inhibitor scaffolds. This eliminated 19.7% of the samples and 25.9% of the actives. This disproportionate filtering of actives relative to the entire dataset is due to the enrichment of compounds with poor physiochemical properties among the most active samples, as discussed in Section 6.1.5. All further operations were performed on this filtered dataset of ∼277 k samples (75 k unique ring hashcodes) including 40 k actives (16 221 unique ring hashcodes). After filtering, a "Top X" selection of 4000 compounds would have reached down to 75.3% inhibition and included 2058 unique ring hashcodes but missed another 36 k actives representing a further 14 163 hashcodes.

Clustering using ring scaffold hashcodes with respect to atom type yielded 7420 clusters with 5 or more compounds, 2796 of which contained at least 2 actives. Although the "active clusters" account for only 3.7% of the total ring hashcodes present in the filtered dataset, 48.8% (∼135 k) of the samples and 61.9% (∼25 k) of the actives are retained. Applying a c.s.p. threshold of 0.05 (see Section 6.2.4.2 for c.s.p. definition) reduces this still further to 972 clusters containing ∼31 k samples and ∼12 k actives (37.9%). Note that these enriched clusters represent only 1.3% of the populated clusters, yet 29.3% of the actives are captured. The mean size of the enriched clusters was 32 and the mean number of actives was 12, although the variation was significant (cluster size maximum 1127, minimum 5, standard deviation 93.6; actives maximum 343, minimum 2, standard deviation 22.5). Thus, the detection of clusters and the application of a c.s.p. threshold results in a 2.6-fold enrichment of actives (37.9%) relative to the background hit rate (14.6%). The 3 most active compounds were selected from each of these enriched clusters, where available, giving 1801 compounds. Taking representatives from clusters helps to dramatically lower the number of actives while retaining the diversity; the enriched clusters contained ∼12 k samples, from which only 1801 compounds were selected (15.3%), covering the same number of ring hashcodes.

To complement this list, the ∼15 k actives falling into clusters of less than 5 compounds were treated separately using diversity analysis, as described in Section 6.2.5. A cell-based selection biased by primary activity was performed from the six-dimensional chemical space defined in DVS. Dividing the chemical space into ∼47 k cells using 6 bins per axis came closest to the desired selection size, giving 3760 compounds; that is, <10% of all possible cells were populated. Of these, the top scoring 2202 were chosen, representing 1990 different ring scaffold hashcodes. The combined selection totaled 4003 compounds representing 2681 ring scaffolds and was submitted for confirmation assays. Note that the full set of 40 k filtered actives contained 16 221 ring hashcodes, so the selected set covers 16.5% of the represented scaffolds. Because of the presence of duplicate samples in the corporate screening collection, 4072 samples were pulled and assayed.

The final selection of compounds was compared with the set that would have been selected using the "Top 4000" method on the entire dataset or on the filtered dataset, in terms of diversity and physical properties. Our selection gave a 30.3 or 41.2% increase in the number of uniquely represented ring hashcodes (2681) as compared with the "Top X" selection (most active 4000 compounds) from either the filtered (2058) or unfiltered

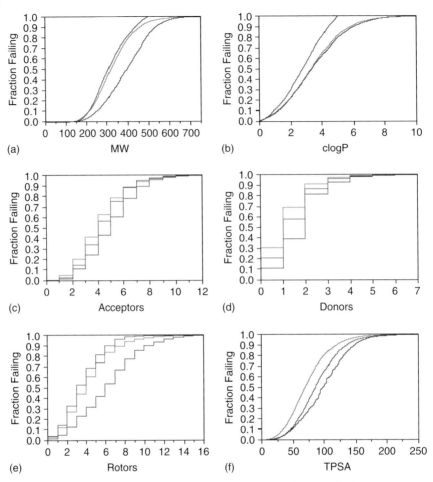

Fig. 6.8 Physiochemical property profiles for the Wyeth screening library (red), the Top 4000 actives (blue) and the selected set (green). (a) MW; (b) clogP; (c) number of hydrogen bond acceptors; (d) number of hydrogen bond donors; (e) number of rotatable bonds; (f) TPSA. Properties were calculated in MOE [41].

datasets (1899) respectively. The physiochemical properties (MW, clogP, number of hydrogen bond acceptors, number of hydrogen bond donors, number of rotatable bonds, and TPSA) of the selected set are improved relative to those of the "Top X" selection. For example, the median MW is 307 versus 398 and median number of rotatable bonds is 3 versus 6. Similarly, the properties of the selected set are similar to or superior to those of the full screening library. Statistical details of the various physiochemical properties for the full screening library, Top 4000, and the selected set are given in Table 6.1, while property profiles are shown in Figure 6.8. The profile of the Top 4000 set is clearly shifted to the right (less druglike) relative to the full library and

Tab. 6.1 Physiochemical properties for the full screening library, Top 4000 most active compounds, and the selected set of 3965 compounds. All properties calculated in MOE [41]

Set		MW	clogP	Acceptors	Donors	Rotors	TPSA
Full screening library	Median	320.4	3.2	4	1	4	69.6
	Mean	329.0	3.2	4.1	1.1	4.2	74.0
	S.D.	95.5	2.0	2.0	1.1	2.6	34.6
	Min.	104	−10	0	0	0	0
	Max.	750	10	26	15	30	416.6
Top 4000	Median	397.5	3.2	5	2	6	98.4
	Mean	394.9	3.2	5.0	1.8	6.2	99.2
	S.D.	108.5	2.1	2.0	1.1	3.2	39.7
	Min.	117.1	−4.6	0	0	0	0
	Max.	746.7	10.7	15	9	19	292.5
Selected	Median	307.4	2.6	4	1	3	85.9
	Mean	315.0	2.5	4.4	1.4	3.6	89.1
	S.D.	81.9	1.5	1.8	1.1	2.1	35.8
	Min.	119.1	−10.8	0	0	0	0
	Max.	499.4	5.0	16	11	16	282.6

MW = molecular weight; TPSA = topological polar surface area; S.D. = standard deviation;
Min. = minimum; Max. = maximum.

the selected set for MW, rotors, acceptors, and donors and shifted to the right relative to the selected set for clogP and TPSA. This supports the suggestion made in Section 6.1.4 that the most active compounds are enriched in compounds with less druglike, and certainly less leadlike, properties.

The confirmation rate for the selected set was dramatically higher than the typical 60 to 80% observed historically, internally, and in the literature [30]. The results showed a 90.5% confirmation rate (9.5% false positive rate) at >30% inhibition, giving 3683 confirmed actives. The confirmation rate rose to 97.4% when considering only compounds selected from enriched clusters and fell to 84.9% when considering compounds selected by diversity. This supports the hypothesis that actives selected from highly enriched clusters are significantly more likely to confirm, that is, be true positives, than other compounds because of the presence of "surrogate replicates".

The confirmation rates for the selected set at different primary and confirmation percent inhibition thresholds appear in Table 6.2. Also shown are confirmation rates for those subsets of the selected that appear in the Top X selection either before or after filtering, which is the union of the selected set and the respective Top 4000 selections. The confirmation rate holds remarkably steady for all three sets at the four different thresholds for the primary inhibition examined if the threshold used to define confirmed actives is kept constant at 30%. If the confirmation threshold is instead increased along with the primary threshold, the confirmation rate drops for all three sets, which might be considered counterintuitive. However, this can be explained by the twin influences of "edge effects" (Section 6.1.5) and the enrichment of false positives at higher activities. This behavior has been reported previously by Fay and Ullmann [31] for both real

Tab. 6.2 Confirmation rates for three sets of compounds at different thresholds for primary and confirmation percent inhibition

Dataset	Primary (%)	Confirmation (%)	Compounds	"Hits"	Confirmation rate (%)
Selected set	30	30	4072	3683	90.5
	50	30	3140	2.690	85.7
	70	30	1334	1232	92.4
	90	30	549	499	90.9
	50	50	3140	2446	77.9
	70	70	1334	924	69.3
	90	90	549	409	74.5
Union of the selected set with the Top 4000, unfiltered	30	30	605	551	91.1
	50	30	602	549	91.2
	70	30	592	539	91.0
	90	30	549	499	90.9
	50	50	602	517	85.9
	70	70	592	485	81.9
	90	90	549	409	74.5
Union of the selected set with the Top 4000, after filtering	30	30	1019	930	91.3
	50	30	1011	923	91.3
	70	30	986	898	91.1
	90	30	497	447	89.9
	50	50	1011	866	85.7
	70	70	986	729	73.4
	90	90	497	360	72.4

data and a theoretical model and was attributed to the increasing dominance of false positives with increasing thresholds. By comparing the confirmation rates at constant and varied confirmation thresholds at the same primary thresholds, we can estimate the maximum extent of the "edge effect". The largest difference occurs at a threshold of 70% inhibition in the primary and either 30 or 70% inhibition in the confirmation assay, where 1232 versus 924 of 1334 confirm. Thus, the edge effect accounts for at most 308 false negatives in the confirmation assay, or 23.1% of the total.

The confirmed actives (3683 of 4072) represented 2337 unique ring hashcodes or 87.2% of those present in the selected set. Among the set selected from clusters, at least 1 member of 967 of the 972 clusters confirmed (99.5%). Only 522 (14.2%) and 961 (26.1%) of these confirmed actives would have been included in a Top X selection of 4000 from the unfiltered or filtered datasets respectively. Secondary assays were limited to ~500 compounds, sevenfold less than the number of confirmed actives. A second round of filtering using additional physiochemical properties (remove MW < 200, rotors > 8, donors > 5, acceptors > 10, TPSA > 140 Å) removed a further 400 compounds. As the set was already diverse and had good physiochemical properties, the remaining compounds were subject to visual inspection to select 528 compounds. Of these, only 152 (28.8%) or 256 (48.5%) would have been identified from a 'Top X" selection of the most potent 4000 actives from either the unfiltered or filtered datasets respectively. Thus, a large number of actives that would otherwise have been missed were successfully "rescued". After secondary assays, eight series were identified as interesting, seven of which came from clustering and one from diversity. Of these series, two had examples with single-digit nM K_i for the target, another five with sub-µM K_i, and the last with single-digit µM K_i. A number of singletons were also identified from the diversity selection with affinities ranging from double-digit nM to single-digit µM.

6.3.2
Protein Kinase

An activity threshold of 25% was chosen after analysis of the primary screening results using both variance of controls and mean + 2 standard deviation (MAD method). The primary HTS assay yielded a high hit rate of 6.1%, or ~23 k actives, fourfold greater than the 8 k throughput limit of the confirmation assay. Selecting the 8000 most active compounds would have disregarded ~15 k compounds with <34.2% inhibition and included representatives from only 4413 unique ring hashcodes compared to 12 160 in the entire set of actives. Furthermore, at the >34.2% level of inhibition, one nonnovel structural family uninteresting to the project team was disproportionately represented – a single relatively specific substructure search yielding 3766 compounds from the entire corporate database (<1%) flagged 1245 or 15.6% of the 8000 most active samples. A strategy was devised to capture all of the most active compounds (exhibiting >40% inhibition) plus a diverse sample of the less active compounds, with appropriate filters to remove undesirables and uninteresting compounds in both cases.

After removing duplicates, 4905 actives remained with percent inhibition >40%. Of these, 992 compounds with $175 > MW > 600$, $clogP > 6$, donors > 5, acceptors > 10, rotatable bonds > 10 or TPSA > 140 Å^2 were flagged and subject to visual assessment in Spotfire by a medicinal chemist. Approximately half of these were retained, leaving 4458 actives from this set for confirmation assays.

In parallel, the entire primary data set was filtered using the same criteria as above plus a substructure search to identify compounds containing a well-known scaffold. This eliminated 22% of the samples and 31% of the actives. This disproportionate filtering of actives relative to the entire dataset is due to the enrichment of compounds with poor physiochemical properties among the most active samples, as discussed in Section 6.1. All further operations were performed on this filtered dataset of ~299 k samples (84 k unique ring hashcodes) including 16 k actives (9254 unique ring hashcodes). After filtering, a "Top X" selection of 8000 compounds would have reached down to 29.6% inhibition and included 4953 unique ring hashcodes but missed a further 8 k actives with percent inhibition between 29.6 and 25% representing another 4301 hashcodes.

Clustering of this filtered set using ring scaffold hashcodes with respect to atom type yielded 8324 clusters containing five or more compounds. Although this represents only 9.9% of the populated clusters, 64.8% (~195 k) of the samples and 60.6% (~10 k) of the actives are retained. Applying a c.s.p. threshold of 0.05 (see Section 6.2.4.2 for c.s.p. definition) reduces this still further to 606 clusters containing ~16 k samples and 3329 actives. Note that this represents <1% of the populated clusters, yet 20.8% of the actives are captured. The average cluster size was 26.4 with 5.5 actives (21%). This represents a modest enrichment over the background hit rate (14.6%). The enriched clusters included 843 actives already selected from the >40% inhibition criteria. Additional actives were chosen to supplement the prior selection as follows. Where a given cluster was already represented by four or more actives with >40% inhibition, no additional compounds were selected. Otherwise, the most active compounds were selected until the cluster was represented by four actives or there were no more actives available. In this manner, another 1270 actives were chosen.

The ~5 k actives with percent inhibition of 25 to 40% and that fell into clusters of less than five compounds were treated separately using BCUT diversity analysis, as described in Section 6.2.5. A cell-based selection biased by primary activity from six bins per each of six axes yielded 1258 compounds. The combined selection from filtering, clustering, and diversity totaled 6986 compounds representing 3337 ring scaffolds and was submitted for confirmation assays. Note that the full set of 16 k filtered actives contained 9254 ring hashcodes, so the selected set covers 36.1% of the represented scaffolds. Because of the presence of duplicate samples in the corporate screening collection, 7275 samples were pulled and assayed.

The results showed a 59% confirmation rate (41% false positive rate), giving 4296 confirmed actives. Of the compounds initially selected with >40% inhibition, 69% confirmed at >40% inhibition. This can be broken down further into compounds that fell into enriched clusters and those that did not, with confirmation rates of 73 and 68% respectively. In this case, there was only a modest increase in the confirmation rate for compounds selected from enriched clusters. This might be attributed to the fact that the enrichment within the active clusters was only about 50% greater than the background hit rate (21 versus 14.7%) in contrast to the peptide hydrolase case where the enrichment was 2.6-fold higher.

The confirmed actives (4296 of 7275) represented 1762 unique ring hashcodes or 52.8% of those present in the selected set. Of the confirmed actives, 3916 (91.2%) and 4097 (95.4%) would have been included in a Top X selection of 8000 from the unfiltered or filtered datasets respectively.

6.3.3
Protein–Protein Interaction

An activity threshold of 25.8% was chosen as the mean + 2 standard deviations of the primary screening data. The primary HTS assay yielded a low hit rate of 1.3%, or 4604 actives, which was less than the throughput limit of the confirmation assay, such that clustering and diversity analysis were not required. Conservative "soft" filtering was applied, where compounds with poor physiochemical properties are flagged and subject to visual assessment in Spotfire [48] by a medicinal chemist. In projects with low hit rates and difficult targets, such as this protein–protein interaction, teams are often interested in actives with physiochemical properties that would be rejected for more tractable targets with larger numbers of hits, and "soft" filtering of this kind has proven useful in these situations.

Actives with $175 > MW > 500$, $clogP > 7$, donors > 5, acceptors > 10, rotatable bonds > 10 or TPSA > 140 Å^2 were flagged and subject to visual assessment in Spotfire by a chemist. After removal of duplicate samples, 3337 compounds remained and were submitted for confirmation assays. These represented 2290 unique ring hashcodes.

The results showed a 46.6% confirmation rate (53.4% false positive rate), giving 1555 confirmed actives (1,096 unique ring hashcodes, 47.9%). A second round of filtering was applied using tighter physiochemical property criteria. Molecules with $200 > MW > 500$, $clogP > 5.0$, rotors > 8 were flagged and visually assessed. The final selection for secondary assays was 1087 compounds (70%).

6.4
Discussion and Conclusion

We have discussed some of the challenges facing HTS in the pharmaceutical industry today and the problems that need to be addressed to overcome those challenges. This is necessary if HTS technology is to live up to its promise and deliver the quantity and, more importantly, the quality of leads that the pharmaceutical industry needs in order to maintain the drug development pipeline. Selection of actives, data visualization, and data management are gaining prominence in this regard.

In order to address the problem of selection of actives from HTS, we have combined rigorous determination of activity threshold, filtering, clustering, diversity analysis, and data visualization. Combinations of these steps offer significant advantages relative to the common "Top X" method, including consideration of all actives with reliably detectable biological activity, the elimination of uninteresting compounds at the earliest possible stage, selection of more diverse sets with improved physiochemical properties, and increased confirmation rates. It achieves this within the practical constraints of confirmation assay throughput limits, and is flexible to meet the needs of individual project teams. We have shown the application of this method to three diverse projects with very different hit rates and requirements.

The greatest advantage of this method is the opportunity it presents to reduce the false negative rate seen with the "Top X" method whenever the hit rate exceeds the confirmation throughput limit. In extreme situations such as for the peptide hydrolase example discussed in Section 6.3.1, this can render the "Top X" method essentially unusable. In contrast, the method discussed here is able to deal with extremely large numbers of actives through clustering and diversity analysis. This provides a rational method for selecting a small percentage of the total number of actives and allows us to "rescue" actives that would otherwise be missed. This was readily apparent for the peptide hydrolase example, where only 152 (28.8%) or 256 (48.5%) of the 528 confirmed actives selected for secondary assays would have been identified from a Top X selection of the 4000 most potent actives from the unfiltered or filtered datasets respectively.

Filtering out uninteresting molecules up front carries a number of significant benefits. It removes large numbers of false positives, leading to an increase in confirmation rate

−90% in the case of the peptide hydrolase example. It creates room for additional, less-potent actives, as mentioned. It also improves the quality of the selected set, as measured by physiochemical properties (Figure 6.8 and Table 6.2) and average binding efficiency per atom (by removing high MW actives).

Clustering enables us to reduce the redundancy seen among the most potent actives, thereby creating room for less-potent actives with more diversity. In the peptide hydrolase example, 972 clusters contained ∼12 k actives, of which only 1801 were selected. At the same time, taking multiple representatives from each cluster ensures adequate coverage to guard against false negatives; for the peptide hydrolase, >99% of the selected clusters contained at least one molecule that confirmed. It also contributes to improved confirmation rates by removing false positives that are detected by their occurrence in clusters otherwise devoid of actives.

Confirmation throughput limits are a practical reality to which HTS analysis and hit selection schemes must conform. If the statistically significant hit rate exceeds the confirmation limit, then a subset of actives must be discarded; the question is how this can be done in such a way as to maximize the return on the screen within the constraints. Removing uninteresting compounds using filters based on functional group, scaffold type, and/or physiochemical properties is an obvious first step that reduces the number of hits without loss of interesting leads. Conservatively, one can also eliminate redundancy without fear of loss of significant compound classes by discarding clusters that are not statistically enriched in activity and retaining only a representative sample from highly represented clusters. If additional reduction in numbers is necessary due to the throughput limits of confirmatory (or secondary) assays, the danger of losing a valuable series becomes more significant. Assuming that all remaining molecules are equally desirable, such that there is no obvious reason for discarding one over another, the fact that "actives" from statistically enriched clusters are less likely to be false positives argues that representatives from these should be the first to be selected. Diversity analysis on the "singletons" provides an alternative means of clustering on chemical space and provides a criterion for the selection of molecules to move forward. There are additional possible criteria that one could use for such a selection that were not discussed here, including synthetic tractability, patentability, and novelty. Additionally, one could weigh any of the selections described above by potency so as to improve the probability of a more-potent molecule being selected rather than a weaker one.

The varying hit rates commonly seen in HTS screens combined with the individual requirements of the project team (stringent or relaxed physiochemical properties, specific scaffolds that represent uninteresting inhibitor classes, and so forth) demand an approach to hit selection that is customizable. This is not true for the "Top X" method but has been shown clearly here for the current method applied to three very diverse projects with hit rates varying by an order of magnitude (1.3 to 14.6%).

The confirmed actives selected from each of the three example projects shown are at varying stages of progression through the secondary assay process. The peptide hydrolase example is the most advanced and so has the most information available. The throughput limit for secondary assays in that case was just ∼500 molecules, and yet 8 interesting and distinct chemical series plus a number of singletons have been identified. As a result of the improved quality of the actives selected by this method, they appear, at first glance, to be more "sustainable" through the follow-up process in a qualitative comparison to historical trends – another distinct advantage.

However, the current method does have drawbacks. The first is related to the limitations of the ring hashcode descriptor for clustering, as discussed in Section 6.2.4.1. However, this is currently being addressed through the development of a new molecular descriptor. The second is the relatively resource-intensive nature of the method; it requires significantly more time to apply than a "Top X" method. This is of considerable importance given the throughput demands placed on HTS groups in modern pharmaceutical companies. However, the method has been streamlined considerably during its development and, furthermore, is amenable to automation.

Once interesting hits have been identified, additional analogs can be quickly identified for biological testing to provide preliminary Structure–Activity Relationship (SAR) data. If the compound came from clusters, the simplest of such searches involves simply looking up the ring hashcode of the hit and selecting all other actives from the same cluster. Further searching can be done using additional molecular descriptors such as fingerprints or, alternatively, by shape-based matching, on either the full screening library or targeted subsets thereof. See Willett [21] for additional information on similarity searching. ROCS (Rapid Overlay of Chemical Shapes, [49, 50]) is an interesting example of shape-based matching.

The reality of HTS is that the true statistical hit rate is often greater than the operational hit rate that can be accommodated by confirmation assays. In such a case, the use of a "Top X" method carries a number of significant drawbacks, the most significant of which is the creation of artificially high false negative rates through the neglect of actives less potent than the cutoff used. This is often overlooked but is of critical importance. Even in projects with lower hit rates, carefully constructed filters should be applied, as was shown in the example of the protein–protein interaction target. Given the huge investment in HTS technology by pharmaceutical and biotechnology companies and the importance of the selection of actives to the drug discovery process, the scheme outlined here offers a practicable alternative to the "Top X" method with tangible and substantial benefits.

Acknowledgments

We would like to acknowledge all of our colleagues in the Computational Chemistry group for helpful discussions, our colleagues in the HTS group for their input and for the data on which this work is based, our colleagues in each of the project teams for their help in shaping the hit selection process to meet the needs of the individual project, Dominick Mobilio for supplying data on purity and integrity rates for random samples of the Wyeth screening library, and Fred Immerman for helpful discussion of statistical issues.

References

1 HERTZBERG, R.P. and POPE, A.J. High-throughput screening: new technology for the 21st century. *Curr. Opin. Chem. Biol.* **2000**, *4*, 445–451.

2 FERNANDES, P.B. Technological advances in high-throughput screening. *Curr. Opin. Chem. Biol.* **1998**, *2*, 596–603.

3 BLEICHER, K.H., BOHM, H-J., MULLER, K., and ALANINE, A.I. Hit and lead generation: Beyond high-throughput screening. *Nat. Rev.: Drug Disc.* **2003**, *2*, 369–378.

4 BAJORATH, J. Integration of virtual and high-throughput screening. *Nat. Rev.: Drug Disc.* **2002**, *1*, 882–894.

5 HANDEN, J.S. High-throughput screening – challenges for the future. *Drug Disc. World* **2002**, 47–50.

6 SMITH, A. Screening for drug discovery: the leading question. *Nature* **2002**, *418*, 453–459.

7 PROUDFOOT, J.R. Drugs, leads, and drug-likeness: an analysis of some recently launched drugs. *Bioorg. Med. Chem. Lett.* **2002**, *12*, 1647–1650.

8 OPREA, T.I., DAVIS, A.M., TEAGUE, S.J., and LEESON, P.D. Is there a difference between leads and drugs? A historical perspective. *J. Comput. Chem. Inf. Sci.* **2001**, *41*, 1308–1315.

9 KUNTZ, I.D., CHEN, K., SHARP, K.A., and KOLLMAN, P.A. The maximal affinity of ligands. *Proc. Natl. Acad. Sci.* **1999**, *96*, 9997–10002.

10 LIPINSKI, C.A., LOMBARDO, F., DOMINY, B.W., and FEENEY, P.J. Experimental and computational approaches to estimate solubility and permeability in drug discovery and development settings. *Adv. Drug. Deliv. Rev.* **1997**, *23*, 3–25.

11 VEBER, D.F., JOHNSON, S.R., CHENG, H-Y., SMITH, B.R., WARD, K.W., and KOPPLE, K.D. Molecular properties that influence the oral bioavailability of drug candidates. *J. Med. Chem.* **2002**, *45*, 2615–2623.

12 HANN, M.M., LEACH, A.R., and HARPER, G. Molecular complexity and its impact on the probability of finding leads for drug discovery. *J. Comput. Chem. Inf. Sci.* **2001**, *41*, 856–864.

13 BOGUSLAVSKY, J. Creating Knowledge From HTS Data. *Drug Disc. Dev.* **2001**, http://www.dddmag.com/feats/0106gsk.asp.

14 WALTERS, W.P., STAHL, M.T., and MURCKO, M.A. Virtual screening – an overview. *Drug Disc. Today* **1998**, *3*, 160–178.

15 LIPINSKI, C.A. Drug-like properties and the causes of poor solubility and poor permeability. *J. Pharm. Toxicol. Methods* **2000**, *44*, 235–249.

16 HANN, M., HUDSON, B., LEWELL, X., LIFELY, R., MILLER, L., and RAMSDEN, N. Strategic pooling of compounds for high-throughput screening. *J. Comput. Chem. Inf. Sci.* **1999**, *39*, 897–902.

17 RISHTON, G.M. Reactive compounds and in vitro false positives in HTS. *Drug Disc. Today* **1997**, *2*, 382–384.

18 OPREA, T.I. Property distribution of drug-related chemical databases. *J. Comput.-Aided Mol. Des.* **2000**, *14*, 251–264.

19 WILLETT, P. Chemoinformatics – similarity and diversity in chemical libraries. *Curr. Opin. Struct. Biol.* **2000**, *11*, 85–88.

20 BAYADA, D.M., HAMERSMA, H., and VAN GEERESTEIN, V.J. Molecular diversity and representivity in chemical databases. *J. Comput. Chem. Inf. Sci.* **1999**, *39*, 1–10.

21 WILLETT, P. Chemical similarity searching. *J. Comput. Chem. Inf. Sci.* **1998**, *38*, 983–996.

22 MASON, J.S. and PICKETT, S.D. Partition-based selection. *Perspect. Drug Disc. Des.* **1997**, *7/8*, 85–114.

23 BROWN, R.D. Descriptors for diversity analysis. *Perspect. Drug. Disc. Des.* **1997**, *7/8*, 31–49.

24 WILLETT, P. Dissimilarity-based algorithms for selecting structurally diverse sets of compounds. *J. Comput. Biol.* **1999**, *6*, 447–457.

25 MENARD, P.R., LEWIS, R.A., and MASON, J.S. Chemistry space metrics in diversity analysis, library design, and compound selection. *J. Comput. Chem. Inf. Sci.* **1998**, *38*, 1204–1213.

26 SPENCER, R.W. High-throughput screening of historic collections: observations on file size, biological targets, and file diversity. *Biotechnol. Bioeng. (Comb. Chem.)* 1998, *61*, 61–67.

27 JOHNSON, M.A. and MAGGIORA, G.M. *Concepts and Applications of Molecular Similarity*. Wiley, New York, 1990.

28 PATTERSON, D.E., CRAMER, R.D., FERGUSON, A.M., CLARK, R.D., and WEINBERGER, L.E. Neighborhood behaviour: a useful concept for validation of "Molecular Diversity" descriptors. *J. Med. Chem.* 1996, *39*, 3049–3059.

29 ZHANG, J-H., CHUNG, T.D.Y., and OLDENBURG, K.R. A simple statistical parameter for use in evaluation and validation of high throughput screening assays. *J. Biomol. Screen.* 1999, *4*, 67–73.

30 SILLS, M.A., WEISS, D., PHAM, Q., SCHWEITZER, R., WU, X., and WU, J.J. Comparison of assay technologies for a tyrosine kinase assay generates different results in high throughput screening. *J. Biomol. Screen.* 2002, *7*, 191–214.

31 FAY, N. and ULLMANN, D. Leveraging process integration in early drug discovery. *Drug Disc. Today* 2002, *7*, S181–S186.

32 MCGOVERN, S.L., CASELLI, E., GRIFORIEFF, N., and SHOICHET, B.K. A common mechanism underlying promiscuous inhibitors from virtual and high-throughput screening. *J. Med. Chem.* 2002, *45*, 1712–1722.

33 MCGOVERN, S.L. and SHOICHET, B.K. Kinase inhibitors: not just for kinases anymore. *J. Med. Chem.* 2003, *46*, 1478–1483.

34 TAMURA, S.Y., BACHA, P.A., GRUVER, H.S., and NUTT, R.F. Data analysis of high-throughput screening results: application of multidomain clustering to the NCI anti-HIV data set. *J. Med. Chem.* 2002, *45*, 3082–3093.

35 GEDECK, P. and WILLETT, W. Visual and computational analysis of structure-activity relationships in high-throughput screening data. *Curr. Opin. Chem. Biol.* 2001, *5*, 389–395.

36 SHEN, J. HAD: an automated database tool for analyzing screening hits in drug discovery. *J. Comput. Chem. Inf. Sci.* 2003, *43*, 1668–1672.

37 NILAKANTAN, R., BAUMAN, N., and HARAKI, K.S. Database diversity assessment: new ideas, concepts, and tools. *J. Comput.-Aided Mol. Des.* 1997, *11*, 447–452.

38 TAYLOR, J.R. The binomial distribution. In *An Introduction to Error Analysis: The Study of Uncertainties in Physical Measurements*. University Science Books, Herndon, VA, 1997.

39 PEARLMAN, R.S. and SMITH, K.M. Novel software tools for chemical diversity. In *3D QSAR and Drug Design: Recent Advances*, KUBINYI, D., MARTIN, Y., and FOLKERS, G. (Eds). Kluwer Academic Publishers, Dordrecht, Netherlands, 1997.

40 FOX, S.J. *The Success of High-Throughput Screening in Developing Clinical Candidates*, HighTech Business Decisions, Moraga, CA, 2000.

41 *MOE*, CCG Inc., Montreal, Canada, 2002.

42 IMMERMAN, F., Personal Communication.

43 TAYLOR, J.R. The normal distribution. In *An Introduction to Error Analysis: The Study of Uncertainties in Physical Measurements*. University Science Books, Herndon, VA, 1997.

44 *RS3 Discovery for Excel*, Accelrys Inc., San Diego, CA, 2001.

45 *Excel*, Microsoft Inc., Redmond, WA, 2000.

46 *Access*, Microsoft Inc., Redmond, WA, 1997.

47 *Diverse Solutions*, Tripos Inc., St. Louis, MO, 1997.

48 *Spotfire Decision Site*, Spotfire Inc., Somerville, MA, 2002.

49 *ROCS*, OpenEye Scientific Inc., Santa Fe, NM.

50 GRANT, J.A., GALLARDO, M.A., and PICKUP, B.T. A fast method of molecular shape comparison: a simple application of a Gaussian description of molecular shape. *J. Comput. Chem.* 1996, *17*, 1653–1666.

7

Molecular Diversity in Lead Discovery: From Quantity to Quality

Cullen L. Cavallaro, Dora M. Schnur, and Andrew J. Tebben

7.1
Introduction

Diversity is an elusive, enigmatic property. In the context of pharmaceutical lead discovery, there are myriad interpretations of this word. Diversity may be viewed in the context of physical properties, such as molecular weight (MW), cLogP [1], polar surface area and the like [2]; or from the perspective of structural features such as pharmacophores [3] or shape [4]; or even with respect to the potential to hit biological targets [5]. However, even without the convoluted semantics, a clear view remains obfuscated, as there are numerous methods available for the assessment of diversity [6]. The underlying fact is that, unlike many scientific measurements, the analysis of diversity is by nature *qualitative* – extremely important, but *qualitative* nonetheless, because diversity only has meaning in a *relative* sense. Sets of compounds can only be measured against other sets or with respect to a fixed number of properties. There is no universal, absolute diversity scale [7].

Scientists may find it uncomfortable to rely on such "soft" data. However, examination of diversity can be critical in a variety of lead discovery endeavors, such as in the design of combinatorial libraries, the construction of biological screening sets, and the acquisition of new screening compounds. As a virtual set of compounds is assembled prior to synthesis or screening, diversity analysis provides a tool that can be used to shape the focus of the set. This exercise is of even greater value if key physical or structural features can be identified in advance. By reducing the redundancy contained within a set of compounds, thus increasing the diversity, smaller subsets can be created that are still representative of the parent collection [8–12]. In other words, fewer compounds can be synthesized, screened, or purchased; yet the subset can still have an impact comparable to the greater set. Research time and materials are saved as efficiency is increased in this manner. The potential of more impact from fewer compounds drives the interest in diversity.

With a variety of tools and techniques available to measure diversity [6], as well as a plethora of properties and derived parameters available to assess it, the approach can be critical. Direction, however, is most frequently provided by the project itself. From one project to the next, the objective of lead discovery remains unchanged: to uncover biologically active compounds that can be developed into a series of leads. However, the

Chemoinformatics in Drug Discovery. Edited by Tudor I. Oprea
Copyright © 2004 WILEY-VCH Verlag GmbH & Co. KGaA, Weinheim
ISBN: 3-527-30753-2

method for pursuing this objective will vary depending on the additional information that is available for a particular project. Often, the size of a library or screening collection will be largest for projects where information is scarce. In this way, a wide net can be cast in an attempt to find active chemotypes or learn something about the target from the hits that are discovered. As more information is gathered, questions about a target become more specific while chemistry becomes more challenging and the number of library members decrease.

Once the objective and appropriate size of a combinatorial library or designed collection are determined, many of the relationships that are important for assessing diversity become more apparent. For example, if a project goal is to design a collection for general screening, then structural diversity might drive the approach. However, the design of a diverse set of 100 analogs of a promising lead compound might utilize parameters and methods that are very target specific, such as a pharmacophore model [13], a CoMFA model [14, 15], or a crystal structure [16]. The relative set against which diversity will be examined will vary as one moves from large collections (often tens to hundreds of thousands) to mid-sized sets to small groups of compounds. As the relative set changes, so does the approach to diversity.

7.2
Large Libraries and Collections

With the emergence of combinatorial chemistry came a boom period of large libraries. Discovery libraries reported in the literature swelled, often reaching 50 000 to 150 000 members [17–19], while peptide libraries eclipsed 1 000 000 members [20–22] on a number of occasions. Size trended from large to enormous as combinatorial technologies advanced and companies worked to build compound decks that quickly dwarfed historical collections they maintained. Diversity, however, was often not a consideration in these early days. When attempting to make 100 000 compounds, the difficulty was often how to obtain enough reagents. If every reasonable amino acid available for purchase is included in a library, how could there be a question of diversity?

Gigantic libraries proved too large and inefficient to carry their own weight, did not meet early expectations in the production of new leads [23], and soon became dinosaurs. One major shortcoming was often a lack of diversity (Figure 7.1). This key deficiency often led to large numbers of compounds with very similar properties that may have been adequately represented by much smaller sets. This approach also led to the production of many promiscuous compounds [24] that would be discovered as hits for many wide-ranging biological targets.

These results led to the design of somewhat smaller diversity-based collections. Initially, this meant structure-driven diversity, especially for deck additions. Typically, this was achieved by comparing structure-based fingerprints such as Daylight [25–27], MACCS [28, 29], UNITY [30, 31], and so on. This eventually gave way to pharmacophore-based diversity [32–34] and diversity within property-based chemistry spaces. The latter used descriptors derived from properties such as connectivity [35, 36], cLog P [37], molecular weight, polar surface area [38], the number of rotatable bonds, hydrogen

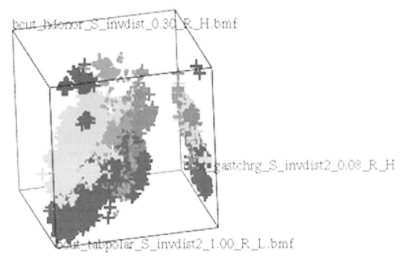

Fig. 7.1   3-D plot of a very large library in BCUT space. As was typically found in very large libraries, the majority of library members are clustered in a fraction of the diversity space. Those areas that are filled are overrepresented and could be covered by far fewer compounds. Many of the library members lie at the very edge of diversity space and do not contribute to filling the broad areas of unfilled space.

bonding, or BCUTs [39, 40]. In addition to assuring diversity among properties, such calculations also allowed for the exclusion of reagents that tended to produce groups of products that consistently fell at the extreme ends of the property scales. Thus, for example, large subsets of highly hydrophobic compounds could be reduced in size or even avoided entirely.

The shortcomings of some of these enormous combinatorial libraries resulted in a new appreciation for diversity and led to the development of a number of new computational approaches to the design of large libraries and collections. Although the number of members contained in such compound sets has steadily decreased, these sets are still employed in situations where there is minimal information about a biological target. By utilizing diversity assessment techniques, a wide sampling of properties can be assured and the probability of discovering a hit can be increased.

7.2.1
Methods and Examples for Large Library Diversity Calculations

The age of ultralarge (50 000–150 000 member) nonpeptide combinatorial libraries [41, 42, 18] is typified by some of the efforts of Merck [43, 44], Pharmacopeia [45], and Houghten Pharmaceuticals [18, 19, 46–48] during the mid-to-late 1990s. While design details often remain unpublished, diversity analyses of several libraries with qualitative comparison to their biological activity exist in the literature [49, 50]. During this era, practical considerations required that ultralarge library design be predominantly based on reagent selection rather than on product selection. Complete enumeration of the huge (million or more compounds) virtual product libraries was impractical, both in terms of CPU time and for most selection methods available at the time. Methodology ranged from experimental design approaches, such as simple multiple property-based partitioning methods based on factorial designs [49] and the D-optimal approaches used at Chiron [51], to maximal diversity-based tools predicated on distances in a PLS (partial least squares)-derived property space as developed by Waldman et al. [52] of Molecular Simulations, Inc. [53] for the Cerius^2 Combinatorial Chemistry Consortium [54]. Other historical diversity methods, including distance-based clustering, Kohenen Maps, and spanning trees, have been reviewed at length [6].

Perhaps the only commercial tool available during the mid-to-late 1990s capable of dealing with diversity-based libraries of this size was Diverse Solutions [55]. This program uses low-dimensional chemistry spaces whose axes are based upon BCUTs [39], a novel type of metric that is based upon both connectivity-related- and atomic properties such as charge, polarizability, and hydrogen bonding abilities. BCUTs appear to correlate with ligand–receptor binding and activity [49, 50, 56]. The program uses a cell-based algorithm that provides reactant-biased, product-oriented combinatorial library design [56] and is used in conjunction with Pearlman's library enumeration software, CombiLibMaker [57], or more recently, Pearlman's Windows-based LibraryDesigner application [58]. Library enumeration, historically a difficult problem, has been greatly facilitated by the introduction of chemically driven enumeration tools such as CombiLibMaker and compact molecular representations such as SMILES [59] strings. As a consequence, very large virtual libraries (millions and even billions) can now be accommodated.

Diverse Solutions accepts a variety of input formats, including SMILES for an enumerated library. A variety of 2-D or 3-D BCUTs are generated for each virtual compound. While the 2-D BCUTs are lowest and highest eigenfunctions of matrices derived from properties on the diagonal and connectivity relationships on the off-diagonals, the matrices for the 3-D BCUTs replace the connectivities with various distance relationships between the atoms using 3-D structures generated on the fly using the algorithms from CONCORD [60].

Once BCUT descriptors of various "flavors" are calculated for the enumerated virtual library, the optimal chemistry space axes that describe the library are determined by finding an orthogonal set of four to six descriptors that distribute the compounds evenly across the space according to a chi-squared algorithm [61, 62] The space is then partitioned into cells, or hypercubes, by subdividing the axes. Since the cell-based space is independent of the compounds evaluated in it once it is derived, comparison of the diversity of various libraries or iterations of libraries in the same space are straightforward and meaningful; thus, it was possible to compare large libraries designed using reagent-based selection methods [49, 50].

In general, larger libraries, other than those of Pharmacopeia and Houghten, have tended to be in the range of 5000 to 15 000 compounds, particularly if the entire library was purified and stored as individual compounds. The same methods and algorithms as for diverse library design apply here. However, it now becomes possible to employ more product-based approaches by filtering reagent sets before enumeration of the virtual library. This includes the incorporation of "druglikeness" [63] or ADME-based constraints such as Lipinski's "rules" [64–66] in addition to synthetic feasibility. Clustering methods and approaches based on pharmacophores, such as those of Mason et al. [67], become feasible for diverse libraries in this size range. Additionally, even though a library may be designed to be diverse, it may also be directed toward one or more targets or target classes. Simplistically, this focus may derive from selection of an intuitively relevant chemotype or one derived from a lead. In this case, the products are selected to find additional active analogs or to assess structure-activity relationships.

Methods to select products range from the cell-based methods, including the concept of receptor relevant subspaces [56], which will be discussed in the context of focused library design, to structure-based combinatorial library design. Other design approaches involve the use of privileged substructures [67–69] or four-point pharmacophores for target-class library design [70]. Additional information provided by privileged substructures, four-point pharmacophores, cell-based descriptors, and other properties can be used to construct profiles of interesting active compounds. These profiles may then be used as a partial constraint on the diversity of the designed library [71]. The simplest example of this would be the use of molecular weight and cLogP profiles to constrain a cell-based library design within ADME-derived cutoffs. Alternatively, the frequency of occurrence of particular substructures in a set of GPCR lead compounds, available in the MDDR [72], could be used to constrain the designed combinatorial library to a similar frequency of occurrence. While the privileged substructures or active leads are used to provide a certain degree of focus, these libraries are usually either intended to be diverse within defined constraints or to contain a percentage of diverse compounds. Methods for combinatorial library design based on target classes or gene families have recently been reviewed [73].

Another approach is the informative library design process, as described by Teig [74]. It is a method that samples the chemical space accessible to a library in a manner that, when combined with assay data, allows the elucidation of a set of pharmacophores consistent with biological activity. The library is selected in a manner that maximizes the Shannon entropy [75] for the possible outcomes being tested by the library [76]. The outcomes most relevant to the pharmaceutical industry are the three-dimensional pharmacophores present in the active molecules because these are scaffold independent and facilitate movement between structural classes.

Miller [77] has presented an intuitive description of the informative design technique in the form of a game of 20 questions (Figure 7.2). The player is asked to guess a number between 1 and 16 using only yes or no questions. If the questions are asked in sequence, the most efficient strategy is uncovered by splitting the space with each subsequent question (i.e., is the number less than eight? Is it less than four? Is it less than two?). While the player does eventually arrive at the answer, one must wait for the answer of each question in order to formulate the next. This strategy becomes ineffective if the player is required to ask all the questions simultaneously, as it will only work if the number is one or two. An informative strategy asks the questions such that when they are answered they will lead the player to any possible outcome. Because the number of outcomes in this simple example is small, it is possible to find the answer

Divide and conquer

1	2	3	4	5	6	7	8	9	10	11	12	13	14	15	16	<9? Y
1	2	3	4	5	6	7	8	9	10	11	12	13	14	15	16	<5? Y
1	2	3	4	5	6	7	8	9	10	11	12	13	14	15	16	<3? Y
1	2	3	4	5	6	7	8	9	10	11	12	13	14	15	16	2? Y

Informative design
pharmacophores   Active?

1	1	1	1	1	1	1	1	1	0	0	0	0	0	0	0	Y
1	1	1	1	0	0	0	0	1	1	1	1	0	0	0	0	N
1	1	0	0	1	1	0	0	1	1	0	0	1	1	0	0	N
1	0	1	0	1	0	1	0	1	0	1	0	1	0	1	0	Y

Fig. 7.2 Informative design – asking the right question: The most efficient means for a player to guess the correct number from 1 to 16 is a "divide and conquer" strategy. However, this requires that the player ask questions in sequence and wait for an answer before asking the next. Informative design asks specific questions so that when they are answered simultaneously, the player is led to the answer. In this example, the questions are compounds either possessing (denoted by 1) or lacking (denoted by 0) a specific pharmacophore. Once the compounds are assayed, a single outcome is found to be consistent with activities and the corresponding pharmacophore is found.

with only a few questions. However, in the context of library design, the player is not attempting to guess a number but to discover those pharmacophores that are most consistent with the biological activities found upon assay by asking "questions" with specific compounds. Because the pharmacophore space of a given library is usually quite large, it becomes difficult to design a set of questions that lead to definitive outcomes within a library of reasonable size. Hence, the real success of the informative method lies in applying it in an iterative fashion [78]. The initial libraries are designed so that they test the pharmacophore space as efficiently as possible. Once these libraries are tested, some fraction of the pharmacophore space can be eliminated, and subsequent libraries can then be used to further narrow the space and, as a consequence, refine the pharmacophore model.

Although no examples of the direct application of the informative process to the design of libraries have been published, a retrospective analysis of the design of a library of potential CDK2 inhibitors has been reported [79]. This analysis compared the rate at which 207 known actives were recovered from a pool of 13 359 compounds using informative design, diverse–similarity design, and random selection. From the initial pool, 1000 compounds were selected via informative design, diverse design, and random selection. The informative selection was carried forward with three subsequent rounds of informative design followed by a single round of similarity design. All of the iterative steps were based upon the actives that appeared in each subsequent round. The diverse design was iterated in four rounds of similarity design, again around the actives discovered. In both cases, the final library consisted of 100 compounds. Ten random selections of 1000 compounds were made as a control. At the end of four rounds, the informative design had recovered 63% of the known actives, the diversity-similarity design had found 32%, and the random selection 24%. It is also noted that the number of pharmacophores was reduced from 1.8 million to 82 000. When this reduced set was used in a similarity search in the final step of the informative design, an enhanced enrichment was achieved.

While many of these methods are still utilized in situations where project information is scarce, new tools for the analysis and design of compound sets have allowed scientists to expect more specific answers from libraries and collections. These events and developments have inevitably led to the design and synthesis of smaller, smarter libraries.

7.3
Medium-sized/Target-class Libraries and Collections

As diversity began to guide library and collection design, the size of compound sets began to trend downward [80]. Attention turned toward a rational analysis of product set diversity, rather than being focused on the availability of reagent sets. Another factor contributing to the reduction in size was the move away from combinatorial libraries containing compound mixtures. Complications in the screening, deconvolution, isolation, and identification of active entities from mixtures proved to be labor intensive, costly, and time consuming.

Simultaneously, it was becoming apparent that drug candidates were unlikely to be directly discovered in combinatorial libraries. Attention shifted to the design of libraries for the generation of potential lead compounds. Analyses of these libraries extended the early property restriction calculations that emerged from the big library era. In addition to monitoring diversity and molecular properties, these calculations were now intended to insure that compound sets displayed leadlike properties [81, 82]. Biological factors also began to play a part in design. The combinatorial shotgun approach – where an attempt was made to hit any and every target available – was tempered by the idea of introducing some biological selectivity or target bias. It was also recognized that related sets of compounds could be used to explore related sets of biological targets. This eventually led to the design of compound sets biased toward particular biological target classes, such as GPCRs, kinases, proteases, and so on. Clearly, the model of choice for design was that of a crystal structure of a ligand bound in the active site of a receptor. For many target classes, such as GPCRs, these are unavailable. However, known ligands for these target classes often share numerous structural features or have similar calculated properties. Thus, an essential starting point for ligand-based designs is the accumulation of the biological data associated with known ligands leading to the development of a target-class knowledge database. This requires the collection of known drugs, analogs, and other active compounds along with their biological response information from the literature, commercial drug databases, and proprietary sources; this is a far-from-trivial undertaking. Commercial drug databases such as the MDDR [72] were not designed for this type of data mining, as they lack primary biological data and often do not indicate a specific target. Nevertheless, an analysis of the MDDR has provided useful information, such as Schuffenhauer's classification of the MDDR ligands with regard to their gene ontology [83]. However, this does not replace the need for comprehensive knowledge databases. A clear indicator of the importance of these databases is the development of commercial products by numerous companies such as Aureus Pharmaceuticals [84], Sertanty [85], Biowisdom [86], and Jubilant Biosystems [87]. Once a target-class knowledge database of ligands has been generated, it may be used to develop computational models for the design of new sets of compounds biased toward the desired targets or for selection of target-class biased compounds from a corporate collection (see e.g., Chapter 12 [88]). The approaches used include the use of four-point pharmacophore descriptors [33, 70], class property profiles [71], privileged substructures [67], DiverseSolutions (DVS) analysis [89, 90], and informative design [91]. This type of analysis can be coupled with property restriction calculations, often based on ADME constraints [64–66], to limit products with undesirable properties, such as very high MW, cLogP, and so on. By linking these approaches, diverse compound sets could be designed with a bias toward a biological target, while simultaneously reducing the probability of producing promiscuous and nondruglike compounds. Libraries could be somewhat focused and remain leadlike. This proves to be an effective strategy for deck building as well, since it allows for focus of libraries in structurally and/or biologically underrepresented areas; this provides a definitive step forward from using large, semirandom collections. Diversity remains critical in these endeavors, although that diversity now operates within the boundaries of biological and property restrictions.

There are three theoretical scenarios for target focused design [73], which can be exemplified by a theoretical 4000-compound library. In the first instance, the common features required for ligand binding are identified/defined for all (or a subset) of the gene family or target-class members. These features are then used to derive a model for selection of the 4000 library products. One or more combinatorial templates or scaffolds may be used for this process. This method might produce weak, nonspecific hits, since the model used for the design emphasizes similarities of the target-class ligands rather than the differences that convey specificity. For the second contrasting approach, a set of specific targets is selected, perhaps four to eight. Separate focused libraries are then designed for each target on the basis of its target-specific computational model. These may be developed using receptor information derived from protein crystal structures or from ligand-based methods. The focused libraries are then combined to form the 4000-compound combinatorial library. This much more computationally intensive method should, if the models used were of high quality, produce some potent hits for the individual targets. It should be pointed out that even the most successful designs produce libraries for which only a small fraction of compounds bind to their intended targets; however, since diversity is still incorporated into the designs and since small changes in structure may favorably or adversely affect binding, they should potentially provide hits for targets in the same gene family that were not explicitly considered in the design effort. The third scenario is, in fact, a combination of these two extremes. The same set of specific targets is selected, but the combinatorial design methodology simultaneously optimizes reagent selection to yield 4000 products made up of subsets that are focused toward those specific targets. Clearly, this type of multivariate optimization would be more efficient synthetically and have the same potential for providing potent leads. It also allows for the possibility of exploring the effect of small structural changes on specificity. However, the price one pays in practice for avoiding the computational and synthetic problems associated with combining multiple small libraries into a single large one is that the resulting library may not be as focused as the separate libraries relative to the individual targets.

In both the second and third approaches, the use of "activity scores" based on the positions in chemistry space of the virtual library products relative to the positions of known ligands of each individual target incorporates all of the information about known ligands, rather than just the common pharmacophoric features. The structural elements that convey specificity are not excluded in the design, as they may be in the first approach.

Needless to state, there are no published examples that compare these three paradigms directly. They do, however, represent some of the driving factors that have contributed to decreases in designed library size over the past decade. Information content, both of models and libraries, has become paramount. A medium or small focused library can more efficiently be directed to gathering SAR (Structure-Activity Relationship) data for a specific target, or small set of related targets, than a big library. Larger libraries, even in a target-class context, tend to be designed for more general lead finding; the computational models tend to be general and run the risk of providing nonspecific weak ligands.

7.3.1
Computational Methods for Medium- and Target-class Libraries and Collections

One of the many methods utilized in the design of target-class-directed libraries is the well-known practice of distance-based similarity searching. In this approach, searches are made in chemistry space around a known active to yield a list of nearest neighbors that are likely to contain other compounds active against the same target. Pearlman and Smith implemented a list-based near-neighbor searching algorithm within Diverse Solutions [92], which can also be used to design combinatorial libraries focused around one or more known lead compounds for a given target. Recent versions of Diverse Solutions and of Pearlman's Windows-based LibraryDesigner application [93] also offer a novel "focused/diverse" library design option that yields products that are focused with respect to receptor-relevant axes of the chemistry space and are diverse with respect to the receptor-irrelevant axes. However, unless all targets in a given target family share identical sets of relevant and irrelevant chemistry space axes (an unlikely situation), a "focused/diverse" library design should not be used to design libraries for an entire target class. Moreover, since simple "focused" library design is distance based (or activity score based), it is well suited for designing libraries for a single target, but not so well suited for target-class libraries because, without reference to absolute position in chemistry space, it is often difficult to ensure that the designed library will cover all members of the target class.

To best address target-class library design, Diverse Solutions and LibraryDesigner also offer a unique, cell-based "fill-in" library design option that is ideally suited for target-class design. A set of known actives from the target-class knowledge database is used to identify a list of "promising cells" in chemistry space (Figure 7.3). The reactant-biased, product-based library design algorithm can then be used to design a large combinatorial or small parallel library that best fills these "promising cells". The degree of target focus is controlled by the number of bins per axis and the number of cell

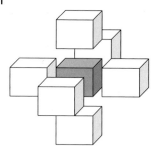

Fig. 7.3 Representation of "interesting" or promising cells. Each cube represents a cell in the partitioned descriptor space. The shaded cell contains one or more active compounds. The virtual compounds that fall into cells that are immediately adjacent to the interesting cells are also likely to be active, so one may add a degree of fuzziness to the design so as to capture these cells. One layer is shown, but the number of layers controls the focus or fuzziness of the region defined as interesting.

radii from the known ligand used to define the size of the "interesting cell" (Figure 7.3). The "focused" design approach uses a set of target ligands to score all the compounds in the virtual library, based on their distance to the actives, and then selects a designed library that optimizes the average virtual activity. An example of using this method to select GPCR compounds for screening to validate the library design approach has been reported by Wang and Saunders [89]. The algorithm also permits the use of externally determined activity scores, such as those from docking, QSAR models, pharmacophore models, or other sources. The "focused/diverse" algorithm combines this approach with the use of "receptor relevant subspaces" [56] to find compounds that are both as similar as possible to the target ligands and as evenly distributed as possible across the nonrelevant descriptor axes.

The concept of receptor-relevant subspaces is so integral to Pearlman's approaches that it needs to be discussed in some detail. Qualitative observations, such as those reported by Schnur [49], about clustering of actives in two- or three-dimensional subspaces of five- or six-dimensional chemistry spaces led Pearlman and Smith [56] to develop a novel algorithm for reducing chemistry space dimensionality. Whereas typical methods that reduce dimensionality discard potentially important information, this algorithm identifies which axes (metrics) convey information that may be related to the affinity for a given receptor and which axes appear irrelevant and therefore may be safely discarded. This is done by identifying the axes that tightly group active compounds on the basis of a cluster-breath normalized value of chi-squared, computed either from a simple count of actives per bin along each axis or from an activity-weighted count of those actives. Multiple binding mode possibilities are addressed by allowing more than one cluster per relevant axis. Receptor-relevant subspaces are useful not only for an easy graphical visualization of active compounds in the chemistry space, and therefore a visual validation of the chemistry space, but more importantly for the calculation of receptor-relevant distances that are essential for identifying near neighbors of actives in focused library design and for comparing libraries.

Several other approaches with the goal of simultaneous optimization of several criteria have been reported. One such approach is the generation of a library that is both focused and diverse via the dual fingerprint metric described by Bajorath [94]. In this method, individual compounds are randomly generated and their similarity to a known inhibitor is evaluated by comparison of their minifingerprints [95] using the Tanimoto coefficient. Those molecules that are above a similarity threshold are then tested for diversity against compounds already selected by evaluating the overlap of MACCS keys [96]. Only those dissimilar to the previously chosen compounds are picked for the library. The selection process is continued until the desired library size has been reached. Since the similarity threshold can be adjusted for both the initial similarity step and the second diversity step, the algorithm can be tuned toward a very focused library or a quite diverse selection. The impact of the initial focusing step was exemplified in the design of two libraries targeted toward known kinase inhibitors. It was shown that when compared to a simple diversity selection the dual-fingerprint method favored scaffolds known to be present in active kinase inhibitors but did not narrow the library to a set of very closely related compounds. While this is a versatile and computationally inexpensive method of library design, it does not attempt to optimize the library.

The nature of combinatorial chemistry can present a considerable challenge because these libraries are generally produced as arrays of compounds and it is often inconvenient to synthesize individual compounds in order to achieve an optimal design. Two methods have been described that attempt to select optimal subset of reagents from a virtual library that has been partitioned into favorable and unfavorable compounds by some method of filtering. The PLUMS algorithm [97] was designed to simultaneously optimize the size of the library based on "effectiveness" and "efficiency". The effectiveness of the library is defined as the ratio between the number of favorable compounds selected for the library and the total number of favorable compounds. Efficiency is the ratio of selected favorable compounds and the total number of products in the library. Sublibraries are scored via a weighted sum of effectiveness and efficiency. The algorithm works by iteratively identifying the monomer that when removed produces a library with the best score for that iteration. Monomer elimination continues until all the unfavorable molecules have been removed, leaving a library of optimal efficiency, but not necessarily of maximal effectiveness. In the case where removal of either of two monomers will result in libraries of identical score, two libraries are generated, one with each monomer removed, and carried forward into the next iteration. Since there is the potential for a very large number of equivalent solutions to propagate forward, the user has the ability to limit the number of solutions that are carried into the next iteration. The PLUMS method was tested on a library consisting of 13 aldehydes, 41 azole aldehydes, and 59 amines. The resulting 31 477 compounds were filtered by molecular weight, ClogP, and ability to fit a Catalyst pharmacophore, leaving 4907 favorable molecules. However, these compounds contained 12 aldehydes, 39 azole aldehydes, and 49 amines, requiring a 22 932-member library to capture all of the favorable compounds. The PLUMS algorithm was used to select the most efficient library from this set of reagents and determined it to be a $6 \times 11 \times 18$ set of 1188 compounds. This library consisted entirely of favorable compounds, but it contained only 24% of the total number of desirables. While this library does represent maximal efficiency, by eliminating 76% of the favored compounds, it lacks balance between efficiency and effectiveness. A more balanced library can be found by selecting the highest scoring iteration rather than the largest library that is composed entirely of favorable compounds. The highest scoring design resulted in an $8 \times 24 \times 35$, or 6720-member library, that captured 86% of the favored compounds representing 63% of the total library. Since PLUMS is a sequential algorithm, there is the danger that it will not find the globally optimal solution. It was compared with both genetic algorithms (GA) [98] and monomer frequency analysis (MFA) [99]. In the design of the above library, it was found that PLUMS found solutions that were not only equivalent to the GA but also outperformed MFA.

Another method of focusing combinatorial arrays has been described by Agrafiotis [100]. This method begins with the selection of individual molecules within a combinatorial array that meet a fitness criterion. The fitness criterion can be as simple as similarity to a reference compound or a multiobjective function composed of several property constraints. A set of subarrays is then randomly selected from the entire combinatorial array and each site is optimized in sequence. For example, the selection of a $5 \times 5 \times 5$ library from a $10 \times 10 \times 10$ array would begin with the generation of ten $1 \times 5 \times 5$ subarrays. These would be evaluated and the five highest scoring R1 monomers would then be used in the construction of ten $5 \times 1 \times 5$ subarrays. The five R2 groups with the highest fitness would then be selected and utilized for ten $5 \times 5 \times 1$ subarrays, resulting in the selection of five R3 groups. Once all the reagents have been selected, the fitness of the resulting sublibrary is evaluated and compared to the previous iteration. If it has a higher score, it is kept and another refinement cycle is carried out; otherwise the process is terminated. The effectiveness of this procedure to generate an array of compounds similar to a reference structure was compared to a simulated annealing algorithm [101] and the selection of discrete compounds. These selection methods were applied to a reductive amination library composed of 300 amines and 300 aldehydes. One member of the library was chosen as the reference structure and simulated annealing was used to select 10×10 arrays and 100 single compounds; 10×10 arrays were also chosen with this algorithm and evaluated on their similarity to the reference, along with the set selected with simulated annealing. It was shown that mean and standard deviation of the calculated similarities for the compounds selected by these three procedures were very comparable. Similar results were obtained in the selection of an optimal set of amines and alkylating or acylating agents for a three-component library described by Cramer et al. [102]. The optimization criteria for this library were not only similarity to a target molecule but also fit to the Lipinski parameters [64]. It was demonstrated that the performance of this algorithm was comparable to simulated annealing and single compound selection when a multiobjective criteria was applied. Although the results of the three methods are comparable, this algorithm has two key advantages. First, it selects optimal combinatorial arrays, alleviating the need for deconvolution of a selection of single compounds. Second, it is extremely fast. The time required for the selection of 25 600 compounds from the 125 000-member, three-component library described above was under 1 s on a standard PC.

In contrast to the stepwise or "greedy" selection algorithms described above, a number of stochastic methods have been applied to the problem of multiobjective library optimization. The most prevalent of these are programs based on either simulated annealing (SA) [103] or genetic algorithms (GA) [104]. GSK developed a SA-based library selection package, PICCOLO (PICking by COmbinatorial Library Optimization), which has been deployed to the synthetic chemists to encourage computational experimentation [105]. While the application of the SA algorithm in the PICCOLO program is not in itself unique, the mechanism by which potential solutions are perturbed is tailored to the selection of monomers for a combinatorial array. PICCOLO biases the random sampling of R-groups by modulating the probability of sampling on the basis of two ratios: the number of reagents available for the R-group over the total number of reagents and the number of reagents selected for an R-group over the total number of selected reagents. This ensures that the more populated R-groups are more frequently sampled while maintaining reasonable sampling of all R-groups. The reagents for each R-group are selected sequentially from randomized lists of reagents until the bottom of the list is reached and then the selection returns to the top. This was found to be a more than random selection of reagents, leading to faster convergence. At each iteration, a reagent is ejected from a given R-group purely at random. Each solution is scored using an objective function that is the weighted sum of several terms: diversity, developability, focusing, and practicality. The diversity term is actually two terms: a measure of the diversity of the reagents at each R-group determined by Tanimoto differences in Daylight fingerprints and product novelty, which is a measure of how much overlap there is between the library and compounds already present in the screening collection. The overlap is determined by generating a low-dimensional cell-based representation of the screening collection and calculating the cell occupancy for the solution. An optimal solution will minimize cell occupancy. Developability is a measure of how well the solution conforms to the Lipinski rules. Focusing is a measure of the similarity to a lead structure either by Daylight fingerprint Tanimoto similarity or Euclidean distance in a principal component space derived from Molconn-Z [106] descriptors. The practicality terms are redundancy in the product molecular weight, which makes identification of individual compounds by mass spectrometry difficult, and cost of reagents. The use of PICCOLO was demonstrated in the design of an 864-member Ugi library that was to be composed of 24 primary amines and 36 aldehydes. The developability and reagent diversity terms were used in the design with the weighting of all other terms set to zero. It was shown that the penalty ascribed to the developability and reagent diversity terms did decrease relative to the starting random library at the cost of product novelty and practicality terms. Since the chemist is given the ability to modify the weightings within the objective function, one can tailor the design to meet specific design goals.

The SELECT program described by Gillet and coworkers [107] also makes use of an aggregate objective function but utilizes a genetic algorithm for optimization. Like PICCOLO, SELECT scores potential solutions with a weighted sum objective function composed of terms that score diversity and physical properties. However, it was found that there are several liabilities with this approach [108]. First, it proved to be difficult to assign appropriate weights to each term in the objective function, requiring significant trial and error to arrive at acceptable parameters. Second, the SELECT program generated only one "optimal" design, while there are certainly several designs that could give rise to an acceptable score. Third, the objectives are often coupled, leading to difficulty in finding an optimal solution. Fourth, it is not always obvious how the fitness function can be composed from disparate objectives (i.e., cost and diversity). Fifth, the weightings in the fitness function determine the regions of space being explored, possibly obscuring other regions.

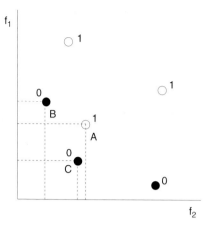

Fig. 7.4 Several solutions to a two objective optimization problem. The dominated solutions are shown as the black circles and the nondominated solutions are the unfilled circles. The dominated solutions are those that are better than every other potential solution in both objectives (e.g., A and B) while the nondominated solutions are those that are bettered in one or both objectives (i.e., C). (Adapted from ref. [103]).

To address these issues, Gillet et al. have devised a multi-objective genetic algorithm (MOGA) in the program MoSELECT [109] that attempts to optimize the overall score of the objective function by simultaneously varying the terms within it. The design of the algorithm also results in multiple solutions, giving a user the ability to select from several high-scoring designs. The MOGA algorithm treats multiple objectives independently and ranks the fitness of a solution by Pareto ranking. The Pareto ranking scheme is based on the principal of dominance (Figure 7.4). A nondominated solution is one in which any one objective cannot be improved without sacrificing one or more of the other objectives, relative to the other solutions in the population. Dominated individuals are those that are deficient in one or more objectives relative to the nondominated individuals. The Pareto ranking method within MoSELECT scores each individual in the population on the basis of their dominance of other individuals. Those individuals that dominate one other individual are given a score of one, those that dominate two are given a score of two, and so on, and the nondominated individuals are given a score of zero. The selection of parents for the subsequent populations is biased toward those solutions with lower ranks. One artifact of this ranking method is the tendency for genetic drift toward a single solution. To ensure the discovery of several unique solutions evenly distributed on the Pareto surface, a niching technique was implemented. Niching is accomplished by defining a radius that is a percentage of the range of values found for each objective found within the nondominated set. The first solution encountered forms the center of the niche, and any individual that is found within the niche radius is penalized by increasing its rank. If a nondominated solution is found outside the niche radius of the others, it forms a new niche center.

The ability of MoSELECT to simultaneously optimize molecular weight and diversity was tested in the selection of 30×30 combinatorial subsets of a 10 000-member virtual amide library. It was shown that as the selection progressed, there was an improvement in both the molecular weight and diversity, as well as a spread of nondominated solutions across the Pareto frontier. The 17 nondominated solutions found after 5000 iterations emphasized the competing nature of these two objectives – those with lower molecular weight tended to be less diverse and those with high molecular weight more diverse. A comparison of SELECT and MoSELECT was also carried out on this library. MoSELECT was run 10 times, as described above, and the nondominated solutions tabulated. Several SELECT runs were carried out: one that optimized molecular weight independently, a second that optimized diversity, 10 runs that weighted diversity and molecular weight equally, and 10 runs that weighted diversity 4 times higher than molecular weight. MoSELECT was able to find nondominated solutions that approached the optimum value of either molecular weight or diversity at the expense of the other objective as well as solutions that balanced the two objectives. The results of the rest of the SELECT runs were distributed within the nondominated solutions found with MoSELECT. This result emphasized the key advantage of the MoSELECT algorithm, that is, the ability to generate several viable solutions without the need for repeat runs and varying weights within the fitness function. In subsequent reports, the MoSELECT procedure has been utilized in the design of focused libraries [109] and the optimization of the size and configuration of combinatorial libraries [110].

By applying some of the lessons learned from the large libraries, target-class and focused library designs can be coupled with property restriction calculations, often based on ADME constraints [64], to limit products with undesirable properties, such as very high MW, cLogP, and so on. By linking these approaches, higher quality diverse compound sets can be designed with the additional feature of bias toward a biological target. Simultaneously, the probability of producing promiscuous and nondruglike compounds can be reduced. Libraries can be somewhat focused and also incorporate relevant biological information. This is a distinct improvement over the production and screening of large, semi-random collections. Diversity remains critical in these endeavors, although that diversity now operates within the boundaries of a series of biological and property restrictions.

7.4
Small Focused Libraries

As a project develops, questions about a lead series or biological target become more specific and one is inevitably drawn into designing focused sets of compounds with greater similarity. Such focused libraries present interesting challenges for diversity-based design. In these projects, there is more information available about the target, or set of leads, and therefore more restrictions placed on the design parameters. The challenge in performing diversity calculations in these cases lies in avoiding redundancy in the face of these increasingly more confining restrictions. These constraints are often structural and may result from available data on biological activity, SAR, target specificity, ADME properties, or some combination of these factors. The underlying diversity of a set of compounds created for these purposes can still be of critical importance, but the relative set to which it is compared has now become severely restricted.

It was quickly realized that parallel synthesis techniques could be used to increase the efficiency of medicinal chemistry in this role [111]. Small optimization libraries were produced using parallel synthesis technologies and they took advantage of the ability to explore multiple variables simultaneously. This multivariate approach provided a more thorough exploration of chemistry diversity space than traditional approaches had allowed, thus increasing the scope of designed libraries and the need for analysis.

In the early days of combinatorial chemistry, Brannigan et al. [11] challenged both traditional medicinal chemistry optimization methods and the full combinatorial library trend with a retrospective study on herbicides that demonstrated that the use of iterative Plackett–Burman designs of acetanilides would have reduced the number of compounds needed to find the optimized herbicidal compound. Additionally, they described a SYBYL SPL program, MOD, that allowed the chemist to choose R-groups at multiple positions on the core based on calculated properties of fragments. The ranges of the selected calculated properties were each divided into three groups (+, 0, −) by adjustable thresholds. The total set of properties created patterns of +'s, 0's, and −'s to describe each fragment, and these patterns were used to group the fragments into sets. A two-level Plackett–Burman design was then used to associate the sets of allowed fragments with each R-position on the core of the molecules to be designed. The chemist then selected fragments for each R-position for each molecule. Synthetic feasibility was incorporated by allowing the chemist to select fragments outside the allowed set either by choosing from the full available list or from a set with similar property ranges (i.e., a − might be replaced by a 0 to access the nearest region of property space). Deviations from the design were recorded for the user.

Other researchers proposed experimental design methods for synthesis, particularly D-optimal designs [1, 8–10] and factorial designs [49, 50]. Wold [8–10] has often pointed out that multivariate designs find optimal experimental conditions and compounds that will be missed by traditional one variable at a time (OVAT) approaches historically used for medicinal chemistry development. Wold developed tools, including MODDE [112], for compound design in multivariate scenarios. It has been demonstrated that this approach can uncover nonadditive SAR effects that would normally be missed by the traditional OVAT [113, 114]. Clearly, these methods provide a huge advantage over OVAT library design approaches for small, diverse libraries. As mentioned in Section 7.2, many new methods have been introduced recently that enable library design, which simultaneously optimize several criteria.

7.4.1
Computational Methods for Small and Focused Libraries

Often, focused small library design draws upon well-known computational approaches, some of which were mentioned in the previous section of this work. These methods include structure-based docking and scoring, CoMFA [14], and Catalyst [115] models, ligand and receptor site–based pharmacophores [67], and various QSAR models. There are many structure-based docking tools, such as DOCK [116], COMBIDOCK [117], LIGANDFIT [118], FLEXX [119], ICM [120], GLIDE [121], and GOLD [122]. Most of these programs have been adapted for screening virtual libraries for combinatorial or parallel synthetic design. While structure-based design has proven to be a powerful method for focusing libraries in those cases where the structure is known, there are weaknesses in the methods. The ability of the scoring functions to predict binding affinity are limited, and no current method adequately handles protein flexibility. Combining docking with CoMFA/QSAR or other methods, including cell-based partitioning methods, may enhance the probability of finding potent compounds, as illustrated in Chapter 16 [123].

Cherry-picking algorithms in cell-based tools such as DiverseSolutions offer the advantage of combining receptor-relevant subspaces [56] with biased property selection. DiverseSolutions allows biasing not only with BCUTs but also with other calculated properties, including those provided by the user [124]. Additionally, the amount of diversity can be controlled by including random selection and/or distance-based selection among the weighted selection criteria. Alternately, compounds may be selected purely on their nearest neighbor distance in chemistry space from a set of known active ligands.

Small, focused libraries for lead discovery and optimization provide a unique design opportunity. In stark contrast to larger, less deliberately designed libraries, there is an abundance of information that may be incorporated in the design. A well thought-out parallel compound set can leverage this knowledge in order to gain a more intimate understanding of the project. The wealth of available information has also led to the development of a wide range of design tools, each with the ability to encompass different types of data. The use of a particular tool is often based upon the data that is available for the project as well as the reliability of that data. Diversity is still important in this environment, but, as informational constraints are added to the design, the relative set from which the diversity is derived begins to shrink.

Small libraries will continue to be used to answer specific biological and medicinal chemistry questions. Their informational output also provides an opportunity for the advancement of library design. By utilizing the information available to pursue iterative library design, these tools will be honed and their effectiveness only improved over time.

7.5
Summary/Conclusion

Although the initial expectations for colossal combinatorial libraries were ultimately not met, important advancements were made in the design of smarter, more efficient libraries and collections. Diversity parameters and ADME constraints were developed to improve the effectiveness of libraries, as well as the quality of the leads obtained from them. As new tools for the analysis and design of compound sets became available, parallel libraries began to focus on more specific chemicals and questions. This work also fostered the development of a large selection of tools for diversity assessment and library design, each with its own particular mixture of benefits, biases, and deficits. Using the project as a guide, these can be weighed to determine the most suitable methodologies for use, but like any other scientific approach, only iterative use and refinement of these tools will allow them to achieve their ultimate potential.

As can be seen from the early combinatorial libraries, chemistry, even in the absence of diversity-oriented design, is capable of producing vast numbers of compounds. The quality and utility of these libraries, however, is not guaranteed [125, 126]. Likewise, design without the consideration of chemistry is nearly futile. The partnering of chemistry and design, however, is powerful, providing the opportunity to produce intelligent libraries with high utility. The data generated by these libraries can then be cycled back into the design loop so that the original ideas and tools can be refined. Ultimately, this partnership provides a sustainable, efficient method by which the science of lead discovery can be advanced.

References

1 MARTIN, E.J., BLANEY, J.M., SIANI, M.A., SPELLMEYER, D.C., WONG, A.K., and MOOS, W.H. Measuring diversity: experimental design of combinatorial libraries for drug discovery. *J. Med. Chem.* **1995**, *38*, 1431–1436.

2 DARVAS, F., DORMÁN, G., and PAPP, A. Diversity measures for enhancing ADME admissibility of combinatorial libraries. *J. Chem. Inf. Comput. Sci.* **2000**, *40*, 314–322.

3 MASON, J.S. and PICKETT, S.D. Partition-based selection. *Perspect. Drug Disc. Des.* **1997**, *7/8*, 85–114.

4 CRAMER, R.D., CLARK, R.D., PATTERSON, D.E., and FERGUSON, A.M. Bioisosterism as a molecular diversity descriptor: steric fields of single "topomeric" conformers. *J. Med. Chem.* **1996**, *39*, 3060–3069.

5 KAUVAR, L.M., HIGGINS, D.L., VILLAR, H.O., SPORTSMAN, J.R., ENGQVIST-GOLDSTEIN, Å., BUKAR, R., BAUER, K.E., DILLEY, H., ROCKE, D.M. Predicting ligand binding to proteins by affinity fingerprinting. *Chem. Biol.* **1995**, *2*, 107–118.

6 LEWIS, R.A., PICKETT, S.D., and CLARK, D.E. Computer-aided molecular diversity analysis and combinatorial library design. *Rev. Comput. Chem.* **2000**, *16*, 1–51.

7 OLSSON, T. and OPREA, T.I. Cheminformatics: a tool for decision-makers in drug discovery. *Curr. Opin. Drug Disc. Dev.* **2001**, *4*, 308–313.

8 ANDERSSON, P.M., SJÖSTRÖM, M., WOLD, S., and LUNDSTEDT, T. Strategies for subset selection of parts of an in-house chemical library. *J. Chemometrics* **2001**, *15*, 353–369.

9 ANDERSSON, P.M., LINUSSON, A., WOLD, S., SJÖSTRÖM, M., LUNDSTEDT, T., and NORDEN, B. Design of small libraries for lead exploration. In *Molecular Diversity in Drug Design*, DEAN, P.M. and LEWIS, R.A. (Eds). Kluwer Academic, Dordrecht, Netherlands, **1999**, 197–220.

10 LINUSSON, A., GOTTFRIES, J., LINDGREN, F., and WOLD, S. Statistical molecular design of building blocks for combinatorial chemistry. *J. Med. Chem.* **2000**, *43*, 1320–1328.

11 BRANNIGAN, L.H., GRIESHABER, M.V., and SCHNUR, D.M. Experimental design in organic synthesis. *ACS Symsp. Series* **1995**, *606*, 264–281.

12 BRANNIGAN, L.H., GRIESHABER, M.V., and SCHNUR, D.M. Use of experimental design in organic synthesis. *CHEMTECH* **1995**, *25*, 29–35.

13 MARRIOTT, D.P., DOUGALL, I.G., MEGHANI, P., LIU, Y.-J., and FLOWER, D.R. Lead generation using pharmacophore mapping and three-dimensional database searching: application to muscarinic M3 receptor antagonists. *J. Med. Chem.* **1999**, *42*, 3210–3216.

14 CRAMER, R.D.I., DEPRIEST, S., PATTERSON, D., and HECHT, P. Comparative molecular field analysis. In *The Developing Practice of Comparative Molecular Field Analysis*, KUBINYI, H. (Ed.). ESCOM, Leiden, Netherlands, **1993**, 443–485.

15 BRUSNIAK, M.-Y., PEARLMAN, R.S., NEVE, K.A., and WILCOX, R.E. Comparative molecular field analysis-based prediction of drug affinity at

recombinant D1A dopamine receptors. *J. Med. Chem.* **1996**, *39*, 850–859.

16 MATTER, H., SCHUDOK, M., SCHWAB, W., THORWART, W., BARBIER, D., BILLEN, G., HAASE, B., NEISES, B., WEITHMANN, K., WOLLMANN, T. Tetrahydroisoquinoline-3-carboxylate based matrix-metalloproteinase inhibitors: design, synthesis and structure-activity relationship. *Bioorg. Med. Chem.* **2002**, *10*(11), 3529–3544.

17 CARELL, T., WINTNER, E.A., SUTHERLAND, A.J., REBEK, J. Jr, DUNAYEVSKIY, Y.M., and VOUROS, P. New promise in combinatorial chemistry: synthesis, characterization, and screening of small-molecule libraries in solution. *Chem. Biol.* **1995**, *2*, 171–183.

18 BLONDELLE, S.E., CROOKS, E., OSTRESH, J.M., and HOUGHTEN, R.A. Mixture-based heterocyclic combinatorial positional scanning libraries: discovery of bicyclic guanidines having potent antifungal activities against candida albicans and cyptococcus neoformans. *Antimicrob. Agents Chemother.* **1999**, *43*, 106–114.

19 BOGER, D.L., JIANG, W., and GOLDBERG, J. Convergent solution-phase combinatorial synthesis with multiplication of diversity through rigid biaryl and diarylacetylene couplings. *J. Org. Chem.* **1999**, *64*, 7094–7100.

20 FERRY, G., BOUTIN, J.A., ATASSI, G., FAUCHERE, J.L., and TUCKER, G.C. Selection of a histidine-containing inhibitor of gelatinases through deconvolution of combinatorial tetrapeptide libraries. *Mol. Div.* **1996**, *2*, 135–146.

21 LUTZKE, R.A.P., EPPENS, N.A., WEBER, P.A., HOUGHTEN, R.A., and PLASTERK, R.H.A. Identification of a hexapeptide inhibitor of the human immunodeficiency virus integrase protein by using a combinatorial chemical library. *Proc. Natl. Acad. Sci. U.S.A.* **1995**, *92*, 11456–11460.

22 SAMSON, I., KERREMANS, L., ROZENSKI, J., SAMYN, B., VAN BEEUMEN, J., and HERDEWIJN, P. Identification of a peptide inhibitor against glycosomal phosphoglycerate kinase of Trypanosoma brucei by a synthetic peptide library

approach. *Bioorg. Med. Chem. Lett.* **1995**, *3*, 257–265.

23 LAHANA, R. How many leads from HTS? *Drug Disc. Today* **1999**, *4*, 447–448.

24 MCGOVERN, S.L., CASELLI, E., GRIGORIEFF, N., and SHOICHET, B.K. A common mechanism underlying promiscuous inhibitors from virtual and high-throughput screening. *J. Med. Chem.* **2002**, *45*, 1712–1722.

25 Daylight software is available from Daylight Chemical Information Systems, Inc., 27401 Los Altos, Suite 360, Mission Viejo, CA 92691.

26 SHEMETULSKIS, N.E., DUNBAR, J.B., Jr., DUNBAR, B.W., MORELAND, D.W., and HUMBLET, C. Enhancing the diversity of a corporate database using chemical database clustering and analysis. *J. Comput.-Aided Mol. Des.* **1995**, *9*, 407–416.

27 LEWELL, X.Q. and SMITH, R. Drug-motif-based diverse monomer selection: method and application in combinatorial chemistry. *J. Mol. Model.* **1997**, *15*, 43–48.

28 MOOCK, T.E., CHRISTIE, B., and HENRY, D. MACCS-3D: a new database system for three-dimensional molecular models. *Chem. Inf. Syst.* **1990**, 42–49.

29 GÜNER, O.F., HUGHES, D.W., and DUMONT, L.M. An integrated approach to three-dimensional information management with MACCS-3D. *J. Chem. Inf. Comput. Sci.* **1991**, *31*, 408–414.

30 POETTER, T. and MATTER, H. Design and evaluation of optimally diverse compound subsets from chemical structure databases. *Book of Abstracts, 213th ACS National Meeting.* San Francisco, **1997**, 1431–1436.

31 BRADLEY, M. Web tools for library design. *Book of Abstracts, 221st ACS National Meeting.* San Diego, **2001**, CINF-016.

32 DAVIES, K. and BRIANT, C. Combinatorial chemistry library design using pharmacophore diversity. *Netw. Sci.* **1995**, *1*. Available at URL: http://www.awod.com/netsci/Issues/July95/feature6.html.

33 PICKETT, S.D., MASON, J.S., and MCLAY, I.M. Diversity profiling and design using 3D pharmacophores:

pharmacophore-derived queries (PDQ). *J. Chem. Inf. Comput. Sci.* **1996**, 1214–1223, COMP–C338996.

34 MAKARA, G.M. Measuring molecular similarity and diversity: total pharmacophore diversity. *J. Med. Chem.* **2001**, *44*, 3563–3571.

35 HOFFMAN, B.T., KOPAJTIC, T., KATZ, J.L., and NEWMAN, A.H. 2D QSAR Modeling and preliminary database searching for dopamine transporter inhibitors using genetic algorithm variable selection of Molconn Z descriptors. *J. Med. Chem.* **2000**, *43*, 4151–4159.

36 REYNOLDS, C.H., TROPSHA, A., PFAHLER, L.B., DRUKER, R., CHAKRAVORTY, S., ETHIRAJ, G., and ZHENG, W. Diversity and coverage of structural sublibraries selected using the SAGE and SCA algorithms. *J. Chem. Inf. Comput. Sci.* **2001**, *41*, 1470–1477.

37 HANSCH, C. and LEO, A. "Calculation of octanol-water partition coefficients by fragments" in Exploring QSAR: fundamentals and applications in chemistry and biology. Am. Chem. Soc. **1995**, 125–168.

38 For example, the descriptor set used in MOE, distributed by Chemical Computing Group, 125 University St., Suite 1600, Montreal, Quebec, Canada H3B 3×3, www.chemcomp.com.

39 PEARLMAN, R.S. Novel software tools for addressing chemical diversity. *Netw. Sci.* **1996**, *2*.

40 STANTON, D.T. Evaluation and use of BCUT descriptors in QSAR and QSPR studies. *J. Chem. Inf. Comput. Sci.* **1999**, *39*, 11–20.

41 NEUSTADT, B.R., SMITH, E.M., LINDO, N., NECHUTA, T., BRONNENKANT, A., WU, A., ARMSTRONG, L., and KUMAR, C. Construction of a family of biphenyl combinatorial libraries: structure-activity studies utilizing libraries of mixtures. *Bioorg. Med. Chem. Lett.* **1998**, *8*, 2395–2398.

42 MORKEN, J.P., KAPOOR, T.M., FENG, S., SHIRAI, F., and SCHREIBER, S.L. Exploring the leucine-proline binding pocket of the Src SH3 domain using structure-based, split-pool synthesis and affinity-based selection. *J. Am. Chem. Soc.* **1998**, *120*, 30–36.

43 ROHRER, S.P., BIRZIN, E.T., MOSLEY, R.T., BERK, S.C., HUTCHINS, S.M., SHEN, D.M., XIONG, Y., HAYES, E.C. Rapid identification of subtype-selective agonists of the somatostatin receptor through combinatorial chemistry. *Science* **1998**, *282*, 737–740.

44 BERK, S.C., ROHRER, S.P., DEGRADO, S.J., BIRZIN, E.T., MOSLEY, R.T., HUTCHINS, S.M., PASTERNAK, A., SCHAEFFER, J.M., and UNDERWOOD, D.J. A combinatorial approach toward the discovery of non-peptide, subtype-selective somatostatin receptor ligands. *J. Comb. Chem.* **1999**, *1*, 388–396.

45 CAVALLARO, C.L. and SCHNUR, D.M. Unpublished observations.

46 NEFZI, A., DOOLEY, C., OSTRESH, J.M., and HOUGHTEN, R.A. Combinatorial chemistry: from peptides and peptidomimetics to small organic and heterocyclic compounds. *Bioorg. Med. Chem. Lett.* **1998**, *8*, 2395–2398.

47 APPEL, J.R., JOHNSON, J., NARAYANAN, V.L., and HOUGHTEN, R.A. Identification of novel antitumor agents from mixture-based synthetic combinatorial libraries using cell-based assays. *Mol. Div.* **1999**, *4*, 91–102.

48 NEFZI, A., OTRESH, J.M., GIULIANOTTI, M.A., and HOUGHTEN, R.A. Solid-phase synthesis of trisubstituted 2-imidazolidones and 2-imidazolidinethiones. *J. Comb. Chem.* **1999**, *1*, 195–198.

49 SCHNUR, D.M. Design and diversity analysis of large combinatorial libraries using cell-based methods. *J. Chem. Inf. Comput. Sci.* **1999**, *39*, 36–45.

50 SCHNUR, D. and VENKATARANGAN, P. Applications of cell-based diversity methods to combinatorial library design. In *Combinatorial Library Design and Evaluation*, GHOSE, A.K. and VISWADHAN, V.N. (Eds). Marcel Dekker, Inc, New York, **2001**, 473–501.

51 MARTIN, E.J. and CRITCHLOW, R.E. Beyond mere diversity: tailoring combinatorial libraries for drug discovery. *J. Comb. Chem.* **1999**, *1*, 32–45.

52 HASSAN, M., BIELAWSKI, J.P., HEMPEL, J.C., and WALDMAN, M. Optimization and visualization of molecular diversity of combinatorial libraries. *Mol. Div.* **1996**, *2*, 64–74.

53 Now Accelrys, Inc., A wholly owned subsidiary of Pharmacopeia, 9685 Scranton Road, San Diego, CA, 92121–93752.

54 Cerius² is available from Accelrys, Inc., 9685 Scranton Road, San Diego, CA, 92121–93752.

55 PEARLMAN, R.S. *Laboratory for Molecular Graphics and Theoretical Modeling*, College of Pharmacy, University of Texas at Austin, Austin TX 78712, distributed by Tripos Associates, 1699 South Hanley Rd., St Louis, MO 63144.

56 PEARLMAN, R.S. and SMITH, K.M. Metric validation and the receptor-relevant subspace concept. *J. Chem. Inf. Comput. Sci.* **1999**, *39*, 28–35.

57 Comb. Libmaker is available in SYBYL from Tripos Associates, 1699 South Hanley Rd., St Louis, MO 63144.

58 Optive Research, Inc., 7000 North Mopac Expressway, Austin, TX 78731, http://www.optiveresearch.com

59 WEININGER, D.J. SMILES: a chemical language and information system. 1. Introduction to methodology and encoding rules. *J. Chem. Inf. Comput. Sci.* **1988**, *28*, 31–36.

60 PEARLMAN, R.S. Rapid generation of high quality approximate 3D structures. *Automation News* **1987**, *2*, 1.

61 PEARLMAN, R.S. and SMITH, K.M. Software for chemical diversity in the context of accelerated drug discovery. *Drugs Future* **1998**, *23*, 885–895.

62 PEARLMAN, R.S. and SMITH, K.M. Novel software tools for chemical diversity. *Perspect. Drug Disc. Des.* **1997**, *9/10/11*, 339–353.

63 WAGENER, M. and VAN GEERESTEIN, V.J. Potential drugs and nondrugs: prediction and identification of important structural features. *J. Chem. Inf. Comput. Sci.* **2000**, *40*, 280–292.

64 LIPINSKI, C.A., LOMBARDO, F., DOMINY, B.W., and FEENEY, P.J. Experimental and computational approaches to estimate solubility and permeability in drug discovery and development settings. *Adv. Drug Deliv. Rev.* **1997**, *23*, 3–25.

65 NAVIA, M.A. and CHATURVEDI, P.R. Design principles for orally bioavailable drugs. *Drug Disc. Tech.* **1996**, *1*, 179–189.

66 VEBER, D.F., JOHNSON, S.R., CHENG, H.-Y., SMITH, B.R., WARD, K.W., and KOPPLE, K.D. Molecular properties that influence the oral bioavailability of drug candidates. *J. Med. Chem.* **2002**, *45*, 2615–2623.

67 MASON, J.S., MORIZE, I., MENARD, P.R., CHENEY, D.L., HULME, C., and LABAUDINIERE, R.F. New 4-point pharmacophore method for molecular similarity and diversity applications: overview of the method and applications, including a novel approach to the design of combinatorial libraries containing privileged substructures. *J. Med. Chem.* **1999**, *42*, 3251–3264.

68 MÜLLER, G. Medicinal chemistry of target family-directed masterkeys. *Drug Disc. Today* **2003**, *8*, 681–691.

69 GUO, T. and HOBBS, D.W. Privileged structure-based combinatorial libraries targeting G protein-coupled receptors. *Assay Drug Dev. Tech.* **2003**, *1*, 579–592.

70 MASON, J.S. and BENO, B.R. Library design using BCUT chemistry-space descriptors and multiple four-point pharmacophore fingerprints: simultaneous optimization and structure-based diversity. *J. Mol. Model.* **2000**, *18*, 438–451.

71 BENO, B.R. and MASON, J.S. The design of combinatorial libraries using properties and 3D pharmacophore fingerprints. *Drug Disc. Today* **2001**, *6*, 251–258.

72 MDL Drug Data Report, MDL Information Systems, San Leandro, CA, USA.

73 SCHNUR, D.S., BENO, B.R., GOOD, A., and TEBBEN, A. Approaches to target class library design. In *Methods in Molecular Biology*, TOTOWA, N.J. (Ed.). Humana Press, Inc., **2004**, *275*, 355–377.

74 TEIG, S.L. Informative libraries are more useful than diverse ones. *J. Biomol. Screen.* **1998**, *3*, 85–88.

75 SHANNON, C.E. A mathematical theory of communication. *Bell Syst. Tech. J.* **1948**, *27*, 379–423, 623–656.

76 BARNUM, D., GREENE, J., and TEIG, S.L. Designing maximally informative

libraries. *ACD National Meeting.* Dallas, TX, **1998**.

77 MILLER, J.L., BRADLEY, E.K., and TEIG, S.L. Luddite: an information theoretic library design tool. *J. Chem. Inf. Comput. Sci.* **2003**, *43*, 47–54.

78 This iterative approach is also used in GA-based selection as illustrated in: WEBER, L., WALLBAUM, S., BROGER, C., and GUBERNATOR, K. Optimization of the biological activity of combinatorial compound libraries by a genetic algorithm. *Angew. Chem., Int. Ed. Engl.* **1995**, *34*, 2280–2282.

79 BRADLEY, E.K., MILLER, J.L., SAIAH, E., and GROOTENHUIS, P.D.J. Informative library design as an efficient strategy to identify and optimize leads: application to cyclin dependent kinase 2 antagonists. *J. Med. Chem.* **2003**, *46*, 4360–4364.

80 DOLLE, R.E. Comprehensive survey of combinatorial library synthesis: 2001. *J. Comb. Chem.* **2002**, *4*, 369–418.

81 TEAGUE, S.J., DAVIS, A.M., LEESON, P.D., and OPREA, T.I. The design of leadlike combinatorial libraries. *Angew. Chem., Int. Ed. Engl.* **1999**, *38*, 3743–3748.

82 OPREA, T.I. Chemoinformatics in lead discovery. In *Chemoinformatics in Drug Discovery*, OPREA, T.I. (Ed.). Wiley-VCH, Weinheim, **2004**, 1–22.

83 SCHUFFENHAUER, A., ZIMMERMANN, J., STOOP, R., VAN DER VYVER, J.-J., LECCHINI, S., and JACOBY, E. An ontology for pharmaceutical ligands and its application for in silico screening and library design. *J. Chem. Inf. Comput. Sci.* **2002**, *42*, 947–955.

84 Aureus Pharmaceuticals, 174, Quai de Jemmapes, 75010 Paris, France, http://www.aureus-pharma.com.

85 Sertanty Inc., 1735 N. First St. #102, San Jose CA, 95112.

86 Biowisdom Ltd., Babraham Hall, Babraham, Cambridge, CB2 4AT, United Kingdom.

87 Jubilant Biosys, 8575 Window Latch Way, Columbia, MD 21045.

88 SAVCHUK, N.P., TKACHENKO, S.E., and BALAKIN, K.V. Rational design of GPCR-specific combinatorial libraries based on the concept of privileged substructures. In *Chemoinformatics in*

Drug Discovery, OPREA, T.I. (Ed.). Wiley-VCH, Weinheim, **2004**, 287–313.

89 WANG, X.-C. and SAUNDERS, J. GPCR library design. *Book of Abstracts, 222nd ACS National Meeting*, Chicago, **2001**, MEDI-012.

90 STEWART, E.L., BROWN, P.J., BENTLEY, J.A., and WILSON, T.M. Selection, application, and validation of a set of molecular descriptors for nuclear receptor ligands. *Book of Abstracts, 222nd ACS National Meeting*, Chicago, **2001**, COMP-182.

91 BRADLEY, E.K., BEROZA, P., PENZOTTI, J.E., GROOTENHUIS, P.D., SPELLMEYER, D.C., and MILLER, J.L. A rapid computational method for lead evolution: description and application to α1-adrenergic antagonists. *J. Med. Chem.* **2000**, *43*, 2770–2774.

92 Diverse Solutions was developed by R.S. PEARLMAN and K.M. SMITH at the University of Texas at Austin and is distributed by Optive Research, Inc. http://www.optive.com, and Tripos Inc., http://www.tripos.com.

93 LibraryMaker and LibraryDesigner are distributed by Optive Research, Austin, Texas, http://www.optiveresearch.com.

94 XUE, L., GODDEN, J.W., STAHURA, F.L., and BAJORATH, J. A dual fingerprint-based metric for the design of focused compound libraries and analogs. *J. Mol. Model.* **2001**, *7*, 125–131.

95 XUE, L., GODDEN, J., and BAJORATH, J. Database searching for compounds with similar biological activity using short binary bit string representations of molecules. *J. Chem. Inf. Comput. Sci.* **1999**, *39*(5), 881–886.

96 MACCS structural keys, MDL Information Systems Inc., 14600 Catalina Street, San Leandro CA, 94577.

97 BRAVI, G., GREEN, D.V.S., HANN, M.M., and LEACH, A. PLUMS: a program for the rapid optimization of focused libraries. *J. Chem. Inf. Comput. Sci.* **2000**, *40*, 1441–1448.

98 BROWN, R.D. and MARTIN, Y. Designing combinatorial library mixtures using a genetic algorithm. *J. Med. Chem.* **1997**, *40*, 2304–2313.

99 ZHENG, W., CHO, S.J., and TROPSHA, A. Rational combinatorial library design.

1. Focus-2D: a new approach to the design of targeted combinatorial chemical libraries. *J. Chem. Inf. Comput. Sci.* 1998, 38, 251–258.

100 AGRAFIOTIS, D.K. and LOBANOV, V.S. Ultrafast algorithm for designing focused combinational arrays. *J. Chem. Inf. Comput. Sci.* 2000, 40, 1030–1038.

101 AGRAFIOTIS, D.K. and LOBANOV, V.S. An efficient implementation of distance-based diversity measures based on k-d trees. *J. Chem. Inf. Comput. Sci.* 1999, 39, 51–58.

102 CRAMER, R.D., PATTERSON, D.E., CLARK, R.D., SOLTANSHAHI, F., and LAWLESS, M.S. Virtual compound libraries: a new approach to decision making in molecular discovery research. *J. Chem. Inf. Comput. Sci.* 1998, 38, 1010–1023.

103 KIRKPATRICK, S., GELATT, C.D. Jr., and VECCHI, M.P. Optimization by simulated annealing. *Science* 1983, 220, 671–680.

104 FORREST, S. Genetic algorithms: principles of natural selection applied to computation. *Science* 1993, 261, 871–878.

105 ZHENG, W., HUNG, S.T., SAUNDERS, J.T., and SEIBEL, G.L. PICCOLO: a tool for combinatorial library design via multicriterion optimization. *Proc. Pacific Symp. Biocomput.* 2000, 5, 588–599.

106 MolConn-Z, Version 3.10; eduSoft, P.O. Box 1811, Ashland, VA 23005.

107 GILLET, V.J., WILLETT, P., BRADSHAW, J., and GREEN, D.V.S. Selecting combinatorial libraries to optimize diversity and physical properties. *J. Chem. Inf. Comput. Sci.* 1999, 39, 169–177.

108 GILLET, V.J., KHATIB, W., WILLETT, P., FLEMING, P.J., and GREEN, D.V.S. Combinatorial library design using a multiobjective genetic algorithm. *J. Chem. Inf. Comput. Sci.* 2002, 42, 375–385.

109 GILLET, V.J., WILLETT, P., FLEMING, P.J., and GREEN, D.V.S. Designing focused libraries using MoSELECT. *J. Mol. Graph. Model.* 2002, 20, 491–498.

110 WRIGHT, T., GILLETT, V.J., GREEN, D.V.S., and PICKETT, S.D. Optimizing the size and configuration of combinatorial libraries. *J. Chem. Inf. Comput. Sci.* 2003, 43, 381–390.

111 DOLLE, R.E. Comprehensive survey of chemical libraries yielding enzyme inhibitors, receptor agonists and antagonists, and other biologically active agents: 1992 through 1997. *Mol. Div.* 1998, 3, 199–233.

112 MODDE is available from Umetrics, Inc., Kinnelon, NJ.

113 LINUSSON, A., GOTTFRIES, J., OLSSON, T., OERNSKOV, E., FOLESTAD, S., NORDEN, B., and WOLD, S. Statistical molecular design, parallel synthesis, and biological evaluation of a library of thrombin inhibitors. *J. Med. Chem.* 2001, 44, 3424–3439.

114 LEE, A. and BREITENBUCHER, J.G. The impact of combinatorial chemistry on drug discovery. *Curr. Opin. Drug Disc. Dev.* 2003, 6, 494–508.

115 Catalyst is available from Accelrys, Inc., 9685 Scranton Road, San Diego, CA, 92121–93752.

116 DOCK, developed and distributed by the Kuntz group, Dept. of Pharmaceutical Chemistry, 512 Parnassus, University of California, San Francisco, USA. URL:http://www.cmpharm.ucsf.edu/kuntz.

117 SUN, Y., SKILLMAN, T.J., and KUNTZ, I.D. CombiDOCK: structure-based combinatorial docking and library design. *J. Comput.-Aided. Mol. Des.* 1998, 12, 597–604.

118 LIGANDFIT is available in Cerius^2 from Accelrys, Inc., 9685 Scranton Road, San Diego, CA 92121–93752.

119 FlexX is available in SYBYL from Tripos Associates, 1699 South Hanley Rd., St Louis, MO 63144.

120 ICM is available from Molsoft LLC, 3366 North Torrey Pines Court, Suite 300, La Jolla, CA 92037.

121 GLIDE is available from Schroedinger, 120 West Forty-Fifth Street, 32nd Floor, Tower 45, New York, NY 10038.

122 GOLD Version 1.2, developed and distributed by the CCDC, 12 Union Road, Cambridge, CB2 1EZ, UK. URL: http://www.ccdc.cam.ac.uk/prods/gold/index.html.

123 TROPSHA, A. Application of predictive QSAR models to database mining. In *Chemoinformatics in Drug*

Discovery, OPREA, T.I. (Ed.).
Wiley-VCH, Weinheim, **2004**,
437–455.

124 PEARLMAN, R.S. *Diverse Solutions
Manual*, Laboratory for Molecular
Graphics and Theoretical Modeling,
College of Pharmacy, University of
Texas at Austin, Austin TX 78712.

125 OPREA, T.I. and MATTER, H. Integrating
virtual screening in lead discovery.
Curr. Opin. Chem. Biol. **2004**, 8,
349–358.

126 HANN, M.M. and OPREA, T.I., Pursuing
the leadlikeness concept in pharmaceutical research. *Curr. Opin. in
Chemical Biology*, **2004**, 8, 255–263.

8

In Silico Lead Optimization

Chris M.W. Ho

8.1
Introduction

In the early 1990s, many of the world's largest pharmaceutical firms spent millions of dollars on hardware and software in their endeavor to make computer-aided *de novo* design a reality. Unfortunately, during that time period, successes were rare. With few exceptions, the *de novo* design was a failure and did not prove to be an effective method to discover lead compounds. The main reasons were limitations in computing power and the lack of useful software functionality. In scientific computing, accuracy and processing time are always a trade-off. Thus, in order to make the calculations run in a finite period of time, a plethora of assumptions, significant approximations, and numerous algorithmic shortcuts had to be utilized. This, in turn, greatly diminished the calculated accuracy of any ligand–receptor interaction. As such, chemists could postulate numerous chemical structures that could potentially complement the active site; however, the calculated binding had little correlation with reality.

This remains the most significant challenge in *de novo* design to this day. Although computers have become exponentially faster, the sheer number of calculations needed to accurately predict the binding of a *de novo* generated ligand to its receptor in a useful time frame still requires significant approximations. In *de novo* design, we are attempting to generate a whole ligand from scratch and dock it within the receptor. As stated above, the difficulty lies in predicting how the chemical structure will behave in real life. A ligand is an inherently flexible structure and can assume a plethora of different conformations and orientations. The key question remains whether the predicted binding structure will mirror the calculated one. Failure in this endeavor has undermined the utility of *de novo* structure generating software.

The second most significant problem in computer-aided *de novo* design is the generation of undesired chemical structures. There are nearly an infinite number of potential combinations of atoms. However, the vast majority of these structures are of no use. As discussed above, undesired structures are rejected because of toxicity, chemical instability, or synthetic difficulty. Nearly all *de novo* design software packages are plagued by this problem, especially with respect to synthetic feasibility. Thus, although such software can postulate potential complementary ligands, a vast majority of them are of questionable value.

Chemoinformatics in Drug Discovery. Edited by Tudor I. Oprea
Copyright © 2004 WILEY-VCH Verlag GmbH & Co. KGaA, Weinheim
ISBN: 3-527-30753-2

The end result of these shortcomings was that computer-aided *de novo* design soon fell out of favor as a means of generating viable lead compounds. By the mid-1990s, there had been a tremendous number of *de novo* software packages released; however, they all suffered the same problems. Gradually, such programs were shelved and investigators looked to other technologies to aid in their drug discovery efforts.

8.2
The Rise of Computer-aided Drug Refinement

At this time, the techniques of mass screening and combinatorial chemistry began to gain widespread acceptance and use. The use of mass screening and combinatorial chemistry allowed researchers to discover lead compounds in a rapid and efficient manner. As such, *de novo* design tools and their associated problems were no longer necessary to generate lead structures. One would surmise that computer-aided drug design technology would have soon ceased to exist. On the contrary, it soon became apparent that computational tools that could *optimize* these lead compounds into more potent molecules were needed.

The concept of lead optimization versus *de novo* design is an important one. The difficulty with *de novo* ligand generation is that an entire structure is being created from scratch. The confidence one has of accurately predicting how this structure will interact and bind within a target receptor is shaky at best. In lead optimization, we begin with a compound whose bound structure within the receptor has been characterized, most likely through X-ray crystallography. Subtle modifications are then performed to generate derivative compounds using structure-based drug design to improve binding affinity. Because we are making much smaller changes, our faith in the validity of the resulting structures is far greater. These derivatives then undergo testing to determine which modifications improve binding. The structures of the best ligands can then be elucidated to verify the accuracy of the modifications. This refinement process continues iteratively until optimal binding ligands are produced.

Since subtle modifications are being made to a common structure, the predictive ability of ligand refinement software is much higher. This is because the effect of a single chemical modification on ligand–receptor binding is far easier to quantitate than that of an extreme change. No longer are we trying to determine the binding affinities of drastically different structures. Instead, we are simply determining the rank order of a list of derivative compounds. This greatly increases the confidence that proposed structures will bind in a manner consistent with our understanding.

In addition, the act of generating chemical derivatives is highly amenable to computerized automation. Consider the application of targeted structure-based combinatorial chemistry as discussed above. Libraries of derivative components are assembled on the basis of the analysis of the active site. Because of the combinatorial nature of this method, an extremely large number of candidate structures may be possible. A computer can rapidly generate and predict the binding of all potential derivatives, creating a list of the best potential candidates. In essence, the computer filters all weak-binding compounds, allowing the chemist to focus, synthesize, and test only the most promising ligands. Thus, utilizing computer-aided drug design software to aid in the refinement of weak-binding lead compounds is the most effective manner in which these tools can be employed. The use of computer modeling to refine structures has become standard practice in modern drug design.

8.3
RACHEL Software Package

RACHEL™ (Real-time Automated Combinatorial Heuristic Enhancement of Lead compounds) is a lead optimization package [1] designed to optimize weak-binding lead compounds in an automated, combinatorial fashion. RACHEL incorporates numerous technological advances that enhance its ability to refine ligands. These technologies compensate for the difficulties inherent to computer-aided drug discovery, and they are discussed below in detail.

8.4
Extraction of Building Blocks from Corporate Databases

RACHEL is a "builder-class" type of program [2] as shown in Figure 8.1. In short, a database of chemical fragments is used to derivatize a lead compound by replacing weak-binding regions (highlighted in gray) with components that will improve receptor complementarity. These compounds are then scored by calculating their affinity for the receptor. Compounds that bind tightly with the receptor are then saved while those that bind poorly are discarded. This new population of compounds is then processed to form the next generation of derivatives. Over time, a lead compound is iteratively refined into a set of tight-binding structures.

When using current commercial ligand refinement packages of this type, the user is dependent on the software for several vital functions that are critical for successful drug discovery. First and foremost, a database of chemical building components is required. All current commercial packages provide this database of components, which allows any researcher to immediately use these tools for drug design. However, this is not always desirable. Pharmaceutical firms are always competing against rival companies. Intellectual property in the form of patented database structures, synthetic know-how, and the biochemical data of characterized lead compounds provides a

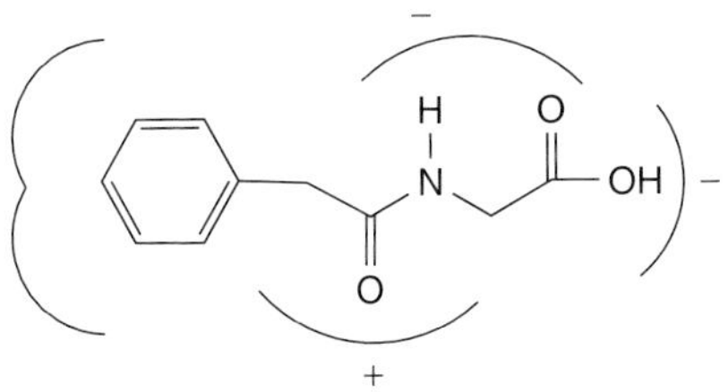

Fig. 8.1   Builder-class ligand optimization program.

competitive edge against other companies. As such, using a program that contains both a standard database and a standard scoring function offers no such advantage over another company.

RACHEL avoids this shortcoming by allowing the user to utilize in-house intellectual property in the design of new ligands. RACHEL pulls building block components directly from the user's structural database. As such, investigators who have invested considerable time and money developing the chemistry to synthesize a particular class of compounds can utilize this knowledge in the design of future drugs. This has the added benefit that generated structures are easier to patent as proprietary chemistry is incorporated.

Figure 8.2 schematically demonstrates this process. On the left, we have the corporate structural database, which may contain hundreds of thousands of compounds. All structures are composed of nonrotatable chemical groups separated by rotatable bonds as defined by the laws of chemistry. These nonrotatable groups represent the fundamental building blocks that will be used to regenerate new derivative compounds. RACHEL first isolates these components by identifying the rotatable bonds in the structure as shown below in Figure 8.3 (arrows). The individual components are then isolated, identified with a unique label that describes their distinct chemical architecture, and stored in

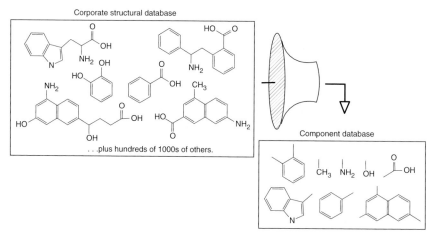

Fig. 8.2 Extraction of components from corporate structural database.

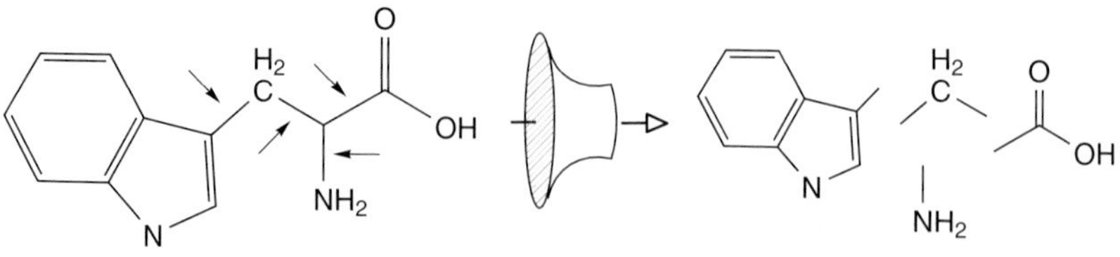

Fig. 8.3 Separation and isolation of components at rotatable bonds.

the component database along with a description of their chemical composition. The component label is very important as it is used to register each fragment and prevent the storage of redundant chemical groups.

There are numerous advantages in extracting components in the manner described above. First, the storage of unique components allows the compression of a massive corporate database into a much smaller and manageable form. Typically, a corporate database containing 1 000 000 structures may be composed of only 50 000 individual components. This is because a few select components, such as methyl, hydroxyl, and amine groups, are utilized over and over again. Second, unique chemical constructions, for which only proprietary synthetic methods are known, are stored and available for use in future ligand design. This allows the user to take advantage of patented corporate chemistry and preserve the competitive edge gained from prior research.

8.5
Intelligent Component Selection System

The goal of builder-type programs is to generate derivatives that are complementary to the active site. Both steric (size and shape) as well as electrostatic forces must be considered. The difficulty in accomplishing this lies in the sheer number of potential component combinations that are possible. As a result, nearly all commercial packages randomly select fragments for assembly. While this ensures an adequate sampling of components, it often leads to the selection of improper fragments. As such, many iterations of structure refinement are wasted generating poor derivatives.

The RACHEL software has a far greater problem. While current builder-type software packages contain databases with 1000 components or less, RACHEL can extract upward of 500 000 components depending upon the size and diversity of the corporate database. Thus, the number of potential fragment combinations is nearly immeasurable. Clearly, a method is needed to rapidly focus on the appropriate combinations that are likely to satisfy binding requirements.

The greatest benefit of RACHEL's component extraction method is that a massive *property index* of the entire corporate database is created. Along with the atomic coordinates of each component, a wealth of chemical information characterizing each building block is stored. Data such as the size of the component, atom composition, connectivity, ring structure, and electrostatic charges are included. As such, a means of rapidly cross-referencing chemical components on demand is available.

Figure 8.4 demonstrates how this property index is generated. On the left, we see a representative component database. Using the stored chemical attributes, the database is sorted and mapped into a multidimensional array, where each axis represents a different descriptor. In this example, we only show size, polarity, and valence (number of connections) for simplicity. Each axis provides a gradient along which the components can be distinguished. Thus, components that are similar with respect to the various descriptors are grouped together, similar to statistical molecular design [3].

This property index offers a powerful means to improve the generation of complementary ligands. Over time, builder-type programs evolve compounds with

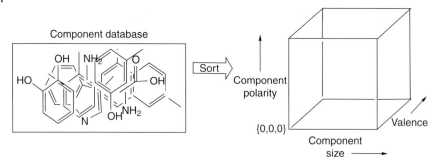

Fig. 8.4 Generation of property index.

improved binding. A moderate affinity structure has reasonable steric and electrostatic complementarity with the active site. However, components can still be added, deleted, or substituted to augment receptor interaction. As stated above, nearly all commercial builder-type programs select substituent fragments at random. Although simplistic, this is absolutely necessary to ensure adequate sampling of the database and generate truly novel solutions. RACHEL implements random sampling in the initial stages of lead compound optimization. Early derivatives that are generated are weak binders. Thus, random component sampling increases the chances of finding the appropriate components to improve receptor interaction.

However, random sampling often diminishes the complementarity of reasonable binding compounds. This is the result of replacing satisfactory components with poor ones. For example, if a small methyl group or a highly charged fragment were to replace a large hydrophobic ring on the ligand, it would ruin interaction with the receptor at that component. Instead, RACHEL incorporates a *heuristic active site mapping algorithm* as shown in Figure 8.5 to determine the optimal chemical characteristics to complement a given region of the active site. This technique records the most complementary substituted components and their loci in order to map the chemical characteristics of the receptor, such as positive charge, negative charge, and steric bulk as a function of distance along the active site axis. Using this active site map, RACHEL can determine the chemical characteristics most likely to complement the receptor at a given component location. RACHEL then determines a list of candidate fragments and substitutes them in a combinatorial fashion.

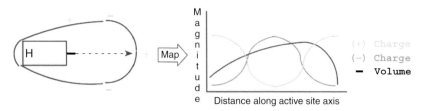

Fig. 8.5 RACHEL active site mapping.

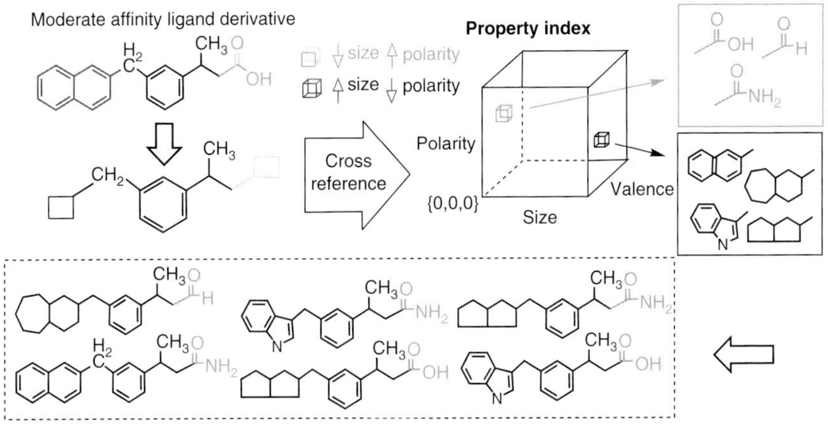

Fig. 8.6 RACHEL intelligent component selection system.

This is diagrammed in Figure 8.6. In this example, RACHEL determines that the naphthalene group (dark gray) and carboxylic acid group (light gray) of a ligand derivative should be replaced with other components to improve binding. The naphthalene group is large and highly nonpolar since it is strictly hydrocarbon. Conversely, the carboxylic acid group is quite small, but highly polar. Using the active site map as described above, RACHEL determines that these characteristics are indeed ideal for complementing the receptor at each respective component. Using the property index, RACHEL can cross-reference other database components that exhibit similar characteristics, as shown in the light and dark boxes on the right. These components are then combinatorially used to generate a new family of derivatives for testing. Each derivative retains the optimal receptor binding characteristics. However, enough variability is generated to potentially improve receptor complementarity.

8.6
Development of a Component Specification Language

The ability to instantly cross-reference components by chemical composition also permits user-directed structure generation, which is the most powerful and unique feature of RACHEL. To our knowledge, no other builder-type ligand design package makes use of this feature so extensively. Essentially, this technology permits the true application of virtual combinatorial chemistry. The inspiration for this technology stems from our previous work [4]. Figure 8.7 demonstrates this with an example. In the middle of this Figure, we see a lead compound scaffold containing an amide bond with various side chains extending from it.

From biochemical characterization of this lead compound, we discover that three chemical groups make up the pharmacophore. The first group, shown in black, must contain a large ring system. Crystallographic analysis reveals that both single and

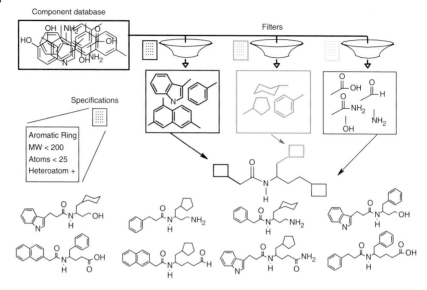

Fig. 8.7 RACHEL user-directed structure generation.

fused bicyclic rings are capable of binding as long as they are planar. Thus, they must be aromatic. Any atom type may be accepted. The second group, shown in gray, has different requirements. Again, a cyclic component is desired. However, the binding pocket in this region is smaller, but more spherical. Thus, only single rings are acceptable, although they need not be aromatic. In addition, this region is very hydrophobic; thus, only hydrocarbon components are acceptable (no polar atoms). The third group, shown in light gray, is quite different from the first two. This region of the active site is highly charged, and requires a small polar group to interact with. Thus, no ring structures are acceptable. Furthermore, heteroatoms (nitrogen, oxygen) are required.

The various chemical requirements of each derivative group are summarized in Table 8.1. In order to implement these requirements, a *component specification language* has been developed. This specification language contains a combination of keywords, target values, and Boolean operators. A brief summary of these commands is listed in Table 8.2 below. The specification language allows the user to control

Tab. 8.1 Chemical requirements for each derivative group in Figure 8.7

Black derivatives	*Dark gray derivatives*	*Light gray derivatives*
(+) Ring (−) aromatic	(+) Ring structure – single	(−) Ring structures
Molecular weight < 200	Molecular weight < 100	Molecular weight < 50
# Atoms < 25	# Atoms < 20	# Atoms < 8
(+) Any atom type	(+) C, X, H = only	(+) N, O = required

Tab. 8.2 RACHEL component specification language

Command	Function
CMPNTS min – max	Number of total components to utilize
ATOMS min – max	Number of atoms in a specific component
R-ATOMS min – max	Number of ring atoms in a specific component
MW min – max	Molecular weight
LINKS atypes ($<$, $>$, $=$) value	Specifies rotatable bond atom types between components
ATYPES (list) ($<$, $>$, $=$) value	Specifies atom type requirements in a specific component
BONDS (list) ($<$, $>$, $=$) value	Specifies bonded atom types within a specific component
PHARM (atype) {x,y,z}	Specifies pharmacophoric group at coordinates {xyz}

these characteristics and many more. Once the chemical requirements are established for each derivative group, RACHEL then filters the master component database and generates individual databases for each subsite.

Using these individual databases, shown in Figure 8.7 as the black, gray, and **light gray** boxes, RACHEL combinatorially generates potential derivatives within the constraints of the active site. In doing so, an immense number of diverse chemical structures may be constructed and tested in a defined and controlled manner.

8.7
Filtration of Components Using Constraints

As illustrated above, the user has considerable control over the chemical species that can be incorporated into structures. Equally important is the fact that RACHEL's component specification language allows the removal of undesired structures. As stated above, this is one of the shortcomings in computer-aided ligand design, especially with builder-type programs. Builders can often generate combinations of components that are neither stable nor synthetically feasible. The terms listed in Table 8.2 may be used in a variety of ways to limit the generation of these unacceptable structures. This is illustrated in Figure 8.8. We see here the various constraints RACHEL can employ and how they relate to a given structure. In this Figure, a hypothetical chemical structure containing seven separate components joined by rotatable bonds is shown. Each component could be potentially replaced using RACHEL to form a derivative compound.

The RATOM descriptor is useful for limiting the generation of linear and branched chains. This can favor the placement of compact, cyclic structures to either fill a subpocket in the active site or bridge a known linear chain to stabilize a given conformer.

Both the LINK and the BOND constraints are ideal for specifying or excluding the chemistry at a particular bond. For example, one could easily exclude the formation of peptide bonds between specific components to generate peptidomimetics. However, the most prevalent use is to eliminate bonded heteroatoms or other unstable chemical species.

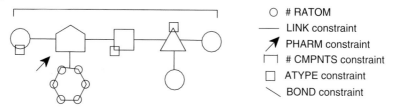

○	# RATOM
——	LINK constraint
↗	PHARM constraint
⌐	# CMPNTS constraint
□	ATYPE constraint
⟍	BOND constraint

Fig. 8.8 RACHEL component constraints.

The PHARM specification signifies that a specific atom type must be present at a precise {xyz|} location in the active site. This is useful when designing ligands that must complement a specific pharmacophoric receptor group.

Other descriptors are useful for implementing ADMET constraints upon the ligand derivatives. For example, using the #CMPNTS descriptor, one can limit the number of rotatable bonds the derivative will possess. The ATYPE descriptor can be used to limit the number of hydrogen bond donors and acceptors that will be included in the derivative. Furthermore, the MW descriptor can be used to insure that the generated compounds fall within acceptable size limits.

8.8
Template-driven Structure Generation

An additional unique feature is an automated method to generate diversity using templates. These templates are an integral part of the component specification language. As is normally the case, a computational chemist is not sure exactly what derivative components might complement the receptor. However, specific chemical groups may be desired at general locations in the active site. These might be pharmacophoric elements or proprietary chemical structures for which synthetic methods are available. This is illustrated in Figure 8.9.

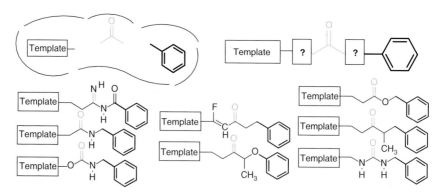

Fig. 8.9 Template-driven structure generation.

In the upper left, we show an active site that contains a portion of a lead compound that has been previously characterized. We also know that two chemical groups, a carbonyl group **(gray)** and a phenyl ring (black), are required to satisfy the pharmacophore for ligand binding. Given these issues, RACHEL allows the user to define a *chemical template*, as shown in the upper right of this Figure, to generate appropriate structures. In this template, the lead compound fragment and the two pharmacophoric groups are separated by *wildcard designations* that denote where chemical variability can occur.

RACHEL will then generate chemically diverse structures using the template as shown in the Figure above. The static portions of the template are left untouched, and they are incorporated into every generated derivative. However, the *wildcard* regions allow RACHEL to creatively insert various components in a random manner to link these pharmacophoric elements together. Obviously, constraints can be placed on these variable regions using the component specifications described above. As such, the use of these templates enables RACHEL to fully explore the chemical diversity within the corporate database and maintain the fundamental groups necessary to achieve receptor binding. The use of templates and chemical descriptors is highlighted later in this chapter with an example.

8.9
Scoring Functions – Methods to Estimate Ligand–Receptor Binding

The calculation of receptor binding affinity for each newly generated derivative ligand remains the most challenging aspect of drug design. Not only is this task very difficult, it also is critical for the success of the program. If the calculated binding affinities have no bearing on reality, then the program might as well generate random structures. Accurate determination of ligand–receptor binding typically involves complex, CPU-intensive quantum chemical calculations, which can take days to weeks on even the fastest computers. On the other hand, a typical ligand refinement program can generate and sample new structures at the rate of several hundred to thousands a minute. Thus, in order to achieve such high throughput, there must be some compromise in the accuracy of the binding calculation. That compromise is in the utilization of *scoring functions*.

Scoring functions estimate ligand-binding affinity using *descriptors* that can be rapidly measured from the ligand–receptor interaction. In essence, a scoring function is an equation that relates measurable descriptors of binding to ligand–receptor affinity. Figure 8.10 reveals how scoring functions are derived. Given a particular ligand and receptor, two things must be done. First, the ligand must be reacted with the receptor to determine its actual binding affinity using a biological assay. Second, the three-dimensional structure of the ligand bound within the receptor must be determined using X-ray crystallography. As we discussed above, the determinants of binding include steric interaction energy, electrostatic interaction energy, and hydrophobicity. Given the three-dimensional structure of a particular compound bound within the active site, we can rapidly calculate the values for these descriptors [5].

For example, to calculate the steric interaction energy, we simply sum up the number of receptor atoms that are within a specific distance (i.e., 5 Å) away from any ligand

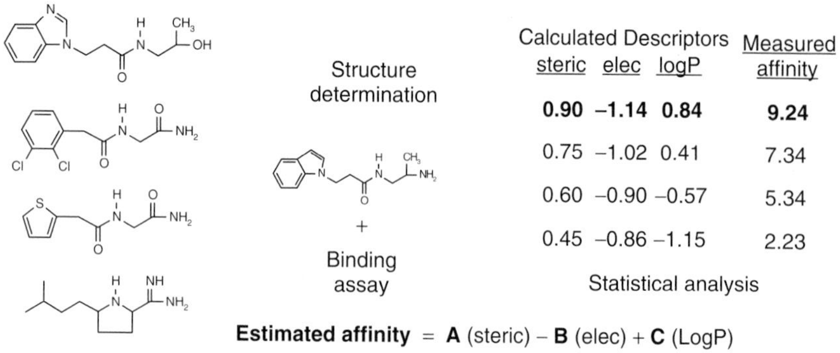

Fig. 8.10 Derivation of scoring function.

atom. The higher the value, the more the interaction between ligand and receptor atoms. Electrostatic interaction energy is computed using Coulomb's law. Hydrophobicity is represented by LogP, the octanol/water partition coefficient [6], which is a measure of the compound's propensity to solubilize in octanol versus water. The higher the value, the more hydrophobic the compound is. These descriptors are simple and very easy to calculate, allowing for the rapid determination of characteristics that relate to ligand-binding strength.

In Figure 8.10, we have four examples whose hypothetical binding affinity is given and whose descriptors have been calculated. Statistical tools, such as partial least squares regression [7], are then employed to generate the equation relating the numerical trends in the descriptors with the corresponding binding affinities. What results is an equation in which the estimated affinity is a function of the calculated descriptors (steric, electrostatic, and LogP) multiplied by their corresponding coefficients (A, B, and C). The coefficients relate calculated descriptors to the binding affinities and are determined by the statistical analysis. For example, we can see that as steric interaction energy (bulk) increases, so does the biological binding activity. Thus, the coefficient (A) is positive. On the other hand, we find that a *negative* electrostatic interaction energy is conducive to tighter binding, although variations in this energy are much smaller. Therefore, the corresponding coefficient (B) is negative, but of lesser importance. LogP follows a trend that is similar to that of steric interaction energy. Thus, coefficient (C) is positive; since LogP values follow the trend in affinity better, (C) is likely to have a higher weight.

Once a scoring function has been derived, it can be employed to estimate binding affinities very rapidly. Given a newly designed ligand or structural derivative that has been docked within the active site, the descriptors of binding are first calculated. These descriptors are then multiplied by the derived coefficients of the scoring function. Once all the terms have been calculated, they are summed to determine the estimated binding affinity of the ligand in question. It is important to note that this example is very simplistic. In reality, some scoring functions contain over twenty terms. Table 8.3 lists the major ligand–receptor binding descriptors employed by RACHEL, based on VALIDATE [8].

Tab. 8.3 RACHEL descriptors used to generate scoring functions

Steric complementarity	Molecular weight
Steric strain	Number of rotatable bonds
Electrostatic interaction energy	LogP estimation
Nonpolar–nonpolar interaction energy	Nonpolar atom fraction

Currently, there are several hundred high quality ligand–receptor complexes in the public domain that can be employed for scoring function development. Pharmaceutical firms have access to far more proprietary structures. However, even with all these structures along with the powerful statistical tools to analyze them, scoring functions still remain mediocre at best. According to the laws of thermodynamics, $\Delta G=\Delta H-T\Delta S$. ΔG is the Gibbs free energy of binding and is the energy that is released when ligand and receptor bind. This is the actual thermodynamic property that we are trying to estimate with the scoring function. ΔH is enthalpy (internal energy) and is grossly approximated by the calculated descriptors. Efforts to improve the accuracy of these approximations often increase calculation time drastically. $T\Delta S$ is an entropy term, and is indicative of the relative gain or loss of disorder when ligand and receptor bind. Perhaps the biggest influence on entropy is the behavior of the water molecules in the active site that are displaced when binding occurs. This is often disregarded, as it is difficult to accurately calculate without considerable computation. Thus, the take-home message is that ΔG is at best very crudely estimated by any scoring function. See [9] for an excellent review.

Compounding this problem is the fact that all current commercial packages utilize a single, proprietary, *generalized* scoring function that has been derived using a wide variety of structures. There are two significant problems with this approach. First, receptor systems vary considerably in their chemical makeup. In some systems, electrostatic interactions dominate ligand binding. In other systems, hydrophobic interactions overshadow the other forces involved. Thus, a master scoring function to estimate binding affinity for all ligand–receptor systems becomes a "jack of all trades, master of none". Using such a variety of ligand–receptor systems in the training set adds considerable noise to the data, which diminishes its predictive power.

The second shortcoming in using a *generalized* scoring function is the loss of competitive advantage. In effect, any laboratory or company that employs these tools utilizes the same predictive function as their closest rival. In addition, considerable resources are spent determining structures and characterizing the activity of candidate structures as a pharmaceutical company hones in on a potential drug. This wealth of structure-activity data could be used to considerably improve the predictive power of the scoring function. However, since current ligand design packages employ their own proprietary scoring functions, this data is lost.

RACHEL offers the unique ability to utilize this structure-activity data and retain the competitive edge gained through research and development. By incorporating the necessary statistical and analytical tools, RACHEL allows the user to easily generate *focused* scoring functions to estimate ligand binding to a specific target receptor using proprietary structure-activity data. This allows companies that have characterized the receptor binding of a number of lead compounds to utilize this knowledge in the design of future drugs.

The advantage of using *focused* scoring functions is significant. By limiting the training set to structures binding within the same receptor, we bias the scoring function toward the interactions that govern ligand association with the target active site. If hydrophobic contacts predominate, the hydrophobic descriptors will be emphasized. Conversely, if electrostatic forces are important to binding, those descriptors will be accentuated. Even something as simple as the size of the active site can have a tremendous impact on the allowable ligands. This is a descriptor that would be lost given a multitude of different training set receptors. As such, focused scoring functions have better predictive power with respect to estimating ligand–receptor binding than generalized scoring functions [10].

8.10
Target Functions

Even with structure-activity data pertaining to a target receptor, difficulties in generating accurate scoring functions may arise. Initially, there must be an adequate number of compounds to make the analysis statistically valid. Suppose the data points in Figure 8.11 are structure-activity data points for individual ligand–receptor complexes. The lines passing through them represent potential scoring functions attempting to describe their distribution. In the graph on the left, we see an ideal distribution of complexes that allows for an easy determination of a best-fit line. This dataset contains a large number of complexes whose activity covers a wide range of values. A scoring function generated from this set thoroughly represents the data. The middle graph is more representative of the situation in academic research. Here, we have far too few compounds to generate an accurate fit of the data. Notice the ambiguity that exists in determining the best-fit line. In essence, any scoring function derived from this dataset has little predictive value. The graph on the right is another scenario that might occur. Here, there is no lack of data. However, given money and time constraints in drug development projects, it can be difficult to justify crystallographic studies on poorly binding compounds. As

Target functions

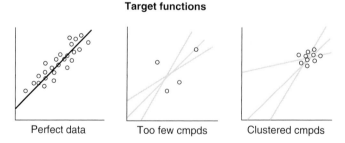

Perfect data   Too few cmpds   Clustered cmpds

Fig. 8.11 Problems in deriving scoring functions.

such, crystal structures of compounds are usually determined only when high-affinity structures have been found by assay. Therefore, a cluster of high-affinity data points is produced. As one can see from the graph, it is also difficult to elucidate an accurate scoring function when the structure-activity data is not broad enough.

In situations when the dataset is either too small or too clustered, RACHEL offers another means of generating a focused scoring system from proprietary structure-activity data. If RACHEL determines that the derived scoring function offers little predictive value, it will revert to a *target function*. A target function is formed by simply averaging the descriptor values of the highest affinity training set complexes. These "ideal" descriptor values are then used as a guide to determine whether the newly generated derivative structures are to be kept or discarded. This is illustrated in Figure 8.12. In this 3-D graph, the axes represent the three classes of descriptors of ligand–receptor binding discussed above. The cube is a plot of the "ideal" descriptor values that have been averaged from the optimal binding ligands in the user's structure-activity data. The black cube represents the target values against which all derivative compounds will be compared. The descriptor values for each derivative structure are then plotted. Those structures whose descriptor values are closest to the target, shown in dark gray, are retained. All other structures, shown in light gray, are rejected.

The primary advantage of using a target function is its ease of implementation. No longer is a large training set of compounds required. Even a single compound can be used as a model for optimal ligand–receptor binding. By simply extracting the descriptor values of the best compounds, we avoid many of the pitfalls in scoring function development that result from data artifacts. In addition, the characteristics of the ligand–receptor association that foster improved binding are allowed to drive the development of future structures. The big disadvantage in using target functions is the lack of extrapolation. In other words, we are constraining the system using the properties of previously characterized ligands. Thus, we are unable to predict whether a new derivative compound can potentially bind better to the receptor than our best compounds. We are also unable to quantitate the binding relative to the other structures in the training set. We are simply building structures that mimic the characteristics of the best compounds. Fortunately, this is often the exact task at hand for pharmaceutical

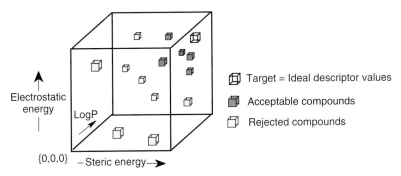

Fig. 8.12 Use of target function to screen compounds.

chemists. By the time a drug development project has reached maturity, the ligands that have been developed are often optimal binding compounds. Therefore, a target function is usually sufficient as it allows the drug designer to construct alternate chemical architecture that retains optimal binding characteristics.

8.11
Ligand Optimization Example

As an example, we use RACHEL™ to explore chemical alternatives for the arginine residue on PPACK, a known peptidic inhibitor [11] of thrombin (PDB entry code 1DWE [12]).

Our goal in this example is to generate ligand derivatives that will substitute the arginine residue of this tripeptide ligand. The region to be replaced is shown in Figure 8.13, labeled ARG3. By studying this crystal complex, we can determine the various ligand–receptor interactions that we must exploit in order to generate appropriate derivatives.

From a steric perspective, the arginine side chain fits tightly into a very defined pocket. This pocket is quite flat, as shown in Figure 8.14, forming a narrow cavity into which the arginine guanido terminus is wedged. One can see that enough room exists for a cyclic system to substitute for the guanido terminus; however, it must be planar. From an electrostatic perspective, it is clear why arginine is an ideal substrate for binding. Numerous hydrogen bonds are formed with the receptor, as shown in Figure 8.15. Interactions occur with the side chain of Asp189, the carbonyl group of Gly219, and possibly with the hydroxy group of Tyr228. This binding pocket is very polar with an abundance of negatively charged functional groups. Thus, the ideal complement for this region should contain hydrogen bond donors.

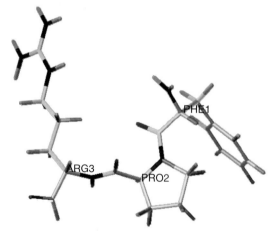

Fig. 8.13　Peptidic inhibitor of thrombin.

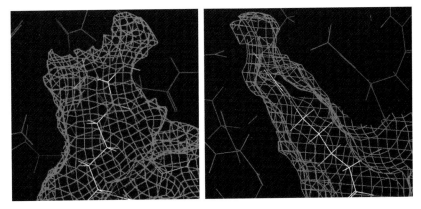

Fig. 8.14 Orthogonal views of arginine binding region.

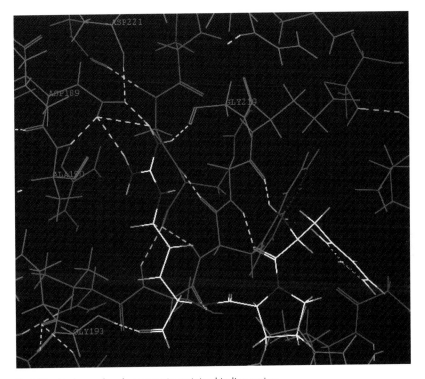

Fig. 8.15 Hydrogen bonds present in arginine binding region.

The carboxy terminus of the arginine residue also resides in a region where growth can occur. In addition, a hydrogen bond is made with the amide nitrogen of Gly193. Any derivative components that are placed in this region should maintain this hydrogen bond as well. Given our knowledge of the ligand–receptor interactions within the

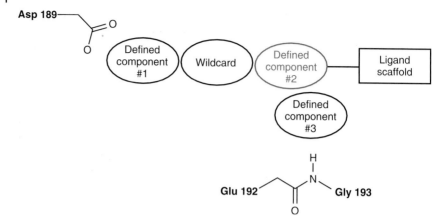

Fig. 8.16 RACHEL template used to generate ARG derivatives.

active site, our first task is to formulate a RACHEL template that will describe the chemistry we would like to implement in our ligand derivatives. The template, shown in Figure 8.16, consists of a user-determined arrangement of defined components and wildcard designations. RACHEL chemical descriptors are then assigned to the defined components. These descriptors, listed in Table 8.4, act as filters to enrich the database for the functionality desired at the various positions in the template.

Defined component #1 is designed to place a cyclic structure with at least one potential hydrogen bond donor within hydrogen-bonding distance of receptor Asp189. Defined component #2 is designed to mimic the alpha carbon of the arginine residue being replaced. As such, a small component is specified with a requirement that an sp3 carbon must attach to the amide nitrogen of the neighboring residue. Because

Tab. 8.4 Chemical descriptors associated with each defined component

Defined Component #1
Must contain ring structures – five to six atoms in size
One or more hydrogen bond donor groups must be present
Component must hydrogen-bond to Asp189

Wildcard
RACHEL will substitute, given the steric and electrostatic environment
RACHEL will add components with varying connectivity as necessary

Defined Component #2
Small component desired to mimic alpha carbon of residue
Number of atoms in this component <6
An sp3 carbon must attach to amide nitrogen of previous residue

Defined Component #3
Component must contain at least one hydrogen bond acceptor. (sp3 or sp2 oxygen)
Component must serve as hydrogen bond acceptor for amide N of Gly193

the distance between component #1 and #2 will vary depending upon the chemical structures selected, a wildcard designation is placed between the two. Finally, descriptors for component #3 specify a group that will accept a hydrogen bond from the amide nitrogen of Gly193. Using the tripeptide ligand and the receptor, a target function is then generated as shown in Figure 8.17. These parameters are used to score the resulting derivatives and guide the structure generation process. The user is given the option to modify parameter values and scalar multipliers if desired.

The search is then performed using components extracted from the user's structural database. Again, granting the user the ability to govern the source of the components is key. Perhaps a database of known inhibitors can be utilized – further enriching the source components for potential solutions. After several minutes, potential derivatives are generated. These initial structures can be analyzed and evaluated for undesired chemical assemblies. The chemical descriptors can then be modified to eliminate such structures and enrich for preferred compounds. Sample hits are shown in Figure 8.18. In each structure, the assembled components satisfy the chemical descriptors specified while maintaining steric and electrostatic complementarity with the active site. The functional groups (in bold) interact with the receptor pharmacophoric elements as described in Figure 8.16.

As we have shown in this example, the use of templates and chemical descriptors allows RACHEL to generate chemically diverse ligand derivatives within specific user constraints. In addition, building structures from enriched corporate database fragments

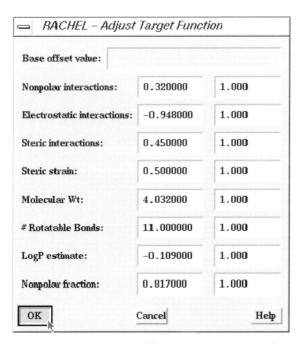

Fig. 8.17 Parameters extracted from ligand–receptor target function.

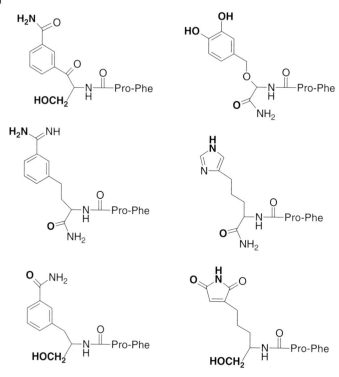

Fig. 8.18 Structures generated according to specification. Functional groups (in bold) interact with receptor pharmacophore.

and allowing the user to generate custom scoring or target functions further enhances the ability of RACHEL to produce potent derivatives that employ intellectual property and afford a competitive edge. As such, *in silico* lead optimization has become incredibly powerful and mainstream as the speed of computers continues to increase and their cost continues to reduce. However, the fact remains that the creativity and genius of the chemist remains paramount. These are simply tools that should provide stimulus and enhance the chemist's ability to think creatively.

References

1 Ho, C.M.W. RACHEL™, RACHEL is developed by Drug Design Methodologies LLC, http://www.newdrugdesign.com and is distributed by Tripos Inc, http://www.tripos.com, **2002**.

2 OPREA, T.I., Ho, C.M.W., and MARSHALL, G.R. De Novo design: ligand construction and prediction of affinity. In *Computer-Aided Molecular Design*, REYNOLDS, C.H., HOLLOWAY, M.K., and COX, H. (Eds). American Chemical Society, Washington, DC, **1995**, 64–81.

3 LINUSSON, A., GOTTFRIES, J., LINDGREN, F., and WOLD, S. Statistical molecular design of building blocks for combinatorial chemistry. *J. Med. Chem.* **2000**, *43*, 1320–1328.

4 Ho, C.M.W. and MARSHALL, G.R. DBMAKER: A set of programs to generate three-dimensional databases based upon user-specified criteria. *J. Comput.-Aided Mol. Des.* **1995**, *9*, 65–86.

5 MUEGGE, I. and ENYEDY, I. Docking and scoring. In *Computational Medicinal Chemistry and Drug Discovery*, TOLLENAERE, J. DE WINTER, H. LANGENAEKER, W., and BULTINCK, P. (Eds). Marcel Dekker, New York, **2004**, 405–436.

6 LEO, A. Estimating $LogP_{oct}$ from structures. *Chem. Rev.* **1993**, *5*, 1281–1306.

7 WOLD, S., RUHE, A., WOLD, H., and DUNN, W.J. The collinearity problem in linear regression. The partial least squares approach to generalized inverses. *J. Sci. Stat. Comput.* **1984**, *5*, 735–743.

8 HEAD, R.D., SMYTHE, M.L., OPREA, T.I., WALLER, C.L., GREENE, S.M., and MARSHALL, G.R. VALIDATE: A new method for the receptor-based prediction of binding affinities of novel ligands. *J. Am. Chem. Soc.* **1996**, *118*, 3959–3969.

9 AJAY and MURCKO, M. Computational methods to predict binding free energy in ligand-receptor complexes. *J. Med. Chem.* **1995**, *38*, 4953–4967.

10 MARSHALL, G.R., HEAD, R.D., and RAGNO, R. Affinity prediction: the sine qua non. In *Thermodynamics in Biology*, DI CERA, E. (Ed.). Oxford University Press, New York, **2000**, 87–111.

11 BANNER, D.W. and HADVARY, P. Crystallographic analysis at 3.0 angstroms resolution of the binding to human thrombin of four active site-directed inhibitors. *J. Biol. Chem.* **1991**, *266*, 20085–20090.

12 The Protein Data Bank is available from http://www.rcsb.org/pdb.

Part III
Databases and Libraries

9
WOMBAT: World of Molecular Bioactivity

Marius Olah, Maria Mracec, Liliana Ostopovici, Ramona Rad, Alina Bora, Nicoleta Hadaruga, Ionela Olah, Magdalena Banda, Zeno Simon, Mircea Mracec, and Tudor I. Oprea

9.1
Introduction – Brief History of the WOMBAT Project

The current paradigm for drug discovery allows a relatively short time period, 6 to 12 months, for the process that modifies an initial active compound – either from high-throughput screening or from publications and patents – into a well-characterized lead molecule. During this time, project team members have relatively little time to familiarize themselves with "prior art", that is, to gather information pertinent to the new biological target, the disease models, as well as active chemotypes on the intended or related targets. The task of gathering background information related to chemotypes is made easier if one has access to chemical databases such as Chemical Abstracts via SciFinder [1], Beilstein [2], and Spresi [3], or to medicinal chemistry related patent databases such as the MDL Drug Data Report, MDDR [4], the World Drug Index, WDI [5], Current Patents Fast Alert [6], or to collections of biologically active compounds such as and Comprehensive Medicinal Chemistry, CMC [7], and the Physician Desk Reference, PDR [8].

These databases are a rich source of information, yet they do not capture an element of interest, namely the biological endpoint: there is no searchable field to identify, in a quantitative manner, what is the target-related activity of a particular compound. Such information is important if one considers that (a) not all chemotypes indexed in patent databases are indeed active – some are just patent claims with no factual basis and that (b) not all chemotypes disclosed as active are equally active, or selective for that matter, on the target of choice. Furthermore, should one decide to pursue a certain interaction "hotspot" in a given ligand–receptor structure (assuming good structure-activity models are available), it would be very convenient to mine structure-activity databases for similar chemotypes to use as potential bioisosteric replacements.

These were some of the practical considerations used by AstraZeneca R&D Mölndal, Sweden, in May 2001, to initiate a data-gathering project centered primarily on the *Journal of Medicinal Chemistry* (JMC) in collaboration with scientists at the Romanian Academy Institute of Chemistry in Timisoara, Romania. The major goal of this project was to capture every chemical structure and associated biological activities that were disclosed in JMC, and an initial goal of 20 000 entries was set for the first year. As a

Chemoinformatics in Drug Discovery. Edited by Tudor I. Oprea

Tab. 9.1 Examples of annotated databases

Database	Description	Homepage
AUR-STORE	Structure-activity information focused on different targets and activities	http://www.aureus-pharma.com
Bioprint	Cerep's proprietary ligand profiling data including target-specific activity and ADME-related properties	http://www.cerep.fr
ChemBank	Structures and biological activity data for over 2000 compounds	http://chembank.med.harvard.edu
Drugmatrix	Pharmacological, pathology and gene expression profiles for benchmark drugs	http://www.iconixpharm.com
GPCR Annotator	Structural and functional information related to GPCRs and their ligands.	http://www.jubilantbiosys.com
Kinase ChemBioBase	Annotated database of over 170 000 molecules covering over 350 kinases	
Sertanty	Annotated database of 22 000 compounds focused on protein kinases	http://www.sertanty.com

Source: Modified from [9].

result, the very first version of this database was available at AstraZeneca R&D Mölndal in May 2002. This version, internally named SADB5 (Structure-Activity DataBase, version 5) contained 21 700 structures (with duplicates) and 36 738 experimental activities measured on 324 targets. The data was indexed from 837 JMC papers (1996–1999) and included the following biological activities: IC_{50} (53.52%), K_i (39.56%), D_2 and EC_{50} (5.54%), as major types.

Because the internal dissemination of this database within AstraZeneca (a company with eleven R&D sites across four continents) was not considered a success, AstraZeneca decided to discontinue the project as of May 2002. Backed by private funding, the database, renamed WOMBAT (WOrld of Molecular BioAcTivity) in 2003, has increased considerably in size, as discussed for WOMBAT 2004.1, below. This chapter summarizes the contents of WOMBAT, some of the problems encountered in appropriately indexing biological activities and correct chemical structures (with focus on machine-readable contents for data mining), and provides some examples of the potential use of WOMBAT data mining. Other annotated databases [9], focused mostly on patent literature, are shown in Table 9.1, together with the on-line references.

9.2
WOMBAT 2004.1 Overview

WOMBAT 2004.1 [10] contains 76 165 entries (68 543 unique SMILES [11, 12]), covering 3039 series from over 3000 papers and $\approx$143 000 activities for $\approx$630 targets. All biological activities are automatically converted to the negative $\log_{10}$ of the molar concentration,

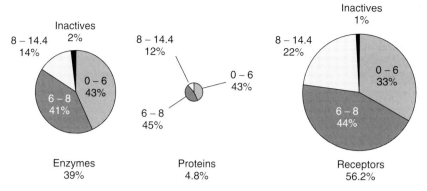

Fig. 9.1 Bioactivity distribution pie charts in WOMBAT 2004.1, classified by target type. The size of the pie chart is proportional to the representation of each target class in the database.

regardless of activity type. Each activity is reported in three different fields; the additional two fields capture the experimental error, when reported[a]. Besides exact numeric values (the vast majority), WOMBAT now captures "inactives" (635), "less than" (8916), "greater than" (259), as well as percent inhibition values (578 single dose experiments). The bioactivity distribution by target type is given in Figure 9.1. Three target types are captured in WOMBAT: *receptors* (which includes GPCRs – G-protein coupled receptors, nuclear hormone receptors, integrins, and ion channels), *enzymes* (all associated with the Enzyme Commission E.C. number [13]) and *proteins* (biological targets that are neither enzymes nor receptors, e.g., transporters, binding proteins, etc.).

A vast majority of the biological activities are related to inhibitors and antagonists: 55.85% of the activities are IC_{50} values (and variations) and 37.1% are K_i values (and variations). Much less frequent are D_2 or EC_{50} values (4.44% of the measurements are

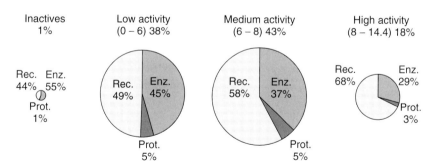

Fig. 9.2 Target-type distribution pie charts in WOMBAT 2004.1, classified by activity value (in the $-\log_{10}$ scale).

---

[a] In the absence of reported errors, the three activity value fields are equal. The decision to index these values for each molecule was taken because "missing values" are given a different interpretation by statistical techniques.

Tab. 9.2 Target class profile for WOMBAT.2004.1. Numbers in the "Compounds" column indicate the number of structures active at least once per each target, and the percentage from the total number of entries is given in the rightmost column

No.	Target class	Compounds	Percent
1	G-protein coupled receptors	28973	38.04
2	Ion channels	9008	11.83
3	Aspartyl proteases	3351	4.40
4	Kinases	2842	3.73
5	Oxygenases	2829	3.71
6	Transporters	2264	2.97
7	Oxidoreductases	2010	2.64
8	Integrins	1772	2.33
9	Phosphodiesterases	1689	2.22
10	Serine proteases	1459	1.92
11	Cysteine proteases	704	0.92
12	Nuclear hormone receptors	688	0.90
13	Others	18576	24.39

for agonists or substrates), binding affinity constants (0.9% K_b and K_d), and in rare cases, we find MIC (0.1%) and ED_{50} (0.04%) values. In WOMBAT 2004.1, enzyme inhibitors populate more of the inactive/low-activity bins, while receptor antagonists populate more of the medium/high-activity bins (Figure 9.2). The target profile of biological activities is given in Table 9.2 with focus on some targets classes of current interest to the pharmaceutical industry. This interest is reflected by the data captured in WOMBAT as well, and the medicinal chemistry publications indexed are listed in Table 9.3. The target-type distribution by activity in Figure 9.2 is, therefore, a reflection of approximately a decade of medicinal chemistry (see also Table 9.2).

Additional data fields provide direct bibliographic links (using the Digital Object Identifier, DOI, format [14]) to PDF files for all literature entries, and SwissProt [15] reference IDs for most targets. The SwissProt ID, in particular, can be challenging,

Tab. 9.3 Medicinal chemistry publications covered in WOMBAT.2004.1

No.	Journal title	Publication year
1	Biochem. Pharmacol.	2001 (partially indexed)
2	Bioorg. Med. Chem. Lett.	2002 (numbers 1–24)
3	Chembiochem	2002 (partially indexed)
4	Eur. J. Med. Chem.	2001 (partially indexed)
5	J. Amer. Chem. Soc.	1975,1992,1993 (partially indexed)
6	J. Healt. Sci.	2003 (partially indexed)
7	J. Med. Chem.	1991 (partially indexed), 1992–2000 (completely indexed), 2003 (partially indexed)
8	Quant. Struct.-Act. Relat.	1998–2000 (partially indexed)

since it is not always clear from a publication what was the exact target used in the assay: sometimes it is unclear what species was the target isolated from (whether recombinant, wild type, or mutant), or even what type of target subtype was used. For example, there are 853 entries in WOMBAT 2004.1 that contain "muscarinic" as the target name, which implies that all muscarinic receptors present in a particular organ (e.g., brain, heart, ileum, or urinary bladder) were tested for agonism and antagonism. For such targets, it is not possible to record a SwissProt ID. A related issue refers to the dynamic nature of the biological targets. As biologists gather more information about a particular target, as our understanding about that target evolves, its exact nomenclature changes as well. For example, there are 211 WOMBAT entries that contain "Flk-1/KDR" as the target name. This enzyme name, which stands for "fetal liver kinase-1" and "kinase insert domain-containing receptor", has recently been reclassified as "VEGFR2", or the vascular endothelial growth factor receptor subtype 2. The VEGFR2 target name is present in all 211 entries, even though some of the older (1998–1999) publications did not contain this particular target name. In an annotated database such as WOMBAT, one has to monitor and update not only changes related to biology but also changes related to chemistry (and chemical errors). This is discussed in more detail below.

9.3
WOMBAT Database Structure

WOMBAT is a dynamic database, which evolves as new data types are included. The database structure is, however, preserved as much as possible from one release to the next. Each root record is identified by a unique number (SMDLID), and is defined by the combination of *one chemical structure* and one or more associated *biological activities* as entered in *one publication* (Figure 9.3). One field, series ID (SID), links all the root records indexed from one reference (article). There are 3039 SID values in WOMBAT 2004.1. At the root level, information about the bibliographic reference (unique SID) that originated the entry is recorded together with several calculated and experimental properties for each structure, for example, counts of miscellaneous atom types, Lipinski's rule-of-five (RO5) parameters [16], including the octanol/water partition coefficient (ClogP) [17] and water solubility [18], polar and nonpolar surface areas (PSA, NPSA), and so on (Figure 9.4). Practical applications based on WOMBAT data mining with these descriptors are given in Chapter 2 [19].

Separate keywords describe structural characteristics related to stereochemistry (e.g., absolute, relative, $+/-$, R/S, "nonchiral" or racemic) and to the salt form. We record the salt separately in order to avoid the salt-removal step that is usually performed in chemoinformatic studies prior to structure computations. For each SMDLID, we define the following biological activity subrecords: AID, the activity identifier, with values from 1 to n, where n is the number of biological activity determinations for one structure; TargetName (the target name on which the activity was measured); ActType (the activity type, e.g., IC_{50}), ValueType, which can be one of five types: exactly ($=$), lower than ($<$), greater than ($>$), percent inhibition at a given concentration ($@$), or inactive; ActValue, the numeric value of the activity, in negative $\log_{10}$ of the molar concentration; Range,

Root	
SMDLID	– structure identifier
SID	– series identifier
Structure	– chemical structure in both formats MOL & SMILES
Reference	– bibliografic reference
Keywords	– structure keywords (chirality, salt information)
Properties	– calculated and experimental properties (MW, RO5, etc.)

Activity #1	
AID	– activity identifier (1, 2, ..., n)
TargetType	– target & experimental determination information
TargetName	– target name
ActType	– activity type (IC_{50}, K_i, EC_{50}, A_2, D_2, ...)
ValueType	– activity value type (=, <, >, @, inactive)
ActValue	– numeric activity value, in log units
Range	– confidence range for the activity value, in log units
BioKeywords	– target & experimental determination information

...

Activity #n	
...	

Fig. 9.3 WOMBAT database schemata (simplified).

the experimental confidence range for the measured activity, also in logarithmic units. For each SMDLID and each AID (Figure 9.5), we also record a number of BioKeywords related to biological activity information (e.g., bio-species, tissue and cell type, etc.) and target-related information (e.g., the E.C. number [13], what radio-labeled substrate or ligand was used, etc.). Thus, for one series (same SID value), each activity block (AID range 1, ..., n) has separate TargetName, ActType, ValueType, and BioKeywords.

9.4
WOMBAT Quality Control

Quality control is performed at the moment of data entry, in particular, with respect to errors present in publications. Chemical structures are checked for structural consistency by matching the molecular weight (MW) and chemical formula with the ones available in the experimental section and/or supporting information – whenever available, and by comparison to prior publications. Whenever in doubt, we also use other sources, such as the Merck Index [20] and free Internet resources. In the instances

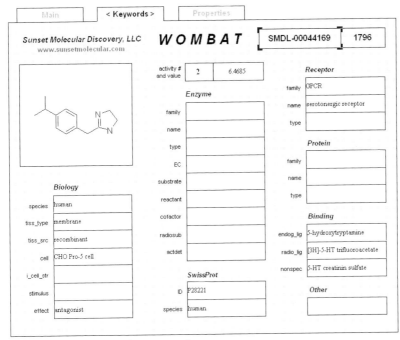

Fig. 9.4 WOMBAT keywords page snapshot.

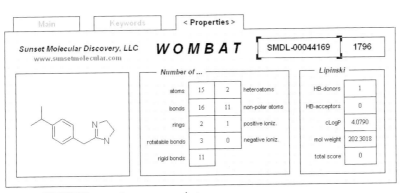

Fig. 9.5 WOMBAT properties page snapshot.

where external and literature data cannot be reconciled, SciFinder [1] is also used. The error rate so far in medicinal chemistry publications is not at all negligible. We find an average of approximately two errors per publication in all the 3030 articles indexed in WOMBAT. Given the median of 25 compounds per series, this implies an overall error rate of 8%. These errors are distributed as follows [21]:

- incorrectly drawn or written structures (3%), incorrect molecular formula or molecular weight (3%);

- unspecified position of attachment of substituents, or ambiguous numbering scheme for the heterocyclic backbone (0.9%);
- structures with the incorrect backbone (0.7%);
- incorrect generic names or chemical names (0.2%);
- duplicates (0.2%);
- incorrect biological activity (0.3%);
- incorrect references (0.2%).

A special attention is given to stereochemistry as some compounds are published without proper chirality representation even though the information is available, for example, for natural compounds and their derivatives. Furthermore, as illustrated in Figure 9.6, compounds published in medicinal chemistry literature are often depicted

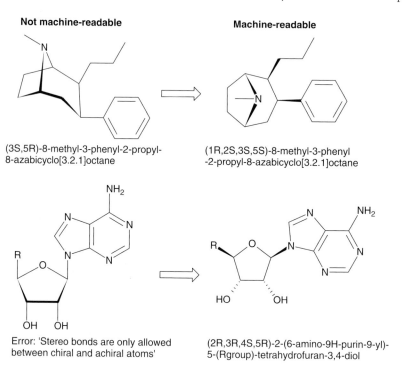

Not machine-readable

(3S,5R)-8-methyl-3-phenyl-2-propyl-8-azabicyclo[3.2.1]octane

Machine-readable

(1R,2S,3S,5S)-8-methyl-3-phenyl-2-propyl-8-azabicyclo[3.2.1]octane

Error: 'Stereo bonds are only allowed between chiral and achiral atoms'

(2R,3R,4S,5R)-2-(6-amino-9H-purin-9-yl)-5-(Rgroup)-tetrahydrofuran-3,4-diol

Cross up/down wedge error

Undefined chirality may be interpreted as both R and S

Fig. 9.6 Human versus machine-readable chemical structure representations. Names based on the depicted structures were interpreted using ACDName [22]. The cross up/down wedge error (middle) causes errors in assigning the absolute chirality.

in a "human-readable" format; that is, structures are drawn in a format that chemists can interpret to reconstruct proper chirality. However, this format is not "machine readable", that is, chemoinformatics software for 3-D structural conversion, or for automatically generating IUPAC nomenclature, cannot perceive the stereo centers correctly if the "above/below plane" convention is not strictly enforced. We illustrate this with ACDName [22] on the structures depicted in Figure 9.6. The software does not perceive two stereo centers for the tropane ring on the left side and returns an error for the sugar structure. The errors are not specific to ACDName – this program is used only to illustrate the problem. Another type of problem in structure conversion is the cross up/down wedge error, when two such bonds emerge from the same chiral center (Figure 9.6). Software cannot assign the proper chirality, since by convention, three atoms are in the "paper plane" and only one is "wedged" (up or down); two wedged bonds are simply not possible according to the convention. Most of these errors can be corrected by checking previous literature. Sometimes, even the cited reference may turn out to be in error; for example, the reported MW is not consistent with the drawn, or named, structure.

9.5
Uncovering Errors from Literature

As the demand for integrated chemical and biological information increases, scientists rely more often on annotated databases that capture medicinal chemistry literature (see Tables 9.1 and 9.3). There is little, if any, error checking downstream from publication time, even though mechanisms for publishing errata have been in place for quite some time. While the responsibility for published data accuracy resides primarily with the author(s), it is also the responsibility of annotated database curators to capture as many of these errors as possible. While ensuring the quality control in WOMBAT, we have found inconsistencies in many of these publications. These errors may have a significant effect on the way we understand the molecular basis of chemical–biological interactions, at least for some particular series used for structure-activity studies. Coats has traced the errors on a known steroid benchmark for QSAR(Quantitative Structure-Activity Relationships) studies to the original publications [23]. Some of these errors are discussed below.

Example 1. The following errors were found in Table 1 of Ref. [24], page 126:

- compounds with molecular names 53 and 56, respectively, appear to be duplicated because all their substituents are identical. On the basis of their activities, 56 (compound 15g in [25]) has the *meta* – C_6H_4–OCH_3 substituent, while 53 (compound 15e in [25]) has the *para* – C_6H_4–OCH_3 substituent;
- the – NH – group is missing from the L substituent in compound 27 (compound 9 in [26]), and the – CH_2 – group is missing from the L substituent in compound 45 (compound 13 in [25]);
- the R substituent of compound 66 is – C_6H_2–2-CO_2CH_3–4,5-$(CH_3)_2$ instead of the correct – C_6H_2–2-CO_2CH_3–4,5-$(OCH_3)_2$ group (compound 5bb in [26]);

- the R substituent of compound 68 is $-C_6H_2-2-CO_2H-4,5-(CH_3)_2$ instead of the correct $-C_6H_2-2-CO_2H-4,5-(OCH_3)_2$ group (compound 7 in [26]);
- compound 44 has a $-\log(IC_{50})$ of 7.67 instead of the correct 7.74 (compound 15d in [25]).

Example 2. In Table 1b of Ref. [27], page 4361, the core structure contains an oxygen atom instead of the correct sulfur atom [28]:

wrong   correct

Thus, 47 structures (where X is the rest of the molecule) are incorrect in [27]. Since the paper illustrates the capabilities of a particular structure-activity method, the consistent error does not influence the validity of the models; it would, however, greatly influence the use of this series/model in a medicinal chemistry project where the goal would be to improve the binding affinity. Starting from the same initial publication [28], other errors were propagated in [29]:

- compound 37 has an incorrect double substitution in the para position of the aromatic ring, 2,4-NO_2,4-OH, while the correct one is 3-NO_2,4-OH;
- the R substituents of compound B.12 is 2,4,6-Cl_2,4-OMe instead of the correct 2,6-Cl_2,4-OMe.

Example 3. Errors could be found also in Chemical Abstracts' SciFinder [1]. All the errors we encountered originate in the primary publications; their appearance in SciFinder illustrates how such errors can propagate (since SciFinder is a very popular resource). For example, the compound RB-380 (CAS# 187454-94-0), published in [30] (original molecule name 24) has a ring size of 14 atoms, instead of the correct 13:

SciFinder structure   Correct structure

$C_{34}H_{42}N_6O_7S_2$ $C_{33}H_{40}N_6O_7S_2$

L-Phenylalaninamide, N-(5-mercapto-1-oxopentyl)-α-methyl-D-tryptophyl-L-homocysteinyl-L-α-aspartyl-, cyclic (1→2)-disulfide (this name is given in CAS)

Cyclo-S,S-[(5-thiopentanoyl)-αMe(R)-Trp-Cys]-Asp-Phe-NH2 (this name is given in the original publication [30], in the experimental section)

The correction we propose is based on the experimental section name and on the following text fragment (p.648, results section [30]): *"...by introducing an additional amide bond (compound 16 or RB 370) or a disulfide bridge (24 or RB 380) into the 13-membered ring (Schemes 2 and 3), and by changing the size of the ring (Table 1, compounds 43 and 45)."* By analyzing the data from Table 1 [30], compound 43 (which is actually 44 – which is another small error) has a 13-atom ring, while compound 45 has a 14-atom ring.

Example 4. Stereochemical ambiguities and structural errors can be encountered in the Merck Index [20] as well, as shown in these two examples:

mpound identifiers and or description	Merck Index structure	Correct structure
30, anagyrine (CAS# 5-89-5): chiral center ersion & cross up/down dge		
1854, carisoprodol AS# 78-44-4): completely ferent structure. All er information about 1854 is correct (name, mula and molecular ight). The formula is rrect in the 9th edition of e Merck Index.		

The examples from SciFinder and the Merck Index are not intended to question the quality of these products, which we consider to be outstanding. They are invaluable resources to many chemists worldwide, and the error rate in these two databases is insignificant if one takes into account the enormous volume of indexed data. One of us has published a structure-activity paper on HIV-protease inhibitors [31] where a modified peptide was present in both the training set and the test set. Al Leo of Pomona College has recently [32] detected 100 chemical and name errors in the printed version of the sixth edition of Burger's Medicinal Chemistry [33], errors that will be corrected in the on-line edition [34]. One can never be too careful in verifying the available information, in particular if one is to invest a significant amount of resources in that area.

9.6
Data Mining with WOMBAT

Example 1. A question that arises often in preclinical pharmaceutical research is how well does an observed biological activity measured in animals match the same activity in humans. WOMBAT 2004.1 contains 29 263 entries with "human" as the biological species and 27 229 entries as "rat". Their intersection, 2339 records, contains measurements from the same research groups, as the query selects only records where only human and rat activities were reported in the same paper. After trimming the dataset to force the same target and the same activity type (e.g., K_i or IC_{50}), we are

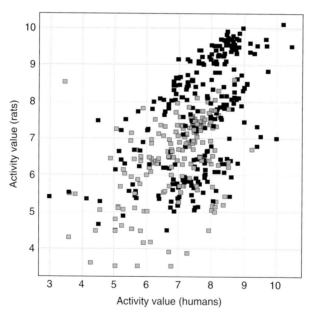

Fig. 9.7 Human versus rat biological activities recorded for the same molecules and reported in the same paper for receptors (black) and enzymes (gray).

Tab. 9.4 Rat versus human bioactivity data comparison using entries from WOMBAT.2004.1; N is the number of compounds, R is the correlation coefficient, and R^2 is the fraction of explained variance

	N	R	R^2
All data	441	0.58	0.34
Enzymes, IC_{50}	173	0.49	0.24
Receptors, IC_{50}	92	0.76	0.58
Receptors, K_i	168	0.48	0.23

left with 441 structures where both rat and human measurements were published in the same paper, possibly performed under the same conditions – see Figure 9.7 and Table 9.4. The correlation coefficient, $r^2 = 0.336$, is rather disappointing, but as listed in Table 9.4, correlations improve if one examines individual subsets: looking only at receptors and IC_{50} values, the trend gains significance. The lack of correlations relates to target (dis)similarity at the primary sequence (in particular, close to the ligand binding site), regardless of functionality. For some targets, rat and human data correlate well only after key aminoacids are mutated from human wild type to the rat sequence [35]. Thus, comparative measurements may be required for each target separately.

Example 2. The concept of leadlikeness [19, 36, 37] and its application in developing leadlike libraries [36, 38, 39] are discussed elsewhere in this book. The reduction of the leadlike concept into practice at Astex [40] resulted in a proposal for fragment libraries in lead discovery called the "Rule of Three": MW < 300, ClogP < 3, number of hydrogen bond donors and acceptors less than or equal to 3, three or less flexible bonds, and PSA $\leq$ 60 Å^2. Using these criteria, WOMBAT 2004.1 returns 4994 entries. Of these, 718 entries contain at least one biological activity better than or equal to 10 nM, and 185 of these contain a generic name. This usually means they are either launched drugs or natural products, or otherwise in an advanced stage of development. The examples given in Figure 9.8 illustrate the chemotype, target and activity diversity that can be found in rule-of-three compliant molecules: neurotransmitter and nuclear hormone receptor agonists (EC_{50}) and antagonists (K_i, IC_{50}, and A_2), neurotransmitter transporters, as well as enzyme inhibitors are present, most of them with multiple activities. On the basis of the WOMBAT 2004.1 entries, it appears that there are a number of interesting chemotypes that are rule-of-three compliant.

9.7
Conclusions and Future Challenges

Being an annotated database, WOMBAT continues to evolve in time – not only with the addition of more entries but also with updates and restructuring of the biological and chemical information, which is subject to change even after the data is captured. The

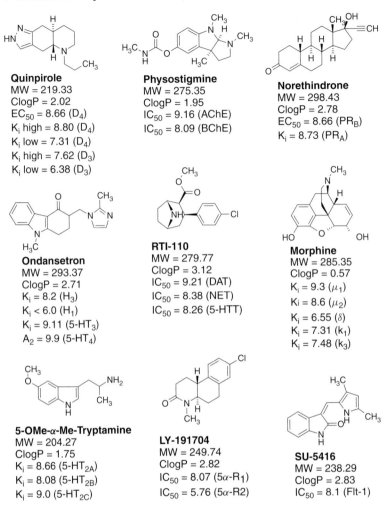

Quinpirole
MW = 219.33
ClogP = 2.02
EC_{50} = 8.66 (D_4)
K_i high = 8.80 (D_4)
K_i low = 7.31 (D_4)
K_i high = 7.62 (D_3)
K_i low = 6.38 (D_3)

Physostigmine
MW = 275.35
ClogP = 1.95
IC_{50} = 9.16 (AChE)
IC_{50} = 8.09 (BChE)

Norethindrone
MW = 298.43
ClogP = 2.78
EC_{50} = 8.66 (PR_B)
K_i = 8.73 (PR_A)

Ondansetron
MW = 293.37
ClogP = 2.71
K_i = 8.2 (H_3)
K_i < 6.0 (H_1)
K_i = 9.11 (5-HT_3)
A_2 = 9.9 (5-HT_4)

RTI-110
MW = 279.77
ClogP = 3.12
IC_{50} = 9.21 (DAT)
IC_{50} = 8.38 (NET)
IC_{50} = 8.26 (5-HTT)

Morphine
MW = 285.35
ClogP = 0.57
K_i = 9.3 (μ_1)
Ki = 8.6 (μ_2)
K_i = 6.55 (δ)
K_i = 7.31 (k_1)
K_i = 7.48 (k_3)

5-OMe-α-Me-Tryptamine
MW = 204.27
ClogP = 1.75
K_i = 8.66 (5-HT_{2A})
K_i = 8.08 (5-HT_{2B})
K_i = 9.0 (5-HT_{2C})

LY-191704
MW = 249.74
ClogP = 2.82
IC_{50} = 8.07 (5α-R_1)
IC_{50} = 5.76 (5α-R2)

SU-5416
MW = 238.29
ClogP = 2.83
IC_{50} = 8.1 (Flt-1)

Fig. 9.8 Examples of rule-of-three compliant molecules that have biological activity better than 10 nM. Under each molecule, the following information is included: molecule name, MW, ClogP, the biological activity type, value and target. Target names are as follows: D_3 and D_4 – dopaminergic receptor types 2 and 3; AChE and BChE – acetyl- and butyryl-choline esterases; PR_A and PR_B – progesterone receptor types A and B; H_1 and H_3, histamine receptor types 1 and 3; 5-HT_{2A}, 5-HT_{2B}, 5-HT_{2C}, 5-HT_3, 5-HT_4 – serotonin receptor subtypes 2A, 2B, 2C, and types 3 and 4; DAT, NET, 5-HTT – dopamine, norepinephrine and serotonin transporter proteins; μ_1, μ_2, δ, k_1, k_3 – opioid receptor types mu-1, mu-2, delta, kappa-1 and kappa-3; 5α-R1 and 5α-R2 – 5-alpha-reductase isozymes 1 and 2; Flt-1–fms-like tyrosine kinase receptor.

inclusion of the precomputed properties panel allows the user to quickly identify rule-of-five or rule-of-three compliant datasets or to constrain the query with respect to, for example, flexible bonds, PSA, and so on. WOMBAT is available in the MDL Isis/Base format and is fully integrated in CABINET (Chemical And Biological Informatics NETwork) [41, 42] as a server. CABINET [42], a federation of high-performance scientific databases, which collaborate via weblike interfaces in order to provide integrated access to diverse chemical and biological information, is described in Chapter 10 [41].

Federated database servers such as CABINET could, for example, bring together WOMBAT and C-QSAR [43], but the challenge goes beyond technical issues related to field correspondence. Data normalization (e.g., ensuring similar treatment regarding chirality, salt information, measured and computed properties) is likely to require on-the-fly data interpreters, which in turn forces lack of ambiguity for all data entries in WOMBAT and other databases. Data transparency is not always possible. For example, most WOMBAT entries related to EGFR (epithelial growth factor receptor) are classified as "TargetType = enzyme", because EGFR is a membrane receptor-linked tyrosine–protein kinase and medicinal chemists target EGFR for kinase inhibition. However, in one instance, "TargetName" was assigned as "receptor" because the endogenous ligand, EGF, was used to test for EGFR antagonism [44]. Thus, restricting data fields to certain value types, which usually is an asset for database indexing, can become a hindrance when the unexpected occurs. Thus, one of the challenges in database federation remains adaptive data normalization for biology-related data fields, since biological phenomena are not always amenable to unambiguous mapping. By successfully addressing this problems, it is quite likely that integrated data mining tools such as CABINET will change the way we conduct everyday research.

Acknowledgments

The authors thank Prof. Hugo Kubinyi (Heidelberg, Germany) for suggestions.

References

1 Chemical Abstracts online and its search module, SciFinder, are available from the American Chemical Society, http://www.cas.org/SCIFINDER/ **2004**.

2 The Beilstein Information Systems is available from http://www.beilstein.com/.

3 The Spresi Database is available from InfoChem GmbH, München, http://www.spresiweb.de/ and from Daylight Chemical Information Systems, http://www.daylight.com/products/databases/Spresi.html.

4 MDDR is available from MDL Information Systems, http://www.mdli.com/products/finders/database_finder/ and from Prous Science Publishers, http://www.prous.com/index.html.

5 WDI, the Derwent World Drug Index, is available from Derwent Publications Ltd., http://thomsonderwent.com/products/lr/wdi/ and from Daylight Chemical Information Systems, http://www.daylight.com/products/databases/WDI.html.

6 The Current Patents Fast Alert database is available from Current Patents Ltd., London, UK, http://www.current-patents.com/.

7 The Comprehensive Medicinal Chemistry database is available from MDL Information Systems, Inc., http://www.mdli.com/products/knowledge/medicinal_chem/index.jsp.

8 The Physician Desk Reference is produced by Thomson Healthcare, **2003**, ISBN 1-56363-472-4, and is available online at http://www.pdr.net/.

9 SAVCHUK, N.P., BALAKIN, K.V., and TKACHENKO, S.E. Exploring the chemogenomic knowledge space with annotated chemical libraries. *Curr. Opin. Chem. Biol.* **2004**, *8*, 412–417.

10 WOMBAT 2004.1 is available from Sunset Molecular Discovery, Santa Fe, NM, http://www.sunsetmolecular.com/ and from Metaphorics LLC, Santa Fe, NM, New Mexico, http://www.metaphorics.com/.

11 WEININGER, D. SMILES 1. Introduction and encoding rules. *J. Chem. Inf. Comput. Sci.* **1988**, *28*, 31–36.

12 WEININGER, D., WEININGER, A., and WEININGER, J.L. SMILES 2. Algorithm for generation of unique SMILES notation. *J. Chem. Inf. Comput. Sci.* **1989**, *29*, 97–101.

13 The Enzyme Nomenclature is recommended by the International Union of Biochemistry and Molecular Biology, and is available at http://www.chem.qmul.ac.uk/iubmb/enzyme/.

14 The Digital Object Identifier (DOI) is a system for identifying and exchanging intellectual property in the digital environment (http://www.doi.org/). An object is directly accessible using the customized address http://dx.doi.org/DOI_VALUE.

15 Swiss-Prot Protein knowledgebase database, http://kr.expasy.org/sprot/.

16 LIPINSKI, C.A., LOMBARDO, F., DOMINY, B.W., and FEENEY, P.J. Experimental and computational approaches to estimate solubility and permeability in drug discovery and development settings. *Adv. Drug Deliv. Rev.* **1997**, *23*, 3–25.

17 LEO, A. Estimating Log P_{oct} from structures. *Chem. Rev.* **1993**, *5*, 1281–1306.

18 RAN, Y., JAIN, N., and YALKOWSKY, S.H. Prediction of aqueous solubility of organic compounds by the general solubility equation (GSE). *J. Chem. Inf. Comput. Sci.* **2001**, *41*, 1208–1217.

19 OPREA, T.I. Chemoinformatics in lead discovery. In *Chemoinformatics in Drug Discovery*, OPREA, T.I. (Ed.). Wiley-VCH, Weinheim, **2004**, 25–41.

20 O'NEIL, M.J. *Merck Index*, 13th edition. Merck & Co. Inc, Rahway, NJ, **2001**.

21 OPREA, T.I., OLAH, M., OSTOPOVICI, L., RAD, R., and MRACEC, M. On the propagation of errors in the QSAR literature. In *EuroQSAR 2002 – Designing Drugs and Crop Protectants: Processes, Problems and Solutions*, FORD, M., LIVINGSTONE, D., DEARDEN, J., and VAN DE WATERBEEMD, H. (Eds). Blackwell Publishing, New York, **2003**, 314–315.

22 ACDName is available from Advanced Chemistry Development Inc., Toronto, Ontario, http://www.acdlabs.com/.

23 COATS, E.A. The CoMFA steroids as a benchmark dataset for development of 3D-QSAR methods. In *3D QSAR in Drug Design*, Vol. 3, Recent Advances, KUBINYI, H., FOLKERS, G., and MARTIN, Y.C. (Eds). Kluwer/ESCOM, Dordrecht, Netherlands, **1998**, 199–213.

24 CHEN, Q., WU, C., MAXWELL, D., KRUDY, G.A., DIXON, R.A.F., and YOU, T.J. A 3D-QSAR analysis of in vitro binding affinity and selectivity of 3-izoxazolylsulfonylaminothiophenes as endothelin receptor antagonists. *Quant. Struct.-Act. Relat.* **1999**, *18*, 124–133.

25 WU, C., CHAN, M.F., STAVROS, F., RAJU, B., OKUN, I., MONG, S., KELLER, K.M., BROCK, T., KOGAN, T.P., and DIXON, R.A.F. Discovery of TBC11251, a potent, long acting, orally active endothelin receptor-A selective antagonist. *J. Med. Chem.* **1997**, *40*, 1690–1697.

26 WU, C., CHAN, M.F., STAVROS, F., RAJU, B., OKUN, I., and CASTILLO, R.S. Structure-activity relationships of N2-aryl-3-(isoxazolylsulfamoyl)-2-thiophenecarboxamides as selective endothelin receptor-A antagonists. *J. Med. Chem.* **1997**, *40*, 1682–1689.

27 So, S.S. and Karplus, M. Three-dimensional quantitative structure-activity relationships from molecular similarity matrices and genetic neural networks. 2. Applications. *J. Med. Chem.* **1997**, *40*, 4360–4371.

28 Burke, B.J. and Hopfinger, A.J. 1-(Substituted-benzyl)imidazole-2(3H)thione inhibitors of dopamine β-hydroxylase. *J. Med. Chem.* **1990**, *33*, 274–281.

29 Vedani, A., McMasters, D.R., and Dobler, M. Multi-conformational ligand representation in 4D-QSAR: Reducing the bias associated with ligand alignment. *Quant. Struct.-Act. Relat.* **2000**, *19*, 149–161.

30 Blommaert, A.G.S., Dhotel, H., Ducos, B., Durieux, C., Goudreau, N., Bado, A., Garbay, C., and Roques, B.P. Structure-based design of new constrained cyclic agonists of the cholecystokinin CCK-B receptor. *J. Med. Chem.* **1997**, *40*, 647–658.

31 Oprea, T.I., Waller, C.L., and Marshall, G.R. 3D-QSAR of human immunodeficiency virus (I) protease inhibitors. II. Predictive power using limited exploration of alternate binding modes. *J. Med. Chem.* **1994**, *37*, 2206–2215.

32 Leo, A. Personal communication, **2004**.

33 Abraham, D. *Burger's Medicinal Chemistry*, 6th edition. Wiley-VCH, Weinheim, **2003**.

34 Abraham, D. Personal communication, **2004**.

35 Parker, E.M., Grisel, D.A., Iben, L.G., and Shapiro, R.S. A single amino acid difference accounts for the pharmacological distinctions between the rat and human 5-HT_{1B} receptors. *J. Neurochem.* **1993**, *60*, 380–383.

36 Hann, M.M., Leach, A., and Green, D.V.S. Computational chemistry, molecular complexity and screening set design. In *Chemoinformatics in Drug Discovery*, Oprea, T.I. (Ed.). Wiley-VCH, Weinheim, **2004**, 43–57.

37 Hann, M.M. and Oprea, T.I. Pursuing the lead likeness concept in pharmaceutical research. *Curr. Opin. Chem. Biol.* **2004**, *8*, 255–263.

38 Goodnow, R.A. Jr., Gillespie, P., and Bleicher, K. Chemoinformatic tools for library design and the hit-to-lead process: A user's perspective. In *Chemoinformatics in Drug Discovery*, Oprea, T.I. (Ed.). Wiley-VCH, Weinheim, **2004**, 381–435.

39 Baringhaus, K.H. and Matter, H. Efficient strategies for lead optimization by simultaneously addressing affinity, selectivity and pharmacokinetic parameters. In *Chemoinformatics in Drug Discovery*, Oprea, T.I. (Ed.). Wiley-VCH, Weinheim, **2004**, 333–379.

40 Congreve, M., Carr, R., Murray, C., and Jhoti, H.A. 'Rule of Three' for fragment-based lead discovery? *Drug Disc. Today* **2003**, *8*, 876–877.

41 Povolna, V., Dixon, S., and Weininger, D. CABINET – chemistry and biological informatics network. In *Chemoinformatics in Drug Discovery*, Oprea, T.I. (Ed.). Wiley-VCH, Weinheim, **2004**, 241–269.

42 CABINET is available from Metaphorics LLC, Santa Fe, NM, http://cabinet.metaphorics.com/.

43 Hansch, C., Hoekman, D., Leo, A., Weininger, D., and Selassie, C.D. C-QSAR database. Available from the BioByte Corporation, Claremont, CA, http://www.biobyte.com/.

44 Furet, P., Gay, B., Caravatti, G., Garcia-Echeverria, C., Rahuel, J., Schoepfer, J., and Fretz, H. Structure-based design and synthesis of high affinity tripeptide ligands of the Grb2-SH2 domain. *J. Med. Chem.* **1998**, *41*, 3442–3449.

10

Cabinet – Chemical and Biological Informatics Network

Vera Povolna, Scott Dixon, and David Weininger

10.1
Introduction

The increasing amount of data that is being generated in chemistry, genomics and biology has necessitated informatics tools to help scientists explore and understand this data. Many efforts are underway to develop such tools. This chapter presents the Cabinet (Chemical And Biological Informatics NETwork) methodology. Cabinet consists of a set of servers that provide information using the HTTP (HyperText Transfer Protocol) and a set of libraries for generation of servers that can collaborate within the Cabinet framework. Each Cabinet server can communicate with other Cabinet servers to provide integrated access to diverse chemical and biological information.

10.1.1
Integration Efforts, WWW as Information Resource and Limitations

Two different, largely independent, dimensions can be used to characterize data integration methodologies: physical integration and logical integration. Physical integration describes the degree to which various data resources are brought together or are used in a distributed fashion. It can vary from completely distributed to completely centralized in a single system. Logical integration describes the degree to which various data sources are brought together under a single data model. This can vary from federated, where each data source may have an individual data model, to completely unified, where all data are subsumed in a single data model.

Physical integration is mainly a matter of technical or practical import. It may be necessary to make local copies of some data sources in order to ensure reliability or to reduce access times. For example, in the WWW (World Wide Web), Internet service providers often use local caches of popular web sites to improve access time. Because of the way certain servers operate, it may be necessary to have local copies of all data.

The far more important dimension of integration is the degree of logical integration. Some current integration efforts along this dimension will be described.

The most familiar highly distributed federation of information resources is the World Wide Web. It is possible to search the Web for text and a few other types of information with search engines like Google. However, since there is no structure to the information

Chemoinformatics in Drug Discovery. Edited by Tudor I. Oprea
Copyright © 2004 WILEY-VCH Verlag GmbH & Co. KGaA, Weinheim
ISBN: 3-527-30753-2

contained in Websites, it is sometimes hard to find sites that precisely match a query. The problems caused by the lack of structure of information in the current Web is the driving force for development of the Semantic Web [1] which adds meta-information and ontologies (structured specifications of the concepts and relationships in a field of knowledge) to provide means to describe the types and relationships between different chunks of data.

In the bioinformatics realm, SRS (Sequence Retrieval System) [2] is a popular system, which uses a centralized collection of data resources primarily in flat text file form and, more recently, handles XML (Extensible Markup Language) files as well. Data resources are treated in a federated manner since each is maintained in its original form. However, SRS contains a large number of cross-references between corresponding fields in various data sources, so that keyword searches can be done across them. SRS thus performs more structured searches across the information than what a simple text search provides (such as web indexes perform, for example). Even though the data model implicit in the cross-reference tables is not very deep, SRS provides a useful way for users to browse and do simple queries across a large number of data sources as well as to integrate results from some computational methods.

There are several efforts being taken to create a tighter logical integration by applying a common query language on top of a set of data resources. Examples include IBM DiscoveryLink [3] and Kleisli and K2 [4]. These systems typically require a wrapper function to be written for each data source, which describes the datatypes returned and provides methods for appropriate data formatting and access. Data sources can be physically distributed, but appear to be logically a part of a single query system. The queries are written in a query language (such as standard SQL) with subqueries for each data source or data type. The facilities of the query language are used to select and combine results.

A common query language provides a powerful mechanism to select and combine results from different data sources but does not necessarily hide the details or peculiarities of each data source. Composing queries can still require considerable expertise for each data source. TAMBIS (Transparent Access to Multiple Biological Information Sources) [5, 6] uses an ontology of biological concepts to control the query specification process. A graphical user interface (GUI) driven directly from the ontology is provided to help users build valid queries. The only queries possible are those that reflect the imposed data structuring.

At the complete unification extreme of the logical integration scale are data warehousing projects. For example, the PlasmoDB project [7] brings together a variety of genomics data on *Plasmodium falciparum*. This effort warehouses data in a relational database system called *Genomics Unified Schema*. It includes "a schema that integrates the genome, transcriptome and proteome of one or more organisms, gene regulation and networks, ontologies and controlled vocabularies, gene expression and interorganism comparisons" [8]. Other recent unification efforts include Integ8 [9] and Genostar [10]. The trade-offs between federated and unified database systems are expanded upon below.

An issue closely related to logical integration is the use of common file formats and shared languages to communicate information between different data sources. The most common language currently in use is the [11] XML, which is widely used in many areas for information interchange. XML is a standard language for defining new languages. As such, it is not by itself sufficient for exchanging chemical or biological information; many XML-based languages have been proposed for such purposes [12]. XML specifies the syntax of a language but not the semantics. Ontologies or other forms of controlled vocabularies or glossaries are necessary to impose semantics on an XML-based language. For example, the Semantic Web developers have proposed several languages to express the structure and relationships of data on the WWW [13].

Another interesting interchange method is Web Services [14], which is an XML-based set of protocols for passing information and services between servers on the Web. An example is the Distributed Annotation System (DAS) [15], which exchanges genomic annotations. These are various types of labels or information that can be attached to particular regions of a genome. While originally fairly limited in scope, the DAS project has led to efforts to extend the Web Services model more generally (e.g., bioMoby [16] in bioinformatics). The Web Services protocols provide means for data or computational servers to describe their inputs and outputs, to register themselves and to discover other Web Services that provide particular services. An older system for implementing distributed services, CORBA (The Common Object Request Broker) [17], has also been used for linking chemical and biological information servers [18].

The problem of unique naming often appears when exchanging or attempting to integrate information. Ontologies and controlled vocabularies help by giving unique names to concepts in related fields. But multiple, conflicting names also appear within a given domain such as sequences, genes and molecules. This has led to the Life Sciences Identifier (LSID) proposed by the Interoperable Informatics Infrastructure Consortium [19].

10.1.2
Goals

Cabinet servers are intended to be easy to learn and use. A user should be able to start with any server and access relevant information from all other servers without needing to learn new concepts or commands. It should be possible to start at any server, depending on what information is available to a user, and to see how that information fits together with information from other servers, even when the domain of information is unfamiliar to the user. By keeping the interfaces simple and by utilizing similarity relationships as a primary means of navigation, Cabinet servers make it possible for users to effectively explore diverse information domains without having to understand the structure of the knowledge before being able to ask questions.

10.2
Merits of Federation Rather than Unification

A philosophical and pragmatic alternative to unification is federation. With respect to information management, the primary difference between these two approaches is how information is organized in different parts of the overall information system.

10.2.1
The Merits of Unification

With unification, databases can become part of the same information system by virtue of their sharing a common underlying data model. Entities in different parts of such a data system are "of the same type" because they have the same semantics (i.e., "meaning").

Consider a molecular structure, which is the most important unifying information model in chemistry. Molecular structures appear in knowledgebases that represent catalogs of commercially available chemicals, pharmacology of named drugs, natural sources of bioactive molecules, protein–ligand interactions, measured molecular bioactivities, metabolic pathways, abstracted research literature, databases of synthetic reactions, and so on.

In a unified system, molecular structures would be stored in the same manner in each of these knowledgebases. When implemented as a relational database management system (RDBMS), each data set is typically represented as one or more tables within what can logically be considered to be a single, large-scale database. This ensures that relations expressed by the database schema are meaningful without having to consider representational issues. Turning this around, the existence of a schema dictates that the semantics of represented data must be consistent; ideally, a particular schema specifically defines such semantics.

The great merits of such an organization are simplicity of the data model and potential efficiency of implementation. One might imagine defining a database operation that accepts the name of a molecule as input and generates a report that lists the substance's natural sources, synthetic reactions, metabolic role, pharmacology, vendors' catalog numbers and price and other information. Since the schema associates one molecule with each record of each table in each dataset and all molecules are represented in the same way, it is possible to collect such data and create such a report. The same data model allows great efficiency in representation and searching, for example, relating information in tables by indexes rather than by their content.

10.2.2
The Merits of Federation

With federation, databases can become part of the same information system because they share common communication language(s). Each database in a federated data system is free to establish and maintain its own data model. The utility of such a system is critically dependent upon the ability of each participating database to process queries using common protocols. It is only required that databases in such a system understand the languages that are appropriate for their local data model. It is not required (nor particularly desirable) that all databases understand all query languages.

Communication between federated databases is much like communication between humans. If you ask someone what they know about "Casablanca", the answer depends on what kinds of things they know about it – even more critically than, for example, how much they know or where they learned it. A video store clerk might answer, "It's available in DVD or VHS, aisle 5". A film buff might answer, "Bogart/Bergman! The most romantic couple in film history". An architect might say, "An ancient walled city in Morocco, with heavy French colonial influences, the Mosque of Hassan II is amazing ...". All this information is valid and potentially useful.

The same phenomenon occurs in the previous example of molecular structure. If one made a query about the Lipitor structure (atorvastatin, a current "blockbuster" drug), one might expect that the information returned from various databases would differ in kind as much as in content: synthetic reactions from one, pharmacology from another, where to buy it and how much it costs from another, how it affects metabolism, and so on. However, if a unified, strictly indexed data system were used, the most common answer would be "nothing found". One reason for this is that atorvastatin occurs in two molecular forms: a linear acid (as sold) and the physiological form, a cyclic lactone. Another common reason is that Lipitor is not commercially available in any molecular form but rather as a salt (usually the calcium salt). In a unified data system, it would be up to the database designers to make sure that one form is linked to all tables in all constituent data sets and up to the data entry people to make sure the data is consistently entered. This sort of diligence is admirable, but in most ways it makes the problem worse rather than better. The underlying issue is one of data model selection. For certain kinds of data sets, only the molecular structure makes sense, for example, databases of synthetic reactions. For a few others, only the salt form makes sense, for example, when you buy Lipitor or dispense it to a patient, what does 100 mg of Lipitor "mean"? For the remaining, the molecular and salt forms are synonymous.

The great merit of a federated informatics system is that each constituent database can have its own data model and information can be stored in ways that are meaningful for that kind of data. One database might store this substance as a "prescription pharmaceutical sold as calcium salt", another as an "HMG-CoA reductase inhibitor", yet another as "a molecule metabolized by Cytochrome P450 3A4", and so on. Each database might represent it as a slightly different molecular structure, that is, appropriate to that database's natural data model. Yet, the federation strategy allows all these databases to form a single integrated data system.

10.2.3
Unifying Disparate Data Models is Difficult, Federating them is Easy

Unifying disparate information sources into a single data model is an intrinsically difficult process. Even when the same kinds of information are collected in different places, it is done so for different reasons, which are reflected in data organization. A common example is extraction of information from scientific literature, where the goal is to make a collection of a particular kind of data. Since the origins of the data are research papers, such efforts often result in "document databases" where the article is the primary entity and the content is represented by various attributes (e.g., lists of referenced molecules). However, a document-oriented database is rarely suitable for the specific requirements of most scientific investigations. Such efforts often "invert" the database, making the molecule the primary entity with attributes that are important to the project, and relegating document-specific information to a citation attribute. Sometimes an effort is made to integrate one or more specific data models into the same database. However meritorious each kind of data organization is in particular, it is certainly true that unifying them into a single data model is very difficult. In almost all such cases, the definitions of corresponding fields in each database start out being nonidentical. One is left with a choice of "joining" fields with similar semantics (thus losing semantic precision) or "splitting" them (thus complicating the data model). Neither choice is ideal.

Even unifying information about a very specific field such as the pharmacology of discrete drugs is a monumental task. The primary reason for this difficulty is the disparity of purpose between various information sources. Consider how different the purposes are of WDI (World Drug Index) [20], PDR (Physician's Desk Reference) [21], MSDS (Material Safety Data Sheets) [22] and REG (Chemical Abstracts Registry) [23], each of which has a different natural data model.

Thinking beyond specific fields, the idea of unifying all human knowledge into a single data model seems nearly impossible.

In contrast, the idea of federating disparate information sources is very straightforward. Each information source can be organized to represent its kind of data in the most natural and expressive manner for that specific kind of information. Given query languages that are adequately expressive, a federation of such databases can appear to be a single integrated information system.

Federated data systems have some characteristic disadvantages. They are less efficient than systems with a predefined organization, which is systematically optimizable. Theoretically, federated data systems should require more computer resources to deliver the same capacity/performance as a unified system. In practice, this does not seem to be the case. The reason for this is largely because the capacity of modern computers has far exceeded the storage requirements of most specific data collections that currently exist. For example, the PDB (Protein Data Bank) [24] was historically considered a very large amount of information, and until a few years ago, it was processed by specialized, dedicated disk-oriented systems. Today, the unabridged PDB fits in the memory of a good consumer-level desktop computer (and being in RAM, data access rates are many orders of magnitude faster). Another example is WWW, insofar as it represents a federated data system. The WWW works remarkably well and represents a huge amount of computational resources – which (almost) always seem available and up to the task.

10.2.4
Language is a Natural Key

The utility of federated systems is dependent upon communication between participating data systems. It is useful to think of this communication as existing at three levels. The lowest level represents the network connection between devices, encompassing the physical connection and network protocols such as TCP/IP. The middle level represents the data transfer protocol, which is logically responsible for reliable transfer of data between processes, for example, HTTP. The highest level represents the specific syntax (format) and semantics (meaning) of the message being communicated.

Although any well-defined computer data format is potentially usable at this level, "languages" are particularly suitable for the highest level of communication in federated information systems.

Although "languages" and "computer data formats" have similar roles and can appear to have similar forms, they differ in their overt purposes. The purpose of a data format is to provide an external representation that will produce a given state in computer programs which "read" that format. Most often, data formats directly represent a computer data structure. An example is the CAS A Distribution Format (ADF), which represents a specific data structure used to process molecular registrations. An even more extreme example is the Wisconsin Molecular Picture (WIMP) format, which was designed to describe a particular way of drawing a molecule. In contrast, the purpose of a language is to represent abstractions; this is true for both artificial languages and natural human languages. One characteristic of languages is that they directly represent a specific abstraction (which may be a model). They are not defined in terms of a particular computer, computer program or program behavior. However, languages can describe a computer as an abstraction itself, for example, Knuth's infamous MIX [25] computer. An example of a molecular language is Simplified Molecular Input Line Entry Specification (SMILES) [26], which specifically represents a valence model of molecules and reactions.

Data systems that participate in a federation can be very heterogeneous both in their content and implementation. In a federated system, it is a great advantage to define communication in terms of external abstraction rather than in terms of internal computer representation. This is not so true for unified systems, where the imposition of a specific data model among components makes it more reasonable to define communication in terms of computer-oriented characteristics and assumptions.

Two communication languages used by Cabinet are SMILES and ECN (Enzyme Commission Number). When one Cabinet server asks another, "What do you know about Lipitor?", it is asking about SMILES:

```
CC(C)c1c(C(=O)Nc2ccccc2)c(c(c3ccc(F)cc3)n1CC[C@@H]4C[C@@H](O)CC
(=O)O4)c5ccccc5
```

When asking, "What do you know about hydroxy-methyl-glutaryl-CoA reductase?", it asks about the EC Number:

```
1.1.1.34
```

Both are entities in well-defined languages that represent specific abstractions: SMILES, the valence model of a molecule; ECN, a classification of enzyme functionality.

10.3
HTTP is Appropriate Communication Technology

The choice of a data transfer protocol (middle layer) is also important to the operation of a federated information system. An ideal data transfer protocol would be architecture-independent, provide reliable data transfer over existing network protocols and, ideally, it would already be globally disseminated. Today (in 2003), the choice seems obvious: HTTP (HyperText Transfer Protocol).

10.3.1
HTTP is Specifically Designed for Collaborative Computing

HTTP was conceived at the European Organization for Nuclear Research (CERN) in 1989 specifically for the purpose of sharing information between scientists working in different universities and institutes all over the world [27]. The emphasis was on providing a means of reliably disseminating many different kinds of information about complex evolving systems to a heterogeneous environment. Many of the most important aspects of distributed communication were addressed from its inception: integration of static and dynamic information, architecture independence through time, efficiency of data exchange versus data security, global network analysis and content indexing.

Originally named the "Mesh", the idea of a distributed information system based on computing elements sharing HTTP as a common data transfer protocol formed the basis for the "World Wide Web" (WWW). Many other critical components were required to create WWW as we know it today. The most important ones include:

- a freely available HTTP/HTML browser (Mosaic, NCSA),
- Uniform Resource Identifiers (URL/URI, NCSA and CERN, 1991–1994),
- extensible message datatyping (MIME, Borenstein/Freed, 1992),
- a client data entry protocol (FORMs, LANL), and
- a standards organization (W3C).

It is a bit of an exaggeration to say that the invention of HTTP is equivalent to the invention of WWW. The specific idea of linked hypertextual information on distributed computer systems for scientific information exchange goes back at least to Vannevar Bush's article *As We May Think* in 1945 [28]. However, it is certain that the emergence of HTTP was the primary enabling technology for WWW and, more generally, it has effectively enabled the use of computers for globally distributed communication.

10.3.2
HTTP is the Dominant Communication Protocol Today

To most end-users, the roles of HTTP and HTML in the WWW are blurred. If they think about it at all, both HTTP and HTML are lumped into the category of "things that make my browser work". The functional distinction between HTTP and HTML is very important when discussing the design of distributed information systems.

HTTP (HyperText Transfer Protocol) is a mechanism to exchange data between computers in a reliable and machine-independent manner [29]. The name "HTTP" is a misnomer. Although named for its historical ability to transfer hypertext, HTTP may be used to transfer any kind of data including information about what kind of data is being transferred. HTTP's datatyping mechanism is formally expressed in the "Content-Type" header of each message; it borrows the extensible MIME (Multipurpose Internet Mail Extensions) specification (originally designed for e-mail) to describe such types [30].

HTML (HyperText Markup Language)is a specific type of information that can be transferred by HTTP [31]. It is transferred in the same way as any other HTTP message; its MIME type is "text/html".

HTML is probably the most visible kind of data transferred over HTTP on WWW. But it is only one of hundreds of kinds of data that are transferred by HTTP on WANs (Wide Area Networks, LANs (Local Area Networks) and internal networks. HTTP offered an appropriate level of technology at the right time and its use has grown exponentially over the past decade. Virtually every computer in the world today contains an HTTP interpreter and generator (as do many dedicated devices which do not look like computers).

10.3.3
HTML Provides a Universally Accessible GUI

The universal availability of HTTP/HTML browsers makes it the GUI of choice.

Ever since FORMs were added to HTML in the mid-90s, browsers on an HTTP network could perform all basic workstation functions (including entry of data and queries) using remote computing resources. FORMs provide a mechanism for textual data entry, selection of text items (menus), graphical pointing (image mapping) and file uploading.

Even so, working HTML FORMs is somewhat primitive compared to other GUIs designed for high bandwidth interactivity. Many of these limitations have been relieved by the availability of Java, JavaScript, plugins and helper applications. Likewise, the availability of "cookies" has eased the limitations of the underlying user-context-free HTTP/HTML client–server model.

In general, a basic HTML-based GUI appears to suffice for most scientific interfaces and the plugin/helper strategy suffices for special purposes. However, a disturbing trend is the recent appearance of browsers that are not extensible via helper applications (e.g., Safari, IE-6). Such browsers are not likely to be suitable for scientific applications.

10.3.4
MIME " Text/Plain" and "Application/Octet-Stream" are Important Catch-alls

The MIME type "application/octet-stream" is the default HTTP type – it describes messages that are composed of 8-bit bytes (octets) but are otherwise uninterpreted binary. The type "text/plain" describes the message content that is composed of simple, printable (7-bit) ASCII text.

In principle, these are extremely important datatypes because it means that the HTTP server–client model subsumes most standard input/output (stdin/stdout) serial application protocols.

A simple HTTP client program (e.g., "fetch") can be pointed at an HTTP server, accept lines of input from a standard input stream, send them as messages to an HTTP server, accept replies from that server and then write them to a standard output stream. The "fetch" program becomes a serial filter for any such HTTP server (which is now acting as a network compute server).

Given that "text/plain" and "application/octet-stream" messages can carry information in arbitrary languages and that database languages exist, HTTP networks have the ability to support distributed information systems. This is very useful.

10.3.5
Other MIME Types are Useful

The "text/plain" example above demonstrates that HTTP networks can support distributed information systems when given appropriate languages, that is, languages that describe abstractions appropriate to that information system. Many other standard MIME types are useful. Most are very specific, for example, "image/gif" is a specific format for bitmapped images, "application/PDF" is a page description format and "application/tar" is a 4.3 BSD archive. Some describe more general abstractions, for example, "application/xml". Private (unauthorized) MIME types are also available, for example, "chemical/x-pdb" and "chemical/x-smiles".

10.3.6
One Significant HTTP Work-around is Required

HTTP is designed beautifully for client–server and server–client communication and the design of HTML follows suit. They were not designed for server–server communications and some deficiencies in this area show up when implementing a distributed information system on top of HTTP/HTML. Fortunately, these are easy to work around.

An important example is the absence of any notion of "encapsulated HTML". Cabinet servers offer *"What's related"* hyperlinks which lead to pages containing lists of related information available on other servers. Static pages of links are easy enough to describe in HTML, and such pages are an important part of most web sites. In a dynamic distributed information system, such pages must be generated on the fly via server–server communication. The server offering a *"What's related"* page needs to ask other servers, "What do you know about *whatever?*" and then generate the page to be sent to the client.

The most natural way of performing this function would be to ask a server for a bit of HTML with links to its pages containing relevant information. Such bits of information would be combined to make the *"What's related"* page. However, there is no support for "encapsulated HTML". This limitation is a bit strange, since one can easily encapsulate other kinds of data, such as images, into HTML. One approach would be to invent a new MIME type, for example, "text/x-ehtml".

Another approach is to transmit the response as a normal HTML page with the encapsulated answer being delimited in some manner. HTML comments are natural delimiters. For instance, Cabinet servers can ask each other, "What do you know about this SMILES?" This is known as a "peeksmi" request. The response is a normal HTML page (or 404 Not Found) in which the encapsulated HTML is delimited by the special HTML comments:

```
<!-- PEEKSMI STARTS -->
```

and

```
<!-- PEEKSMI ENDS -->
```

The encapsulated portion of the peeksmi response is then extracted into the *"What's related"* page.

10.4
Implementation

Implementation of a federated information system falls into three parts: deciding on an implementation strategy, developing tools to make this strategy practical, then actually implementing the components. In an ideal universe, strategic decisions are easy for a project with well-defined goals – one simply picks technologies that support the goal set. In practice, technologies are often not available to support all project goals, and in many cases, available technologies conflict with each other.

In this case, the availability of HTTP and HTML resources greatly simplifies the strategic design problem. The tactical decisions of how to implement HTTP services for use by HTML clients (and other services) are discussed here. The primary goals of universality, flexibility and longevity led to an unexpected but very usable result.

10.4.1
Daylight HTTP Toolkit

A number of widely available HTTP servers were examined, two of which were used in experimental Cabinet implementations (NCSA's httpd [32] and Apache [33]). Both are robust, easy to set up, easy to customize, highly standards-compliant and are available for a wide variety of platforms. The Apache server also provides a limited ability to link to run-time libraries and good support for secure services. However, both were designed to provide support for general-purpose websites with primarily static content and limited (Common Gateway Interface (CGI)-like) dynamic content. They are also designed to be the primary HTTP service on a given machine. Neither was designed to support multiple instances of large, interoperating services needed by a federated data system.

Alternatives are to use an HTTP programming library or to "roll your own" HTTP server, that is, write one from scratch. Daylight Chemical Information Systems (CIS) produces a commercial object-oriented HTTP Toolkit [34], which allows a great deal of flexibility in generating HTTP servers and clients. In particular, it is possible to build multiple interoperating servers with large amounts (4 GB+) of data embedded in memory, an ideal architecture for Cabinet. Therefore, production versions of Cabinet servers are implemented as Daylight HTTP toolkit programs rather than built into multiple instances of standard Apache servers. One factor which simplified this decision is that the Daylight HTTP Toolkit provides an architecture-independent API (Advanced Programming Interface) with a strict data interface, which means that the HTTP layer implementation can be easily changed to some other mechanism if alternative technologies appear in the future.

10.4.2
Metaphorics' Cabinet Library

The strength of a federated data system is that it provides access to many different data sources. On one hand, this implies a profound requirement to process data in a flexible manner because the nature of the data is so variable. On the other hand, this implies that there will be a large number of utilities that are duplicated in many places, for example, management of help, options, login, language processing, HTML and graphics generation, interserver communication, and so on.

This is an archetypical situation for a special-purpose programming library. Starting with about a dozen fully functional data servers, a "cabinet library" was written to subsume common functionality. First, essential communication services were implemented, including common language processing and interservice communication utilities. Next, common utility services such as help and preferences via cookies were implemented to minimize duplicated code and allow uniform interfaces. Finally, convenience libraries were implemented to reduce the task of programming a new server, for example, clean APIs for gif and PDF generation. As the library grew, individual servers shrank. By the time Metaphorics' first Cabinet production servers were released, each Cabinet server essentially contained only the code that is specifically needed to support the kind of data served by that server.

10.5
Specific Examples of Federated Services

The number of Cabinet servers is growing. Most of them are destined for production; some serve as utility servers. All have the same type of interface and navigation. Each server has a home page with a navigation bar on top and a search entry form. Databases can be searched by text (a word or a part of a word), molecular structure, SMILES, EC number, pathway, pattern (in Wombat) or protein alignment (in Planet). Wherever possible, the output text is hyperlinked to indexes of associated information in the database.

All of the databases have two common search features: *similar* and *related*. *Similar* searches within the current database, whereas *related* searches other Cabinet databases. The *related* search lists links to all the databases where a particular structure, keyword or EC number appears. New relationships can be discovered while going from one database to another.

10.5.1
Empath – Metabolic Pathway Chart

Empath is a Cabinet database of metabolic pathways that models a metabolic pathway chart. It initially models the Boehringer Mannheim wall chart [35] but other pathway layouts are possible. It currently includes 1462 steps (metabolic reactions). The Empath database consists of more than 8000 live objects. Every object has an exact geometric location, that is, x, y coordinate, which is optionally indicated by visible hotspots. The chart is clickable everywhere and the image recenters itself around the selected point. The current object is the one closest to the center of the image and is identified by a bull's eye. A summary of the object is given (e.g., its structure, EC number, SMILES, reaction stoichiometry, etc.). Empath provides navigational features such as zoom in/out, wider, thinner, taller and shorter.

Empath maintains ten object classes. *Landmarks* indicate a general area of the chart where a particular metabolism occurs. *Compounds* are primary reactants and products. *Cofactors* are additional reactants and products that usually drive steps energetically. *Enzymes* are the primary catalysts of steps; most of them have an associated EC number. *Agents* are secondary catalysts, usually metal ions. *Regulators* are substances that enhance or suppress a particular reaction. *Steps* are reactions that include all of the above object types. *Pathways* are sequences of life essential steps or steps that occur across all species. *Diseases* as well as *Notes* are associated with steps. *Abbreviations* are a separate category for abbreviated names on the chart and have a link to the area of primary synthesis.

A fingerprint similarity search is available within Empath by selecting *similar* in the summary of objects that includes structures. The *related* button finds similar structures in other Cabinet databases.

A directed graph search is available where compounds are nodes and metabolic steps are edges. The server finds the shortest allowable path between one node and another. One can limit the search to steps that occur in a certain kingdom and can also choose to display only pathways visually connected on the metabolic pathway chart. There are thousands of connections that are not drawn on the chart but can be found by using this search.

Empath provides a view of a reaction in the context of metabolic pathways. As such, it serves as an index for other Cabinet databases.

10.5.2
Planet – Protein–ligand Association Network

Planet is a Cabinet server that provides information about protein–ligand interactions. Data are read in the PDB format for observed or hypothetical interactions. Each data object is a relationship between a protein and one or more ligands and other small molecules (cofactors, etc.). Several specific relationships are explored: the relationships between small and big molecules, between protein chain residues and relationships based on EC numbers.

Planet offers a 2-D representation (similar to that used in LIGPLOT [36]) of the protein–ligand interaction called a PLAID (Protein–Ligand Accessibility and Interaction Diagram, Figure 10.3). It is automatically calculated from the PDB data and shows van der Waals interactions, hydrogen bonds and solvent accessible areas. The interaction can be also viewed in 3 D or as the original PDB entry.

Planet interacts with itself and with other Cabinet servers in different ways depending on what relationship is desired. Searching is available by protein chain alignment and ligand structural similarity within Planet itself by selecting *similar*. Enzyme functionality and ligand-structure search across all Cabinet databases is available with the *related* button.

10.5.3
EC Book – Enzyme Commission Codebook

EC book is an index server that provides access to enzymes and EC numbers. EC numbers are assigned by the Enzyme Commission on the basis of enzyme functionality [37]. Each EC number consists of four levels of function. There are six main classes of enzymes. The EC book database is usually updated once a year (but more frequently in recent years).

EC book is usefully connected to Wombat, Empath, Planet and WDI, where many EC numbers appear.

10.5.4
WDI – World Drug Index

WDI is a comprehensive database of named drugs. These include marketed drugs, drugs in development, new chemical entities and trial preparations [38]. The data come from journal articles and conference reports. WDI is updated quarterly.

WDI includes molecular structures and 2-D coordinates for most entries. Thirteen kinds of names are distinguished such as Trade Name, USANs and CAS numbers. There are seven types of pharmacological data: pharmacological activity, mechanism of action, indications and usage, precautions and warnings, contraindications, adverse effects and interaction data. Two kinds of enzymes appear in the database: the enzymes that act as drugs and enzymes that are the drug targets. The Cabinet version of WDI includes EC numbers for both kinds.

The navigation through the database starts at the WDI home page. It is designed to be simple and a few clicks away from anything in the database. Text, SMILES and EC number searches are available. The search results are sorted alphabetically with the typed query highlighted. The WDI home page also has hyperlinks to A–Z index, introductory examples, pharmacological activities, keywords, indication and usage classes, popular drugs and enzymes.

All WDI entries include *similar* and *related* functions. *Similar* shows structures within WDI and *related* gives a list of other databases where this substance or related structures appear.

10.5.5
WOMBAT – World of Molecular Bioactivity

WOMBAT [39] is a Cabinet database that provides information about biological activity of small molecules [40]. The dataset comes from publications in the Journal of Medicinal Chemistry and other periodicals. WOMBAT is updated twice a year. The data consist of series of compounds that were observed and compared for a specific activity, for example, K_i, IC_{50}, D_2 and EC_{50}. The database also includes calculated LogP and Log S_w values, as well as other descriptors related to flexibility, size, and so on.

Each series has a set of keywords that specify the series' properties. The database can be searched by author name, citation, compound name, SMILES or keywords. A search of a substructure or a generic pattern is also possible. The pattern of molecules in a series is highlighted. Compounds in most series include the same or parent structures with different functional groups. Many series also include existing drugs or other lead compounds. A possible correlation can be derived between the structure and the measured bioactivity.

Each substance in the database has its unique ID number. With every structure, there are two associated functions: similar and related. *Similar* searches WOMBAT and *related* searches all Cabinet databases. Some series involve an enzyme with an EC number listed and can be searched by selecting the *related by EC* button.

10.5.6
TCM (Traditional Chinese Medicines), DCM (Dictionary of Chinese Medicine), PARK (Photo ARKive) and zi4

TCM is a Cabinet database that attempts to bridge Eastern and Western medicines and their use. The TCM data of molecules come from the book "Traditional Chinese Medicines" [41] and has two kinds of data. Molecular entries are structures contained in TCM medicinals. Biological entries are the sources of these medicinals (plants, minerals etc.). Many relationships can be discovered within the database on the basis of structure (SMILES) or language (English, Chinese, Latin and Pinyin).

TCM is loosely coupled with DCM (Dictionary of Chinese Medicine) and PARK (Photo ARKive) and establishes relationships with Planet and WDI.

DCM is a Cabinet server that provides encyclopedic information of Chinese medicines and medical conditions from the book, *A Practical Dictionary of Chinese Medicine* [42]. It has over 220 000 searchable fields in four different languages (English, traditional Chinese, Pinyin and Latin). DCM is a tool that includes Chinese and Western medical terminology and gives medication as well as acupuncture treatment descriptions. DCM is implemented for finding definitions without leaving the TCM database. Many medicinal sources are identified by pictures generated from PARK.

PARK is a Cabinet server that provides photographs for the purpose of plant identification. All photographs are annotated. The database is searchable in multiple languages (English, Latin and Chinese). Photographic sources are images from collections such as the *Atlas of the Chinese Materia Medica* [43].

zi4 is a noninteractive Cabinet server that provides various Chinese character image and mapping services to other Cabinet servers. zi4 is tightly connected with TCM and DCM. It outputs transparent gifs of Unicode, Traditional and Simplified Chinese characters. It is also able to translate Traditional Chinese to Simplified Chinese.

10.5.7
Cabinet "Download" Service

It has been proved to be useful to provide universal availability of consistent browsers and helper applications via Cabinet servers. The *download* server is a convenient place for browser, PDB viewer and PDB helper downloads. The combination of Netscape [44] as a browser and PyMOL [45] as a molecular viewer seems to work well. Other options are available.

10.5.8
Cabinet Usage Example

A nearly infinite number of relationships can be found using Cabinet. The starting point may be a structure, name, biological activity, and so on, which occur in any database. The existence of relationships does not need to be known beforehand but can be discovered while surfing Cabinet. Here is an example of one way Cabinet can be used.

Lipitor is currently one of the best selling drugs for lowering cholesterol. Searching in WDI for "Lipitor" produces the page shown in Figure 10.1 [46]. The preferred name is "Atorvastatin Calcium". WDI has information about the pharmacology, indications, interactions, and so on, hyperlinked for keyword search.

When the *related* button above the structure is selected, a list of similar structures in other databases is shown. Empath, Planet, TCM and WOMBAT are found. In Planet, the closest structure is 1HWK. Selecting it brings up the Planet version of the PDB

wdi: **ATORVASTATIN CALCIUM**	DERWENT
	WORLD DRUG INDEX

WDI A-Z search help preferences fedora prev next

structure (similar, related)   **approved names**

DXRN: CI981

DPN: ATORVASTATIN CALCIUM

CAS: 134523-03-8

DRN: CI-981

USAN: ATORVASTATIN-CALCIUM

Substance description: 2:1 ratio

Pharmacology

Pharmacological activity: ANTIARTERIOSCLEROTICS; HMG-COA-REDUCTASE-INHIBITORS; (EC-1.1.1.34);
Mechanism of action: HMG-COA-REDUCTASE-INHIBITOR; (EC-1.1.1.34);
Indications & usage: LIPID-METAB.DISORDER; as adjunct to diet in heterozygous familial-hypercholesterolemia; LIPID-METAB.DISORDER: dysbetalipoproteinemia; LIPID-METAB.DISORDER: elevated serum trigluceride levels; LIPID-METAB.DISORDER: as adjunct to diet in primary-hypercholesterolemia; LIPID-METAB.DISORDER: as adjunct to diet in homozygous familial-hypercholesterolemia; LIPID-METAB.DISORDER: as adjunct to diet in mixed-hyperlipemia
Precautions & warnings: Children; perform hepatic function test before and during treatment; adequate female contraception; history of alcohol-addiction or hepatopathy
Contraindications: Acute-hepatopathy; pregnancy; persistent elevated hepatic enzymes; lactation
Adverse effects: GASTROENTEROPATHY; constipation; GASTROENTEROPATHY; diarrhea; GASTROENTEROPATHY: dyspepsia; HEPATOPATHY; MISC.: headache; MISC.: nausea; MYOPATHY: myalgia; MYOPATHY; MISC.: asthenia; GASTROENTEROPATHY: gastralgia; GASTROENTEROPATHY: flatulence; MISC.: insomnia
Interactions: Antacids; ciclosporin; colestipol; drugs metabolized by cytochrome P450 3A4; fibrates; oral-contraceptives; warfarin; niacin; erythromycin; digoxin; azole-antifungals

Registered trade names

trade name	company	country
LIPITOR	PARKE-DAVIS	United States of America
LIPITOR	WARNER-LAMBERT	United Kingdom
		Ireland
		Japan
		United States of America
		Australia
LIPITOR	YAMANOUCHI	Japan
SORTIS	WARNER-LAMBERT	Germany
		Croatia
		Slovenia
ZARATOR	PHOENIX	Argentina

Other names

ATORVASTATIN CALCIUM
CALCIUM ATORVASTATIN

Update information and references

Appeared: 1994.1
Updated: 2003.2

peek

Metaphorics	*Metaphorics LLC*	wdi 4.82

Fig. 10.1 WDI entry for Lipitor.

entry. The protein that binds atorvastatin is HMG-CoA reductase (Figure 10.2) [47]. Other ligands of HMG-CoA reductase are also depicted.

Figure 10.3 shows an enlarged image of the PLAID diagram from Figure 10.2 [48]. It demonstrates the interaction of the protein and the ligand in 2 D. Underneath the

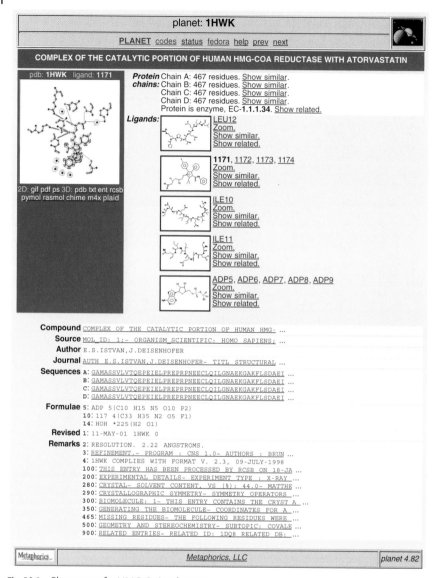

Fig. 10.2 Planet entry for HMG-CoA reductase.

PLAID image are options for seeing the 3-D structure (Figure 10.4 [49]) and a link to the official PDB website. It is possible to search for similar proteins and see their alignment by selecting *show similar* (Figure 10.2). There is also a link to enzymes of similar function, *related by EC*. Clicking on it shows all places in Cabinet where the same or similar EC number occurs. Reference sources are listed at the bottom of Figure 10.2.

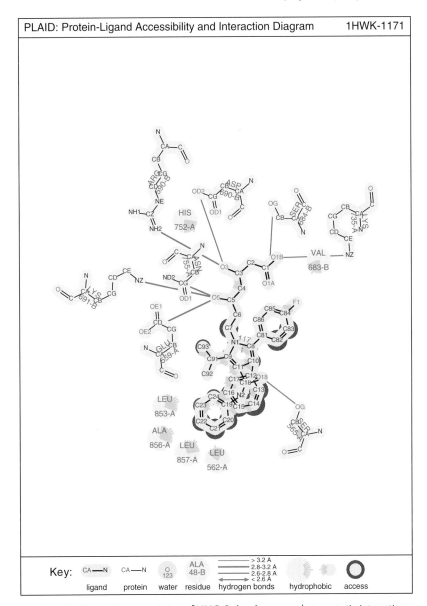

Fig. 10.3 PLAID – 2-D representation of HMG-CoA reductase and atorvastatin interaction.

In our example, EC book and Empath find an exact match to HMG-CoA reductase. The Empath link shows the metabolic step that the enzyme catalyzes (Figure 10.5 [50]). The reaction is between *S*-3-hydroxy-3-methylglutaryl-CoA and Mevalonate. The step summary on the right side of the chart image shows activation and regulation of the enzyme, its biological scope, direction, reversibility and stoichiometry. A pathway search

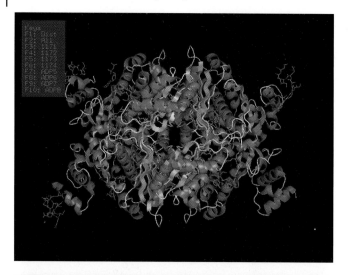

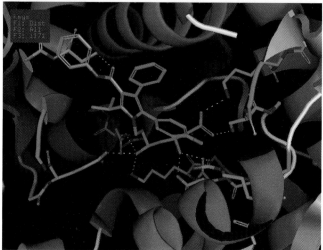

Fig. 10.4 Two PyMOL pictures of HMG-CoA reductase (whole molecule and atorvastatin up close).

from HMG-CoA reductase to cholesterol is available via the *path* link. Cholesterol is entered as the end product, the animal kingdom is selected and a pathway is generated (Figure 10.6 [51]).

Clicking on the product of the reaction, mevalonate, other *related* information such as 3-D structure with Ruby (Rule-based invention of conformations), a connection to TCM, WDI and WOMBAT are found. TCM depicts several similar compounds, one of which is citric acid with a 0.87 similarity to mevalonate. Clicking on it brings a list of 12 TCM sources in which citric acid appears (Figure 10.7 [52]). All the sources come

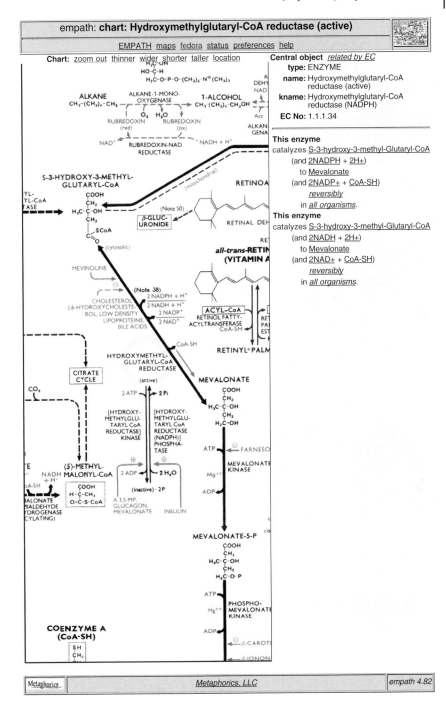

Fig. 10.5 Empath entry for a metabolic reaction catalyzed by HMG-CoA reductase.

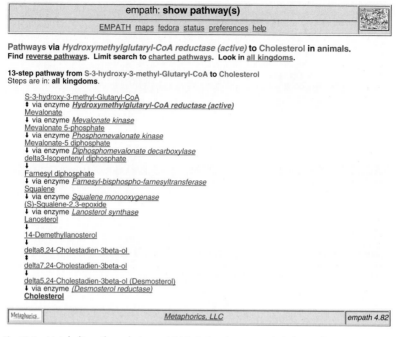

Fig. 10.6 Metabolic pathway between HMG-CoA reductase and cholesterol (altered by Lipitor).

from plants and are illustrated by photographs from PARK. Family name, effects and indications are hyperlinked to DCM.

Biological activities of structures similar to mevalonate can be found in WOMBAT. The closest similar molecule is in series 1062, which is (according to the paper it was published in) a novel thiol-containing citric acid analog (Figure 10.8 [53]). It has a measured K_i value that can be compared to other molecules in the series. The figure in bold indicates the common pattern for the majority of compounds in this series.

10.6
Deployment and Refinement

Implementation of Cabinet was done under Solaris using various Sun workstations. Development included simultaneous porting to Red Hat Linux and Macintosh OS-X. The initial deployment of Cabinet consisted of 14 servers delivering about 2 GB of in-memory data to WWW and was piggybacked on a corporate web server. Although the performance was generally acceptable, its operation generated some fear that critical network services such as mail and ftp might be interrupted. Two small server solutions were evaluated for a stand-alone Cabinet implementation – current versions of Dell/Red Hat Linux and Macintosh/OS-X dual processor desktop machines. Both

| tcm: **Citric acid** |
| chemical entry: substance structure similarity related... |
| TCM bio pin chem help preferences fedora prev next |

English name: Citric acid
TCM ID: M1124
Formula: C6H8O7
Mol. wt: 192.14
CAS No: 5373-11-5

The following 6/12 TCM medicinal sources contain the above substance.

Seabuckthorn Fruit   51 known compounds

Elaeagnaceae   *(HIPPOPHAE RHAMNOIDES L.)*

CU LIU GUO

effects: To quicken blood and dissipate stasis, transform phlegm and loosen the chest, invigorate the spleen and fortify the stomach.
indications: Knocks and falls, stasis swelling, cough with profuse phlegm, difficult breath, indigestion.

Chinese Ephedra   38 known compounds

Ephedraceae   *(EPHEDRA SINICA STAPF)*

MA HUANG

effects: To induce diaphoresis, relieve panting and facilitate the lung-qi and disinhibit urine.
indications: Syndrome of exterior repletion shown as aversion to cold, fever, headache, nasal congestion, anhidrosis, floating and tense pulse; cough; edema.

Narrowleaf Cattail Pollen   37 known compounds

Typhaceae   *(TYPHA ANGUSTIFOLIA L.)*

PU HUANG

effects: To cool the blood and stanch bleeding, quicken the blood and disperse stasis.
indications: Various kinds of bleeding, including hemoptysis, spontaneous external bleeding, blood ejection, bloody stool, bloody urine, flooding and spotting, and traumatic bleeding; cardiac or abdominal pains, menstrual pain and afterpains; blood strangury with inhibited pain.

Ginseng   124 known compounds

Araliaceae   *(PANAX GINSENG C. A. MEY.)*

REN SHEN

effects: To reinforce qi and restore pulse from collapse, to supplement the lung and the spleen, to promote secretion of body fluid and relieve mental stress.
indications: Prostration syndrome due to deficiency of qi; deficiency syndrome of the spleen; syndrome of deficiency of the lung-qi; thirst and diabetes due to impairment of body fluid; irritability, insomnia, dreaminess, palpitation induced by fright, forgetfulness, listlessness, lassitude and others due to deficiency of both the heart and spleen; syndromes of deficiency of the blood or both the qi and blood.

Chinese Hawthorn   47 known compounds

Rosaceae   *(CRATAEGUS PINNATIFIDA BGE.)*

SHAN ZHA

effects: To remove food stagnancy and blood stasis.
indications: Indigestion and retention of food, abdominal distention, anorexia, abdominal pain and diarrhea; postpartum abdominal pain and lochiorrhea due to blood retention; hypertension; hyperlipemia; angina pectoris of coronary heart disease.

Tall Gastrodia   20 known compounds

Orchidaceae   *(GASTRODIA ELATA BLUME)*

TIAN MA

effects: To subdue the exuberant yang of the liver, calm the internal wind, and relieve convulsion and fainting.
indications: Epilepsy induced by fright and convulsion ; headache and light-headedness; rheumatic impediment, numbness of the limbs or tetraplegia.

| Metaphorics | *Metaphorics LLC* | *tcm 4.82* |

Fig. 10.7   TCM list of plants in which citric acid occurs.

Single bioactivity entry.

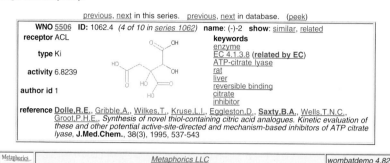

Fig. 10.8 WOMBAT entry for a molecule similar to Mevalonate.

can be configured with the RAM needed to manage all data in memory (4 GB) and both offer outstanding performance in this configuration.

The only major technical difficulty encountered during initial deployment was providing access to Cabinet's unusual TCP/IP ports to clients behind firewalls via WWW. This was solved by promoting the *cabinet* Cabinet server from its original role as a server index to a reverse proxy. Later, its role was further extended to provide login management services as described below.

10.6.1
Local Deployment

Although Cabinet is fundamentally network-oriented, it is also usable locally as a stand-alone system, that is, server(s) and client(s) run on the same machine. In fact, Cabinet was developed in such an environment. TCP/IP networks are typically configured with a nonroutable "local loopback" pseudonetwork with the IP address 127.0.0.1 ("localhost"). To demonstrate the ultimate form of local deployment and for use in mobile environments, Cabinet was operated in stand-alone mode on Intel ×86 laptops and Apple Gx PowerBooks. Since laptops can currently (2003) be configured with only 1–2 GB RAM, they were run with abridged versions of the largest datasets (e.g., PDB). With that one caveat, both Intel/Linux and G4/OS-X laptops provided excellent Cabinet performance as single-user platforms.

10.6.2
Intranet Deployment

Modern intranets provide very high network performance, and as one might expect, Cabinet performance over such networks is excellent. Overall Cabinet performance over 1000 BT (1 Gbs) networks is only slightly better than that over 100 BT (100 Kbps) networks.

Even though this was expected, there were two performance surprises on intranets. First, offloading heavily used services to independent machines was not particularly useful. For example, the zi4 server provides a glyph for each translated traditional/simplified Chinese character and is often required to process 100's or 1000's of HTTP requests for a complex page of multilingual output. One might imagine that offloading this task to a separate machine might be necessary to provide high performance data delivery. Similarly, generating *"What's related?"* pages requires that all participating servers do a search and return the results to the requesting service, which might strain a single machine. Cabinet is designed to make coarse-grained parallelization easy to configure and test. Robotic clients were used to simulate increasing numbers of simultaneous users making independent, random requests. A Cabinet network can eventually be overloaded (failures started at between 50 and 100 simultaneous users in the simulation). However, when using the current generation of computers and networks, there was no evidence that coarse-grained parallelization is of any performance benefit at all compared to "pumping up" a single machine with more processors and memory. In the test environment, this was primarily due to improved server–server communication within a single machine (local loopback latency is apparently very small), while client–server communication overhead was fairly constant.

The second surprise was to discover that putting the client on a different machine from the server(s) significantly enhanced overall performance (compared to stand-alone operation) as observed in the time to complete a variety of complex pages. Given the above result, this seems counterintuitive. This was most dramatically observed when comparing a G4 PowerBook used for servers versus both client and servers: despite the absence of network overhead in stand-alone mode, average time to complete complex pages was almost twice than that in client–server mode. The most plausible explanation is that on a stand-alone machine, the time spent on context-switching between internal processing and GUI I/O exceeds the network overhead required to offload the GUI I/O to another machine. This result might be specific to the hardware used (the G4 PowerBook has only a 1 × 1 GHz CPU but an excellent 1 Gbs network implementation).

The bottom line is that the current crop of workstations/small servers can deliver 4 GB of high-performance scientific data to about 50 intranet users. Today, that hardware costs about US$5000 configured with 4-GB RAM, 1 Gbs networking, DVD-R backup and other features.

10.6.3
Internet Deployment

Fourteen Cabinet servers were deployed on the Internet (WWW) over a dedicated 2 × 2 GHz 4 GB G5DP over a corporate T1 connection. As with most Internet servers, performance is connection-limited, so sheer server performance is not the limiting factor. However, operation over the Internet has its own special issues, some of which Cabinet ran into.

The production Internet deployment of Cabinet has proven to be robust and reliable. The first two months of operation were analyzed carefully. Average usage was 182 000 HTTP requests processed per day. Service was interrupted four times: twice due to external network glitches, once due to a power failure that exceeded the UPS backup capacity, and once due to a botched server upgrade (which affected only one service for 5 min). Overall uptime was 99.85% during this period.

One problem encountered in preproduction testing was that the Cabinet strategy of using unusual TCP/IP port numbers for various services does not work well for many corporate users. Such users are typically behind firewalls that allow IP transport only through a few ports such as 80. This is fine in an intranet setting but not over the Internet (it is not deemed acceptable to require reconfiguration of the client-side firewall). The solution was to create a reverse-proxy server that accepts requests on the standard HTTP port 80, handles all communication with internal Cabinet services on the client's behalf and passes responses back to the client. Reverse-proxying was implemented in the *cabinet* library and requires no changes to individual servers. Performance degradation in the Internet environment is negligible (performance remains connection-limited).

10.6.4
Online Deployment

"Online deployment" is the same as "Internet deployment" except that there is a requirement for user management, that is, user registration, user authorization and usage monitoring. Some of the Cabinet databases are proprietary and must be licensed, so user management is required for Internet deployment. Since the *cabinet* reverse-proxy service described above handles all Cabinet Internet traffic, it is an ideal place to implement user management. User management services were implemented using "Netscape-style cookies" and server-side DES encryption in a manner similar to "shopping cart" systems with which most users are already familiar.

One fortuitous aspect of implementing user management for WWW deployment is that it solves the "robot problem" cleanly. Web robots are programs that automatically index WWW by following links, for example, for Google (and other less well-behaved robots). Such robots work well for websites that have a limited number of static pages (thus a limited number of different links). However, when robots attempt to "index" the content of a Cabinet compute server, they can run on forever (or until they hit a large, predefined limit). Mechanisms exist to limit robot access via hints, but these are not uniformly followed. A user registration/authorization scheme effectively stops them.

10.7
Conclusions

A federated database system based on web technology has been designed, implemented and deployed in local, intranet and Internet environments. A number of constituent databases have been deployed that represent a wide variety of data models, for example, protein–ligand interactions, Western allopathic and traditional Chinese pharmacologies, photographic collections, a metabolic pathway chart, and so on. Each database implements a different data model that is appropriate for the nature of its information. These seemingly disparate data sources are cleanly federated into a user-friendly and integrated data system.

User accessibility is extremely high due to the exclusive usage of the common web browser GUI. The same GUI delivers the highest levels of scientific rigor and citation. "Full strength" searching of unabridged technical databases is provided during interactive time. This means that a user can work in their area of expertise and yet explore related information areas, discovering informatic relationships on the fly.

A lot of technology is required to achieve effective federation of disparate information sources. The Cabinet design is such that the computer does a tremendous amount of data processing to deliver a simple and powerful interface to the user. This is contrary to traditional database design that focuses on processing efficiency at the cost of flexibility and simplicity. Fortunately, current computers are more than adequate to do this job well. By the standards of the last few years, current desktop computers are supercomputers (and they are still getting faster and more capable each year).

The difficulties of globally federated data systems are primarily social rather than technical. In many ways, the nature of information federation is at odds with the way our society has evolved to control the dissemination of information and to protect intellectual property. For instance, many of the most important Cabinet datasets can only be licensed *in toto* on an annual basis, typically with a negotiated price, encouraging long-term but restricted/predefined usage. This very traditional approach evolved from disseminating information in print. It is not well suited to integrative usage, where the value of information is discovered rather than proscribed in advance. This phenomenon is evident in our society far beyond the scientific realm: it pervades all forms of information exchange and control, for example, telecommunication, dating, book and music publishing, to name a few. Like biological evolution, the nature of social evolution is fundamentally an adaptation to changing conditions. Our current social mechanisms such as copyrights are used to harness and control information flow. They are largely adaptations to the advent of the printing press, which was the "great extinctor" of scribes, private libraries and the religious control of knowledge. Perhaps the next "great extinctor" will prove to be computer networks. Will federated information systems form the basis of society's informatic nervous system in the near future? Given that the technical capability is obviously here and the equally obvious success of the WWW, it does not seem too far-fetched.

References

1 BERNERS-LEE, T., HENDLER, J., and LASSILA, O. The semantic web. *Sci. Am.* **2001**, 279–287.

2 ZDONBNOV, E.M., LOPEZ, R., APWEILER, R., and ETZOLD, T. The EBI SRS server – new features. *Bioinformatics* **2002**, *18*, 1149–1150.

3 HAAS, L.M., SCHWARZ, P.M., KODALI, P., KOTLAR, E., RICE, J.E., SWOPE, W.C. DiscoveryLink: A system for integrated access to life sciences data sources. *IBM Syst. J.* **2001**, *40*, 489–512.

4 DAVIDSON, S.B., CRABTREE, J., BRUNK, B.P., SCHUG, J., TANNEN, V., OVERTON, G.C., STOECKERT, C.J., Jr. K2/Kleisli and GUSS: experiments in integrated access to genomic data sources. *IBM Syst. J.* **2001**, *40*, 512–531.

5 BAKER, P., BRASS, A., BECHOFER, S., GOBLE, C., PATON, N., and STEVENS, R. *TAMBIS – Transparent Access to Multiple Biological Information Sources*, **1998**.

6 GOBLE, C.A., et al. Transparent access to multiple bioinformatics information sources. *IBM Syst. J.* **2001**, *4*, 532–551.

7 BAHL, A., et al. PlasmoDB: the Plasmodium genome resource. An integrated database that provides tools for accessing, analysing and mapping expression and sequence data (both finished and unfinished). *Nucl. Acids Res.* **2002**, *30*, 87–90.

8 Genomics Unified Schema, **2004**, http://www.gusdb.org/

9 Integ8, http://www.ebi.ac.uk/integr8/

10 Genostar, http://www.genostar.org/english/index.html

11 XML, http://www.w3.org/XML

12 ACHARD, F., VAYSSEIX, G., and BARILLOT, E. XML, bioinformatics and data integration. *Bioinformatics* **2001**, *17*, 115–125.

13 Semantic Web, **2001**, http://www.w3.org/2001/sweb

14 Web Services, http://www.w3.org/TR/2002/WD-ws-arch-20021114/

15 DOWELL, R.D., JOKERST, R.M., DAY, A., EDDY, S.R., and STEIN, L. The distributed annotation system. *BMC Bioinform.* **2001**, *2*, 7.

16 WILKINSON, M.D. and LINKS, M. BioMOBY: an open source biological web services proposal. *Brief. Bioinform.* **2002**, *3*, 331–341.

17 CORBA, http://www.omg.org

18 CORBA, http://www.omg.org/lsr/home.html

19 LSID (Life Sciences Identifier), http://www.i3c.org/pub/mar03clark.asp

20 WDI (World Drug Index), http://www.derwent.com/products/lr/wdi/

21 PDR (Physician's Desk Reference), http://www.pdr.net/pdrnet/librarian

22 MSDS (Material Safety Data Sheets), http://hazard.com/msds/index.php

23 REG (Chemical Abstracts Registry), http://www.cas.org/

24 PDB (Protein Data Bank), http://www.rcsb.org/pdb/

25 KNUTH, D. (MIX) *The Art of Computer Programming*. Addison-Wesley, New York, **1973**.

26 WEININGER, D. SMILES, A chemical language and information system. 1. Introduction to methodology and encoding rules. *JCICS* **1998**, *28*(1), 31–36.

27 BERNERS-LEE, T. *Information Management: A Proposal*. CERN, **1989**.

28 BUSH, V., *Atlantic Monthly* **1945**, *176*(1), 101–108, http://www.theatlantic.com/unbound/flashbks/computer/bushf.htm

29 RFC 2616, Hypertext Transfer Protocol – HTTP/1.1, June **1999**, http://www.w3.org/Protocols/

30 RFC 1341, Multipurpose Internet Mail Extensions, http://www.w3.org/Protocols/rfc1341/0_Abstract.html

31 HTML 4.01 Specification, December **1999**, http://www.w3.org/TR/html401/

32 NCSA's httpd, http://hoohoo.ncsa.uiuc.edu/

33 Apache, http://httpd.apache.org/

34 HTTP Toolkit, http://daylight.com/products/

35 MICHAL, G. *Biochemical Pathways*, 3rd edition. Boehringer, Mannheim, Germany, GmbH, Biochemica, **1993**.

36 LIGPLOT,
http://www.biochem.ucl.ac.uk/bsm/
ligplot/ligplot.html

37 Enzyme Commission,
http://www.chem.qmw.ac.uk/
iubmb/enzyme/

38 WDI (World Drug Index),
http://www.derwent.com/
products/lr/wdi/

39 OLAH M., MRACEC M., OSTOPOVICI L., RAD
R., BORA A., HADARUGA N., OLAH I.,
BANDA M., SIMON Z., MRACEC M., and
OPREA T.I. WOMBAT: world of molecular
bioactivity. In *Chemoinformatics in Drug
Discovery*, OPREA, T.I. (Ed.). Wiley-VCH,
Weinheim, **2004**, 223–239.

40 Wombat (World of Molecular Bioactivity),
http://www.sunsetmolecular.com/
products/

41 YAN, X., ZHOU, J., and XIE, G. *(TCM)
Traditional Chinese Medicines – Molecular
Structures, Natural Sources, and
Applications.* Ashgate Publishing,
Aldershot GB, UK, **1999**.

42 WISEMAN, N. and YE, F. *(DCM) A
Practical Dictionary of Chinese Medicine*,
2nd edition. Paradigm Publications,
Brookline MA, **1998**.

43 XUETIAN, P., SHICHEN, Z., YINQING, C.,
and JIGUAN, Z. (Eds). *A Coloured Atlas of
the Chinese Materia Medica Specified in
Pharmacopoeia of the People's Republic of
China*, 1995 edition. Joint Publishing
Co., Hong Kong, **1996**.

44 Netscape, http://channels.netscape.com/
ns/browsers/download.jsp

45 PyMOL, http://pymol.sourceforge.net/

46 http://cabinet.metaphorics.com/
wdi/dxrn/CI981.html

47 http://cabinet.metaphorics.com/
planet/pdb/1HWK.html

48 http://cabinet.metaphorics.com/
planet/pdb/1HWK.html

49 http://cabinet.metaphorics.com/
planet/pdb/1HWK.html

50 http://cabinet.metaphorics.com/
empath/chart/cw:479/ch:1132/cs:1/
cx:2321/cy:5196/chart.html

51 http://cabinet.metaphorics.com/
empath/paths/on:0/in:4/1490/
474/showpaths.html

52 http://cabinet.metaphorics.com/
tcm/mol/M1124.html

53 http://cabinet.metaphorics.com/
wombatdemo/wno/5506.html

11
Structure Modification in Chemical Databases

Peter W. Kenny and Jens Sadowski

11.1
Introduction

In chemical database applications, preserving the integrity of structures is normally of paramount importance. Accurate and efficient structure retrieval is the main function of a chemical database system and structures may not be registered in forms appropriate to molecular modeling. Here, we describe methods for controlled modification of chemical structures and their application to a range of cheminformatics problems, in particular, generation of appropriate ionization, tautomeric and configurational states for virtual screening.

Virtual screening [1, 2] is generally taken to mean a process of docking large numbers of structures into binding sites of proteins and scoring the resulting poses to assess probability of binding. Typically, both docking and scoring procedures aim to model interactions between ligand and protein. Electrostatic interactions between ligand and protein molecules are largely a consequence of nitrogen and oxygen atoms and the hydrogen atoms bonded to them, although the involvement of aromatic rings [3, 4] and halogens [5] should not be ignored. Interaction potential of these atoms is determined by formal charge and connectivity, so it is vital that these be appropriately represented for meaningful virtual screening.

Drug molecules frequently contain functional groups such as amines and carboxylic acids that are ionized under normal physiological conditions. Typically, these are registered in databases in their neutral forms that are perfectly reasonable when the main objective is efficient retrieval of structures. Representation of tautomers [6, 7], which differ only in bond type and hydrogen atom connectivity, raises similar issues while measurement of the relevant equilibrium constants is significantly more difficult.

Ionization and tautomerism present challenges for virtual screening because the processes reverse polarity of atoms. Balanced equilibria pose additional problems because both forms must be included for a meaningful representation of molecular structure. Provided that the virtual screening program provides the capability, these problems can be addressed to some extent by defining generic atom types although this situation is the exception rather than the norm. It should also be noted that atom typing, even with expressive line notations such as SMARTS (SMILES arbitrary target specification) [8] or SLN (Sybyl line notation) [9], does not address the problem

Chemoinformatics in Drug Discovery. Edited by Tudor I. Oprea
Copyright © 2004 WILEY-VCH Verlag GmbH & Co. KGaA, Weinheim
ISBN: 3-527-30753-2

of correlation between atom types. For example, if one nitrogen atom in pyrazole is a hydrogen bond donor, the other must be an acceptor.

A related problem is the configuration of trivalent nitrogen atoms. In many chemical environments, this has to be seen as an additional conformational freedom, that is, a continuum of states between the pyramidal and planar configurations. Because of the static view on conformations which many molecular modeling programs take, a decision for the "right" configuration has to be made in advance. In many cases, this means checking a multitude of states against the target.

Three examples taken from crystal structures of protein–ligand complexes from the protein data bank [10] shall illustrate the importance of correct hydrogen placements for structure-based design. Normally, protein crystal structures do not contain information on hydrogens or formal charges. In order to perceive this information, the ligand has to be analyzed in the context of the protein binding site. (Extending this to ligand databases for virtual screening, a list of reasonable hydrogen placements has to be considered since this context is missing.)

The first example in Figure 11.1 shows a pyrimidine trione compound bound to the active site of stromelysin. In order to get the interaction of one of the acidic pyrimidine nitrogens with the zinc cation correctly, the deprotonated negatively charged isomer has to be considered.

The second example in Figure 11.2 shows guanine bound to a phospho-ribosyl transferase. Only one of the two possible tautomers is able to form the correct hydrogen bond to an active site aspartate.

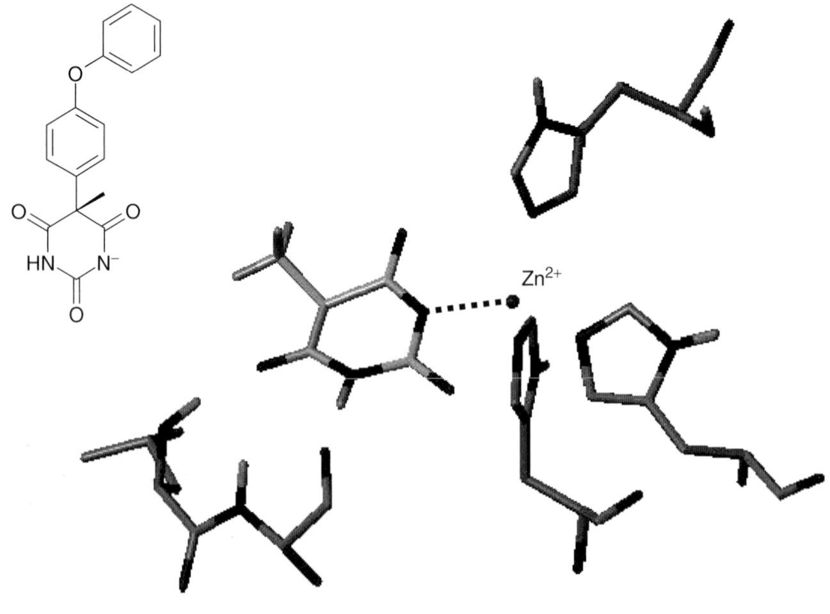

Fig. 11.1 Protonation state example: pyrimidine trione compound bound to stromelysin (PDB 1g4k). Only the pyrimidine trione ring of the ligand is shown in 3D for clarity.

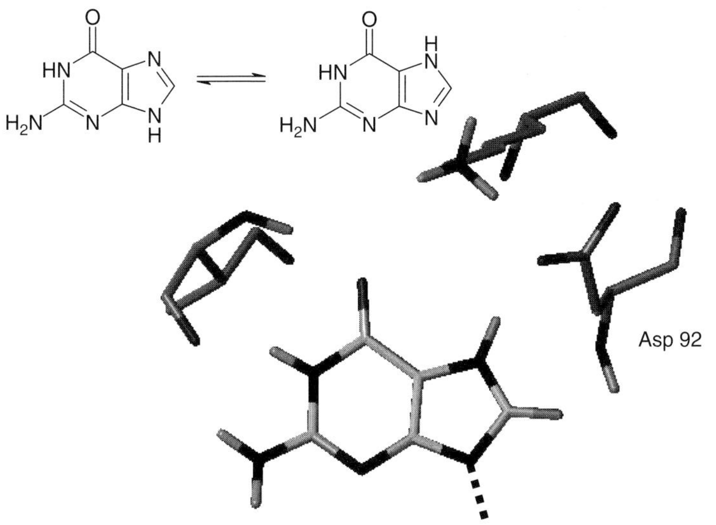

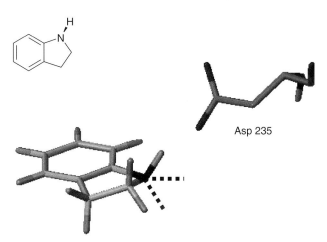

Fig. 11.2 Guanine bound to a phospho-ribosyl transferase (PDB 1a95).

Fig. 11.3 Nitrogen configuration example: Indoline/cytochrome C peroxidase complex (PDB 1aek). Alternative hydrogen positions indicated by dotted lines.

The last example (Figure 11.3) underlines the importance of the correct nitrogen configuration. In the indoline/cytochrome C peroxidase complex, only one of the three sketched hydrogen positions allows for an interaction with an active site aspartate.

Normally, programs for virtual screening, for example, docking software, do not take multiple hydrogen placements of these three types into account. There are, however, a few partial solutions. The docking program FlexX [11] can generate stereoisomers on the fly. This can be applied to the pseudochirality of pyramidal nitrogens but necessarily would also change all true stereo centers in the ligand. This approach would still not be able to generate planar configurations as well. The docking program DOCK [12] seems to have a module for superprotonation of nitrogen heterocycles, by this leaving multiple choices for special hydrogen placements to the docking stage.

It is unlikely that vendors of commercial software for virtual screening will all implement satisfactory solutions to the problems that we have outlined. Our favored approach is to preprocess the virtual screening library before converting to the format appropriate to each virtual screening program, and we describe two complementary approaches. Requiring a minimum of user intervention, Permute allows extensive enumeration of ionization, tautomeric states and nitrogen configurations in 3-D structures. In contrast, Leatherface modifies connection tables according to user-specified rules prior to 3-D coordinate generation. As a general-purpose 2-D structural editor, Leatherface finds application in areas other than virtual screening, and some of these will be discussed.

Recently, two more approaches that will add similar preprocessing functionality to commercial virtual screening software have been reported [13, 14].

11.2
Permute

Permute has been written to exhaustively generate reasonable isomers of druglike molecules with respect to protonation and formal charge, tautomerism, and nitrogen configurations. The approach applies a set of empirical rules for possible sites of hydrogen placements, for restricting the number of isomers, and for the prevention of duplicates. It is optimized for isomer generation for large databases of druglike molecules for virtual screening.

The program has been implemented on top of the Corina [15] toolkit library. All chemical substructures have been hard-coded for performance reasons. The program can handle 2-D or 3-D SDFiles and will, in the latter case, produce correct 3-D coordinates for all hydrogens being changed in the process.

11.2.1
Protonation and Formal Charges

A list of functional groups along with their allowed formal charges is given in Table 11.1. Note that there are a number of groups listed with only one possible charge state. These are groups put into a "permanent charge" state. The remaining groups are referred to as "chargeable" since they can exist in both charged and neutral state.

By default, Permute would generate all possible combinations of protonation states at all of these functional groups when contained in a molecule. A number of additional rules are used to prevent unreasonable combinations and combinatorial explosion:

Tab. 11.1 Functional groups and their associated formal charges

Group	Formal charges	Remarks
$-COOH$	-1	Optional uncharged
$-PO_3^{2-}$	-2	
$-SO_3^-$	-1	
Guanidine, amidine, aminopyridine	$+1$	
Tetrazole	$0, -1$	
Thiole	$0, -1$	
Hydroxamic acid	$0, -1$	
Acidic amide nitrogen	$0, -1$	$O=[C,S]N[C,S]=O$
Acidic enol	$0, -1$	$O=CC(O)=C$
Amine	$0, +1$	
Imidazole	$0, +1$	
Triazole	0	
Pyridine	0	Pyridine type N, optional $+1$

- Maximum number of "permanent charges"   3
- Maximum "chargeable" groups   5
- Maximum total charge ± 2
- Maximum number of zwitterions   1
- Maximum charges on nitrogens in the same ring   1
- No neighboring charges of the same sign allowed

This set of rules stops isomer generation whenever one rule is violated. It implicitly restricts the total number of isomers generated also.

11.2.2
Tautomerism

Permute is restricted to one individual type of tautomerism that is seen as most important for protein–ligand interactions. This is the shift of one hydrogen between two oxygen or nitrogen atoms along a path of alternating single/double bonds following a concept introduced by Sayle [16] and illustrated by the following example:

Scheme 11.1

This covers the most common tautomer pairs including, for example, imidazole and tetrazole. No other tautomerism is considered, for example, keto-enol. Two additional rules prevent combinatorial explosion:

- Maximum number of bonds between N/O   6
- Maximum number of tautomeric pairs   12

11.2.3
Nitrogen Configurations

This module generates multiple configurations for flexible ("flappy") nitrogen atoms under the following conditions:

- Uncharged
- Nonaromatic
- No bridgehead atom, that is, not locked in fused or bridged rings
- Pseudochiral center, that is, three distinguishable neighbors
- Maximum number of "flappy" nitrogens is four

For nitrogen atoms fulfilling these conditions, the two possible pyramidal configurations are generated. For nitrogens conjugated to aromatic rings, for example, aniline nitrogens, a planar configuration is generated in addition.

11.2.4
Duplicate Removal

Since there are no symmetry considerations embedded in the isomer generation approaches described above, duplicate isomers may be generated for symmetric parts of a molecule.

A postprocessing step takes care of these. For this, the concept of molecular hash codes introduced by Ihlenfeldt [17] is used. A molecular hash code is a unique integer number directly calculated from the connection table of a molecule. The hash code takes into account augmenting properties such as formal charges or stereo descriptors. Therefore, a hash code is perfectly suited to reflect all the differences between different isomers. It is independent of the atom numbering and is guaranteed to be identical for identical molecules. Duplicate isomers are detected by comparing their hash codes to all previously generated isomers for a given molecule.

This approach is sufficiently fast and needs sufficiently moderate memory in order to be applicable as a postprocessor.

11.2.5
Nested Loop

All three approaches to isomerism – formal charges, tautomers, and nitrogen configurations – and the duplicate removal are put together in a nested loop as shown in the following piece of pseudocode:

```
Uncharge all atoms
Apply permanent charges
For all tautomers{
        For all protonation states{
```

```
For all nitrogen configurations{
            Calculate hash code
            If not duplicate hash code{
                        Write isomer
                        Save hash code
            }
      }
  }
}
```

Observe that the loop nesting results in protonation states getting generated individually for all possible tautomers and nitrogen configurations getting generated individually for all protonation states of all tautomers. This is in clear contrast to the Leatherface concept, which applies all types of transformations simultaneously and allows for extra isomers, for example, protonation states that can be generated only for a certain tautomer. This approach exhaustively generates isomers up to a maximum number of 1024 per molecule.

11.2.6
Application Statistics

In order to give a feeling of the overhead added to typical virtual screening databases by the described algorithm for isomer generation, we give here the statistics for two commonly used public domain databases: Available chemicals directory (ACD) [18] and MDL drug data report (MDDR) [19]. Table 11.2 shows the results. All numbers refer to prefiltered versions of the databases after taking away unwanted compound types (data not shown).

In general, the pregeneration of isomers using Permute results in a two- to threefold increase in the number of compounds. The higher rate of compounds with isomers and also the higher average of isomers per compound in MDDR reflect the higher density of functional groups in typical druglike compounds compared to the more general chemicals in ACD.

Tab. 11.2 Permute statistics for common virtual screening databases

Database	Compounds (k)	With isomers (%)	Isomers (k)	Isomers/cpd.
ACD	328	36	703	2.2
MDDR	380	57	1236	3.3

11.2.7
Impact on Docking

In order to monitor the importance of isomeric hydrogen placements for docking and structure-based design, a large dataset of protein–ligand complexes was automatically derived from the PDB [20] and converted into files suitable for the docking program FlexX [11]. In detail, 2193 complexes of druglike molecules under certain restrictions with respect to crystallographic quality were obtained (data not shown).

In the course of the automatic file conversion, the hydrogen bonding interactions between ligand and protein were optimized by permuting hydrogen placements for the protein side chains by using the protein ensemble approach implemented in FlexE [21] and by generating multiple isomers for the ligands by using Permute. Optimal combinations were identified by the FlexX scoring function.

Table 11.3 gives the statistics for the PDB dataset. 41% of the ligands had more than one isomer according to Permute. Almost half of them (19%) were cases where the first isomer generated was different from the one identified later to form the best interactions with the protein. In other words, these 19% complexes are cases where a single default protonation pattern would not have led to the optimal hydrogen bonding interaction with the target.

For further considerations, only the subset of 419 (19%) complexes with the above-described "isomeric effect" was used. The impact of multiple hydrogen placements on docking was studied by carrying out two different attempts to redock the ligands into the protein binding sites by using FlexX. The first round docked the first isomers available for the ligands. The second round docked the isomers forming the best interactions with the proteins. The success rate was measured by counting the percentage of complexes having one solution better than 1.5 Å amongst the top ten solutions.

Table 11.4 shows the summary for the docking experiments. There is significant statistical evidence that the chances to reproduce the experimentally observed binding mode increase drastically when taking multiple hydrogen configurations into account. The reproduction rate goes up from 37 to 51%. This is a clear illustration of how sensitive docking programs like FlexX are to hydrogen placement and of how important it therefore is to search explicitly for the correct isomer.

Tab. 11.3 PDB dataset statistics

Total from PDB	19 220
Cleaned	2193
Isomers per ligand	2.3
Ligands with isomers	903 (41%)
Complexes with best isomer other than the first	419 (19%)

Tab. 11.4 Summary of docking experiments on the 419 complexes showing an "isomeric effect" for the ligand hydrogen configuration

Isomers	*Top ten pose(s) <1.5 Å (%)*
First	37
Best	51

11.3
Leatherface

Leatherface is a 2-D molecular editor that modifies properties of atoms and bonds in molecular connection tables according to rules specified by the user. Unlike Permute, Leatherface encodes no chemical knowledge and neither processes nor generates 3-D structures. Its real strength is that it allows the user to impose a very detailed and precisely specified chemical view on large numbers of connection tables.

Leatherface was built using Daylight programming toolkits [22] and connection tables are read and written as SMILES (simplified molecular input line entry specification) strings [23–25]. The highly expressive SMARTS [8] notation is used to direct structural editing, allowing substructural rules to be encoded with exquisite precision. Leatherface was originally created as a bond deletion tool for identification of ring systems in large chemical databases, and its unusual name reflects this original molecular fragmentation role. Atom and bond modification, needed for preparation of virtual screening databases, were introduced primarily to address limitations of pharmacophore searching software. Enumeration capability was added in order to apply our ionization and tautomer model that uses multiple forms to represent balanced equilibria.

The capabilities of Leatherface are best illustrated with examples. While a number of these will be typical of substructural modifications involved in building virtual screening databases, we will also show how a 2-D structure editor can be used to establish relationships between molecules and perform unorthodox substructural searches.

11.3.1
Protonation and Formal Charges

Each molecular transformation applied by Leatherface is specified by a SMARTS definition followed by a series of instructions that specify how the substructure matched by the SMARTS is to be modified. The first example shows how neutral carboxylic acids are converted to their more physiologically relevant anionic forms. The Prt instruction indicates that a single proton is to be removed (-1) from the atom matching the third atom in the SMARTS.

```
O=C[OH]
Prt     3      −1
```

Hypervalent nitrogen can be represented as ylid or pentavalent forms and both of these may be encountered, occasionally even in the same database. The problem is easily dealt with by converting the ylid species to the form with pentavalent nitrogen using the Chg and Bnd keywords. The pentavalent nitrogen convention simplifies atom typing in pharmacophore searching and scoring with functions that discriminate between hydrogen bonds involving neutral and charged species. The syntax of the Chg keyword is similar to that of Prt while Bnd indicates that the single bond between the first and second atoms in the substructure matching the SMARTS be changed to a double bond.

```
[n,N;+][O−]
Chg     1     0
Chg     2     0
Bnd     1     2     2
```

Leatherface allows definition of atom types that can be used to specify more complex SMARTS patterns. This is done in a separate input file and can lead to a more readable specification of substructure. Neutral aliphatic amines can be defined as follows:

```
Csp3          [CX4]
PrimAmin      [N;H2;X3][$Csp3]
SecAmin       [N;H;X3]([$Csp3])[$Csp3]
TerAmin       [N;X3]([$Csp3])([$Csp3])[$Csp3]
AllAmin       [$PrimAmin,$SecAmin,$TerAmin]
```

This definition for aliphatic amines can be used to direct protonation of these species although it should be noted that our ionization model uses more complex definitions to prevent simultaneous protonation of both nitrogen atoms in a piperazine ring.

```
[$AllAmin]
Prt     1     1
```

11.3.2
Tautomerism

Tautomer conversion involves nonlocal structural editing, and substructural patterns must generally be defined more carefully. 2-Pyridones are consistently registered in the AstraZeneca database as their typically less stable 2-hydroxypyridine tautomers. The conversion can be carried out by first defining an atom type corresponding to a nonfused aromatic carbon atom.

```
car     [$([cH]),$(c-A)]
```

A SMARTS definition can be built using this atom type to direct structural editing.

```
[OH]c1[nX2][$car][$car][$car][$car]1
Hyd     1     −1
Hyd     3     1
Bnd     1     2     2
```

In the examples described so far, Leatherface operates in its normal mode in which each editing operation is applied to the molecule until no match is found for its defining SMARTS pattern. The program can also be used to enumerate alternative forms as is appropriate to represent balanced equilibria where no single form is favored. The generation of the two alternative tautomers of asymmetrically substituted pyrazoles is specified by first defining an atom type to prevent an inappropriate substructural match against indazoles and analogous species in which the pyrazole is fused with an aromatic ring.

cx   [$([cH]),$(c-A)]

This is used to move the hydrogen from N1 to N2 in pyrazoles or 1, 2, 4-triazoles.

[nH]1[nX2][$cx][$cx,n&X2][$cx]1
Hyd   1   −1
Hyd   2   1

The transformation used above to enumerate tautomers would lead to identical products when applied to symmetrically substituted pyrazoles. The set of structures generated in the enumeration process is converted to a sorted list of canonical SMILES [23] from which duplicates are easily eliminated. Structures registered in alternative tautomeric forms are converted to identical lists of SMILES that can each be represented by their common first member. This effectively extends the definition of canonical SMILES to cover an ensemble of tautomeric forms and makes it possible to check for duplicate structures without having to register multiple forms [16, 26].

11.3.3
Ionization and Tautomer Model

Preparation of virtual screening databases starts with standardization of the input SMILES. This procedure was originally developed to deal with databases from commercial suppliers. Preferred tautomeric forms are generated in this step and ionized species are neutralized. Ionization states are set in the second step for biased equilibria and multiple forms are enumerated in a third step to represent balanced equilibria. The model treats an equilibrium as balanced if the equilibrium constant associated with its defining rule is likely to be less than about 1.5 log units.

Our current ionization and tautomer model encodes biased and balanced equilibria with 108 and 35 rules respectively. Extensive use is made of atom typing to specify chemistry in as generic a manner as possible, which reduces the number of rules required. For example, a number of heteroaromatic nitrogen types, including selected 2-aminopyridines and imidazoles, are aggregated into a definition of bases with pK_a

close to physiological pH. Other examples include piperazines with two basic nitrogen atoms, for which both singly protonated species are generated, and sulfonamides, for which three classes of ionization behavior are defined.

Corina [15] is used to generate 3-D coordinates and enumerate configurations if these are unspecified as is the case for racemates. Processing databases in this manner typically increases the number of 2-D structures by 10 to 15%. Although our treatment of ionization and tautomerism was developed for docking and scoring, we have found that it greatly simplifies atom typing for 3-D pharmacophore searching. These protocols for building databases were used in virtual screening for inhibitors of Checkpoint Kinase-1 [27].

11.3.4
Relationships between Structures

In database mining, it is useful to be able to search for objects that are related in a defined manner. Relationships between structures can be established by applying the appropriate structural transforms and searching for the products of these in the original database. We term this matched molecular pair analysis and have found it to be widely applicable in relating structure to properties. For example, the effect of a substituent can be quantified by converting the substituent to hydrogen and matching the resulting canonical SMILES [23–25] against the database. Leatherface is particularly useful in this regard because its enumeration capability can be used to ensure that single substituent deletions are selected. The effects of a number of substituents on aqueous solubility are quantified (see Table 11.5) for a selection of neutral compounds from the AstraZeneca solubility database.

Tab. 11.5 Matched molecular pair analysis of the effect of substituent on aqueous solubility

X	Mean	Standard error of mean	Number of pairs	Standard deviation	Significant
			$log(S_X/S_H)$		
Bromo	−1.19	0.14	15	0.53	Yes
Chloro	−0.75	0.06	111	0.61	Yes
Fluoro	−0.15	0.05	171	0.66	Yes
Methoxy	−0.08	0.07	87	0.67	No
Methyl	−0.22	0.08	66	0.66	Yes

Solubility measured at $25\,^{\circ}$C in phosphate buffer (pH $= 7.4$).

11.3.5
Substructural Searching and Analysis

2-D structural editors are particularly useful in retrosynthetic analysis [28]. A good example of this class of problem is identification of fragments in molecules. Typically, the full set of fragments is unknown prior to analysis or too large for manual substructural searching to be a practical option. This is the case in the problem of identifying the set of cyclic fragments present in a structural database [29, 30]. 3-D databases of rings are particularly useful in scaffold-hopping [31] applications where the target of the search is often a template presenting appropriately oriented bonds [32]. Leatherface identifies rings in a database by breaking exocyclic and other selected single bonds. One feature of the program particularly relevant to bond breaking is the ability to set isotopic mass which allows the nature of discarded fragments to be encoded in a generic manner such as electron withdrawing or hydrogen bond donor.

Even when fragments are known beforehand, setting up substructural searches manually is laborious, especially if hydrogen counts have to be specified, and unlikely to be a practical option for more than a small number of fragments. One example of this type of application is ranking reagents by counting occurrences of compounds derived from them in a database representing medicinal chemistry space. While this can often be achieved by bond deletion, transformation of reagent SMILES into a SMARTS substructural query is also an option. Specification of hydrogen count is the key problem as illustrated by the example of the SMILES (NC) for methylamine matching any aliphatic amine when used as a SMARTS. One way to force hydrogen to be written in the SMILES is to break normal valency rules, for example, by adding three hydrogen atoms to aliphatic carbon atoms already bonded to hydrogen.

[C;!H0]
Hyd   1   3

Methyl groups occur in the output SMILES as [CH6], which is easily converted to [CH3] using the UNIX stream editor to give the required SMARTS. Aromatic carbon forces a different tactic because adding hydrogen destroys aromaticity. In this case, hydrogen count is decremented causing the atom to be written out as [c] which can be converted to [cH] in analogous manner.

[cH]
Hyd   1   −1

11.4
Concluding Remarks

We have described two complementary tools for controlled modification of structures and their application to building databases for virtual screening. It is likely that virtual screening tools will integrate this capability to an increasing extent in the future.

We have also shown how 2-D molecular editing can be used to identify relationships between structures and perform other analysis that would be difficult using conventional substructural perception.

Acknowledgments

It is a pleasure to acknowledge helpful discussions with David Cosgrove, James Damewood, James Empfield, Charles Lerman, Sorel Muresan, Tim Perkins, Roger Sayle, George Stratton, Peter Taylor, Dave Timms, David Weininger and Jeremy Yang.

References

1 WALTERS, W.P., STAHL, M.T., and MURCKO, M.A. Virtual screening – an overview. *Drug Disc. Today* **1998**, *3*, 160–178.

2 LYNE, P.D. Structure-based virtual screening: an overview. *Drug Disc. Today* **2002**, *7*, 1047–1055.

3 HUNTER, C.A., LAWSON, K.R., PERKINS, J., and URCH, C.J. Aromatic interactions. *J. Chem. Soc., Perkin Trans.* **2001**, *2*, 651–669.

4 ZACHARIAS, N. and DOUGHERTY, D.A. Cation–π interactions in ligand recognition and catalysis. *Trends Pharm. Sci.* **2002**, *23*, 281–287.

5 LOMMERSE, J.P.M., STONE, A.J., TAYLOR, R., and ALLEN, F.H. The nature and geometry of intermolecular interactions between halogens and oxygen or nitrogen. *J. Am. Chem. Soc.* **1996**, *118*, 3108–3116.

6 ELGUERO, J., MARZIN, C., KATRITZKY, A.R., and LINDA, P. *The Tautomerism of Heterocycles*. Academic Press, London, **1976**.

7 KENNY, P.W. Handling heterocyclic tautomerism. Mug-99. Santa Fe, NM, 23–26 February **1999**, http://www.daylight.com/meetings/mug99/Kenny/kenny_mug99.htm.

8 SMARTS Manual, Daylight Chemical Information Systems Inc., 27401 Los Altos - Suite 360, Mission Viejo, CA 92691, http://www.daylight.com/dayhtml/doc/theory/theory.smarts.html.

9 SLN Manual, Tripos Inc., 1699 South Hanley Road, St. Louis, MO 63144-2917.

10 BERMAN, H.M., WESTBROOK, J., FENG, Z., GILLILAND, G., BHAT, T.N., WEISSIG, H., SHINDYALOV, I.N., and BOURNE, P.E. The protein data bank. *Nucleic Acids Res.* **2000**, *28*, 235–242.

11 RAREY, M., KRAMER, B., LENGAUER, T., and KLEBE, G. A fast flexible docking method using and incremental construction algorithm. *J. Mol. Biol.* **1996**, *261*, 470–489.

12 MAKINO, S. and KUNTZ, I.D. Automated flexible ligand docking method and its application for database search. *J. Comput. Chem.* **1997**, *18*, 1812–1825.

13 PEARLMAN, R.S., KHASHAN, R., WONG, D., and BALDUCCI, R. ProtoPlex: User-control over tautomeric and protonation state. *224th ACS National Meeting*, Boston, MA, **2002**.

14 KLICIC, J. Automated database ionization. *224th ACS National Meeting*. Boston, MA, **2002**.

15 GASTEIGER, J., RUDOLPH, C., and SADOWSKI, J. Automatic generation of 3D-atomic coordinates for organic molecules. *Tetrahedron Comput. Methods* **1990**, *3*, 537–547.

16 SAYLE, R. and DELANY, J. Canonicalization and enumeration of tautomers. EuroMug-99. Cambridge, UK, 28–29 October **1999**, http://www.daylight.com/meetings/emug99/Delany/taut_html/sld001.htm.

17 IHLENFELDT, W.D. and GASTEIGER, J. Hash codes for the identification and classification of molecular structure

elements. *J. Comput. Chem.* **1994**, *15*, 793–813.

18 ACD: Available Chemicals Directory, Version 1/02, MDL Information Systems, **2002**.

19 MDDR: MDL Drug Data Report, Version 1/02, MDL Information Systems, **2002**.

20 SADOWSKI, J., BUNING, C., RAREY, M., and CLAUßEN, H. An automatic clean-up procedure for protein-ligand complexes from the PDB. *European QSAR Meeting*, Bournemouth, UK, **2002**.

21 CLAUßEN, H., BUNING, C., RAREY, M., and LENGAUER, T. FlexE: Efficient molecular docking considering protein structure variations. *J. Mol. Biol.* **2001**, *308*, 377–395.

22 Toolkit Manual, Daylight Chemical Information Systems Inc., 27401 Los Altos - Suite 360, Mission Viejo, CA 92691, http://www.daylight.com/dayhtml/doc/man/refman.html.

23 WEININGER, D. SMILES, a chemical language and information system. 1. Introduction to methodology and encoding rules. *J. Chem. Inf. Comput. Sci.* **1988**, *28*, 31–36.

24 WEININGER, D., WEININGER, A., and WEININGER, J.L. SMILES. 2. Algorithm for generation of unique SMILES notation. *J. Chem. Inf. Comput. Sci.* **1989**, *29*, 97–101.

25 SMILES Manual, Daylight Chemical Information Systems Inc., 27401 Los Altos - Suite 360, Mission Viejo, CA 92691. http://www.daylight.com/dayhtml/doc/theory/theory.smiles.html.

26 TREPALIN, S.V., SKORENKO, A.V., BALAKIN, K.V., NASONOV, A.F., LANG, S.A., IVASHCHENKO, A.A., and SAVCHUK, N.P. Advanced exact structure searching in large databases of chemical compounds. *J. Chem. Inf. Comput. Sci.* **2003**, *43*, 852–860.

27 LYNE, P.D., KENNY, P.W., COSGROVE, D.A., DENG, C., ZABLUDOFF, S., WENDOLOSKI, J.J., and ASHWELL, S. Identification of compounds with nanomolar binding affinity for checkpoint kinase-1 using knowledge-based virtual screening. *J. Med. Chem.* **2004**, *47*, 1962–1968.

28 LEWELL, X.Q., JUDD, D.B., WATSON, S.P., and HANN, M.M. RECAP – Retrosynthetic combinatorial analysis procedure: a powerful new technique for identifying privileged molecular fragments with useful applications in combinatorial chemistry. *J. Chem. Inf. Comput. Sci.* **1998**, *38*, 511–522.

29 (a) BEMIS, G.W. and MURCKO, M.A. The properties of known drugs. 1. Molecular frameworks. *J. Med. Chem.* **1996**, *39*, 2887–2893. (b) BEMIS, G.W. and MURCKO, M.A. Properties of known drugs. 2. Side chains. *J. Med. Chem.* **1999**, *42*, 5095–5099.

30 LEWELL, X.Q., JONES, A.C., BRUCE, C.L., HARPER, G., JONES, M.M., MCLAY, I.M., and BRADSHAW, J. Drug rings database with web interface. A tool for identifying alternative chemical rings in lead discovery programs. *J. Med. Chem.* **2003**, *46*, 3257–3274.

31 SCHNEIDER, G., NEIDHART, W., GILLER, T., and SCHMID, G. Scaffold-Hopping by topological pharmacophore search: a contribution to virtual screening. *Angew. Chem., Int. Ed.* **1999**, *38*, 2894–2896.

32 BARTLETT, P.A., SHEA, G.T., TELFER, S.J., and WATERMAN, S. CAVEAT: a program to facilitate the structure-derived design of biologically active molecules. *Spec. Publ.- R. Soc. Chem.* **1989**, *78*, 182–196 (Mol. Recognit.: Chem. Biochem. Probl.).

12

Rational Design of GPCR-specific Combinational Libraries Based on the Concept of Privileged Substructures

Nikolay P. Savchuk, Sergey E. Tkachenko, and Konstantin V. Balakin

12.1
Introduction – Combinatorial Chemistry and Rational Drug Design

Owing to historic inefficiency of mass random bioscreening, the current paradigm suggests that target-specific and pharmacokinetic properties of small molecule libraries should be addressed as early as possible in the discovery process. Computational medicinal chemistry can address this problem at the level of presynthetic library design. A number of advanced *in silico* methods have recently been developed and applied to combinatorial templates to enhance their target-specific informational content. Appropriate strategies for the design of combinatorial libraries are developed in accordance with the target, disease area, resources on hand and the specific project goals. In a general sense, combinatorial synthesis can be defined as the process of production of all possible combinations of appropriate reagents using a given reaction. Obviously, the number of possible reaction products greatly exceeds the synthetic and screening resources at even the largest pharmaceutical companies. For example, consider the reaction of anilide generation using the available reagents from the MDL® Available Chemical Directory (ACD; MDL Information Systems, Inc., URL: http://www.mdl.com) online catalog. One can select 4749 anilines (Ar–NH$_2$) and 34 637 carboxylic acids (R–COOH). Combining these two reagent sets would result in a total of over 164 million (!) products. Both the production and the screening of such a library would be economically infeasible. This task is even more staggering when we consider the number of targets now identified to assay. Therefore, the problem of constraining the size of such virtual combinatorial libraries appears. This problem is solved via selective combinatorial library design aimed at selection of a "rational" compound subset that would meet the predefined goals of a bioscreening program.

   In this work, we will describe a practical approach to the design of target-biased combinatorial libraries. As an illustration, we will discuss the design of combinatorial libraries targeted against different G-protein coupled receptors (GPCRs). The goal of the combinatorial synthesis planning strategy presented here is to construct an algorithm utilizing simple, automated procedures for designing combinatorial libraries that are expected to show GPCR activity. "GPCR activity" is assumed here as the ability of a compound to be a successful ligand for a GPCR. We will demonstrate the

Chemoinformatics in Drug Discovery. Edited by Tudor I. Oprea
Copyright © 2004 WILEY-VCH Verlag GmbH & Co. KGaA, Weinheim
ISBN: 3-527-30753-2

practical significance of the concept of privileged substructures in the identification of combinatorial building blocks for synthesis of libraries rich in target-specific structural motifs. We will also illustrate the novel and interesting possibilities associated with the property-based selection of reactants and products using advanced machine-learning strategies.

The superfamily of seven-transmembrane-domain G-protein-coupled receptors is a diverse group of transmembrane proteins involved in signal transduction [1, 2]. GPCRs are of extraordinary importance for the pharmaceutical industry; in that space nearly 40% of marketed drugs act through modulation of the GPCR functions [3].

All the calculations described in this work were performed with the use of the ChemoSoft™ cheminformatics software system (Chemical Diversity Labs, Inc.).

12.2
Rational Selection of Building Blocks Based on Privileged Structural Motifs

In a most general way, a virtual library can be regarded as a computed array of products resulting from every combination of initial reactants. On the basis of this definition, two principal types of molecular entities, *the reagents* (or *"building blocks"*) that may undergo a chemical reaction and the *products* resulting from all possible combinations of those reagents, can be specified. Accordingly, there are two principal possibilities for selection of the rational compound subset. *Reagent-based selection* implies selection of reagent molecules only, which are further used for synthesis of a full combinatorial library of products. The second possibility is *product-based selection*, which is related to direct selection among the whole virtual library. In general, reagent-based selection is much faster and the results of calculation are easier to reproduce in the laboratory. On the other hand, product-based selection is usually more accurate and provides more information per molecule synthesized [4]. However, the superiority of the product-based design has to be balanced against its increased synthetic cost, which can be unacceptable when the resulting library is intended for repeated screening, using many different primary assays [5]. These considerations suggest that different strategies for manipulation of both product and reagent subsets should be present in the arsenal of a contemporary combinatorial library designer.

12.2.1
Privileged Structures and Substructures in the Design of Pharmacologically Relevant Combinatorial Libraries

The term "privileged structure" was proposed by Evans et al. [6] in 1988 to describe particular structural types (originally benzodiazepine-2-ones) that bind to multiple, unrelated classes of protein receptors as high-affinity ligands. It was envisioned that the use of such molecules would accelerate the discovery of promising drug candidates with activity against novel "druggable" biological targets identified through genomics initiatives. Many examples illustrating the concept of privileged structures have been published in the last 16 years. The recent reviews devoted to this research area [7–9] describe a large number of privileged scaffolds, and novel examples are continuously appearing in scientific literature [10, 11]. As an example, we mention the peptidomimetic squaric acid scaffold with a diverse portfolio of biological activities (Figure 12.1), including inhibition of cell adhesion; antagonism of neuropeptide Y1 binding; antagonism of IL-8, integrins VLA-4 and LPAM-1 as well as cation and anion channel regulation. Its close structural relationship to amino acids suggested its inclusion in the general family of privileged structures.

It is now recognized that it would be advantageous to synthesize one library on the basis of a privileged core structure and screen it against a variety of different receptors, yielding several active compounds each time. However, the resulting compounds can possess a low IP potential, as their biological activity encourages rapid patenting.

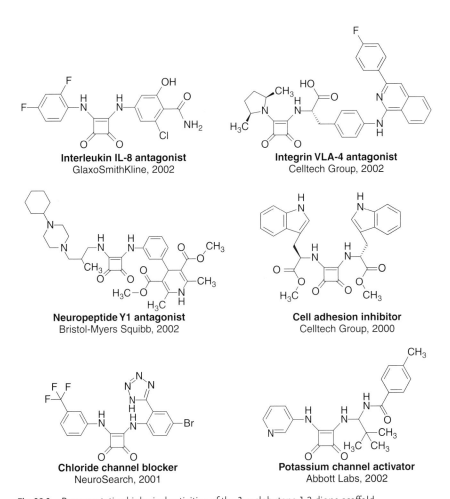

Fig. 12.1 Representative biological activities of the 3-cyclobutene-1,2-dione scaffold.

This potential problem should be taken into account while planning synthetic projects around the privileged core structures.

Another problem associated with the use of privileged structures is that the broad spectrum of their biological activity can lead to the lack of selectivity and, as a result, to limited practical applicability of such molecules [12]. In addition, they can be a source of false positive results in biological screening [12, 13]. It has been suggested that the nonselective action of several compounds containing the privileged motifs is due to aggregation (micelle or vesicle formation) [12, 13]. Though the aggregation phenomena can really occur in several particular cases, for example, flavonoid quercetin, bisindolylmaleimide IX, indirubin, and so on, the great majority of privileged structures do not appear to form aggregates and possess a very high level of selectivity against particular specific receptors. Thus, an atypical antipsychotic drug, Olanzapine (launched in 1996), which is a member of a classical benzodiazepine family of privileged structures, shows nanomolar activity against more than 15 different CNS receptors (Figure 12.2) [14]. Thus, Olanzapine displays the greatest selectivity for Dopamine D2 and 5-HT2A receptor binding (100–1000-fold) compared to its HERG channel IC_{50} [15]. At the same time, Olanzapine does not reveal any side effects due to nonspecificity. In fact, Olanzapine can be considered as a "concrete privileged structure".

Though the term "privileged structure" has been widely used in medicinal chemistry literature, it usually does not mean, as in the case of Olanzapine, the "concrete privileged structure". As a rule, a "concrete privileged structure" refers to a distinct structural motif typical of a particular family of active molecules and determining their "privileged" status. Recently, a more correct term, "privileged substructure", has been used more frequently [8, 16]. The privileged substructures are typically rigid, polycyclic or heterocyclic systems capable of orientation to varied substituents in a well-defined three-dimensional space. Because of its 3-D spatial parameters, such a fragment is able to nonspecifically bind to a large number of biotargets, while substituents attached to the scaffold may be responsible for its target specificity [8]. Biphenyl fragment is an example of such a well-known privileged substructure. Hajduk et al. performed an analysis of NMR-derived binding data and observed that the biphenyl framework bound to multiple proteins as a preferred motif [17]. We have found that biphenyl fragment, even without taking into account its heteroatom-containing bioisosteres, is a part of

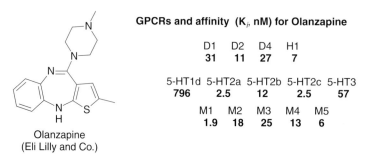

GPCRs and affinity (K_i, nM) for Olanzapine

D1	D2	D4	H1
31	11	27	7

5-HT1d	5-HT2a	5-HT2b	5-HT2c	5-HT3
796	2.5	12	2.5	57

M1	M2	M3	M4	M5
1.9	18	25	13	6

Olanzapine
(Eli Lilly and Co.)

Fig. 12.2 Atypical neuroleptic Olanzapine with a multiple binding pattern.

more than 3% of all described GPCR ligands from the MDDR database. This fact is in agreement with a hypothesis that the active site geometry of GPCRs is complementary to the architecture of this privileged substructure [18].

It is worth noting that the term "privileged substructure" can have alternative meanings and interpretations [19, 20]. The specificity of privileged substructures in its classical medicinal chemistry context is in the presence of properties that allow for a specific molecular recognition depending on electrostatic and steric surface complementarity of receptor and the "privileged substructures". On the basis of this definition, it would be interesting to discover novel "privileged substructures" having an increased selectivity against receptors belonging to particular families (e.g., GPCRs). In this work, we attempted to solve this problem using special cheminformatics tools and data-mining algorithms.

12.2.2
Analysis of Privileged Structural Motifs: Structure Dissection Rules

This part of our work is related to analysis of an existing database of GPCR active agents. Our goal is to identify the privileged substructures possessing a specific affinity to particular GPCRs. Several techniques have been developed to identify structural motifs that specifically interact with biotargets or target families using retrosynthetic analysis of existing knowledge bases and generation of specific molecular fragments [21–24]. Thus, a RECAP (Retrosynthetic Combinatorial Analysis Procedure) method was described on the basis of fragmenting molecules around bonds that are formed by common chemical reactions [23]. The main advantage of this approach is that the initial molecules are fragmented at several predefined bond types, all of which are amenable to combinatorial chemistry. Therefore, the resulting fragments represent direct precursors of building blocks for combinatorial library synthesis. Though the RECAP technique is very useful in the design of combinatorial libraries, this pure retrosynthetic approach can lead to ungrounded simplification of some privileged scaffolds. For example, according to RECAP rules, biphenyl fragment is dissected. In our work, we use a modified approach that takes into consideration not only the chemistry-derived rules but also the distinctive structural features of some privileged scaffolds. Along with the cleavage rules, we specified several bond types, which are left intact. Thus all mono- and biheterocyclic structures, benzylheterocycles, spirocyclic fragments, biphenyl and diarylmethane fragments and their heterocyclic bioisosteres, as well as all ring fragments are considered as indestructible.

The main chemical bond types at which a molecule is to be cleaved are shown in Figure 12.3. Hydrazides, hydrazones, ketoximes, ureas, uretanes, esters of hydroxamic acids are cleaved in a similar manner. If the terminal fragment to be cleaved contains only small functional groups with molecular weight less than 45, the fragment is left uncleaved. The nonterminal cleaved fragments with molecular weight less than 45 are eliminated. The main reasons for this are to avoid generation of very simple fragments and to obtain more "druglike" fragments. An example of a typical cleavage is given in

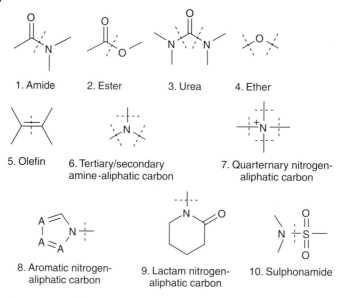

1. Amide   2. Ester   3. Urea   4. Ether

5. Olefin   6. Tertiary/secondary amine-aliphatic carbon   7. Quarternary nitrogen-aliphatic carbon

8. Aromatic nitrogen-aliphatic carbon   9. Lactam nitrogen-aliphatic carbon   10. Sulphonamide

Fig. 12.3   Main chemical bond cleavage types.

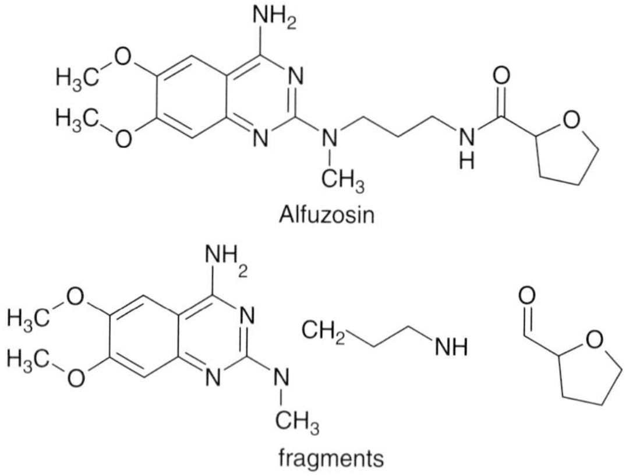

Alfuzosin

fragments

Fig. 12.4   Results of dissection of α1-adrenoceptor antagonist Alfuzosin.

Figure 12.4, where an initial molecule, Alfuzosin, with three cleavage points and three resulting fragments are shown.

It is important to note that the applied rules are not the pure retrosynthetic rules. The described structure dissection procedure also takes into account the structures of typical privileged motifs and is directed to the search of structural chemotypes with selective action against particular receptors.

The structure cleavage algorithm is an integrated module of the ChemoSoft™ software suite. The program allows the user to define chemical bond types to be cleaved and thus to vary content of the resulting database of fragments.

12.2.3
Knowledge Database

As mentioned in the previous section, our method is based on extracting information from large reference databases of active agents that show any target-specific activity. A total of 12 540 known GPCR ligands belonging to more than 100 different GPCR target-specific groups were used in this work as a reference knowledge database (Table 12.1). Compounds in the stages of preclinical and clinical trials, marketed drugs and compounds with experimental data on GPCR activity were included in this data set. All compounds were selected from the Ensemble database of biologically active compounds (Prous Science, URL: http://www.prous.com), which is a licensed database

Tab. 12.1   Reference database of GPCR ligands

Number	GPCR targets	Ligand type	Compounds
1	5-HT1A/1B/1D	Agonists/antagonists	1749
2	5-HT2A/2B	Agonists/antagonists	463
3	5-HT4	Agonists/antagonists	164
4	α1/α2-Adrenoceptor	Agonists/antagonists	713
5	β-Adrenoceptor	Agonists/antagonists	1000
6	Bradykinin B2	Agonists/antagonists	126
7	Cannabinoid CB1/CB2	Agonists/antagonists	93
8	CCKB	Antagonists	200
9	CRF	Antagonists	372
10	δ-Opioid	Agonists	195
11	Dopamine Autoreceptor	Modulators	189
12	Dopamine D1-D4	Agonists/antagonists	1551
13	Endothelin ETA/ETB	Antagonists	319
14	κ-Opioid	Agonists	164
15	μ-Opioid	Agonists/antagonists	86
16	Muscarinic M1	Agonists	630
17	Neuropeptide Y	Antagonists	101
18	Oxytocin	Antagonists	209
19	PGE2	Antagonists	129
20	Tachykinin NK1/NK2	Antagonists	1734
21	Vasopressin V1/V2	Antagonists	126
22	Approximately 80 minor GPCR-specific groups with number of compounds < 50	Agonists/antagonists	2700
Total number of compounds in training set[a]			12 540

a)   The total number of compounds is not equal to the sum of the shown values, as some compounds are not selective and manifest activity against more than one target.

Tab. 12.2 Diversity parameters of the studied database

Parameter	Value
Total number of compounds	12 540
Number of screens[a]	13 503
Diversity coefficient[b]	0.822
Number of core heterocycles	1131

a) Screens are simple structural fragments, centroids, with the topological distance equal to 1 bond length between the central atom and the atoms maximally remote from it.

b) Cosine coefficients are calculated, and the sums of nondiagonal similarity matrix elements are used in ChemoSoft™ program as a diversity measure; the diversity coefficient can possess the value from 0 to 1, which correspond to minimal and maximal possible diversity of a selection.

of known pharmaceutical agents compiled from the patent and scientific literature. Structures were extracted according to the assigned activity class, where the class indicates a single GPCR target. Some structures manifest activity against more than one GPCR. In this case, the molecule was included in all the corresponding activity classes. Molecules were filtered on the basis of molecular weight range (no more than 800) and atom type content (only C, N, O, H, S, P, F, Cl, Br, and I allowed).

Diversity parameters for this reference database are shown in Table 12.2. As evident from the number of screens, the number of core heterocyclic fragments, and the diversity coefficients (all these parameters are calculated using the *Diversity* module [25] of the ChemoSoft™ software tool), the studied compound database has high structural diversity and can be considered to be a good representation of known GPCR-active compounds.

The reference database of GPCR-active agents was fragmented using a structure splitting package of the ChemoSoft™ software and the cleavage rules described above. The procedure is automatic for each particular rule, and it results in a database of final fragments for which no further cleavage is possible. In this database, target-specific activity of a parental molecule is indicated for each fragment. Table 12.3 shows the number of fragments obtained from the initial database and for three arbitrary receptor-specific groups of ligands.

Tab. 12.3 The number of fragments obtained as a result of structure dissection procedure for the initial 12 540-compound reference database

	Fragments	Unique fragments
Total fragmented database	32 756	7425
Fragmented set of Dopamine D2 agonists	1839	452
Fragmented set of Tachykinin NK1 antagonists	5211	1400
Fragmented set of Cannabinoid agonists/antagonists	294	120

12.2.4
Target-specific Differences in Distribution of Molecular Fragments

In this part of our work, we will show that the privileged substructures, which are probably associated with target-specific activity of the uncleaved compounds, are present in fragmented GPCR-specific groups of ligands. As an example, we will use the fragmented database of Dopamine D2 agonists. This database, obtained by fragmenting 89 intact Dopamine D2 agonists, contains 1839 nonunique and 452 unique final retrosynthetic fragments. It was clustered using a method based on the distance matrix derived from Tanimoto similarity coefficients. As a result, several clusters containing more than five structures were generated, which represent the most frequently occurring fragments in the Dopamine D2 database.

To correctly address the problem of identification of target-specific privileged motifs, one should take into account the phenomenon of bioisosterism [26]. Thus, several different bioisosteric structures can constitute only one distinct privileged structural motif. In order to include all possible bioisosteric analogs into one cluster, we use a special algorithm of ChemoSoft™ based on a collection of rules for bioisosteric conversions described in literature. All bioisosteric analogs are considered similar with similarity coefficient 1 if they have identical substituents around the central bioisosterically transformed fragment.

To facilitate analysis of the association of specific fragments (possible privileged motifs) with a given target-specific dataset, we constructed a *characteristic occurrence* metric, as described below. For each privileged motif obtained after the cleavage procedure, we determined its occurrence in each GPCR-specific data set, and then compared this to the frequency of occurrence in the entire fragmented database. To construct the characteristic occurrence (CO) metric for a fragment in a particular set, we calculated the percentage ratio of the fragment's occurrence in the set to the total number of compounds in this set. To quantitatively assess an enrichment of a particular activity set with a specific fragment, we used the CO of this fragment in a particular compound set relative to its CO in the whole fragmented data set. The ratio of these characteristic occurrences of any fragment can serve as a measure of uniqueness of the fragment's distribution in the corresponding receptor-specific fragment base compared to the total database.

As an illustration, Table 12.4 shows four heterocyclic structures that were present in the two data sets, the fragmented database of Dopamine D2 receptor antagonists and the total fragmented database, with at least fivefold difference in the CO values, $CO_{Dop\_D2}/CO_{tot} > 5$. Such a difference gives a reasonable indication of whether a fragment is specific for this particular activity group or whether it is widely distributed in many unrelated groups. In fact, the fragments shown in Table 12.4 represent privileged substructural motifs for Dopamine D2 receptor antagonists.

In a similar manner, such privileged motifs can be identified for each GPCR-specific activity group. The typical number of privileged substructures per group lies in the range of 5 to 20 for the studied GPCR-specific compound sets.

The *N*-arylpiperazine fragment shown in Table 12.4 represents an interesting structural motif with an expressed mixed type of receptor-specific activity. Compounds

Tab. 12.4 Some privileged structural motifs of Dopamine D2 agonists

Fragment	CO_{tot}, %	CO_{Dop_D2}, %	CO_{Dop_D2}/CO_{tot}
	0.42	10.18	24.2
	0.59	3.34	5.7
	0.21	2.41	11.5
	0.13	6.21	47.8

containing this fragment can be active against Dopamine D2, Tachykinin NK1, Vasopressin V1A/V2 and other GPCRs (see also Table 12.5). Such activity with respect to the entire GPCR family substructures represents very valuable objects in combinatorial synthetic strategies. Their ability to serve as selective ligands against different receptors can be modulated with the proper selection of other parts of the molecule.

12.2.5
Privileged versus Peripheral Retrosynthetic Fragments

Analysis of GPCR ligands allows us to identify two principal categories of retrosynthetic fragments. The main category is the privileged fragments, described in the previous section. In most cases, the occurrence of a privileged motif is crucial for the target-specific activity of a compound. It should be noted that identification of privileged target-specific motifs is rather a technical problem in the sense that whenever a large-enough reference database of active agents and the appropriate chemical database management tools are available (such as chemical database, clustering package module, bioisosteric similarity module etc.), privileged target-specific substructures can be identified.

Peripheral structural motifs are a second principal category of retrosynthetic fragments. The presence of such fragments usually does not substantially influence

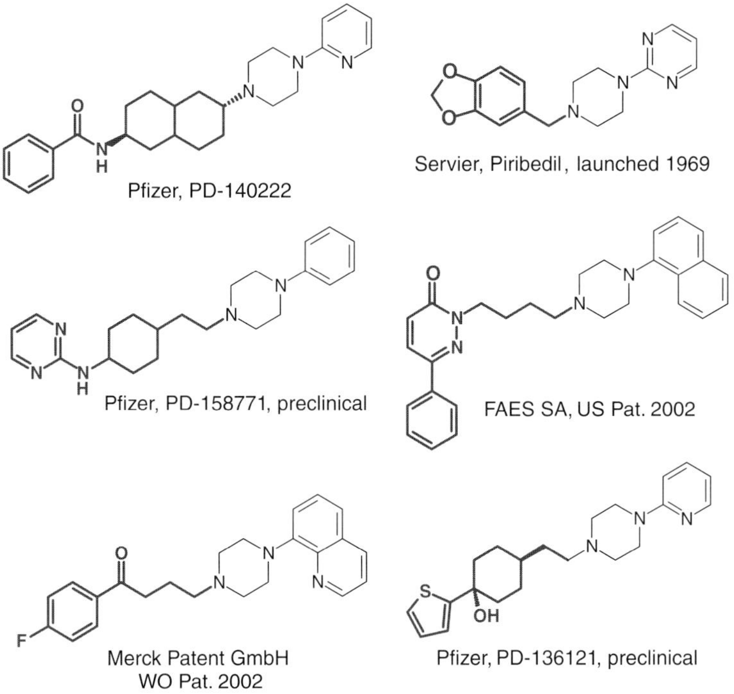

Fig. 12.5 Dopamine D2 agonists having a distinct *N*-arylpiperazine privileged motif and structurally different peripheral fragments.

the target specificity of a compound (with the exception of molecules containing the privileged core motifs with multiple target selectivity) but can seriously affect the protein–ligand binding affinity as well as its pharmacokinetic properties. In contrast to privileged motifs, the peripheral fragments belonging to a particular target activity class usually do not possess any structural similarity. Thus, several selective Dopamine D2 agonists have a distinct privileged substructure motif (*N*-arylpiperazine moiety) and different peripheral fragments (shown in bold in Figure 12.5).

Nevertheless, the structural peripheral fragments shown in Figure 12.5 have similar size, molecular topology, lipophilicity, number of H-bond accepting groups and number of rotatable bonds. It can be concluded that peripheral structural motifs, being structurally different molecular fragments, can exhibit similar physicochemical and spatial properties.

12.2.6
Peripheral Retrosynthetic Fragments: How to Measure the Target-specific Differences?

In this stage of our study, we seek to answer the following important question: whether we can quantitatively discriminate between different target-specific combinations of peripheral cleaved fragments on the basis of their molecular properties, rather than on the basis of their structural dissimilarity. To address this question, we used an advanced data-mining method based on artificial neural networks and unsupervised learning approach.

In recent years, the methods based on neural networks have become popular in focused compound libraries design because of their efficiency whenever a large-enough training database is available. Neural networks have been used successfully for segregation of pharmaceutical compounds into categories based on different properties. Recently, we applied neural network classification methodology for property-based design of GPCR [32, 33] and serine protease-targeted [34] libraries. We have shown that with proper combinations of specific physicochemical parameters, neural network models allow successful differentiation of GPCR or serine protease ligands from compounds active against other target-specific classes. Recently, we extended neural network approach to the design of combinatorial libraries targeted for other therapeutically significant classes of biotargets, namely, protein kinases and nuclear receptors [35].

In this work, we applied an advanced method of neural network quantitative SAR and data visualization based on Kohonen self-organizing maps [36]. A similar strategy was recently used and described in our work devoted to prediction of cytochrome P450-mediated metabolism of organic compounds [37]. The Kohonen type of network is based on a single layer of neurons arranged in a 2-D or 3-D plane with a well-defined topology. Each object, or molecule, can be represented as a vector, the components of which are variables with a definite meaning (molecular descriptors). The Kohonen neural network automatically adapts itself to the input data in such a way that similar input objects are associated with the topologically close neurons in the neural network. In the Kohonen approach, the neurons learn to determine the location of the neuron in the neural network that is most "similar" to the input vector. This means that objects located physically close to each other on the map have similar properties. In this work, we used a 20×20 Kohonen net with a 2-D organization of the network neurons arranged in a rectangular grid. This algorithm allows us to create a 2-D image of a multidimensional property space of the studied data sets of cleaved peripheral fragments. As an alternative method, 3-D Kohonen maps can be used for the same purposes. However, in our opinion, 2-D maps provide the comparable quality of visualization of the studied multidimensional property space; at the same time, 2-D maps are much more convenient to analyze and test.

In these experiments, we used the entire 7452-compound database of unique cleaved fragments. Each fragment in this database is characterized by a defined profile of target-specific activity of its active compound-precursor, focused against 1 of more than 100 different GPCR targets. Molecular features encoding the relevant physicochemical properties of compounds were calculated from 2-D molecular representations of the molecular fragments. Fragment size, topological complexity, H-binding capacity and lipophilicity were the main contributors to the models generated.

A self-organizing Kohonen map of the total database of cleaved retrosynthetic fragments generated as the result of an unsupervised learning procedure (data not shown) indicates that the cleaved fragments occupy a wide area on the map, characterized

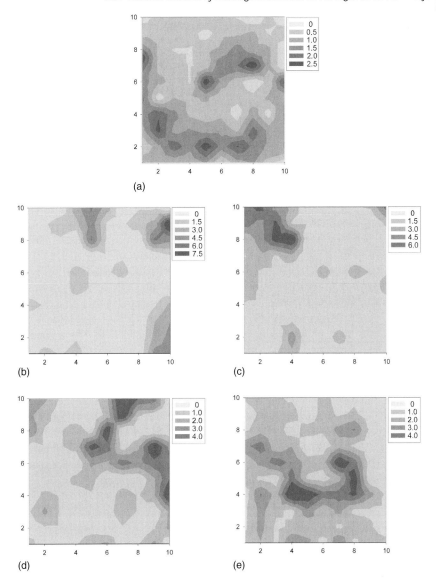

Fig. 12.6 Distribution of five different target-specific groups of peripheral fragments within the Kohonen map: (a) Tachykinin NK1 antagonists (521 fragments), (b) Dopamine D2 agonists (113 fragments), (c) Cannabinoid CB1/CB2 agonists/antagonists (89 fragments), (d) β3-Adrenoceptor agonists (294 fragments), (e) 5-HT1A agonists (354 fragments). The data are in percent (the total number of peripheral fragments in a receptor-specific group corresponds to 100%).

as the area of potential building blocks for combinatorial synthesis. Studying the distribution of various target-specific groups of peripheral structural fragments in the Kohonen map yielded interesting results consistent with our intuitive hypothesis about similarity of physicochemical properties between peripheral retrosynthetic fragments belonging to a particular target-specific category. Most of the groups have distinct locations in specific regions of the map (Figure 12.6a–e). The differences in location sites allow us to formulate the underlying principle, which can be used for selection-preferred peripheral fragments for each particular receptor-specific category – every group of peripheral structural fragments associated with defined target specificity can be characterized by a distinct and sometimes unique combination of physicochemical parameters. One possible explanation of this observation is that receptors of one type tend to share a structurally conserved ligand binding site. The structure of this site dictates the bundle of properties that a receptor-selective ligand should possess to properly bind the site. These properties include specific spatial, lipophilic, and H-binding parameters, as well as other features influencing pharmacodynamic behavior. On the other hand, the observed difference of physicochemical properties for particular target-specific groups of peripheral fragments can result from different pharmacokinetic requirements for compounds acting on specific GPCR.

The observed differences create a basis for a rational selection of building blocks for synthesis of combinatorial libraries enriched in target-specific motifs. The quantitative structure-activity discrimination function found at this stage of our study can be used for effective search of reactive monomers possessing the desired physicochemical and spatial parameters.

To summarize this part of our work, the privileged motifs can be considered, in a general case, as a main category of molecular fragments playing an essential role in the target-specific activity of a compound. On the other hand, the peripheral structural motifs are less important for target specificity and usually do not have any structural similarity, but, nevertheless, they are important for protein–ligand interactions and for a compound's pharmacokinetic profile. Modern chemical database management tools in combination with advanced methods of data mining permit effective identification of both privileged and peripheral molecular fragments. After these fragments are identified for each target activity group, they can be readily transformed into chemical building blocks for generation of a virtual target-biased combinatorial library.

12.2.7
Selection of Building Blocks

Combinatorial building blocks containing the privileged target-specific or nonselective structural motifs constitute the main category of reagents for synthesis. On the basis of statistical analysis of the total fragmented database, we created platforms of "privileged" core building blocks for all receptor-specific areas studied. For each privileged substructure, we selected a set of closely related compounds containing this privileged fragment (or its bioisosteric analog) and one or more "points of diversity" for introduction of peripheral building blocks. Examples of privileged structural motifs and the related core building blocks are shown in Table 12.5. A significant step in the selection of core building blocks is the estimation of IP potential of the resulting compounds using the Beilstein database based on a number of known structures for

Tab. 12.5 Examples of core and assembling building blocks structurally related to privileged motifs belonging to different target-specific ligand groups

Privileged substructures (unselective/target selective)	Core building blocks IP potential (Beilstein score)	Assembling building blocks Example of reaction
Unselective[a)] Dopamine D2 agonists, Tachykinin NK1 antagonists, α1 Adrenoceptor antagonists, PGE2 antagonists, 5-HT1D agonists, Vasopressin V1A/V2 antagonists, CCKB antagonists, 5-HT2A antagonists, Muscarinic M1 agonists, etc.	Low IP – 852 examples High IP – 3 examples High IP – 8 examples	[27] [28]

Tab. 12.5 (*Continued*)

Privileged substructures (*unselective/target selective*)	Core building blocks IP potential (*Beilstein score*)	Assembling building blocks Example of reaction
	 Low IP – 142 examples	
Unselective Dopamine D2 antagonists, Tachykinin NK3 antagonists, Neurokinin NK3 antagonists, etc.	 High IP – 2 examples	[29]

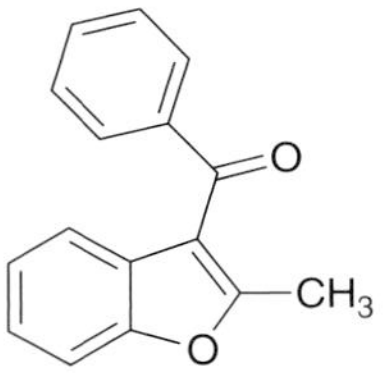

Target specific
Tachykinin NK1
antagonists

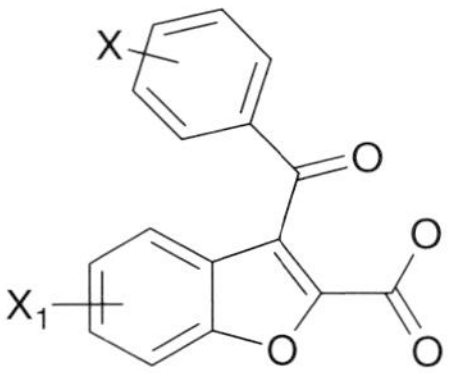

[30]

High IP – 11 examples

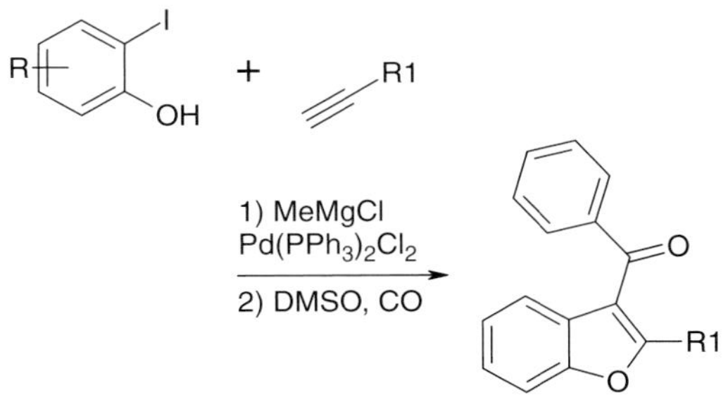

a) *N*-Arylpiperazines are typical privileged substructures in a broad sense and have been used frequently in combinatorial synthesis [31]. MDDR (August 2003 issue) contains 3449 physiologically active compounds containing *N*-arylpiperazine moiety, including 91 structures in clinical trials (55 – Phase I, 35 – Phase II, and 1 – Phase III), across more than 20 therapeutic areas.

reported active compounds containing the particular substructure. The next step in our selection procedure is related to assessment of synthetic accessibility (the third column in Table 12.5) and in generation of sets of assembling building blocks. It is important to note that such reagents sets, which can be used for synthesis of core building blocks, cannot be formed with the use of pseudoretrosynthetic automatic approaches similar to RECAP. In most cases, the trivial simple cleavage rules used in such algorithms do not correspond to the practical methods of synthetic assembling of complex privileged structures.

As discussed above, the peripheral building blocks determine pharmacokinetic and pharmacodynamic properties of compounds as well as their target specificity in the case of compounds with unselective privileged cores. The selection of peripheral building blocks for the design of GPCR-targeted combinatorial libraries is based on application of Kohonen neural networks. It is important to note that before this experiment, all reagent structures should be reduced to their "normalized" representations to allow correct comparison with the structures of retrosynthetic fragments. For example, all carboxylic acid derivatives, such as acid chlorides, anhydrides or activated esters are transformed into their reduced radical form identical to that obtained after dissecting the amide bond; all alkyl halides and alcohols are transformed into alkyl radicals; and so on. Such a "normalized" database of building blocks has been used in all neural network experiments. This database of available building blocks was processed on a Kohonen map. For illustration, we show the results of the selection procedure for Cannabinoid CB1/CB2 receptor ligands (Figure 12.7). The hashed zone restricts the area of preferable peripheral fragments for Cannabinoid CB1/CB2 agonists/antagonists. A total of 10 221 building blocks falling into the restricted area were selected for synthesis planning.

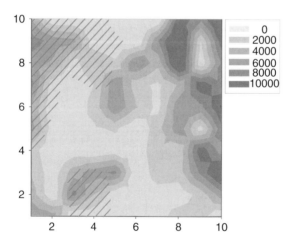

Fig. 12.7 Distribution of reagents within the Kohonen map. The hatched zone restricts the area of preferable peripheral fragments for Cannabinoid CB1/CB2 agonists/antagonists.

Selection of peripheral building blocks using the described method usually results in a relatively high number of candidates for synthesis. To reduce their number, additional, more stringent selection criteria can be applied. For instance, the structural similarity to the peripheral fragments found in the structures of active agents can be used. Optimization of structural diversity is another natural way to restrict the size of the initial selection. Additional filtering is related to exclude monomers that contain reactive chemical functions incompatible with the complementary functions of the privileged building blocks. These algorithms and filtering procedures usually allow selection of 200 to 300 peripheral monomers for the generation of virtual combinatorial library targeted against a particular GPCR.

At the stage of virtual target-biased combinatorial library generation, the reagents containing the privileged structural motifs ("privileged building blocks") are categorized according to their reactive chemical functionality. Then, the reagents with peripheral structural motifs ("peripheral building blocks") are divided into the chemical classes complementary to the corresponding privileged building blocks. After such a categorization is done, the automated procedure of virtual combinatorial library generation can be performed. Typically, it results in several tens of combinatorial libraries targeted against a particular GPCR containing a total of 10^4 to 10^6 virtual compounds.

12.2.8
Product-based Approach: Limiting the Space of Virtual Libraries

The total number of structures contained in the virtual libraries generated as described in the previous section is relatively large (up to 1 million). Therefore, there is still a serious need for additional product-based filtering procedures that will allow reduction of the size of such selections. Here, we will elucidate some important aspects of product-based combinatorial library design and describe a promising strategy of virtual library screening based on application of self-organizing Kohonen maps.

A general approach to limiting the space of virtual libraries of combinatorial reaction products consists of implementation of a series of special filtering procedures.

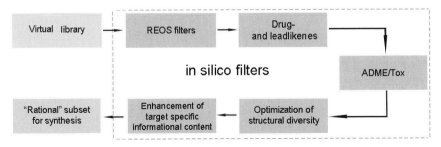

Fig. 12.8 General procedures of selection of a rational target-specific subset within an initial virtual combinatorial library.

The typical filtering stages are summarized in Figure 12.8. A variety of "Rapid Elimination of Swill" (REOS) filters is used to eliminate compounds that do not meet certain criteria [38]. These criteria can include: (1) presence of certain nondesirable functional groups, such as reactive moieties and known toxicophores; (2) molecular size, lipophilicity, the number of H-bond donors/acceptors, the number of rotatable bonds. At the next stage, the design focuses on "lead" and "druglikeness" of combinatorial molecules [39]. The ADME/Tox properties of screening candidates should be taken into consideration as early as possible [40, 41]. Additional filters are therefore used for *in silico* prediction of some crucial ADME/Tox parameters, such as solubility in water, logD at different pH values, cytochrome P450-mediated metabolism and toxicity, and fractional absorption. Optimization of structural diversity is another natural and very important way to constrain the size of combinatorial libraries (reviewed in [42]).

A particular challenge in compound selection lies in choosing structures that are most likely to have a predefined biological property of interest from the vast assortment of structurally dissimilar molecules. This challenge has been tackled with powerful computational methodologies, such as *in silico* docking of available structures into the receptor site and pharmacophore searches for particular geometric relations among the elements thought to be critical for biological activity [43, 44]. Both methodologies focus on conformational flexibility of both target and ligand, which is a complex and computationally intense problem. The limits of computational power and time restrict the practical library size selected by these methods. Another popular approach to virtual screening is based on ligand structure and consists of selecting compounds structurally related to hits identified from the initial screening of existing commercial libraries, or to active molecules reported in research articles and patents. Although broadly used in the development of structure-activity relationship (SAR) profiling libraries, ligand structure–based methods usually perform poorly when it comes to the discovery of novel lead chemotypes. In general, the target and ligand structure–based technologies cannot adequately address all the real problems of rational drug design, particularly those connected with virtual screening of large compound databases or the discovery of novel lead chemotypes.

12.2.9
Alternative Strategy: Property-based Approach

An alternative strategy for the design of target-specific libraries draws on similarity of molecular physicochemical properties including specific steric, lipophilic and H-binding, and other features influencing the pharmacodynamic requirements [45]. The theory is realized in computation models for quantitative discrimination between the ligand groups. Whenever a large set of active ligands is available for a particular receptor, the mean values of the key molecular properties are considered as optimal and characteristic of this group of ligands. On the basis of these values, one can generate a quantitative discrimination function that permits selection of a series of compounds to be assayed against the target.

It can be shown that the unsupervised learning methodology based on Kohonen self-organizing maps algorithm can be effectively used for differentiation between various receptor-specific groups of GPCR ligands. The method is similar to that described in Section 12.2.6.

As a reference database, we used the same database of 12 540 GPCR-active agents with demonstrated experimental activity against over 100 different GPCRs (Table 12.1). Our Kohonen neural network models were built using a set of seven molecular descriptors selected by principal components analysis. These descriptors are molecular weight, logarithm of partition coefficient in 1-octanol/water at pH 7.4, the number of H-bond donors and acceptors, the number of rotatable bonds, total polar surface area and partial positive surface area. The chosen descriptors are readily computable and, in combination, provide a reasonable basis for the assessment of the particular GPCR activity potential. This set of descriptors defines a bundle of the most relevant factors affecting the ability of a compound to possess GPCR activity: lipophilicity, molecular surface area and size, H-binding potential and surface charge properties.

12.2.10
Kohonen Self-organizing Maps

We used the internally developed program included in CDL's proprietary ChemoSoft™ software suite for unsupervised learning and generation of Kohonen maps. A 15 × 15 node architecture was chosen in order to provide the studied molecules with optimal distribution space. The reference database was used for neural network training and Kohonen map generation (Figure 12.9). The GPCR ligands are widely distributed throughout the map as irregularly shaped islands with a trend toward the bottom of

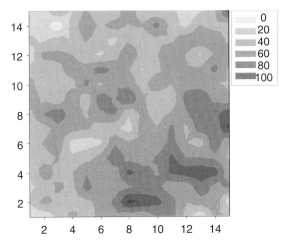

Fig. 12.9 15 × 15 Kohonen network trained with seven selected descriptors for total reference database of GPCR-active agents (12 540 compounds). The data have been smoothed.

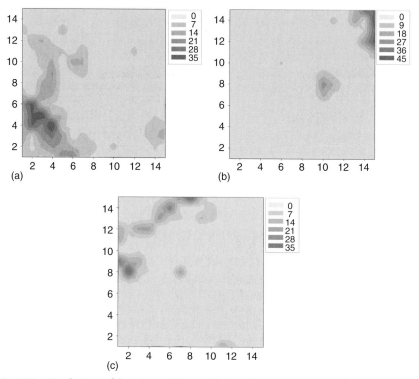

Fig. 12.10 Distributions of three large GPCR-specific ligand groups within the Kohonen map:
(a) Tachykinin NK1 antagonists (1400 compounds), (b) Muscarinic M1 agonists
(563 compounds), (c) β3-Adrenoceptor agonists (433 compounds).

the map. The area occupied by the GPCR ligands is relatively large, reflecting their significant diversity.

At the next step, we studied the distribution of different receptor-specific groups of GPCR ligands within the generated Kohonen map. These ligand groups appeared to be clustered at distinctly different areas of the map. As an illustration, Figures 12.10a–c show the distributions of three large GPCR-specific ligand groups. Three-dimensional distribution plots of these ligand groups on the same map (Figure 12.11a–c) demonstrate statistically significant differences in their location. Interestingly, active agents presently in clinical trials and approved launched drugs usually gravitate toward the central parts of the corresponding receptor-specific sites on the map, while the peripheral positions are occupied by compounds at earlier stages of development (data not shown). The Kohonen maps for particular receptor-specific groups of ligands can be used for predicting potential receptor-specific activity. Thus, the processing of a diverse exploratory compound library on this Kohonen map allows us to distinguish between the specific compound subsets falling into particular receptor-specific areas. We suggest that the molecules from these subsets are more likely to be active against the corresponding receptors.

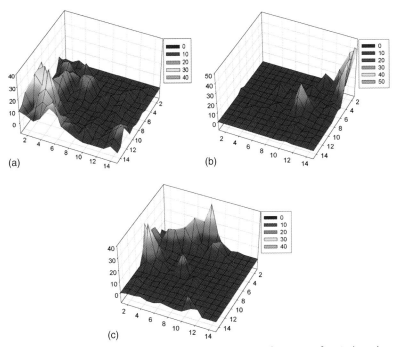

Fig. 12.11 3-D diagrams of distribution of three target-specific groups of GPCR ligands on the Kohonen map: (a) Tachykinin NK1 Antagonists (1400 compounds), (b) Muscarinic M1 Agonists (563 compounds), β3-Adrenoceptor Agonists (433 compounds).

The computational algorithm described here is useful in constraining the size of virtual libraries of potential GPCR active agents. It can be applied as an *in silico* filter to assist in the product-based design and planning of novel combinatorial libraries. For example, when applied to the total virtual library of 200 000 potential Dopamine D2 receptor ligands generated with the use of reagent-based approaches described in the previous sections, this filter permits to select approximately 6000 compounds having the highest probability of activity against Dopamine D2 receptor.

The described methodology can be generalized to aid in the selection of an optimal methodology for any arbitrary target-specific library design; it is not restricted to the GPCR-targeted libraries studied here. In addition, the results can be used for profiling the bioactivity of compounds on the basis of comparison with the structures of known agents possessing a certain biological activity.

12.3
Conclusions

In this paper, we elucidated some theoretical and practical aspects in the design of target-specific combinatorial libraries. In the studies we described to illustrate a real process of focused library design, two principal strategies were considered: reagent-based selection and product-based selection. In the framework of reagent-based strategy, we described a multistep procedure of selection of privileged structural motifs, which are the key structural elements of target-biased libraries. We also demonstrated the practical significance of the target-specific peripheral structural fragments concept and suggested a method for their rational selection based on application of advanced machine-learning algorithms.

In general, reagent-based selection is much faster and more convenient to execute in the laboratory as compared with the product-based selection. On the other hand, the latter strategy usually provides more accurate results. There exists a potential to combine both approaches to achieve more optimal results, particularly in the case of large exploratory virtual combinatorial libraries, for which mass random synthesis and screening are not economically feasible. In this article, we demonstrated the usefulness of property-based approach for selection of optimal GPCR ligands.

The approach we described combines reagent- and product-based selection procedures and results in generation of a compact virtual compound library (in the general case, several thousand compounds) targeted against a particular GPCR target. Usually, such a library consists of several tens of distinct medium-sized combinatorial sublibraries (50–200 compounds each) rich in target-specific structural motifs and possessing optimized physicochemical properties. Such libraries represent a very useful tool at early stages of drug discovery and development, as a valuable source of primary hits easily amenable to further hit-to-lead optimization. Examples of particular GPCR-, protein kinase- and ion channel–targeted libraries generated using the described strategy can be found among the commercial products currently available at Chemical Diversity Labs, Inc.

The statistics-based approach described consists of a series of automated procedures, is easily reproducible, and can be recommended for the practical design of compound libraries targeted against biotargets, for which sufficiently large number of selective ligands is available. It is important to note that all the procedures described are computationally inexpensive and permit real-time calculations with moderate hardware requirements.

Primary bioscreening of large exploratory libraries of small molecules produced by combinatorial synthesis remains a key element of modern drug discovery. The problem of enhancement of bioscreening effectiveness necessitates more serious attention to the quality of screening compound libraries. In this context, advanced cheminformatics technologies, aimed at selection of the proper screening candidates, are of great industrial demand. The further evolution of such technologies will result in the development of integrated cheminformatics platforms, where all the issues related to selection of a rational pharmaceutically relevant screening candidate, having good synthetic feasibility, a desirable profile of target-specific action, druglikeness, unexploited IP position, favorable ADME/Tox profile, compatibility with assay protocol, and so on, will be solved with maximal quality, time- and cost-effectiveness.

References

1 GUDERMANN, T., NURNBERG, B., and SCHULTZ, G. Receptors and G proteins as primary components of transmembrane signal transduction. Part 1. G-protein-coupled receptors: structure and function. *J. Mol. Med.* **1995**, *73*, 51–63.

2 HAMM, H. The many faces of G protein signaling. *J. Biol. Chem.* **1998**, *273*, 669–672.

3 DREWS, J. Drug discovery: a historical perspective. *Science* **2000**, *287*, 1960–1964.

4 GILLET, V.J., WILLETT, P., and BRADSHAW, J. The effectiveness of reactant pools for generating structurally-diverse combinatorial libraries. *J. Chem. Inf. Comput. Sci.* **1997**, *37*, 731–740.

5 FERGUSON, A.M., PATTERSON, D.E., GARR, C.D., and UNDERINER, T.L. Designing chemical libraries for lead discovery. *J. Biomol. Screen.* **1996**, *1*, 65–73.

6 EVANS, B.E., RITTLE, K.E., BOCK, M.G., DIPARDO, R.M., FREIDINGER, R.M., WHITTER, W.L., LUNDELL, G.F., VEBER, D.F., ANDERSON, P.S., CHANG, R.S., LOTTI, V.J., CERINO, D.J., CHEN, T.B., KLING, P.J., KUNKEL, K.A., and SPRINGER, J.P. Methods for drug discovery: development of potent, selective, orally effective cholecystokinin antagonists. *J. Med. Chem.* **1988**, *31*, 2235–2246.

7 PATCHETT, A. and NARGUND, R. Priviledged structures – an update. *Annu. Rep. Med. Chem.* **2000**, *35*, 289–298.

8 HORTON, D.A., BOURNE, G.T., and SMYTHE, M.L. The combinatorial synthesis of bicyclic privileged structures or privileged substructures. *Chem. Rev.* **2003**, *103*, 893–930.

9 HORTON, D.A., BOURNE, G.T., and SMYTHE, M.L. Exploring privileged structures: the combinatorial synthesis of cyclic peptides. *J. Comput.-Aided Mol. Des.* **2002**, *16*, 415–430.

10 BLEICHER, K., WUTHRICH, Y., ADAM, G., HOFFMANN, T., and SLEIGHT, A. Parallel solution- and solid-phase synthesis of spiropyrrolo-pyrroles as novel neurokinin receptor ligands. *Bioorg. Med. Chem. Lett.* **2002**, *12*, 3073–3076.

11 VOURLOUMIS, D., TAKAHASHI, M., SIMONSEN, K., AYIDA, B., BARLUENGA, S., WINTERSA, G., and HERMANN, T. Solid-phase synthesis of benzimidazole libraries biased for RNA targets. *Tetrahedron Lett.* **2003**, *44*, 2807–2811.

12 MCGOVERN, S.L., CASELLI, E., GRIGORIEFF, N., and SHOICHET, B.K. A common mechanism underlying promiscuous inhibitors from virtual and high-throughput screening. *J. Med. Chem.* **2002**, *45*, 1712–1722.

13 MCGOVERN, S.L. and SHOICHET, B.K. Kinase inhibitors: not just for kinases anymore. *J. Med. Chem.* **2003**, *46*, 1478–1483.

14 BYMASTER, F.P., CALLIGARO, D.O., FALCONE, J.F., MARSH, R.D., MOORE, N.A., TYE, N.C., SEEMAN, P., and WONG, D.T. Radioreceptor binding profile of the atypical antipsychotic olanzapine. *Neuropsychopharmacology* **1996**, *14*, 87–96.

15 KONGSAMUT, S., KANG, J., CHEN, X.L., ROEHR, J., and RAMPE, D. A comparison of the receptor binding and HERG channel affinities for a series of antipsychotic drugs. *Eur. J. Pharmacol.* **2002**, *450*, 37–41.

16 MASON, J., MORIZE, I., MENARD, P., CHENEY, D., HULME, C., and LABAUDINIERE, R. New 4-point pharmacophore method for molecular similarity and diversity applications: overview of the method and applications, including a novel approach to the design of combinatorial libraries containing privileged substructures. *J. Med. Chem.* **1999**, *42*, 3251–3264.

17 HAJDUK, P.J., BURES, M., PRAESTGAARD, J., and FESIK, S.W. Privileged molecules for protein binding identified from NMR-based screening. *J. Med. Chem.* **2000**, *43*, 3443–3447.

18 ABROUS, L., HYNES, J., FRIEDRICH, S., SMITH, A., and HIRSCHMANN, R. Design and synthesis of novel scaffolds for drug discovery: hybrids of beta-D-glucose with

1,2,3,4-tetrahydrobenzo[e][1,4]diazepin-5-one, the corresponding 1-oxazepine, and 2- and 4-pyridyldiazepines. *Org. Lett.* **2001**, *3*, 1089–1092.

19 ZEFIROV, N. and PALYULIN, V. Fragmental approach in QSPR. *J. Chem. Inf. Comput. Sci.* **2002**, *42*, 1112–1122.

20 MERLOT, C., DOMINE, D., CLEVA, C., and CHURCH, D. Chemical substructures in drug discovery. *Drug Disc. Today* **2003**, *8*, 594–602.

21 GASTEIGER, J., MARSILI, M., HUTCHINGS, M.G., SALLER, H., LOEW, P., ROESE, P., and RAFEINER, K. Models for the representation of knowledge about chemical reactions. *J. Chem. Inf. Comput. Sci.* **1990**, *30*, 467–476.

22 RIDINGS, J.E., BARRATT, M.D., CARY, R., EARNSHAW, C.G., EGGINGTON, C.E., ELLIS, M.K., JUDSON, P.N., LANGOWSKI, J.J., and MARCHANT, C.A. Computer prediction of possible toxic action from chemical structure: an update on the DEREK system. *Toxicology* **1996**, *106*, 267–279.

23 LEWELL, X.Q., JUDD, D.B., WATSON, S.P., and HANN, M.M. RECAP – retrosynthetic combinatorial analysis procedure: a powerful new technique for identifying privileged molecular fragments with useful applications in combinatorial chemistry. *J. Chem. Inf. Comput. Sci.* **1998**, *38*, 511–522.

24 SCHNEIDER, G., CLÉMENT-CHOMIENNE, O., HILFIGER, L., SCHNEIDER, P., KIRSCH, S., BÖHM, H., and NEIDHART, W. Virtual screening for bioactive molecules by evolutionary de novo design. *Angew. Chem., Int. Ed.* **2000**, *39*, 4130–4133.

25 TREPALIN, S.V., GERASIMENKO, V.A., KOZYUKOV, A.V., SAVCHUK, N.P., and IVASHCHENKO, A.A. New diversity calculations algorithms used for compound selection. *J. Chem. Inf. Comput. Sci.* **2002**, *42*, 249–258.

26 PATANI, G.A. and LAVOIE, E.J. Bioisosterism: a rational approach in drug design. *Chem. Rev.* **1996**, *96*, 3147–3176.

27 SINGH, U., STRIETER, E., BLACKMOND, D., and BUCHWALD, S. Mechanistic insights into the Pd(BINAP)-catalyzed amination of aryl bromides: kinetic studies under synthetically relevant conditions. *J. Am. Chem. Soc.* **2002**, *124*, 14104–14114.

28 WOLFE, J. and BUCHWALD, S. Palladium-catalyzed amination of aryl triflates. *J. Org. Chem.* **1997**, *62*, 1264–1267.

29 ALBERT, J., AHARONY, D., ANDISIK, D., BARTHLOW, H., BERNSTEIN, H., BIALECKI, R., DEDINAS, R., DEMBOFSKY, R., HILL, D., KIRKLAND, K., KOETHER, G., KOSMIDER, B., OHNMACHT, C., PALMER, W., POTTS, W., RUMSEY, W., SHEN, L., SHENVI, L., SHERWOOD, S., WARWICK, S., and RUSSELL, K. Design, synthesis, and SAR of tachykinin antagonists: modulation of balance in NK(1)/NK(2) receptor antagonist activity. *J. Med. Chem.* **2002**, *45*, 3972–3983.

30 ORRU, R.V.A. and DE GREEF, M. Recent advances in solution-phase multicomponent methodology for the synthesis of heterocyclic compounds. *Synthesis* **2003**, *10*, 1471–1499.

31 BARALDI, P., DEL CARMEN NUNEZ, M., MORELLI, A., FALZONI, S., DI VIRGILIO, F., and ROMAGNOLI, R. Synthesis and biological activity of N-arylpiperazine-modified analogues of KN-62, a potent antagonist of the purinergic P2X7 receptor. *J. Med. Chem.* **2003**, *46*, 1318–1329.

32 BALAKIN, K.V., TKACHENKO, S.E., LANG, S.A., OKUN, I., IVASHCHENKO, A.A., and SAVCHUK, N.P. Property-based design of GPCR-targeted library. *J. Chem. Inf. Comput. Sci.* **2002**, *42*, 1332–1342.

33 BALAKIN, K.V., LANG, S.A., SKORENKO, A.V., TKACHENKO, S.E., IVASHCHENKO, A.A., and SAVCHUK, N.P. Structure-based versus property-based approaches in the design of G-protein-coupled receptor-targeted libraries. *J. Chem. Inf. Comput. Sci.* **2003**, *43*, 1553–1562.

34 LANG, S.A., KOZYUKOV, A.V., BALAKIN, K.V., SKORENKO, A.V., IVASHCHENKO, A.A., and SAVCHUK, N.P. Classification scheme for the design of serine protease targeted compound libraries. *J. Comput.-Aid. Mol. Des.* **2002**, *16*, 803–807.

35 NIKOLSKY, Y., BALAKIN, K.V., IVANENKOV, Y.A., IVASHCHENKO, A.A., and SAVCHUK, N.P. Intelligent machine learning technologies in pre-synthetic combinatorial design. *PharmaChem.* **2003**, *4*, 68–72.

36 ANZALI, S., GASTEIGER, J., HOLZGRABE, U., POLANSKI, J., SADOWSKI, J., TECKENTRUP, A., and WAGENER, M. The use of self-organizing neural networks in drug design. In *3D QSAR in Drug Design, Vol. 2*, KUBINYI, H., FOLKERS, G., and MARTIN, Y.C. (Eds). Kluwer/ESCOM, Dordrecht, Netherlands, **1998**, 273–299.

37 KOROLEV, D., BALAKIN, K.V., NIKOLSKY, Y., KIRILLOV, E., IVANENKOV, Y.A., SAVCHUK, N.P., IVASHCHENKO, A.A., and NIKOLSKAYA, T. Modeling of human cytochrome p450-mediated drug metabolism using unsupervised machine learning approach. *J. Med. Chem.* **2003**, *46*, 3631–3643.

38 WALTERS, W.P., STAHL, M.T., and MURCKO, M.A. Virtual screening – an overview. *Drug Disc. Today* **1998**, *3*, 160–178.

39 CLARK, D.E. and PICKETT, S.D. Computational methods for the prediction of 'drug likeness'. *Drug Disc. Today* **2000**, *5*, 49–58.

40 EDDERSHAW, P.J., BERESFORD, A.P., and BAYLISS, M.K. ADME/PK as part of a rational approach to drug discovery. *Drug Disc. Today* **2000**, *5*, 409–414.

41 OPREA, T.I. and MATTER, H. Integrating virtual screening in lead discovery. *Curr. Opin. Chem. Biol.* **2004**, *8*, 349–358.

42 VALLER, M.J. and GREEN, D. Diversity screening versus focused screening in drug discovery. *Drug Disc. Today* **2000**, *5*, 286–293.

43 KRUMRINE, J., RAUBACHER, F., BROOIJMANS, N., and KUNTZ, I. Principles and methods of docking and ligand design. *Methods Biochem. Anal.* **2003**, *44*, 443–476.

44 RAREY, M., LEMMEN, C., and MATTER, H. Algorithmic engines in virtual screening. In *Chemoinformatics in Drug Discovery*, OPREA, T.I. (Ed.). Wiley-VCH, Weinheim, **2004**, 59–115.

45 BALAKIN, K.V. Pharma ex machina. *Mod. Drug Disc.* **2003**, *8*, 45–47.

Part IV
Chemoinformatics Applications

13

A Practical Strategy for Directed Compound Acquisition

Gerald M. Maggiora, Veerabahu Shanmugasundaram, Michael S. Lajiness, Tom N. Doman, and Martin W. Schultz

13.1
Introduction

With the development of many new methodologies for rapidly screening large numbers of compounds – high-throughput and ultrahigh-throughput screening, (HTS and UHTS), respectively – the generation of compound collections of sufficient magnitude to effectively support these screening methodologies has become an important concern within the pharmaceutical industry. As is well known, combinatorial chemistry methods, while useful, tend to produce compound libraries that lack sufficient diversity. Thus, other means must be sought to provide the large numbers of diverse compounds needed. Currently, the acquisition of existing compounds from various sources, primarily commercial vendors, has been the method of choice throughout the pharmaceutical industry. Thus, a variety of computational approaches were devised to provide a rational means for augmenting proprietary compound collections in the most effective manner possible. These approaches, however, raise a number of issues that must be dealt with before practical compound acquisition strategies can be developed. Of these issues, compound diversity is certainly one of the most important.

Molecular diversity has a relatively brief history, which began in the late eighties and somewhat parallels the development of combinatorial chemistry [1]. Unlike molecular similarity [2–4], which is a pairwise measure, molecular diversity is a measure of the *similarity distribution* over a population of molecules. Alternatively, molecular diversity can be assessed in terms of the *dissimilarity distribution* over a population of molecules since the dissimilarity of two molecules i and j is the complement of their similarity, that is $D(i, j) = 1 - S(i, j)$.

Molecular diversity is thus plagued not only with the problems inherent in molecular similarity/dissimilarity [5, 6] but also with those problems associated with molecular populations [7]. One of the foremost problems is that computed molecular similarity values are not invariant to the molecular representation and to the similarity measure used [5]. Nearest-neighbor (NN) relationships, which are employed extensively in many aspects of HTS, are thus problematic, and it is difficult, and in many cases impossible, to obtain consistent subsets [8]. The structure of chemistry space can also be altered significantly in a global sense. As molecular diversity also depends on these factors, it too can be problematic and inconsistencies will no doubt arise.

Chemoinformatics in Drug Discovery. Edited by Tudor I. Oprea
Copyright © 2004 WILEY-VCH Verlag GmbH & Co. KGaA, Weinheim
ISBN: 3-527-30753-2

Another important and related problem is the scope and resolution of a corporate compound collection. Whether it is better to have a compound collection that covers chemistry space broadly but with a relatively low average density of compounds ("sparse model") as opposed to one that covers less of chemistry space but at a much higher average density of compounds ("dense model") is moot. Early on, the "Similarity Principle" that suggested that similar compounds generally have similar activities or properties [3] led to the view that activity or property landscapes in chemistry space are much like the gently rolling hills of Kansas shown in Figure 13.1. In such cases, the sparse model would generally be sufficient. More recently, however, the Similarity Principle has been shown to be an oversimplification. Many examples now exist that show that the activities of even quite similar compounds may differ significantly and thus the activity landscapes would then tend to look more like the rugged topography of Bryce Canyon shown in Figure 13.2. In such cases, a greater density of compounds is needed to ensure that chemistry space is adequately covered. This is especially important since this situation is likely to represent the usual case rather than the exception [9, 10].

The implication of these problems is daunting and has led researchers to overlook some of the "nasty details" and to develop what might be called "practical strategies". This situation is just another example of the heuristic nature of many of the computational methods employed in biological research in general and drug research in particular. What is described in this chapter is a practical procedure based on a soft, heuristic approach to the problem of compound acquisition that has been implemented and used at Pharmacia over the last several years.

Fig. 13.1 An ideal activity landscape. In such landscapes, which are conceptually similar to the gently rolling Flint Hills of Kansas depicted in the figure, it is generally assumed that similar molecules have similar biological activities [3]. It is clear from the figure that even relatively sparse sampling of chemistry space may be sufficient to construct a reasonable estimate of the activity landscape.

Fig. 13.2 A typical activity landscape. In such landscapes, which are conceptually similar to the rugged terrain of Bryce Canyon, Utah depicted in the figure, small changes to structure in many cases lead to large changes in biological activity. It is clear from the figure that in this case considerably large sampling is required to obtain a suitable estimate of the underlying activity landscape.

13.2
A Historical Perspective

Before the formation of Pharmacia through the merger of Pharmacia & Upjohn with Monsanto-Searle in 2000, each company had its own approaches to augment and diversify their compound collections, and these approaches set the stage for the procedure implemented at Pharmacia. Thus, it was felt that a brief description of the historical background to the present work would be informative and would afford additional insights that bear directly on the development of the compound acquisition strategy described here.

The first application of a computational method to select structurally diverse compounds for purchase started in 1992 at the Upjohn Company, which predated the formation of Pharmacia & Upjohn by about three years. The basic approach selected compounds using a method based upon maximum dissimilarity and was implemented using SAS software [11]. This later evolved into the program Dfragall, which was written in C and is described in Section 13.6.3. Basically, a set of compounds that is maximally dissimilar from the corporate compound collection is chosen from the set of available vendor compounds. Early versions of the process relied solely on diversity-based metrics but it was found that many nondrug like compounds were identified. As a result, structural exclusion criteria were developed to eliminate compounds that were considered unsuitable for purchase from consideration [12]. Over the years, the number of compounds acquired through this strategic purchase program increased from several hundred to tens of thousands, which was a direct consequence of its success in acquiring structurally unique compounds that resulted in important leads in a number of therapeutic programs.

Around 1995, the Monsanto-Searle Company devised its own computational methods for selecting compounds for acquisition. These methods centered around the use of the proprietary clustering algorithm known as "algorithm5" [13], which clusters compounds using the molecular fingerprints developed by Daylight Chemical Information Systems [14] to compute Tanimoto similarity (TS) between all pairs of molecules. Interestingly, in nearly all cases, algorithm5 generates clusters with average intracluster Tanimoto similarity of 0.85 or greater. In addition to the Monsanto compound collections investigated, a number of other collections, including some commercial ones such as the ACD (Available Chemicals Directory, MDL Information Systems, San Leandro, CA), showed essentially the same behavior. Moreover, since algorithm5 does not require user-supplied clustering parameters, it is an especially good technique for carving up compound collections into clusters in an unbiased way such that compounds within a given cluster have a high probability of sharing similar bioactivities. This is consistent with activity landscapes that follow the hills of Kansas analogy in Figure 13.1.

The Monsanto-Searle cluster-based selection method proceeded by coclustering the corporate compound collection with the set of commercially available compounds and then looking for those clusters containing only commercially available compounds. Choosing one compound from each of these clusters ensured that the sample thus obtained consisted of a minimal set of diverse compounds to augment the collection. In practice, it was found that the cluster-based method was complementary to the cell-based method utilized in our current work (*vide infra*). In a typical case, significantly more compounds are selected using the cell-based method, since each cell of interest is, in general, populated with more than one compound. This approach, in distinction to that employed with algorithm5, is in keeping with the notion that the topography of activity landscapes is more similar to that of the Bryce Canyon depicted in Figure 13.2 than to that of the hills of Kansas (see Figure 13.1). Nevertheless, the algorithm5 cluster-based method is useful in situations in which a minimum number of compounds are needed and there is a desire to choose these in a very unbiased fashion.

13.3
Practical Issues

A number of ancillary issues must also be addressed in addition to those dealt with in the compound selection procedure itself. One of these is compound attrition; each year, about 1% of the compounds in a given collection are, for a variety of reasons, no longer available. Some of these may be rare or unusual compounds that are difficult to replace. If and how these are replaced is a subject for discussion that will not, however, be dealt

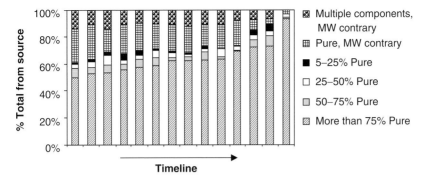

Fig. 13.3 The quality of commercial compounds. In general, quality control by the compound brokers has improved significantly over the past decade. Sample identity and purity is now less of an issue, but still bears watching. The graph shown above summarizes the historical data gathered over a period of time over which vendor performance was monitored. In recent years, more than 90% of preferred vendors have provided accurate and highly pure samples.

with here. Other, more mundane but nevertheless important, issues include the number of compounds to be acquired, the overall budget, the amount and cost per compound and vendor choice. A number of issues are associated with vendor choice. Two of the most important issues are reliability in providing compounds in an expeditious manner and sample purity. Issues of sample purity are addressed here only indirectly through vendor performance. Only vendors who provide reasonably pure samples based on historical records are included in our Preferred Vendor list. In recent years, however, the quality of commercially available compounds has improved significantly as shown in Figure 13.3. Another factor that is unrelated to vendor performance is availability of rare or unusual compounds. Many compounds that appear in the catalogs of a number of large vendors are identical. Small vendors, on the other hand, may be potential sources of novel compounds not typically found in the catalogs of large vendors, as illustrated in Figure 13.4. Thus, we have chosen a mix of large-, medium-, and small-size vendors. The procedure described here was used at Pharmacia for the purpose of augmenting our corporate compound collection. It excluded purchases of libraries derived using combinatorial chemistry methods or libraries constructed for specific target classes (e.g., kinase libraries). These were dealt with using strategies other than those described in this work.

The ultimate goal in the present work was to acquire on the order of 40 000 compounds. Because of a number of factors, such as changes in the available amount or quality of an ordered compound, obtaining all of the compounds ordered is generally not possible. In our earlier efforts, only about 50% of the ordered compounds were actually delivered. By requesting that vendors indicate which compounds in their catalogs were available in amounts of at least 50 mg, more than 80% of the compounds ordered were delivered, significantly improving the effectiveness of the compound acquisition process.

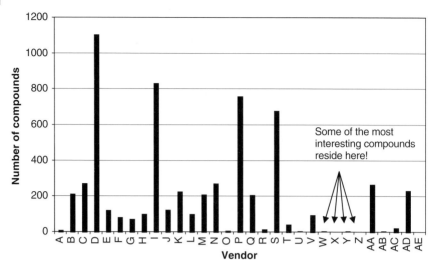

Fig. 13.4 A logistical nightmare. Processing orders from numerous vendors is very challenging. Shown here are 6111 compounds purchased from 30 different vendors. To alleviate this kind of ordering in the current process, only compounds from the preferred vendor list and a set of "small vendors" that have potential sources of novel compounds (compounds not found in the catalogs of larger vendors) were considered. A vendor prioritization scheme that assigns compounds to the top vendors (if compounds were available from multiple sources) was also put in place.

13.4
Compound Acquisition Scheme

The basic heuristic procedure used in this work can be divided into four main activities:

1. Preprocessing Compound Files
2. Diversity Assessment and Initial Compound Selection
3. Compound Reviews
4. Final Selection and Compound Purchase

The process is schematically depicted in Figure 13.5. Detailed summaries describing the various methodologies applied in this work are found at the end of the chapter.

13.4.1
Preprocessing Compound Files

Before compound selection can begin, it is necessary to preprocess the set of vendor compound files and the Pharmacia Compound Collection (PCC) to ensure that they both represent the most current compound information. In the former case, it is necessary to generate a single file for all vendor compounds free of redundancies, that is, each compound should appear only once in the file. The set of vendors supplying each

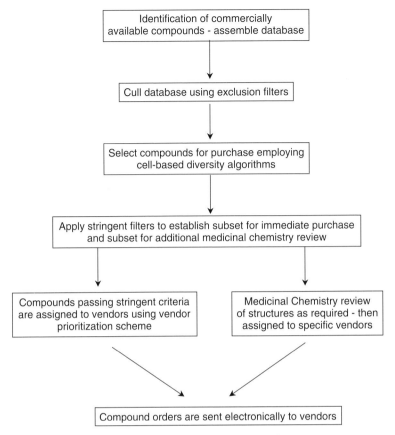

Fig. 13.5 The compound acquisition scheme. The overall process depicted in the figure provides an efficient procedure for the selection and acquisition of compounds that optimally populate the PRCC with diverse, druglike molecules for suitable screening.

compound, ranked by vendor preference, was also included. Vendor preferences were determined by vendor performance in previous compound acquisitions and are updated yearly. The compound catalogs for the approximately 30 large and small vendors used in this study contained more than three million compounds that, according to the vendors, were readily available in 50 mg or greater quantities. The 50-mg sample size was chosen for three reasons: (1) cost, (2) sample lifetime – 50 mg of sample would last a number of years, and (3) hit validation – 50 mg of sample would provide sufficient sample for numerous follow-up hit validation experiments.

Removing redundancies reduced the number of compounds in this set to about 1.8 million. An additional set of criteria based upon properties associated with "druglikeness" were used to further reduce the set, which we called the Commercial Research Chemicals Directory (CRCD), to just above one million compounds. The

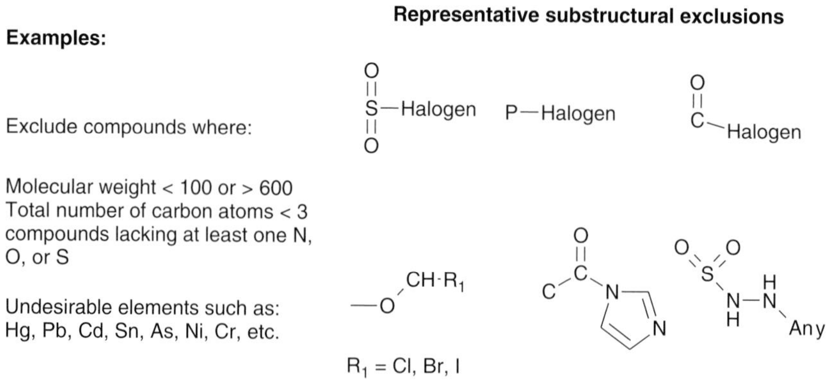

Examples:

Representative substructural exclusions

Exclude compounds where:

Molecular weight < 100 or > 600
Total number of carbon atoms < 3
compounds lacking at least one N,
O, or S

Undesirable elements such as:
Hg, Pb, Cd, Sn, As, Ni, Cr, etc.

R_1 = Cl, Br, I

Fig. 13.6 Preprocessing compound files. Representative substructural exclusion criteria and compound filters used to prune the CRCD database are shown. These were designed to eliminate reactive compounds, toxic materials, and so on. The complete list of exclusion criteria was developed over a number of years in close collaboration with experienced medicinal chemists.

druglike properties considered include molecular weight, octanol–water partition coefficient, number of hydrogen-bond donor and acceptor groups, number of rotatable bonds, and exclusion of compounds containing highly reactive or otherwise undesirable functional groups. This is depicted in Figure 13.6; Section 13.6 should be consulted for further details and discussion.

Preprocessing the PCC presented somewhat different problems. The PCC contained many different types of compounds including a substantial fraction of compounds derived from combinatorial chemical methods as well as "chemically undesirable" compounds, which contain highly reactive or otherwise unsuitable structures as potential drug leads. Peptides, compounds containing heavy atoms (e.g., mercury) and compounds with improper or incorrect structures were also removed. Removing all of these compounds reduced the size of the PCC substantially. Choosing only compounds with inventory, that is, compounds that could actually be screened, reduces the size of the PCC even further. It should be noted here that the significant drop in the number of screenable compounds is due to the fact that only dry powder inventory was considered. Compounds in DMSO were not considered, as generally only less than 5 mg was available in such cases. This was certainly enough for screening but the quantities were not sufficient for follow-on validation and characterization experiments. The remaining compounds are taken to be the Pharmacia Reference Compound Collection (PRCC). All compounds selected from the CRCD for acquisition are referenced against the PRCC to ensure that it is augmented in a way that enhances its diversity overall.

Interestingly, the PRCC is relatively free of compounds that violate Lipinski's "Rule of Five" (ROF) [15] as shown in Table 13.1. Similar ROF behavior has also been observed by Oprea [16] for other compound databases including the ACD and the MDDR (MDL Drug Data Report, MDL Information Systems, San Leandro, CA).

As seen in the table, less than 3% of the PRCC violated two or more of the rules, which is slightly less than the approximately 3% of cases found in the entire PCC. Thus,

Tab. 13.1 Percent ROF violations in the PRCC

ROF violations	0	1	2	3	≥4
% ROF Violations	70.6	26.7	2.4	0.3	~0.0

ROF violators were not removed from the PRCC. Although removing them could lead to some "holes" or "voids" within the PCC, it was felt that this would not be significant.

13.4.2
Initial Compound Selection and Diversity Assessment

In the initial phase of compound selection, the strategy was to select compounds that filled "diversity voids". Diversity voids are defined as regions of a chemistry space that are not populated with compounds from the PRCC. It should be noted, however, that regions of chemistry space might not be populated by molecules of interest in drug design or, for that matter, by molecules of any type since chemistry space is, in general, sparse.

The program DiverseSolutions (DVS) developed in Professor Bob Pearlman's group at the University of Texas is a tool to create models of chemistry spaces. This software was used to accomplish the task of defining the PRCC chemistry space and identifying diversity voids. DVS represented the chemistry space of the PRCC in terms of six 3-D BCUT descriptors [17]. Two BCUTs characterize positive and negative partial-charge properties, two characterize hydrogen-bond donor and acceptor properties, and two characterize polarizability properties, which are associated with hydrophobicity. These define the 6-dimensional chemistry space used in this work that is derived from the PRCC. A brief discussion of the method is found in Section 13.6.2.

To facilitate compound selection, chemistry space is partitioned into cells, as depicted by a simplified scheme shown in Figure 13.7. This was accomplished by dividing each BCUT axis into ten equal-sized bins yielding a total of one million cells. About 91% of the cells were unoccupied by the PRCC, leaving a significant amount of chemistry space available for compound selection. The histogram in Figure 13.8 shows the distribution of cell occupancies in the PRCC. The selection criterion is based on filling all cells containing fewer than five compounds with up to five new compounds from the CRCD. Other selection criteria were also investigated, but the above criterion was chosen to ensure that the number of compounds selected was sufficient in number to ultimately meet our goal of approximately 40 000 acquired compounds. This yielded slightly more than 140 000 compounds that filled about 73 000 cells that originally contained fewer than five compounds. Even though this selection has filled some very low-occupancy regions of chemistry space, the overall density of compounds remains rather low.

Although the number of cells in chemistry space is large, the hyper volume of each individual cell is also very large. Thus, compounds within a given cell may be quite similar or quite dissimilar. The latter could occur, for example, for compounds located

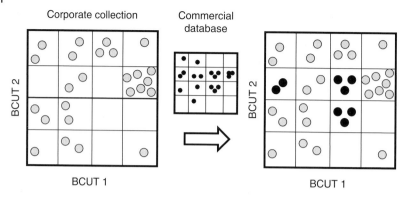

Fig. 13.7 A schematic depiction of the compound selection procedure used in this work, which is based on *DiverseSolutions* [18] and a cell-based representation of chemistry space. Commercial compounds are selected to fill unoccupied and low-occupancy cells. This ensures that new regions of chemistry space are given precedence, which extends the scope of coverage of chemistry space.

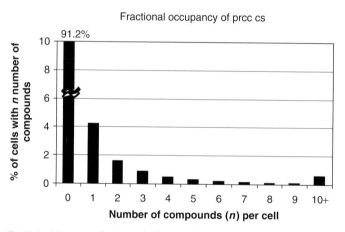

Fig. 13.8 Histogram depicting the fractional occupancy of PRCC chemistry space. A significant fraction of the PRCC chemistry space (91.2%) is unoccupied and is available for compound selection. Only a very small percent (0.59%) is densely populated (10+ compounds per cell).

in the "corners" of the cell that are maximally separated (i.e., that lie along the main diagonal of the hyper-dimensional cell). Boundary effects can also occur: compounds located close to but on either side of a boundary between two cells will be more similar than compounds at either end of the main diagonal within the same cell. Because of these types of idiosyncratic behaviors, an additional level of similarity analysis is performed to ensure that the newly selected compounds remain as dissimilar as possible from the PRCC.

This analysis is carried out using a program called "Dfragall", which is based on a maximum dissimilarity algorithm [19]. Dfragall computes the Tanimoto similarity among all pairs of compounds using a bitstring representation of molecules based on the molecular fragment scheme employed in our chemical information system, *ChemLink*. This allows identification of the most structurally diverse molecules contained in a set of compounds [19]. In the present application, it allows one to eliminate compounds that are highly similar to those previously selected.

Approximately 62 000 molecules were picked from the set of 140 000 molecules obtained earlier using the cell-based procedure in DVS. The number 62 000 was chosen based on previous experience to ensure that after additional processing the number of molecules remaining would be sufficient to meet our compound acquisition goal of ~40 000 compounds – recalling that only a fraction of the ordered compounds are ever actually received. No compound was chosen from the initial selection set of about 140 000 compounds that had a Tanimoto similarity value, based upon ChemLink fingerprints, greater than 0.96 to any of the compounds in the PRCC. This may at first seem rather large, but it should be borne in mind that such high similarity values will inevitably occur in the compound acquisition process, since commercial vendor collections such as the CRCD are unlikely to fill entirely new regions of chemistry space but rather will mainly add additional compounds to already occupied regions. This is not necessarily bad because, as illustrated in Figure 13.2, activity landscapes can be rather rough with many cliffs so that proper sampling requires a significant density of compounds even in regions of active compounds. The analysis just described outlines a cell-based selection procedure. In earlier compound selection campaigns, we also employed a procedure based on clustering using algorithm5 (*vide supra*).

13.4.3
Compound Reviews

Before making final compound selections, a last round of compound reviews was carried out. In previous years, this has usually been carried out by medicinal chemists using the time-honored method of visual inspection. This can be quite daunting and subject to the usual inconsistency problems humans experience when considering large datasets of any kind. Thus, datasets of 50 000 or more compounds need to be handled in a more automated fashion. To expedite this process, a set of "No-Review" (NR) filters was developed based upon the drug-design knowledge of experienced medicinal chemists. The filters are designed so that molecules medicinal chemists "like" are allowed to pass through. As will be seen, these filters add some additional criteria as well as increase the stringency of already existing criteria. Some of the criteria are as follows: number of rings, 2 to 4; MW, 200 to 400; number of rotatable bonds, 0 to 5; ratio of hetero- to carbon atoms, 0.1 to 1; ClogP, -1 to 2; and so on. In a number of evaluations carried out at Pharmacia, it was found that $\sim 99.7\%$ of the compounds that passed through the NR filters were acceptable to medicinal chemists. In the current application, the NR filters passed approximately 65% (~40 000) of the ~62 000 selected by the diversity-based procedures described earlier. The remaining 35% that did not pass the NR filters (~22 000 compounds) were then handled using "traditional" visual inspection by medicinal chemists.

In an effort to assess the reliability and consistency of visual inspection by medicinal chemists, the following process was implemented. The approximately 22 000 compounds that did not pass the NR filters were broken up into 11 lists of approximately 2000 compounds each. Two chemists were assigned to each of the lists, and each chemist reviewed two lists. A list of the "acceptable" compounds was compiled from the reviews. A detailed analysis of the study is being prepared for publication [18].

13.4.4
Final Selection and Compound Purchase

A list of approximately 50 000 compounds was chosen from the 62 000-compound set discussed above, which was about 10 000 compounds greater than the number of 40 000 that was the initial acquisition target. The "overselection" of 10 000 compounds was needed to compensate for compounds that simply may not be available at the time of ordering. Two steps were taken to obtain the intermediate 50 000-compound set. First, an analysis of the ring systems present in the selected compounds was carried out, which resulted in the elimination of around 750 compounds possessing ring systems already well represented within the PRCC. A brief discussion of this analysis is found in Section 13.6.4. The ring-systems analysis was followed by a proportional sampling of the compounds that passed the NR filters and those that passed visual inspection. Initially, the intention was to immediately order ~32 000 of the 40 000 compounds that passed the NR filters. This would jump-start the ordering process and allow us to obtain the compounds as quickly as possible. The remaining 18 000 compounds would then come from compounds that passed visual inspection. As it turned out, however, all of the compounds that passed the NR filters as well as those that passed visual inspection, a total of about 59 000 compounds, were ordered because of a late infusion of additional funds for compound acquisition. To summarize, the set of ~59 000 compounds ordered consisted of the ~40 000 that passed the NR filters and ~19 000 compounds that passed visual review. The ~750 compounds that did not pass the ring-review (*vide supra*) were, of course, not included in the set of compounds to be ordered. Approximately 83% of the compounds ordered (~49 000 compounds) were received from vendors, which far surpassed the usual ~50% rate of past compound acquisitions. This is no doubt due, in large measure, to our effort to assess the vendors' ability to provide sufficient samples for our needs (*vide supra*).

13.5
Conclusions

Compound collections play a crucial role in the search for new leads in drug research at large pharmaceutical companies. Hence, it is essential that compound collections be augmented through acquisition of new compounds, as well as by other means, to ensure that chemistry space is adequately covered, thus enhancing the chance that new and interesting leads will be identified through HTS. The present work, which describes the compound acquisition process at Pharmacia, represents a refinement of the processes carried out for a number of years at Pharmacia & Upjohn and Monsanto-Searle.

As the nature of chemistry space depends upon the way in which compounds are represented, an absolute or universal chemistry space does not exist. Thus, any procedure that utilizes chemistry space may be subject to considerable uncertainty, and the results obtained in different chemistry spaces are likely to differ, sometimes in quite significant ways. This rather daunting circumstance has necessitated the use of practical, heuristic approaches that, while imperfect, have nevertheless performed in a reasonably satisfactory manner over the last three years. During this period about 120 000 diverse, quality compounds have been added to our corporate compound collection. This does not include the many compounds obtained from combinatorially derived libraries and special target-directed (e.g., kinase) libraries.

Preprocessing both the vendor and Pharmacia compound collections has resulted in a substantial improvement in the quality of the compounds selected and in the performance of vendors in supplying the compounds on a timely basis. In 2002, more than 83% of the nearly 60 000 compounds ordered were obtained, exceeding our earlier performances that were usually in the neighborhood of 50%. This is partially due to the preselection of vendors who had the ability to provide an appropriate amount (~50 mg) of sample. Unfortunately, we are unable to *directly* compare our approach with those tried by others or with an entirely random sampling approach. Nevertheless, it is clear that the compound acquisition process described here does, indeed, provide an effective means for enriching our corporate compound collection.

13.6
Methodologies

13.6.1
Preprocessing Filters

Several druglike properties including molecular weight, octanol–water partition coefficient, and the number of rotatable bonds were considered for preprocessing compound collections. Threshold levels for each of these properties were conservatively chosen so as not to exclude compounds that might serve as useful, albeit slightly less druglike, leads for drug discovery. We preferentially kept compounds with molecular weight 100 to 600, with Daylight clogP −2.00 to +6.00 and with 15 or fewer rotatable bonds. Our algorithm defines a rotatable bond as any single bond except: (a) an amide N–C bond, (b) a single bond adjacent to a triple bond, or (c) a single bond in a ring, which is formally assigned as half a rotatable bond (all rings were considered regardless of size, but since the number of 3- and 4-membered rings is small, the approximation did not significantly affect the overall statisitics).

Finally, we employed a list of 56 substructures that are either required or forbidden to be in compounds we select. These represent decades of medicinal chemistry experience at the legacy Pharmacia & Upjohn and Monsanto-Searle companies and are conservatively chosen so as to avoid obviously reactive or otherwise undesirable compounds. Figure 13.6 shows some of these filters, including substructure filters for several reactive structural types, as well as filters for heavy metals and for compounds with too few carbon atoms or heteroatoms.

13.6.2
Diverse Solutions (DVS)

The program DiverseSolutions (DVS) was developed in Professor Bob Pearlman's laboratory at the University of Texas, Austin [20]. DVS is designed for addressing a variety of diversity-related chemoinformatic tasks, including assessing the diversity of a collection of compounds, selecting representative subsets from a larger compound collection, comparing the diversity of two or more different compound collections and filling "diversity voids" within one collection by selecting compounds from a second collection, and so on.

For a given collection of compounds, DVS generates chemistry-space descriptors called BCUTs that describe atomic properties relevant to protein–ligand interactions (interpreted as charge, polarizability and H-bonding ability) using the connectivity information (bond types, interatomic distances, etc.) contained in these compounds. DVS then selects those descriptors that best distinguish the structural differences between compounds and creates an approximately orthogonal chemistry space. This can be used to represent the diversity of the given compound collection.

DVS offers a cell-based method for assessing the diversity of compound collections. The orthogonal chemistry space is partitioned into cells, and the occupancy of each cell is easily determined. "Diversity voids", or unoccupied regions of chemistry space, are identified with unoccupied cells or cells of low occupancy, which are also easily recognized. Enhancing diversity of a compound collection thus consists of filling empty and low-occupancy cells (see Figure 13.5).

13.6.3
Dfragall

Dfragall computes the Tanimoto similarity among all pairs of compounds using a bitstring representation that allows one to identify the most structurally diverse compounds contained within a set of compounds. In applications at Pharmacia, the bitstring representation used was one available in our chemical information system *ChemLink* and is composed of 320 bits associated with a set of molecular fragments [21]. Given a starting "seed compound", the program computes Tanimoto similarity values between the seed and all remaining compounds selecting that compound with the lowest similarity to the seed. The process iterates selecting those compounds that have the lowest similarity (highest diversity) with any of the previously selected compounds until the desired number of compounds is obtained.

13.6.4
Ring Analysis

A common complaint by medicinal chemists when reviewing lists of compounds, generated by the compound acquisition process or by, for example, the results of HTS, is that there are too many compounds with a given ring pattern. It is well known that two compounds that possess the same ring system may in fact have low similarity due to differences in side chains and substituents. This problem was addressed by identifying those ring-pattern classes contained in each molecule and by choosing only those compounds from under-represented ring classes. The ring-pattern classes from both the potential purchases and the PRCC were obtained using Structural Browsing Indices formalism developed by Xu and Johnson [22, 23]. The frequency of occurrence of each ring class in the PRCC was determined and a calculation was made as to how many new compounds could be added. If necessary, a random selection was used to identify the compounds from the over-represented classes out of which compounds are selected for purchase.

Acknowledgments

The authors would like to acknowledge the support provided by Gordon Bundy, Mary Torkelson, Dick Thomas, Chic Spilman, Bob Manning, Mario Varasi, and Michael Clare.

References

1 MARTIN, Y.C. Diverse viewpoints on computational aspects of molecular diversity. *J. Comb. Chem.* **2001**, *3*, 231–250.

2 WILLETT, P. *Similarity and Clustering in Chemical Information Systems.* Research Studies Press, Letchworth, **1987**.

3 JOHNSON, M.A. and MAGGIORA, G.M. (Eds). *Concepts and Applications of Molecular Similarity.* John Wiley & Sons, New York, **1990**.

4 DEAN, P.M. (Ed.). *Molecular Similarity in Drug Design.* Chapman & Hall, Glasgow, **1994**.

5 MAGGIORA, G.M. and SHANMUGASUNDARAM, V. Molecular similarity measures. In *Chemometrics: Methods and Protocols*, BAJORATH, J. (Ed.). Humana Press, Totowa, NJ, **2003**, 1–50.

6 WILLETT, P., BARNARD, J.P., and DOWNS, G.M. Chemical similarity searching. *J. Chem. Inf. Comput. Sci.* **1998**, *38*, 983–996.

7 WALDMAN, M., LI, H., and HASSAN, M. Novel algorithms for the optimization of molecular diversity of combinatorial libraries. *J. Mol. Graph. Model.* **2000**, *18*, 412–426.

8 SHERIDAN, R.P. and KEARSLEY, S.K. Why do we need so many chemical similarity search methods? *Drug Discov. Today* **2002**, *7*, 903–911.

9 MARTIN, Y.C., KOFRON, J.L., and TRAPHA-GEN, L.M. Do structurally similar molecules have similar biological activity? *J. Med. Chem.* **2002**, *45*, 4350–4358.

10 SHANMUGASUNDARAM, V. and MAGGIORA, G.M. Characterizing property and activity landscapes using an information-theoretic approach. *Abstr. Papers ACS*, 32-CINF, Part 1 August **2001**, *222*.

11 Lajiness, M.S., Johnson, M.A., and Maggiora, G.M. Implementing drug-screening programs using molecular similarity methods. In *QSAR: Quantitative Structure-Activity Relationships in Drug Design*, Fauchere, J.L. (Ed.). Alan Liss Inc., New York, USA, **1989**, 173–176.

12 Lajiness, M.S. Dissimilarity-based compound selection techniques. *Perspect. Drug Discov. Des.*, **1997**, *7/8*, 65–84.

13 Doman, T.N., Cibulskis, J.M., Cibulskis, M.J., McCray, P.D., and Spangler, D.P. Algorithm5: A technique for fuzzy similarity clustering of chemical inventories. *J. Chem. Inf. Comput. Sci.* **1996**, *36*, 1195–1204.

14 Daylight Chemical Information Systems, Inc., Mission Viejo, California, http://www.daylight.com/.

15 Lipinski, C.A., Lombardo, F., Dominy, B.W., and Feeney, P.J. Experimental and computational approaches to estimate solubility and permeability in drug discovery and development settings. *Adv. Drug. Deliv. Rev.* **1997**, *23*, 3–25.

16 Oprea, T.I. Property distribution of drug-related chemical databases. *J. Comput. Aided Mol. Des.* **2000**, *14*, 251–264.

17 Pearlman, R.S. and Smith, K.M. Novel software tools for chemical diversity. *Perspect. Drug Disc. Des.* **1998**, *9*(11), 339–353.

18 Lajiness, M.S., Maggiora, G.M., and Shanmugasundaram, V. An assessment of the consistency of medicinal chemists in reviewing compounds lists. *J. Med. Chem.*, in press.

19 Lajiness, M.S. Dissimilarity-based compound selection techniques. *Perspect. Drug Disc. Des.* **1997**, *7/8*, 65–84.

20 Pearlman, R.S. *Diverse Solutions User's Manual.* University of Texas, Austin, TX, **1995**.

21 Hagadone, T.R. and Lajiness, M.S. Integrating chemical structures into an extended relational database system. In *Proceedings of the Second International Chemical Structures in Chemistry Conference*, Warr, W.A. (Ed.). Springer-Verlag, Berlin, Germany, **1993**, 257–269.

22 Johnson, M.A. and Xu, Y.-j. Using molecular equivalence numbers to visually explore structural features that distinguish chemical libraries. *J. Chem. Inf. Comput. Sci.* **2002**, *42*, 912–926.

23 Xu, Y.-j. and Johnson, M.A. Algorithm for naming molecular equivalence classes represented by labeled pseudographs. *J. Chem. Inf. Comput. Sci.* **2001**, *41*, 181–185.

14

Efficient Strategies for Lead Optimization by Simultaneously Addressing Affinity, Selectivity and Pharmacokinetic Parameters

Karl-Heinz Baringhaus and Hans Matter

14.1
Introduction

The increasing pressure on efficiency and cost of research in the pharmaceutical industry has caused a technological paradigm shift [1, 2] in order to bring promising candidate molecules earlier to the market [3, 4]. The total research and development costs for a novel compound to enter this market have recently been estimated as US$600 to 800 million [5, 6], while continuously increasing expenses are mainly attributed to the high attrition rate in later development phases. The major reason for failure in phase II and III of clinical trials is the inadequate understanding of pharmacokinetic behavior of drugs [7–9] and what constitutes a pharmacokinetic profile for candidate drugs. This again underscores the necessity to improve on the success rates in order to maintain economic viability of the pharmaceutical drug discovery process on longer terms.

Several technological advances [10] along the drug discovery value chain have entered the field and become indispensable tools for identifying novel lead compounds. However, the advent of technologies like combinatorial chemistry [11, 12], automated synthesis technologies and high-throughput screening [13] also caused an exponential increase in the number of single data points for analysis, while it is debatable to which extent these technologies contributed to the launch of new chemical entities (NCEs) to the market [2, 3, 14, 15]. The current challenges in drug discovery are related to increased regulatory hurdles, more effective integration of technological advances, the extraction of relevant information and knowledge from primary data to support data-driven decisions and – at the same time – the cost pressure, prompting for increased efficiency and lower attrition rates.

Improving on the low success rates prompts for a balanced compound progression driven by clearly defined knowledge-based decisions to advance or discontinue particular lead series as early as possible. Any failure to identify promising lead series has severe implications on time and resources within a disease-related program in a pharmaceutical company. A clear process, rigorous and relevant metrics for compound progression are essential here [2, 16]. This requires early attention to stringent quality criteria for compound series, reflecting the complex multitude of biological, pharmacological, pharmacokinetic and physicochemical parameters, which a project team has identified as key drivers for the chemical optimization and for candidate selection.

Chemoinformatics in Drug Discovery. Edited by Tudor I. Oprea
Copyright © 2004 WILEY-VCH Verlag GmbH & Co. KGaA, Weinheim
ISBN: 3-527-30753-2

Although the stage of lead optimization on the first look is not very spectacular, the efficient conversion of molecules with promising biological activity into viable drug candidates fulfilling a multitude of requirements appears to be one of the most challenging steps [17, 18] with a very high impact on the successful continuation of efficient drug discovery programs [2]. Conversion of a biologically active chemical compound into an effective and safe drug adds substantial value in the drug discovery process. Consequently, the improvement of a compound profile toward a clinical candidate is one of the essential skills in integrated drug discovery teams. Those candidate requirements include not only desirable selectivity against related or diverse "antitargets" but also favorable physicochemical and pharmacokinetic properties, leading to oral administration and acceptable half-life of the final candidate. To arrive at candidates with suitable pharmacodynamic and pharmacokinetic properties, hence, requires a simultaneous optimization of multiple parameters in carefully planned iterations. Here, it is interesting to note that recent comparisons of molecular properties of launched drugs revealed only limited differences compared to the original lead compound as starting point [19].

The necessity to shorten the discovery process has prompted for an early integration of pharmacokinetic and drug development efforts to rapidly identify those molecules that are unlikely to become drugs and those lead series with significantly lower chemical optimization potential. It is of high importance to initiate lead optimization programs only for those molecules that intrinsically have the potential to be converted into drugs. Hence, this lead optimization stage has entered into a new phase of complexity caused by advances in modern technologies like protein crystallography, assay technology, ADME (absorption, distribution, metabolism, excretion) assays, medicinal chemistry automation and others. Although rational approaches during this phase to manipulate molecules are typically guided by QSAR (quantitative structure-activity relationship) and structure-based design, it now becomes a tight interplay between various disciplines: medicinal chemistry, structural biology, pharmacology and pharmacokinetics. It is vital for success to conceive lead optimization as simultaneous multidimensional optimization rather than to address one parameter at a time (Figure 14.1).

The review focuses on a discussion of novel strategies, processes and computational tools with an impact on improving the poor industry success rates during the chemical optimization phase of lead series. The ultimate goal of any chemical optimization program is to convert quality lead structures into high value drug candidate molecules prior to clinical development. First, the starting point for lead optimization is analyzed, namely the origin of lead and current selection criteria for the promising series. Knowledge-based decisions are essential at this stage to circumvent time and resource investments for later stages. The emphasis of subsequent sections is to discuss several components in this lead optimization process toward refinement of series with improved druglike properties. This includes a summary of the understanding and predictive models for biological affinity, selectivity versus closely related targets as well as clearly defined antitargets and pharmacokinetic problems.

Traditionally, in this optimization process, the affinity is optimized first, while pharmacokinetic and ADME issues are investigated later in this process [20, 21]. This approach, however, showed only limited success, as optimizing for affinity only

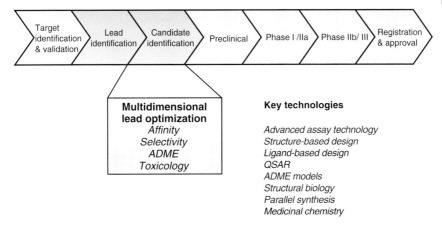

Fig. 14.1 Drug discovery value chain and selected key technologies and activities [2].

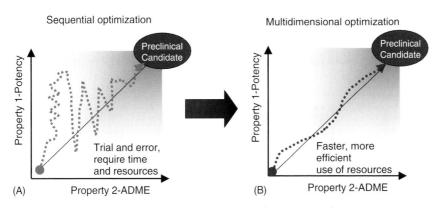

Fig. 14.2 Different drug discovery strategies: (A) Sequential optimization (historic approach) – first optimization of affinity, while ADME properties are treated at a later stage. (B) Multidimensional scenario – combined optimization of affinity and ADME properties, simultaneous monitor changes in relevant properties [20].

can result in chemical classes, where subsequent optimization for ADME properties becomes difficult, if not impossible. The efficiency of the drug discovery process is expected to improve if both aspects are considered simultaneously. Hence, we discuss different approaches and case studies on lead optimization using such a tight integration of different parameters by means of simultaneous optimization of a variety of relevant molecular properties, sometimes referred to as "multidimensional optimization" (MDO) [2, 16, 22]. Consequently, we will use this term throughout the following discussion. This multidimensional optimization requires a clear risk assessment prior to initialization of a resource-intensive research program to optimize compounds toward a balance of such a multitude of properties [23]. General considerations to multidimensional lead optimization are discussed with an emphasis on computational approaches to assist in this simultaneous task. Data integration is mandatory for extraction of knowledge and reuse in following iterations in the design cycle, while *in silico* approaches to estimate ADME parameters are crucial to focus only on promising molecules to address the increased complexity in lead optimization. These concepts will be illustrated using literature and in-house examples (Figure 14.2).

14.2
The Origin of Lead Structures

This chapter summarizes strategies to identify lead structures with a particular emphasis on modern technologies complementing high-throughput screening in the pharmaceutical industry, while recent reviews discuss some aspects in detail [22] and provide a summary of the individual processes and interfaces between disciplines involved within a pharmaceutical setting [16].

The success of many drug development projects seems to be fundamentally limited by the nature of the target. Experience in medicinal chemistry and historic drug discovery programs suggests that small-molecule organic compounds can modulate more readily some privileged protein target classes. Hence, the selection of tractable targets in addition to careful selection following therapeutic requirements is essential to guide drug discovery at an early stage. All drugs that are currently on the market are estimated to modulate less than 500 targets, like nucleic acids, enzymes, GPCRs (G-protein-coupled receptors) or ion channels [9]. Consequently, the "druggability" of protein targets including small-molecule binding sites has been discussed [24] following the completion of the Human Genome Project, with an attempt to estimate a reasonable number of targets that indeed provide an opportunity for future therapeutic intervention. While earlier estimations discuss a total of 5000 to 10 000 potential targets according to the number of disease-related genes [9], a focus on properties of orally bioavailable, small drug molecules that could potentially interact with those targets results in a lower number of targets (600–1500) [24]. Those physical–chemical properties of "druglike" molecules include the well-established "rule-of-five" [25] or related knowledge-derived guidelines [26]. Those summarize simple, intuitive parameters for medicinal chemists to give a warning for compounds that are unlikely to be orally absorbed via the passive intestinal route. The increasing understanding of druglike properties hence might additionally refine our understanding of complementary properties for drug targets.

Hit finding for a selected target typically starts with a collection of actives from high-throughput screening (HTS) [13, 27, 28], which today is the most widely applicable technology, while quality of results is critically dependent on assay type and quality [16]. In addition, the recent years have witnessed an emerging number of complementary biophysical and *in silico* hit-finding approaches like fragment-based SAR-by-NMR [29–31], surface plasmon resonance binding [32], high-throughput X-ray crystallography [33, 34], protein structure–directed *de novo* design [35] and structure- and ligand-based library design and virtual screening [36, 37].

The costs associated with hit identification using diverse assay systems and technologies as well as, alternatively, biophysical methods and approaches are regarded as minor compared to the finally required clinical development of novel candidate molecules [38]. The validation and exploration around lead series is significantly more resource-intensive and dependent upon the selection criteria to progress on promising series. Hence, the series of hits are thoroughly evaluated and validated as early as possible for collecting as much information on individual structures and entire structure classes. Any systematic method to evaluate and validate results from hit finding requires a clear understanding on which terms and criteria to monitor in order to decide upon the fate of a series [2, 16].

Several criteria for selecting promising lead structures and tailoring focused libraries have been proposed in the literature. The early awareness of liabilities within a lead series with respect to the desirable compound profile in terms of selectivity, physicochemical and ADME properties is important for choosing the series with potential for starting a successful lead optimization program, while none of these knowledge-driven decisions, however, can prevent unpredictable problems like animal toxicity and others. Generally speaking, it is the overall characteristics of a particular compound class, considering a multitude of properties including synthetic feasibility and patentability, which makes it attractive for starting a lead optimization program. This potential, however, has to be systematically explored within the frame of a limited synthetic program dedicated to explore the structure-activity relationship as well as obtain as much information on key drivers in order to identify those series with improved druglike properties.

Recent comparisons of physicochemical properties of lead structures to drug molecules also enhanced the understanding of which criteria to apply for selecting a promising series. Those studies were reported by Oprea, Teague et al. [39–41] based on a dataset of 96 lead-drug pairs and a variety of calculated physicochemical properties. Lead structures tend to exhibit lower molecular weight, lower number of rings rotatable bonds and lower logP, suggesting that the subsequent process of lead optimization in medicinal chemistry tends to add complexity, mainly by means of less directional hydrophobic interactions.

The impact of molecular complexity on the probability to detect hits was studied by Hann et al. [42] on the basis of statistical arguments. Using simple models of ligand–receptor interactions, the probability of useful binding events of increasing complexities (number of potential protein–ligand interactions) was estimated. Low-complexity compounds have an increased probability of being detected as hits in screening and thus might offer better starting points for drug discovery. If systems become more complex, the chance of observing a useful interaction for a randomly picked ligand falls dramatically, based on statistical reasons.

One possible route to increase binding affinity of low-complexity fragments is to link two fragments binding to different subpockets. From an analysis of high-resolution X-ray data for fragments and corresponding larger molecules, the loss of rigid body translational and rotational entropy, which forms a significant barrier of protein–ligand recognition, was estimated as 15 to 20 kJ mol^{-1}, that is, 3 orders of magnitude in affinity at 298 K [43].

These studies consistently suggest that less complex lead structures are better to detect and exhibit a favorable optimization potential, which might influence attempts to design more "leadlike" [44] compound libraries for screening. They have also lead to alternative lead-finding strategies directed toward the initial discovery of small fragments as starting points for optimization [45]. Recent examples include the discovery of DNA gyrase inhibitors by means of 3-D structure-based fragment screening combined with biophysical assays and protein structure–based optimization [46] and the discovery of IMPDH (inosine 5′-monophosphate dehydrogenase) inhibitors [47] from a virtual screening protocol tailored to discover small, alternative warheads to known phenyloxazole anilines.

The combination of alternative screening approaches, virtual screening, parallel medicinal chemistry in combination with an early profiling on the multitude of relevant compound properties will hopefully generate an improved basis for proper decisions about which promising lead series to take forward.

14.3
Optimization for Affinity and Selectivity

14.3.1
Lead Optimization as a Challenge in Drug Discovery

This chapter summarizes several approaches with and without knowledge of the target protein 3-D structure to optimize lead structures for affinity and selectivity as one prerequisite for multidimensional optimization. The lead optimization phase to efficiently convert the lead structures with promising biological properties into clean drug candidates fulfilling a multitude of criteria is a challenging task [17, 18] with a high impact on the successful continuation of efficient drug discovery programs [2]. Consequently, the improvement of a compound profile toward a clinical candidate is one of the essential skills in integrated project teams. Those candidate requirements include not only desirable selectivity against related or diverse "antitargets", but also favorable physicochemical and pharmacokinetic properties, leading to oral administration and acceptable half-life of the final candidate. To arrive at candidates with suitable pharmacodynamic and pharmacokinetic properties hence requires a simultaneous optimization of multiple parameters in carefully planned iterations.

Owing to the lower dimensionality of the ADME space, ADME properties should be easier to be predicted than biological receptor affinity [48], while in practice this is more difficult, as many experimental screens for ADME properties are multi-mechanism rather than single mechanism systems. In contrast, biological assays for the majority of pharmacological targets are typical single-mechanism systems, for which computational models to correlate structural descriptors are easier to develop and resulting predictions tend to be more robust. Many experimental screens for these ADME properties include multiple mechanism systems, however. Computational models, on the other hand, for data with multiple underlying mechanisms tend to get worse if more data for structurally more diverse compounds are experimentally obtained and included into the training set. This is mainly due to the fact that the increase of assay data relates to an increase of underlying mechanisms on which those data have been obtained and the noise level rises for each individual mechanistic component [48]. For smaller, structurally, and thus probably, mechanistically homogeneous datasets, acceptable correlations are obtained, while the ability of descriptors to capture a more diverse experimental dataset is limited. Although the same descriptors might still have some statistical significance and thus explain global trends for less homogeneous data, their predictivity is often too low for a valid *in silico* prediction. This only offers the possibility to construct statistically significant rules or filters based on descriptors and property distributions and ranges. Hence, it is mandatory to carry out high-quality single-mechanism ADME experimental data and use them to derive single-mechanism models [48]. In contrast, assays and models to correlate structural properties to biological activity for a biomolecular target are mainly based on a single mechanism, namely, favorable protein–ligand interactions upon the molecular recognition event.

Although the present chapter does not summarize the pharmacophore [49, 50] and 2-D/3-D-QSAR [51–55] approaches toward ligand-based optimization in the absence of any 3-D receptor structure, these methods are extremely important in today's lead optimization settings, as for the majority of targets, it might be difficult to obtain relevant X-ray structures of protein–ligand complexes. This is true for membrane-bound proteins, like ion channels, G-protein-coupled receptors (GPCRs) and others. Hence, validated methods in ligand-based design to extract knowledge from molecules in a series, build statistical models and use them for further design are useful. The focus here is on the use of 3-D protein structure information *per se* and of the tight integration of ligand and structure-based design approaches toward a more reliable affinity prediction and understanding of selectivity. However, both ligand- and structure-based design approaches require the close monitoring of ADME and physicochemical properties of ligand series for successful optimization.

14.3.2
Use and Limitation of Structure-based Design Approaches

Over the past few years, there has been an exponential increase in the number of available 3-D structures in the public domain [56, 57], which is increasingly being used within the drug discovery process in the lead finding and lead optimization phase. However, there are still other factors that add up to the complexity of guided lead optimization by structure-based design [58–62]. The reality of protein–ligand interactions at the molecular level is still far too complex to provide a correct *in silico* approach for accurate affinity prediction, either based on the knowledge of the target protein 3-D structure or even without that information. The effect of entropy and the dielectric constant are only two examples, which are controversially discussed within the literature. Other challenging factors in structure-based design are the existence of multiple ligand binding modes [63, 64], the accessibility of conformational states for both ligand and receptors [65, 66], the influence of structurally conserved water [67], pH [68] and others. One important point to address is the well-documented protein flexibility [65] as direct consequence of amino acid chemical properties. Furthermore, it is of critical importance in structure-based design process to reflect about limitations of the use of X-ray protein structures [69], which also is partially due to significant errors in some reported structures [70, 71] and experimental limitations that are difficult to overcome. These complicating influences turn the reality of lead optimization for affinity into a difficult task.

Some important reviews summarized the current status of our understanding of protein–ligand interactions [72, 73]. This knowledge prompted some groups to derive scoring functions for guiding flexible docking algorithms and rank-ordering series of docked molecules. Several publications indicate that substantial progress has been made in both fields of docking and scoring. A variety of methods exist to estimate how strongly a ligand interacts with a protein binding site, while today's existing functions belong to three main categories: force field–based methods, empirical scoring functions and knowledge-based methods [62, 73–77]. In an intersection-based consensus scoring approach, Charifson et al. [78] and Clark et al. [79] combined a range of different functions to rank protein–ligand geometries, which resulted in an increased performance with respect to hit rates. The first paper [78] also discusses one of the possible limitations for earlier scoring functions, namely, that those are derived from X-ray structures of extremely potent ligands, while information about less active analogs is often not available.

14.3.3
Integration of Ligand- and Structure-based Design Concepts

The global scoring functions described above are derived from a more or less balanced collection of protein–ligand complexes from the public domain, leading to empirical functions, which might be able to globally provide some guideline on whether a compound could potentially bind to a particular binding site. However, their accuracy in ranking compounds on the basis of experimental binding affinities is typically limited. This caused studies to tailor scoring functions in lead optimization-stages using a narrow training set for only the protein target of interest. In addition, this need in lead optimization prompted others toward a tight integration of ligand- and structure-based design approaches to model and understand biological affinities by means of 3-D-QSAR (quantitative structure-activity relationship) methods, which is based on an alignment rule derived from reliable docking modes.

Gohlke et al. developed the AFMoC procedure [80] to derive tailored scoring functions for protein–ligand complexes based on DrugScore [81] knowledge-based pair-potentials, which are adapted to a single protein binding site by incorporating ligand information (Figure 14.3). The formalism is similar to CoMFA (comparative molecular field analysis) [82, 83], while the fields in their approach originate from the protein environment and not from the ligands alone. A regular-spaced grid is placed into the binding site and knowledge-based pair-potentials between protein atoms and ligand atom probes are mapped onto the grid intersections resulting in "potential fields". In a PLS (*partial least squares*) [84] analysis, these atom-type specific interaction fields are correlated to the actual binding affinities of the docked or experimentally known ligands, resulting in individual weighting factors for each field value. They described significant improvements of the predictive power for affinity prediction compared

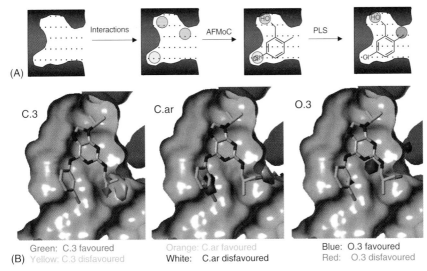

(A)

| C.3 | C.ar | O.3 |

| Green: C.3 favoured | Orange: C.ar favoured | Blue: O.3 favoured |
| (B) Yellow: C.3 disfavoured | White: C.ar disfavoured | Red: O.3 disfavoured |

Fig. 14.3 (A) Deriving a tailored scoring function using AFMoC [80], which integrates protein- and ligand-derived descriptors and produces a statistical model on protein–ligand interactions relevant for affinity. On a predefined grid, favorable protein interactions with ligand atom types are computed. These are adapted by incorporating actual information from a series of docked or crystallized ligands. A statistical analysis led to a model with favorable and unfavorable contributions. (B) AFMoC interpretation for 86 substituted purines as CDK-2 inhibitors. The underlying PLS model (q^2: 0.521, r^2: 0.800, 5 PLS components) indicates favorable and unfavorable regions for atom types, as shown for C.3, C.ar and O.3 and the CDK-2/purvalanol A complex (PDB 1CKP). The CDK-2 binding site is represented as MOLCAD solvent accessible surface.

to global knowledge-based potentials by considering a sample set of only 15 known training ligands.

The conceptually similar COMBINE approach was developed by Wade et al. [85] on the basis of the analysis of force-field energy contributions per amino acid to describe interaction differences to a congeneric set of ligands. The resulting predictive regression equations were also reported for applications in structure-based design projects [86–88].

Other attempts are based on tailor-made empirical functions, mainly by adding additional descriptors and determine weights for these, plus the known terms using appropriate statistical methods and a training set of several ligands exhibiting a range of biological affinities to only one protein cavity. Approaches to understand the structure-activity relationship in chemotypes by a determination of the importance of weights of different coefficients of individual terms in scoring functions are described by Rognan et al. [89] and Murray et al. [90], both using the ChemScore global scoring function [91] as a start.

Structure-based design is focused on understanding protein–ligand interactions but does not always lead to predictive models for ligand series. In contrast, 3-D-QSAR with acceptable statistics does not always reflect topological features of the protein structure, as those are not necessarily built using alignment rules that reflect the bioactive conformation. Hence, several groups have successfully combined both approaches. On the basis of the protein X-ray structure and a series of analogs with a potentially similar binding mode, consistent and highly predictive 3-D-QSAR models were derived, which could be mapped back to the protein topology. This leads to a better understanding of important protein–ligand interactions and provides guidelines for ligand design and a predictive model for scoring novel synthetic candidates. Docking calculations based on already available 3-D structures often might result in an alignment for all other compounds by superimposing them onto the template and relaxing them within the cavity for consistent 3-D-QSAR models.

These receptor-based 3-D-QSAR models represent a strategy to integrate ligand-based design and structure-based design approaches. Comparative molecular field analysis (CoMFA) [82, 83, 51] and comparative molecular similarity index analysis (CoMSIA) [92] are used to derive relationships between molecular property fields and biological activities. Electrostatic and steric interaction energies are computed between each ligand and a probe atom located on predefined grid points for CoMFA, while for CoMSIA those interaction fields are replaced by fields based on similarity indices between probe atoms and each molecule. The PLS [84] method is used to derive a linear relationship, while cross-validation [93] is used to check for consistency and predictiveness. The resulting contour maps from 3-D-QSAR models enhance the understanding of electrostatic, hydrophobic and steric requirements for ligand binding, guiding the design of novel inhibitors to those regions where structural variations altering steric or electrostatic fields reveal a significant correlation to biological properties. 3-D-QSAR results then allow focusing on those regions where steric, electronic or hydrophobic effects play a dominant role in ligand–receptor interactions.

One of the earlier and most influential applications of receptor-based CoMFA studies was carried out by Marshall et al. on a series of HIV-protease inhibitors [94, 95]. The bound conformation of several ligands was known from X-ray crystallographic studies and was shown to provide the most consistent and predictive QSAR models, while the authors also pointed out that conformational energy and entropic effects are not adequately included within these datasets. Hydrophobic field types used within CoMSIA [92] or HINT [96] might be partially useful to overcome the second limitation. Other successful applications include estrogen receptor ligands [97], acetyl cholinesterase inhibitors [98] and – from the authors' own work – receptor-based QSAR models for matrix metalloproteinase [99, 100], CDK-2 [101] and factor Xa inhibitors [102]. These models collectively found applications in lead optimization projects, as they provide an enhanced understanding of important effects in protein–ligand interactions and reliable affinity predictions.

14.3.4
The Selectivity Challenge from the Ligands' Perspective

Protein target families are the fundamentals to build a framework for matching biological motifs with chemical scaffolds. Many structurally related targets find implications in different therapeutic areas, while exhibiting similar protein–ligand interactions. A key issue for this interdisciplinary field is how to generate and efficiently use this information to guide drug design across targets [2]. The knowledge of structurally related proteins for a particular target is of interest in early drug discovery stages to identify leads by analogy within a target family [2, 22] from focused libraries enriched with privileged motifs of importance for that family [103] or by annotation-based similarity searching across target families [104, 105]. This latter approach allowed by linking chemical structure databases, biological target sequences, and structure activity data (screening results) to perform similarity searching to identify appropriate ligands for new targets within the same target family.

However, selectivity toward only a single biological target is an essential requirement for drug candidates to minimize side effects. Hence, it is mandatory to involve selectivity considerations early in the drug discovery process and to monitor experimentally and by *in silico* approaches closely related or structurally unrelated "antitargets" for the potential drug candidate. The following section will briefly summarize methods to take this requirement into account from both the perspective of ligand and the protein 3-D target structures, if available.

While most chemistry-driven approaches rely on trial and error, often 2-D and 3-D-QSAR approaches have been used to correlate biological affinities against both target proteins in order to identify those structural determinants or favorable spatial regions around the ligands gaining selectivity on the target enzymes. This approach requires a series of biologically characterized chemical analogs, which are typically available at this stage within the drug discovery value chain.

Some earlier examples on the use of 3-D-QSAR to understand selectivity differences include the work of Wong et al. to unveil structural requirements for selective binding of ligands to the diazepam-insensitive benzodiazepine receptor [106]. The chemical interpretation of the resulting CoMFA model on a larger series of 1,4-benzodiazepines highlighted the marked influence of particular substitutions in only a few scaffold positions on the ratio of affinities against the diazepam-insensitive (DI) and diazepam-sensitive (DS) benzodiazepin receptor binding site. Interestingly, individual 3-D-QSAR models have been derived for binding affinities against both individual sites as well as against the ratio, defined as $K_i(DI)/K_i(DS)$, as additional dependent variable for statistical analysis. The use of K_i values for this mathematical transformation is preferable because they are independent of concentration effects, while practically IC_{50} values for closely related targets or sites at one receptor are often determined at very similar conditions, thus also allowing the use of similar ratios as dependent variables for statistics. However, PLS can also be used for two (or more) dependent Y-variables in only one model [84].

A similar approach has been taken to explain selectivity differences between the matrix metalloproteinases MMP-8 and MMP-3 [100] and the serine proteases factor Xa and thrombin [107]. However, in both cases, the availability of experimental 3-D protein structures also allowed the interpretation of selectivity differences by careful inspection of protein binding site requirements and additional statistical models tailored to highlight key differences in potential protein–ligand interactions from binding sites only [100, 101, 107, 108]. In both cases, the selectivity differences from the analysis of ligand series nicely correspond to binding site requirements. Furthermore, the availability of 3-D structures for both the target and the closely related "antitarget" might lead to the use of structure-based design approaches to qualitatively identify amino acid differences, which might allow for a more focused interaction toward only one target. The combined use of both types of information greatly enhanced the understanding of selectivity requirements for these series of target enzymes.

Other quantitative approaches to address selectivity rely on generating and consistently interpreting individual statistical models, which again provide very detailed information on desirable scaffold substitution pattern and spatial areas around superimposed ligands, which allow to selectively interact only with one binding site. There are multiple interesting applications of this concept in the literature (e.g., [109–112]).

14.3.5
Selectivity Approaches Considering Binding Site Topologies

The availability of protein 3-D structures considerably simplifies the search for selective ligands, as these structures allow the comparative analysis of favorable protein–ligand interactions with a particular focus on those that are only possible in one member of a protein family. There are several computational tools that assist in an unbiased characterization of protein–ligand interactions, for example, the force field–based approach GRID [113] and the knowledge-based approaches SuperStar [114, 115] and DrugScore [81, 116]. In the recent GRID/CPCA (consensus principal component analysis) approach [108], the GRID descriptors derived for multiple protein binding sites are analyzed using a consensus principal component analysis (CPCA [117]) as statistical method, which allows in a straightforward way to identify possible modifications in potential ligands to improve their selectivity toward a chosen target. This statistical analysis evaluates the relative importance of individual molecular interaction fields from particular probe atoms for the final (PCA) principal component analysis model. Because the input data matrix is structured in chemically meaningful blocks, hierarchical and consensus multivariate analysis tools like CPCA provide information about the relative importance of individual blocks on the stage of an intermediate level of the analysis, while the final model is identical to a regular PCA. Plotting of the individual PCA and CPCA scores then allows to visualize selectivities between different members of the protein family, while the structural reason in terms of potential protein–ligand interactions that allows discrimination can be deduced from CPCA loadings plots and interactive variations thereof. Those "active plots" help to focus ligand designs toward binding site residues, which are essential for discriminating between subfamilies.

This approach was first applied toward an understanding of discriminating interactions in the serine proteases factor Xa, thrombin and trypsin [108] and provided selectivity information for all important serine protease subpockets, which are in agreement to experimental selectivities of typical protease inhibitors. This approach was complemented by a 3-D-QSAR selectivity analysis on a series of 3-amidinobenzyl-1H-indole-2-carboxamides [107], which points, from the viewpoint of the ligands, to similar main interactions driving selectivity between key enzymes in the blood coagulation cascade: factor Xa and thrombin. Other applications of GRID/CPCA include the interpretation of structural differences in human cytochromes P450, 2C8, 2C9, 2C18, and 2C19 [118] from homology models using the mammalian CYP2C5 X-ray structure [119], matrix metalloproteinases from X-ray and homology modeling [120], the classification of Eph Kinases [121] and a comparison of the binding characteristics of the family of lipid-binding proteins [122]. For these cases, the selectivity regions were in good agreement to available experimental information and inhibitor structure-activity relationships. Recent extensions include FLOGTV toward a simplified visualization of differences between related receptor sites based on a trend vector analysis [123].

Many structurally related protein targets find implications in different therapeutic areas, while exhibiting similar protein–ligand interactions. It is essential to efficiently use this information to guide drug design across targets. This encompasses the knowledge about selectivity regions in binding sites, while, on the other hand, a general entrance by less specific interactions into such target family is of particular interest for shaping a compound library at the lead-finding stage. To this end, GRID/CPCA was successfully applied to arrive at the "target family landscape" concept [101] to study features for general interactions to an entire target family and to identify determinants for interactions to only one member, which was of particular interest for broader sets of kinases [101] and serine proteases [107] with experimentally known 3-D crystal structures. The applications to major protein target families produced models in agreement with in-house ligand SAR models and thus provide guidelines toward rapid optimization in a library format for family-specific scaffolds.

To use a set of consistent descriptors for docking and selectivity studies, the replacement of the force field GRID by DrugScore [81, 116] derived knowledge-based potentials was successfully investigated on the factor Xa/thrombin/trypsin problem [124] as well as for matrix metalloproteinases. It was shown that this change produces models of qualitatively similar interpretation. Predictive submodels for all protein pockets (S1, S2, S4) were obtained. Here, CPCA scores discriminate factor Xa from thrombin and trypsin on PC1 in Figure 14.4 (Cl probe for S2 submodel). Differential plots reveal the structural reason for this discrimination: the piperidinyl moiety in the thrombin selective inhibitor NAPAP can only favorably interact with this subsite in thrombin. The detailed interpretation for other pockets is consistent and in agreement with internal series.

Lapinsh et al. [125] developed the proteochemometrical strategy combining ligand and protein information for a comparative analysis of series of receptors and ligands. Proteochemometrics exploits affinity data for series of diverse organic ligands binding to different receptor subtypes, correlating it to descriptors and cross terms derived from the amino acid sequences of the receptors and the structures of the small ligands. Statistically valid models resulted in all cases with good external predictive ability, while evaluation of these models gave important insight into the mode of interactions of the G-protein-coupled receptors with their ligands. This method was successfully applied to other GPCR receptor subtype analyses without experimental knowledge on binding site topologies, that is, melanocortin receptor subtypes [126],

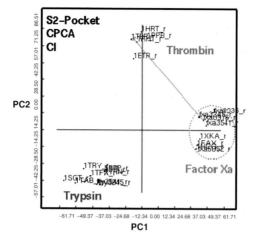

Cl: Factor Xa / Thrombin   **C.3:** Factor Xa / Thrombin

NAPAP X-ray
FXa: 7900 nM
Thr: 6 nM
Try: 690 nM

Fig. 14.4

serotonin, dopamine, histamine, adrenergic receptor subtypes [127], and α1a-, α1b- and α1d-adrenoceptors [125]. This approach combining information from ligands and receptors provided more detailed information about receptor–ligand interactions and determinants for receptor subtype selectivity than ligand-based QSAR studies on individual ligand series alone.

These presented examples of quantitative descriptions of protein–ligand interactions remain to be a very promising area to address affinity and the selectivity challenge in the future. Those approaches are collectively seen as interesting for a more integrated lead optimization on a simultaneous consideration of numerous balanced descriptors and models for optimization.

14.4
Addressing Pharmacokinetic Problems

14.4.1
Prediction of Physicochemical Properties

Pharmacokinetics and toxicity have been identified as important causes of costly late-stage failures in drug development. Hence, physicochemical as well as ADMET properties need to be fine-tuned even in the lead optimization phase. Recently developed *in silico* approaches will further increase model predictivity in this area to improve compound design and to focus on the most promising compounds only. A recent overview on ADME *in silico* models is given in Ref. [128].

The physical properties of a compound also determine its pharmacokinetic and metabolic behavior in the body. Poor biopharmaceutical properties are often linked to poor aqueous solubility, as recently summarized by Lipinski [129]. Thermodynamic solubility measurements consider the crystal packing energy of the solute as also its cavitation and solvation energy. The crystal packing energy accounts for disruption of the crystal to bring isolated molecules into the gas phase. The cavitation energy is the energy required to disrupt water for creation of cavity in which the solute is to be hosted. Finally, the solvation energy is the sum of favorable interactions between solvent and solute. Solubilization is largely kinetically driven, and the effect of crystal packing energy and polymorphic crystal forms are lost.

Currently available solubility models based on turbidimetry as well as nephelometry are not very predictive and are limited in their broad applicability, because they make use of training data from different laboratories determined under varying experimental conditions [130]. However, the access to many aqueous solubilities measured under standardized conditions is expected to greatly improve currently available models.

Fig. 14.4 DrugScore/CPCA for ligand selectivity in structure-based design. The binding sites of serine protease X-ray structures (factor Xa/thrombin/trypsin) are profiled using DrugScore for favorable interactions of 6 atom types (C.3, O.2, N.am, C.ar, O.3, Cl). Individual CPCA models are derived for each subsite, allowing the identification of regions for discrimination. The interpretation for the S2 pocket and NAPAP as thrombin selective inhibitor (PDB 1ETS) highlights yellow regions, where hydrophobic groups increase thrombin selectivity, while blue regions should favorable effect factor Xa selectivity. The interpretation is in good agreement with internal SAR data.

14.4.2
Prediction of ADME Properties

Poor intestinal absorption of a potential drug molecule can be related to poor physicochemical properties and/or poor membrane permeation. Poor membrane permeation could be due to low paracellular or transcellular permeability or the net result of efflux from transporter proteins including MDR1 (P-gp) or MRP proteins situated in the intestinal membrane. Cell lines with only one single efflux transporter are currently engineered for *in vitro* permeability assays to get suitable data for reliable QSAR models. In addition, efforts to gain deeper insight into P-gp and ABC on a structural basis are going on [131, 132].

Discrimination of efflux, active or passive transport is already feasible by suitable *in vitro* experiments. For instance, the PAMPA assay detects passive transport only, while Caco-2 cells include transporters. A comparison between transport in PAMPA and Caco-2 cells by a calibration plot reveals compounds with greater or less transport in Caco-2 cells than in PAMPA. These compounds should be tested in uptake and efflux transport assays in order to gain deeper insight into absorption fate.

Several *in silico* models for prediction of oral absorption are available [133–136]. Simple models are based on only few descriptors like logP, logD, or polar surface area (PSA), while they are only applicable if the compounds are passively absorbed. In case of absorption via active transporters or if efflux is involved, prediction of absorption is still not successful.

GastroPlus [137] and iDEA [138] are absorption-simulation models based on *in vitro* input data like solubility, Caco-2 permeability and others. They are based on advanced compartmental absorption and transit (ACAT) models in which physicochemical concepts are incorporated. Both approaches were recently compared and are shown to be suitable to predict the rate and extent of human absorption [139].

Problems related to poor systemic exposure are also tied to volume of distribution, which is indirectly related to plasma protein binding. Human plasma contains more than 60 different proteins of which the major components are serum albumin (HSA, 60%) and glycosylated proteins (AGP). Eighteen different variants arising from single amino acid mutations have been identified, accounting for different protein binding. Allelic variation makes data consistency difficult and hence modeling of the resultant data less reliable [140].

Prediction of bioavailability from molecular structure is quite difficult, since bioavailability depends on absorption and first-pass clearance [141]. By applying "fuzzy adaptive least squares", Yoshida and Topliss generated a QSAR model using logD at pH 7.4 and 6.5 as input for physicochemical properties and the presence/absence of certain functional groups as structural input. They achieved a classification of drugs into one of four bioavailability categories with an overall accuracy of 60% [142].

First-pass clearance can be tracked to gut-stability or metabolism by phase I and then either direct clearance or clearance of the metabolite by phase II enzymes or biliary, renal or plasma clearance. Metabolite stability by phase I enzymes include inhibition, induction, regiospecificity, lability or affinity toward several cytochrome

P450 enzymes or Flavin monooxygenase (FMO). Inhibition might cause drug–drug interactions and can be related to competitive, noncompetitive, uncompetitive or mechanistic inhibition [143]. Metabolite-based inhibition of the P450 enzymes also needs to be considered if toxicity aspects are related to these enzymes. Assays for differentiating these various types of mechanisms of inhibition are necessary and need to be applied for a reasonable number of compounds in a series in order to apply molecular modeling techniques to help in designing proper molecules with the preferred inhibition properties. Typical fluorimetric assays do not differentiate substrates that can be competitively bound or those that can covalently modify and inactivate the P450 enzymes. Only by different experimental approaches (time, substrate concentration, inhibitor concentration, NADPH dependent inhibition), these types of compounds can be identified. Sometimes, the identification of metabolites helps to decipher the binding modes in the Cytochrome P450 responsible [144]. Despite these problems, some successful applications of modeling to reduce drug–drug interactions, especially for CYP2D6 and CYP2C9, have been reported (see below).

Assays for FMO and most phase II enzymes are typically not included as part of a standard eADMET profile and therefore, modeling related to clearance by phase II enzymes has not been attempted although structural information for some sulfotransferases is available [145].

In silico methods to predict metabolism are based on QSAR, 3-D-QSAR or protein and pharmacophore models [146]. Early predictions of metabolism within a particular compound are now feasible. One approach by Korzekwa et al. uses, for example, reaction energetics to develop a predictive model for CYP3A4 mediated metabolization [147]. Sheridan et al. [148] described a model to predict likely sites of CYP3A4-mediated metabolism based on the energy necessary to remove a hydrogen radical from each site, as estimated by AM1 semiempirical MO calculations, and the surface area exposure of the hydrogen atom. The development and validation of a general quantitative structure-property model of metabolic turnover rates in human S9 homogenate based on uniform biological data of 631 diverse compounds proprietary to GlaxoSmithKline has also been described [149]. This model was able to classify 83% of test compounds correctly for their metabolic stability. Other approaches are based on databases to predict metabolism, for example MetabolExpert (Compudrug), META (MultiCASE), Meteor (Lhasa) and the MDL database Metabolite [150].

14.4.3
Prediction of Toxicity

Approximately 20 to 40% of drug failures are attributed to toxicity problems. Hence, *in silico* predictive toxicology tools are necessary to reduce attrition. Most of the current software packages available are dealing with carcinogenicity and mutagenicity [151] and are based on available public-domain toxicological data. *In silico* tools for other important endpoints such as hepatotoxicity, QT prolongation [152] and phospholipidosis are emerging and are expected to improve the design and optimization of novel compounds.

14.4.4
Physicochemical and ADMET Property-based Design

Property-based design supplements successful activity-based strategies to produce drug candidates [153]. Property screening in parallel with activity screening allows medicinal chemists to simultaneously optimize both biological activity and druglike properties, which is a widely accepted and implemented approach within the pharmaceutical industry. ADMET filters are now applied even at very early stages of drug design, for instance, in virtual screening. The first generation of predictive ADMET tools allows focusing on compounds with a high potential of the required pharmacokinetic and safety criteria and should reduce compound attrition. However, all currently available models are based on a limited set of data and are therefore restricted in their applicability. Hence, a continuous improvement and refinement of these models is required by taking into account more data that is also of high quality. Thereby, ADME might be considered as "automated decision making engine" in the early discovery paradigm and as absorption, distribution, metabolism and excretion in the regulatory phase [154].

14.5
ADME/Antitarget Models for Lead Optimization

14.5.1
Global ADME Models for Intestinal Absorption and Protein Binding

This section will provide an overview on ADME models from our group to illustrate our approach for building predictive models on structurally diverse training sets. Datasets for intestinal human absorption and human serum albumin binding are discussed, while models for other relevant ADME properties have also been obtained. Those models, however, do not stand alone but are used in combination with those models tailored for affinity and selectivity in the frame of multidimensional lead optimization.

Recently, a set of alignment-free 3-D descriptors named VolSurf was developed by Cruciani et al. [155], referring to molecular properties extracted from 3-D molecular interaction fields of various probe atoms. VolSurf transforms the information present in GRID [113] derived 3-D fields for particular probes into quantitative descriptors (Figure 14.5), which carry information related to pharmacokinetic properties like polarity, hydrogen bonding, lipophilicity, size, polarizability and others. The descriptors are easy to understand and to interpret, thus providing guidelines for chemical optimization after a linear model has been established. In most cases, the GRID water, hydrophobic (DRY) and carbonyl oxygen probes were utilized, as interaction of molecules with biological membranes is mediated by surface properties like shape, electrostatic forces, hydrogen bonding and hydrophobicity.

These descriptors have been reported in the literature to correlate with bioavailability, blood-brain partitioning, membrane transport and other properties [156–159]. They are also correlated to relevant physicochemical properties and were also successfully applied to many internal and public data. For example, we derived PLS models [160] using 72 VolSurf descriptors for human serum albumin (HSA) binding using 95 drugs on a

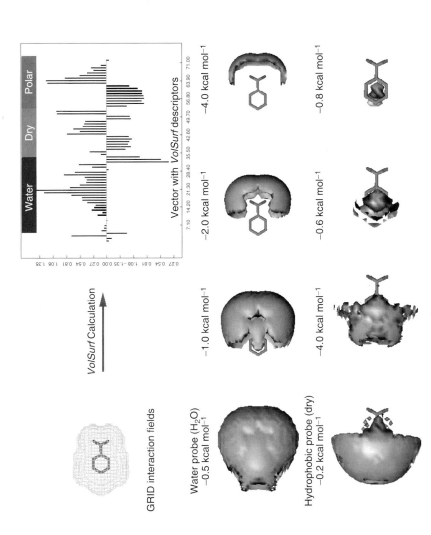

Fig. 14.5 Computation of VolSurf descriptors [155, 156] derived from GRID molecular interaction fields. Interactions of the example molecule with a water and dry probe at different contour levels are used to compute a vector of 72 volume-, size- and surface-based descriptors.

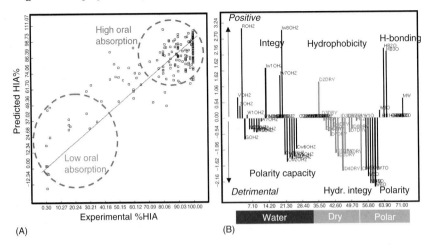

Fig. 14.6 Correlation of VolSurf descriptors with human intestinal absorption for 169 drug molecules. (A) Predicted versus experimental %HIA (human intestinal absorption) from final PLS model with q^2: 0.662, r^2: 0.709 and 4 PLS components. (B) PLS loadings for four-component model showing the importance of individual VolSurf descriptors to the prediction of human intestinal absorption.

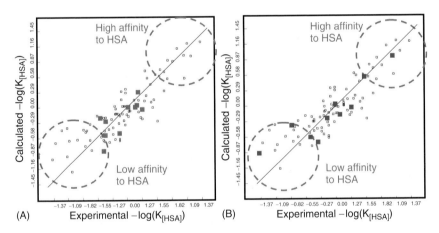

Fig. 14.7 Correlation of VolSurf descriptors with human serum albumin binding affinity. Two submodels indicate their predictive ability to external test sets. Ten compounds were removed by either experimental design on PCA scores (A) or literature proposals (B). Model A: q^2: 0.668, r^2: 0.763, 6 PLS components, SDEP for prediction 0.223 for 10 cpds. Model B: q^2: 0.646, r^2: 0.745, 6 PLS components, SDEP for prediction 0.274 for 10 cpds.

dataset from Colmenarejo et al. [161] ($r^2 = 0.76$, $q^2 = 0.67$, 6 PLS components), and for human intestinal absorption using data of 169 drugs compiled by Zhao et al. [133] ($r^2 = 0.71$, $q^2 = 0.66$, 4 PLS components). Those models are obtained using a canonical 3-D structure of neutral molecules.

As an example, the correlation of VolSurf descriptors to the human intestinal absorption for 169 drugs is shown in Figure 14.6. The conclusions for factors influencing permeability and absorption are in agreement with earlier findings, pointing to the positive impact of hydrophobicity, integy moment, shape and hydrogen bonding, while polarity derived using the GRID water and carbonyl O probes as well as the capacity factors from polar interactions on the entire surface are detrimental for intestinal absorption. However, because of the nature of these descriptors, this approach led to an enhanced understanding of physicochemical requirements for a pharmacokinetic effect. For example, the balance between lipophilic and hydrophilic parts in combination with size, volume and other effects guides the design of new compounds.

The interpretation for human serum albumin also is in accord with literature findings and X-ray information on warfarin binding to the first drug binding site in HSA [162]. For instance, hydrophobicity, volume, shape and molecular weight increase HSA binding, while factors like polarity, integy moment, hydrogen bonding and favorable interactions to water are detrimental. Typically, hydrophobicity and geometrical factors like shape are reported as essential in literature [161]. These models are always validated by external test datasets, which is schematically shown in Figure 14.7, where 10 compounds have been removed from the HSA training dataset by two different approaches and predicted using two only slightly differing models. In both cases, the agreement between experiment and prediction was very good.

Alifrangis et al. [158] reported a structure-property model for membrane partitioning for 20 peptides with data from two chromatographic systems with phospholipides as stationary phase, immobilized artificial membranes (IAM) and immobilized membrane chromatography (ILC). The relationship between these measures and three different sets of calculated descriptors (molecular surface area, MolSurf and VolSurf) were analyzed using PLS, showing that VolSurf-derived models are superior to both others [158]. Especially, the VolSurf critical packing descriptor to describe interactions of amphiphilic molecules with membranes is important to explain the membrane partitioning ability. This agrees to internal VolSurf models derived for PAMPA membrane transport [163] to understand passive transcellular transport across membranes. One of our internal models based on 29 compounds characterized by immobilized artificial membrane chromatography by Salminen et al. [164] shows an r^2 of 0.81 and $q^2 = 0.70$ for two PLS components derived using VolSurf descriptors. This is one of the rare examples where ionized starting molecules led to slightly better PLS statistics, while the general chemical interpretation is not affected.

VolSurf was also successfully applied in the literature to predict absorption properties [156] from experimental drug permeability data of 55 compounds [165] in Caco-2 cells (human intestinal epithelial cell line derived from a colorectal carcinoma) and MDCK cell monolayers (Madin-Darby canine kidney). In this interesting case, it was shown that models including counterions for charged molecules clearly show significantly better quality and overall performance. The final model was also able to correctly predict, to a great extent, the relative ranking of molecules from another Caco-2 permeability study by Yazdanian et al. [166].

14.5.2
Selected Examples to Address ADME/Toxicology Antitargets

For ADME and toxicology properties, which are related to a particular target protein, approaches similar to those for affinity and selectivity are used in our group, while most of the ligand-based studies have recently been augmented by recent X-ray structures and/or validated homology models. Again, both approaches led to significant results in combination, when applied in the context of lead optimization in internal projects.

The K^+ channel encoded by the human ether-à-go-go related gene (hERG) is one of the many ion channels that are crucial for normal action potential repolarization in cardiac myocytes [152]. This hERG channel is of considerable pharmaceutical interest, as it is connected to drug-induced long QT syndrome and therefore cardiac toxicity of a wide range of pharmaceutical agents. This undesirable side effect for noncardiac drugs has caused recent withdrawals of drugs from the market and stimulated many studies to establish a structural hypothesis for hERG and a structure-activity relationship of hERG channel blockers. Given the wide range of chemical structures as hERG inhibitors and the observation that different compounds are reported to bind to different states (open or closed) of the channel, multiple binding sites have been proposed [152]. On the basis of a homology model of the closed form of the tetrameric pore derived from the X-ray structure of the bacterial K^+ channel KcsA [167] combined with site-directed mutagenesis studies, several amino acids were identified as important for high-affinity binding of the methansulfonanilide MK-499 and thus provided a structural basis for drug-induced long QT syndrome [168]. Interestingly, two aromatic residues Tyr652 and Phe656 lining the inner pore as unique structural motif in hERG compared to other K^+ channels (except EAG) are proposed to interact with aromatic groups in ligands.

A recent review [169] summarizes the current status of *in silico* models developed in the past two years to explain the hERG structure-activity relationship and hopefully eliminate undesirable features from potential drug candidates. While some of the models are intended to be used as early filters in compound design [170], only the 3-D-QSAR studies of Cavalli et al. [171] and the combination [172, 173] of 3-D-QSAR from consistent biological data with homology models of the closed and open form of hERG built from the low-resolution crystal structure of the MthK K^+ channel [174] allow us to understand structural reasons for hERG blocking and thus provide sufficient details to apply them during the multidimensional lead optimization of a candidate molecule. These models have been internally derived from consistent biological data on literature compounds and close analogs, while they are constantly updated using novel molecules and biological data from lead optimization programs. Many internal applications have proven the value in using such a combined approach. The essential elements of the hERG pharmacophore, the 3-D-QSAR and the homology model are: (1) the hydrophobic feature optimally consists of an aromatic group that is capable of engaging in π-stacking with Phe656. Optionally, a second aromatic or hydrophobic group may contact an additional Phe656 side chain. (2) The basic nitrogen appears to undergo a π-cation interaction with Tyr652. (3) The pore diameter and depth of the selectivity loop relative to the intracellular opening, act as constraints on the conformation-dependent inhibitor dimensions [172]. However, undesirable structural features for hERG are not sufficient in the context of a lead optimization project. It is important to understand which variations influence hERG binding, but are still tolerated by the desired biological target. Both properties were recently used as lead optimization parameters by Friesen et al. [175] for PDE-4 inhibitors in a systematic approach to identify structural features that only affect hERG binding.

Toxicity caused by drug–drug interaction also resulted in withdrawal of several drugs from the market. Compounds that are potent inhibitors of the major metabolizing enzymes can potentially affect the metabolism of other molecules and thus lead to toxicity. Early information on cytochrome P450 inhibition thus is useful to develop structure-activity relationships and minimize the potential toxicity of development candidates. Inhibition of CYPs can be substrate dependent, which especially is reported for CYP3A4 with a very large active site that is able to bind multiple substrates in different orientations [176]. The growing knowledge about CYP substrates has increased the understanding of active site requirements either by protein- or ligand-oriented studies [177]. However, understanding the nature of substrate specificity of each CYP requires knowledge of the interaction of drugs with each CYP active site. A detailed analysis of substrate binding affinities and selectivities for the CYP2 family is given by Ridderström and Lewis et al. [118, 178].

Pharmacophore-based models for many cytochrome P450 isoforms have been reported for CYP2B6 substrates [179], CYP3A4 substrates [180] and inhibitors [181], CYP2C9 inhibitors [182] and CYP2D6 inhibitors [183]. A 3-D-QSAR model for CYP2C9 inhibitors for prediction of drug–drug interactions was reported by Rao et al. [184] and compared with appropriate homology models. Afzelius et al. [185] used alignment-independent GRIND descriptors implemented in the program Almond [186] to obtain qualitative and quantitative predictions of CYP2C9 inhibitors for a series of structurally very diverse molecules. De Groot et al. reported a combined protein and pharmacophore model approach to understand and predict CYP2D6 mediated drug metabolism [187, 188]. A model for 40 different CYP2D6 substrates (hydroxylation, *O*-demethylation) was obtained by combination of pharmacophores, protein models and molecular orbital calculations. This model was extended by a second pharmacophore explaining 14 less-common *N*-dealkylation transformations. The final model was in agreement with additional substrate and site-directed mutagenesis data. Zamora et al. developed an interesting novel approach named MetaSite to predict the metabolization site of substrates in CYP2C9 on the basis of a combined application of alignment-independent descriptors to describe the protein binding site taken from a CYP2C9 homology model and a distance-based representation of individual substrates [189].

The recently solved X-ray structures of rabbit microsomal CYP2C5 in complex with diclofenac [190] and a sulfaphanazole derivative [191] provided additional evidence on how the molecular recognition of structurally diverse substrates takes place. Comparisons of the complex with apo CYP2C5 [119] indicates that the protein closes around the substrate and prevents open access of water from bulk solvent to heme Fe. Multiple substrate binding models of the sulfaphenazole derivative are in agreement

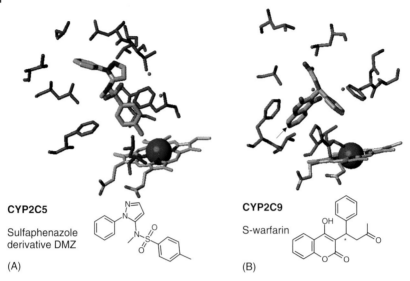

CYP2C5

Sulfaphenazole
derivative DMZ

CYP2C9

S-warfarin

(A)

(B)

Fig. 14.8 Experimental ligand interactions with cytochrome P450 2C family. (A) X-ray structure of the sulfaphenazole derivate DMZ in rabbit CYP2C5 at 2.3 Å resolution (PDB 1N6B) from Wester et al. [191]. Only one of the two-ligand orientations for DMZ in accord with electron density is shown placing the benzylic methyl group in a 4.4 Å distance to the heme iron. (B) X-ray structure of S-warfarin in human CYP2C9 at 2.55 Å resolution (PDB 1OG5) from Williams et al. [192]. The substrate is situated in a predominantly hydrophobic pocket. This binding mode places the 6- and 7-hydroxylation sites 10 Å from the iron (arrow).

with the experimentally derived ligand electron density maps and the finding that the substrate is not tightly constrained in the active site. For diclofenac, a single binding mode is consistent with the observation of a highly regiospecific hydroxylation at the distal ring in 4′ position. This large active site and the striking enzyme flexibility upon binding of both ligands underlines the ability of the drug-metabolizing enzymes to work on structurally diverse substrates of different sizes.

This work was complemented by the X-ray structure of human CYP2C9 in complex with warfarin by Williams et al. [192] (Figure 14.8). This report provides the first human CYP protein structure, while the warfarin binding mode reveals previously unanticipated interactions in a new binding pocket. From the binding mode, the authors conclude that CYP2C9 may undergo an allosteric mechanism during its function and that two molecules for warfarin could be accommodated in the very large CYP active site. Collectively, these X-ray structures provide insights into relevant protein–ligand interactions for particular human CYP subfamilies and thus can be used for docking studies and building scaffold-specific 3-D-QSAR models driven by a protein-derived alignment rule. Those models provide guidelines on where to optimize a molecule with a CYP liability. Hence, a higher resolution picture of drug-CYP interactions begins to emerge for some subfamilies, allowing the use of this information as one optimization parameter in multidimensional optimization projects.

14.6
Integrated Approach

14.6.1
Strategy and Risk Assessment

While previous sections discussed computational approaches toward the understanding and analysis of individual properties like affinity, selectivity, and ADME parameters, here we show how those tools should be combined to arrive at a lead optimization strategy that is able to manage and design for multiple properties in lead optimization cycles. While in early phases of discovery without knowledge of multiple analogs for a particular hit "global" models for ADME properties are applied for library design and selection of promising hits, this focus is shifted toward "local" models based on information about analogs during the subsequent optimization phase. The lack of consistent publicly available data on properties like oral absorption prompted us to optimize pharmacokinetic properties based on consistent biological data for one series.

Any significant improvement of the lead optimization process requires predictive *in vitro* assays, models and computations for ADME and toxicological properties to be incorporated into the iterative cycle of compound design and synthesis. In this manner, compounds can be optimized in parallel by a multidimensional optimization strategy, which integrates the evaluation and optimization for affinity, selectivity, physicochemical and ADME properties. This approach, however, requires a proper experimental design of compounds related to a specific chemotype prior to any synthesis. Such a design should take into account a broad coverage of the chemotype-specific chemical space. The reliability of any subsequent statistical model strongly depends on the information content of the training set data. If experiments are planned inefficiently, resulting in higher number of experiments, less information and sometimes misleading data are generated. The systematic variation of one factor at a time is only appropriate if the response surface is not influenced by interactions between factors, while cooperative effects may require advanced design approaches [193]. Thereby, appropriate chemotype-specific local models can be built and subsequently applied in the design of new compounds.

Needless to say that reliable experimental data based on single mechanism should be preferred. Such single-mechanism data provide not only a deeper understanding of the underlying molecular mechanisms but also allow us to derive better, less noisy and thus more predictive *in silico* models with information on directions to be taken for further design. Furthermore, this approach allows an early identification of properties to be improved in lead optimization and, in addition, whether two, three or more variables can be optimized in parallel. For example, if the optimization of the biological properties Y1 and Y2 requires increasing hydrophobicity for better activity, while this descriptor is detrimental for the improvement of a third property Y3, a detailed statistical assessment is necessary whether it is possible at the end to optimize all three dependent variables in parallel to derive a compound with a suitable target product profile.

In particular, parallel optimization of affinity/selectivity and pharmacokinetic properties are difficult to achieve, especially if several PK parameters need to be optimized (e.g., absorption and volume of distribution). The number of variables to be optimized as well as their optimization potential should always be assessed in order to deal with a minimal set of properties to work with. It is our experience over the last couple of years in multiple lead optimization projects that as more variables need to be addressed in parallel, a more difficult, time-consuming and exhaustive lead optimization needs to be performed. In general, the probability of a successful lead optimization of a certain chemotype resulting in a development candidate decreases substantially with the need to improve more than three properties in order to fulfill the target property profile.

This requires risk estimation by an early assessment of the optimization potential for any novel chemistry series in order to get early information on ADME and antitarget properties as well as potential toxicity issues, which are known to become increasingly important for drug failures. Unfortunately, only a few ADME antitargets could be addressed already (e.g., hERG activity, see above), while others need to be improved. Toxicity, however, is quite often not related to a single mechanism and therefore very difficult to address [194]. Certainly, more reliable and more predictive SAR for target affinity as well as for antitarget and PK properties will strengthen the multidimensional optimization of compounds. The integration of these tools into connected processes will certainly raise modern drug discovery to a new level (Figure 14.9).

Experimental compound profiling
Affinity, selectivity, safety, physicochem, ADME incl. CYP450 inhibition, bioavailability

	Affinity	Selectiv	CYP1	CYP2	CYP3	hERG
Cpd1	z	z	z	z	z	z
Cpd2	z	z	z	z	z	z
Cpd3	z	z	z	z	z	z
Cpd4	z	z	z	z	z	z

Turning data into knowledge   **Next design cycle**
Statistical models to predict properties

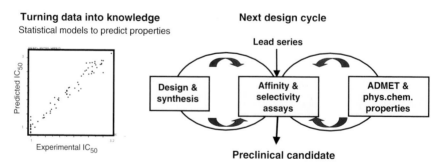

Fig. 14.9 Individual components of multidimensional optimization. This approach requires experimental compound profiling against key properties, which should be done on a designed compound subset to maximize information with a minimum number of molecules. These data are used to derive models for key properties, which are applied during the next design cycle. The results then led to augmented models. The process is characterized by a tight integration of *in vitro* and *in silico* tools for profiling compound series to guide chemical optimization.

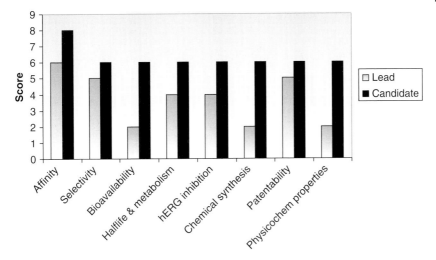

Fig. 14.10 Example for multidimensional optimization on relevant properties during the lead optimization phase from leads to a development candidate. After some iteration, compound properties are either improved or show no further optimization potential.

14.6.2
Integration

The ideal case during lead optimization is to use a consistent set of descriptors for multiple biological variables. This concept is followed wherever applicable, while this is not always possible from a practical point of view. If such a single set of informative descriptors, like, for example, VolSurf, to model binding affinity in structure- or ligand-based design and ADMET properties at the same time [21] is used to build appropriate models for all dependent (Y) variables, an early assessment of the optimization potential is feasible. This approach described by Zamora et al. [21] is the promising integration of both binding affinity and passive pharmacokinetic properties on the descriptor level.

However, if different models and descriptors are used and applied, certain criteria, decision trees or scoring functions need to be used to deal with the multidimensional optimization. Decision trees refer to approaches similar in spirit to the well-established rule-of-five and are often project-specific variations and improvements thereof. The chemical meaning and interpretability of the descriptors is essential here. Scoring models, as a slightly different approach, condense each *in silico* estimated compound property into a range between 0 and 1 based on project team–specific thresholds. Some degree of uncertainty could be considered by introducing a linear increase in this score contribution from 0 (undesirable) to 1 (desirable) within a narrow area around the threshold. Those individual scores are compiled into a weighted sum and used to evaluate novel synthesis proposals, often biased by the knowledge of the protein 3-D structure (Figure 14.10). Step-by-step application of decision trees or a set of single models, however, might ultimately sort out almost all compounds depending on the sequence used and are therefore less valuable. Scoring functions, however, offer the opportunity to fine-tune the function in terms of the most important variables without neglecting any of the other properties. The thoughtful integration of *in silico* tools in this multidimensional optimization process will certainly improve candidate quality in the next decade.

14.6.3
Literature and Aventis Examples on Aspects of Multidimensional Optimization

Although the optimization of promising lead compounds toward clinical candidates is one of the essential skills within the pharmaceutical industry, there are not many reports on the entire multidimensional optimization strategy or selected aspects. Here, we discuss examples from the recent medicinal chemistry literature and projects at Aventis in which either the chosen optimization strategy or the computational tools are interesting and agree with the proposed multidimensional optimization strategy and its requirements. Important prerequisites and milestones for lead optimization that are essential to successfully manage a drug discovery program in an industrial setting are often not discussed in those contributions. These include the early assessment of the optimization potential for a lead series, the risk assessment on how many parameters have to be optimized in a multidimensional context while all the others should remain in the positive range, where they have been from the beginning. The literature reports address one or only a few additional parameters in addition to biological affinity at the target enzyme to progress a series. This could be seen as typical examples for one single optimization cycle within multidimensional lead optimization.

Linusson et al. at AstraZeneca described the application of statistical molecular design for planning and analysis of a parallel synthesis lead optimization library of thrombin inhibitors [195]. This structurally well-characterized serine protease is involved in the blood coagulation cascade and was targeted by many industrial drug discovery projects. The report described how for a central scaffold with three vectors directed toward individual thrombin subpockets S1, S2 and S4 building blocks were selected. This was done on the basis of a quantitative statistical analysis of known experimental information such as biological affinities for related thrombin inhibitors at AstraZeneca. The focus for the planned library was to replace benzamidine, which provides important polar interactions within the protease S1 pocket, while it has been recognized for a while that benzamidine substituents are detrimental for oral bioavailability, mainly due to their low intestinal membrane permeability, which could be attributed to their high pKa value. Hence, parameters like pKa and membrane permeability are seen as equally important in the planning phase for multidimensional optimization. There are many reports on benzamidine-based thrombin inhibitors, which are all endpoints in lead optimization and led to potent inhibitors in terms of affinity, selectivity and anticoagulation effect, while significant oral bioavailability could not be demonstrated, except for prodrugs like ximelagatran [196]. Considering this property for a development candidate very late in the discovery phase, however, often results in a significant loss of affinity when replacing the S1 benzamidine. These considerations led the authors to include topologically similar S1 directed substituents with a potential for lower pKa into their statistical selection procedure of appropriate building blocks. The resulting parallel synthesis library was analyzed with respect to affinity, membrane permeability, pKa and trypsin selectivity. To this end, the authors derived multivariate QSAR models for all key biological data (affinity, selectivity, pKa, permeability). As the library has been carefully designed, the final models, although only based on a limited series of molecules, are significant and predictive and might guide the direction for further optimization. The SAR information for key properties, which were only found using statistical molecular design in combination with multivariate analysis, could now be applied to focus a second follow-up library. This interesting study combines the concepts of multidimensional optimization and suggests statistical approaches to obtain informative models. As this study presents only one cycle in lead optimization, it is only a "snapshot" in the search for a development candidate.

Sugano et al. studied the membrane permeation of 51 benzamidine-based thrombin inhibitors in a rat everted sac permeability model [197]. They reported significant membrane permeabilities in this *in vitro* model, which they attributed to passive paracellular transport, a different absorption mechanism to transcellular permeability. On the basis of their evaluation and our internal predictive VolSurf model [160] for this series (r^2: 0.81, q^2: 0.60, 4 PLS components), it can be concluded that factors like size and shape, which had previously been reported to affect paracellular permeability, are indeed important in the VolSurf PLS model to explain the local structure-permeability relationship of one particular scaffold. Hence, local statistical models provide a qualitative ranking of candidates, and thus are valuable for optimization of pharmaceutically relevant compounds, especially if combined with additional models to understand affinity, selectivity or any particular pharmacokinetic behavior.

Burgey et al. at Merck described their approach toward metabolism-directed optimization of 3-aminopyrazinone-based thrombin inhibitors [198] to result in an orally bioavailable series. Several research groups have now successfully replaced the benzamidine moiety in S1 by less-basic or nonbasic substituents with different protein–ligand interactions to overcome the specific benzamidine problems described above, resulting in series with increased oral bioavailability and pharmacokinetics. Starting from an aminopyrazinone–acetamide scaffold with moderately basic 2-aminopyridine as S1 directed benzamidine mimetic (Figure 14.11), they discovered three main sites of metabolism, which they related to the observed insufficient pharmacokinetic behavior. The optimization was dictated by metabolic considerations in concert with required target affinity. On the basis of the available experimental information on binding modes for this chemotype and the detailed DMPK studies to unravel the mechanistic origin of metabolic instability, they were able to design a series of metabolically more stable variations of the original lead structure with similar enzyme affinity and selectivity. The introduction of metabolically more stable substituents at this scaffold led to a final compound with improved pharmacokinetic properties. Two key observations are essential in this and other successful studies: (a) the careful balance of decreasing thrombin affinity by metabolically more stable substitutions in one subpocket by compensation within another pocket, which could have been achieved only using the wealth of structural knowledge and understanding of thrombin protein–ligand interactions; (b) the careful correlation of *in vivo* PK observations to

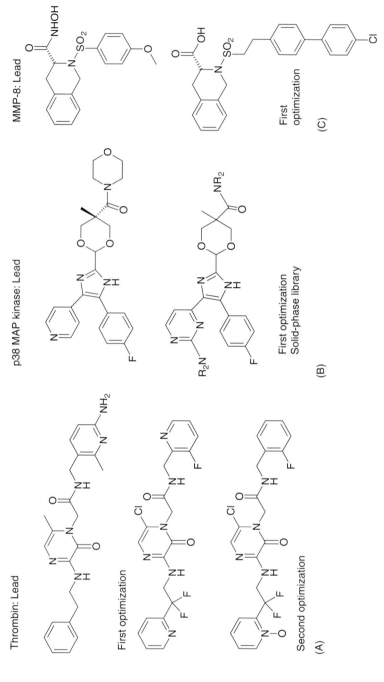

Fig. 14.11 Selected examples for lead optimization under consideration of multiple parameters simultaneously: (A) thrombin inhibitors, (B) p38 MAP kinase inhibitors, (C) MMP-8 inhibitors.

their *in vitro* origin and the subsequent monitoring of each metabolically less labile substituent in appropriate *in vivo* dog PK studies.

In a second contribution from the same group, a subsequent optimization cycle for the previous scaffold was directed toward improving solubility while reducing the number of chemical steps in the overall synthesis of S1 replacements. The readily available S1 directed building blocks then led to establish the structure-activity relationships of three key parameters besides thrombin affinity, namely oral bioavailability, half-life and human liver microsome stability. Hence, the optimization strategy of the S1 subpocket was again guided by pharmacokinetic considerations, which were related by *in vitro* solubility assays. Finally, the introduction of S4 pyridine N-oxides with increased solubility led to a new orally bioavailable series. An expedited investigation of the P1 SAR incorporating several S1 *N*-benzylamides with respect to oral bioavailability, plasma half-life, and human liver microsome stability subsequently revealed an interesting candidate for advanced pharmacological evaluation after careful monitoring and optimization of a variety of affinity, selectivity, physicochemical and pharmacokinetic parameters.

Hasegawa et al. from Roche [199] described a local model to understand structure-pharmacokinetic relationships for 107 benzofurans as *N*-myristoyltransferase inhibitors, based on cassette dosing PK studies and rat elimination half-lifes as dependent variables. They obtained a relatively simple, yet effective, statistical model, which they describe as useful and which give a direction for designing new inhibitors having good PK profiles. Similar local QSAR models for relevant PK properties could be used in combination with affinity prediction models or docking and scoring calculations, if applicable, to select novel synthesis candidates from a series of feasible proposals in a multidimensional fashion.

McKenna et al. at Aventis described the design and synthesis of a solid-phase library to optimize a pyridine-imidazole-based p38 MAP kinase inhibitor toward target affinity and rat oral bioavailability [200]. The authors describe a computational approach toward the design of a library with 570 analogs on two attachment points. Those are selected using a Monte Carlo monomer selection (MCMS) strategy on the basis of combinatorial synthetic efficiency on one hand, while maximizing the estimated bioavailability, as considered by *in silico* descriptors like PSA and a modified rule-of-five definition. MCMS applies a scoring function controlled by a series of weights for individual components to drive monomer selection toward solutions that satisfy the design constraints [201]. From this focused lead-optimization library, ten relatively structurally diverse compounds with improved potency and acceptable bioavailability and PK parameters were rapidly identified as follow-up to the previous candidate.

In their search for nonbenzamidine-based inhibitors of the serine protease factor Xa, Choi-Sledeski et al. [202] described the discovery of an orally efficacious inhibitor that incorporates a neutral substituent directed toward the protease S1 pocket as a result of an optimization strategy guided by structure-based rational design and a qualitative consideration of bioavailability. The conflicting and nonconclusive SAR of initial derivatives on that scaffold could only be resolved by X-ray structure analysis of a relevant protein–ligand complex [64]. The X-ray structure revealed a strong preference for a neutral substituent directed toward this S1 pocket for the explored ketopiperazine scaffold, which offered a new perspective on designing nonbasic novel factor Xa inhibitors with increased potential for oral bioavailability. This striking preference for neutral S1 substituents was earlier reported for bioavailable thrombin inhibitors as well [203]. By minimizing the size and lipophilicity of the S4-directed substituents and by incorporating hydrogen-bonding groups on the N-terminus or on the 2-position of the S1 directed ring system, a series of active thrombin inhibitors with good bioavailability profiles was reported that showed a unique binding mode after X-ray structure analysis in thrombin. Both optimizations on serine protease inhibitors were supported by structure-based design techniques, while the latter, including favorable protein–ligand interactions, alone is not sufficient to arrive at bioavailable inhibitors. Here, the careful analysis of available X-ray structures helps to identify substituent positions that will not affect enzyme activity but will modulate other properties favorably. Replacing a high-affinity substituent requires, on the other hand, a careful monitoring of the loss of affinity, possible substituents in other areas that might compensate for this loss and the favorable change in the desired ADME or PK property.

A similar situation was experienced in the search for potent and bioavailable inhibitors of the matrix metalloproteinase MMP-8 [99, 204]. For the explored tetrahydroisoquinoline scaffold, hydroxamic acids for zinc binding in 3-position are essential for MMP affinity in early inhibitors, while those showed insufficient PK properties and low oral bioavailability. Driven by X-ray and 3-D-QSAR, alternative Zn binding groups like carboxylates were investigated, while the expected loss in binding affinity could be compensated by optimally filling the proteinase S 1′ pocket close to the catalytic zinc. The design and SAR for this series is in good agreement with those protein requirements and, moreover, monitored multiple properties including selectivity against the undesirable MMP-1 [204]. For several MMP-8 inhibitors, the oral bioavailability from rabbit studies could be correlated again to VolSurf descriptors (r^2: 0.65, q^2: 0.42, 4 PLS components), which led to a semiquantitative model [160], used in conjunction with structure-based docking and scoring, 3-D-QSAR based affinity and selectivity predictions and *in silico* ADME models to estimate membrane permeability, solubility and other key properties for the optimization process in this series. Hence, in this as well as in other series, multiple models are collectively utilized to ranking and prioritizing novel synthesis candidates and focused virtual libraries (Figure 14.12).

In the search for bile-acid resorption inhibitors (BARI), a predictive 3-D-QSAR pharmacophore model for the ileal Na^+/bile acid cotransporter was derived, which enhanced the understanding of binding and transport properties [205]. This model was then also successfully explored to search for potential substitution sites, which are not relevant for the SAR of this series, while they allow the addition of additional substituents to minimize the oral uptake of inhibitors.

The approach discussed to use VolSurf derived *in silico* models to understand structure-PK relationships for pharmacokinetic properties was also applied to one series of selective cardiac KATP channel blockers [160]. It was found that compounds fulfilling the predefined selectivity profile exhibit only less-optimal pharmacokinetic properties because of a short plasma mean residential time (MRT). Consequently, the MRT for 28 compounds from rabbit iv studies for one series was used as dependent variable to derive a VolSurf PLS model in addition to ligand affinity SAR data. The chemical

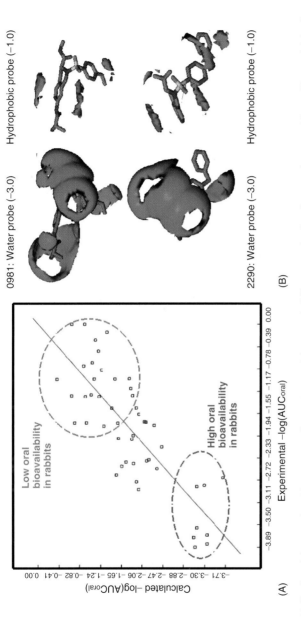

Fig. 14.12 VolSurf model to correlate 49 matrix metalloproteinase inhibitors with different zinc-binding functionalities to rabbit oral bioavailability for metabolically stable compounds. (A) Semiquantitative PLS model (q^2: 0.424, r^2: 0.646, 4 PLS components) to rank novel synthesis candidates. Main factors influencing absorption, that is, lower polarity, capacity factors and increased hydrophobicity, are in agreement with global models for human intestinal absorption. (B) Distribution of polar and hydrophobic surfaces for two molecules with low (0981) and higher (2290) rabbit AUC from oral pharmacokinetic studies.

interpretation of the VolSurf model ($r^2 = 0.89$, $q^2 = 0.76$, 3 PLS components) reveals hydrophobic interactions and the hydrophobic integy moment to be strongly correlated to MRT. This model was then used to prioritize a novel set of promising synthesis candidates from a virtual library of potential products accessible via parallel synthesis.

Those examples illustrate typical workflows, namely, the use of a few predictive models in parallel to rank order very focused synthesis proposals that have been derived on the basis of chemical feasibility and ligand SAR data. Each iteration and biological testing of a well-defined, information-rich set of compounds enriches our knowledge about the problem and is, of course, used to update the statistical models in order to guide the next optimization cycle. On the other hand, it might also happen that models for different properties reveal that key parameters influence different properties in opposite directions. This indicates that optimization might be extremely difficult, if not impossible, because of the very narrow balance and possibilities in the design of novel analogs.

14.7
Conclusions

Optimization of early lead compounds into promising drug candidates for pharmaceutical development is one of the key technologies in today's drug discovery efforts. Several medicinal chemistry approaches have been successfully applied to this problem, while the flexible integration of approaches to form an adaptive project team strategy is essential. The high complexity of this task prompts for an early identification of critical success factors and a risk assessment before individual series are progressed. High throughput alone, in this phase, does not provide a solution, as only the scientific understanding of critical success factors will enable a project team to focus on the most promising series. Therefore, a clear understanding of factors controlling affinity, selectivity, ADME and physicochemical compound properties of lead structures is essential to direct the medicinal chemistry strategies.

This chapter discussed approaches to simultaneously consider multiple aspects in optimization with the challenging goal to improve the efficiency of the drug discovery process. Some literature case studies on lead optimization illustrate the value of tight integration of parameters by means of simultaneous optimization. Data integration is mandatory for reusing the knowledge in subsequent design cycles toward an enhanced lead optimization and drug candidate selection process. Typical problems are, in particular, the improvement of affinity at the desired target, sufficient selectivity against closely related proteins and ADME and toxicology antitargets (e.g., hERG). Furthermore, a parallel optimization of compound properties toward favorable physicochemical and related ADME/PK properties is essential. At the end, it is important that no single approach will solve all problems for a series, but the challenge lies in the effective integration of these concepts depending on the individual problem. Only a flexible and data-driven integration of those tools will enable drug discovery project teams to process interesting lead series and cancel less promising ones early.

The paradigm shift from critical activities from later drug development to earlier discovery phases some years ago has effectively led to a change in lead optimization and added a new dimension of complexity, while it is envisioned that from a multidimensional, data-driven process more suitable candidates in accord with the therapeutic target product profiles may emerge for the treatment of currently unmet medical needs.

Acknowledgment

The authors would like to thank Thorsten Naumann, Roy Vaz (Aventis Pharma), Gabriele Cruciani (Univ. Perugia) and Gerhard Klebe (Univ. Marburg) for stimulating discussions on several aspects of this chapter.

References

1 LLOYD, A. Modifying the drug discovery/drug development paradigm. *Drug Disc. Today* **1997**, *2*, 397–398.

2 WESS, G., URMANN, M., and SICKENBERGER, B. Medicinal chemistry: challenges and opportunities. *Angew. Chem., Int. Ed. Engl.* **2001**, *40*, 3341–3350.

3 WESS, G. How to escape the bottleneck of medicinal chemistry. *Drug Disc. Today* **2002**, *7*, 533–535.

4 LAWRENCE, R.N. Sir Richard Sykes contemplates the future of the pharma industry. *Drug Disc. Today* **2002**, *7*, 645–648.

5 SHAH, N.D., VERMEULEN, L.C., SANTELL, J.P., HUNKLER, R.J., and HONTZ, K. Projecting future drug expenditures-2002. *Am. J. Health Syst. Pharm.* **2002**, *59*, 131–142.

6 DIMASI, J.A. Risks in new drug development: approval success rates for investigational drugs. *Clin. Pharmacol. Ther.* **2001**, *69*, 297–307.

7 PRENTIS, R.A., LIS, Y., and WALKER, S.R. Pharmaceutical innovation by the seven UK-owned pharmaceutical companies (1964–1985). *Br. J. Clin. Pharmacol.* **1988**, *25*, 387–396.

8 KENNEDY, T. Managing the drug discovery/development interface. *Drug Disc. Today* **1997**, *2*, 436–444.

9 DREWS, J. Drug discovery: a historical perspective. *Science* **2000**, *287*, 1960–1964.

10 HILLISCH, A. and HILGENFELD, R. (Eds). *Modern Methods of Drug Discovery.* Birkhäuser, Basel, **2003**.

11 (a) GORDON, E.M. and KERWIN, J.F. (Eds). *Combinatorial Chemistry and Molecular Diversity in Drug Discovery.* Wiley, New York, USA, **1998**; (b) JUNG, G. (Ed.). *Combinatorial Chemistry.* Wiley-VCH, Weinheim, **1999**.

12 (a) GALLOP, M.A., BARRETT, R.W., DOWER, W.J., FODOR, S.P.A., and GORDON, E.M. Applications of combinatorial technologies to drug discovery. 1. Background and peptide combinatorial libraries. *J. Med. Chem.* **1994**, *37*, 1233–1251; (b) GORDON, E.M., BARRETT, R.W., DOWER, W.J., FODOR, S.P.A., and GALLOP, M.A. Applications of combinatorial technologies to drug discovery. 2. Combinatorial organic synthesis, library screening strategies, and future directions. *J. Med. Chem.* **1994**, *37*, 1385–1399; (c) MADDEN, D., KRCHNAK, V., and LEBL, M. Synthetic combinatorial libraries: views on techniques and their applications. *Perspect. Drug Discov. Des.* **1995**, *2*, 269–285; (d) ELLMAN, J.A. Design, synthesis and evaluation of

small-molecule libraries. *Acc. Chem. Res.*
1996, *29*, 132–143; (e) GORDON, E.M.,
GALLOP, M.A., and PATEL, D.V. Strategy
and tactics in combinatorial organic
synthesis. Application to drug discovery.
Acc. Chem. Res. **1996**, *29*, 144–154;
(f) FRUCHTEL, J.S. and JUNG, G. Organic
chemistry on solid supports. *Angew.
Chem., Int. Ed. Engl.* **1996**, *35*, 17–42;
(g) THOMPSON, L.A. and ELLMAN, J.A.
Synthesis and applications of small
molecule libraries. *Chem. Rev.* **1996**, *96*,
555–600; (h) CHOONG, I.C. and ELLMAN,
J.A. Solid-phase synthesis: applications
to combinatorial libraries. *Annu. Rep.
Med. Chem.* **1996**, *31*, 309–318;
(i) BALKENHOHL, F., VON DEM BUSSCHE-
HÜNNEFELD, C., LANSKY, A., and ZECHEL,
C. Combinatorial synthesis of small
organic molecules. *Angew. Chem., Int.
Ed. Engl.* **1996**, *35*, 2288–2337; (j) NEFZI,
A., OSTRESH, J.M., and HOUGHTEN, R.A.
The current status of heterocyclic
combinatorial libraries. *Chem. Rev.*
1997, *97*, 449–472; (k) BRISTOL, J.A.
(Ed.). Applications of solid-supported
organic synthesis in combinatorial
chemistry. *Tetrahedron* **1997**, *53*,
6573–6706; (l) GEYSEN, H.M., SCHOENEN,
F., WAGNER, D., and WAGNER, R.
Combinatorial compound libraries for
drug discovery: an ongoing challenge.
Nat. Rev. Drug Disc. **2003**, *2*, 222–230.

13 LANDRO, J.A., TAYLOR, I.C.A., STIRTAN,
W.G., OSTERMAN, D.G., KRISTIE, J.,
HUNNICUTT, E.J., RAE, P.M.M., and
SWEETNAM, P.M. HTS in the new
millennium: the role of pharmacology
and flexibility. *J. Pharm. Toxicol.
Methods* **2000**, *44*, 273–289.

14 LAHANA, R. How many leads from HTS?
Drug Disc. Today **1999**, *4*, 447.

15 ZALL, M. The pricing puzzle. *Mod. Drug
Disc.* **2001**, *4*, 36–38.

16 ALANINE, A., NETTEKOVEN, M., ROBERTS,
E., and THOMA, A.W. Lead generation
– enhancing the success of drug
discovery by investing in the hit to lead
process. *Comb. Chem. High Throughput
Screen.* **2003**, *6*, 51–66.

17 HILL, S.A. *Drug Discov. Word Spring*
2001, 19–25.

18 BRENNAN, M.B. *Chem. Eng. News* **2000**,
78(26), 63–74.

19 PROUDFOOT, J.R. Drugs, leads, and
drug-likeness: an analysis of some
recently launched drugs. *Bioorg. Med.
Chem. Lett.* **2002**, *12*, 1647–1650.

20 OPREA, T. Virtual screening in lead
discovery: a viewpoint. *Molecules* **2002**, *7*,
51–62.

21 ZAMORA, I., OPREA, T., CRUCIANI, G.,
PASTOR, M., and UNGELL, A.-L. Surface
descriptors for protein-ligand affinity
prediction. *J. Med. Chem.* **2003**, *46*,
25–33.

22 BLEICHER, K.H., BÖHM, H.J., MÜLLER, K.,
and ALANINE, A.I. Hit and lead
generation: beyond high throughput
screening. *Nat. Rev. Drug Disc.* **2003**,
369–378.

23 HODGSON, J. ADMET – turning
chemicals into drugs. *Nat. Biotechnol.*
2001, *19*, 722–726.

24 HOPKINS, A.L. and GROOM, C.R. The
druggable genome. *Nat. Rev. Drug Disc.*
2002, *1*, 727–730.

25 LIPINSKI, C.A., LOMBARDO, F., DOMINY,
B.W., and FEENEY, P.J. Experimental and
computational approaches to estimate
solubility and permeability in drug
discovery and development settings.
Adv. Drug Deliv. Rev. **1997**, *23*, 3–25.

26 VEBER, D.F., JOHNSON, S.R., CHENG,
H.Y., SMITH, B.R., WARD, K.W., and
KOPPLE, K.D. Molecular properties that
influence the oral bioavailability of drug
candidates. *J. Med. Chem.* **2002**, *12*,
2615–2623.

27 SUNDBERG, S.A. High-throughput and
ultra-high-throughput screening:
solution- and cell-based approaches.
Curr. Opin. Biotechnol. **2000**, *11*, 47–53.

28 THIERICKE, R. High-throughput
screening technologies. In *Modern
Methods of Drug Discovery*, HILLISCH, A.
and HILGENFELD, R. (Eds). Birkhäuser,
Basel, **2003**, 71–85.

29 SHUKER, S.B., HAJDUK, P.J., MEADOWS,
R.P., and FESIK, S.W. Discovering
high-affinity ligands for proteins:
SAR by NMR. *Science* **1996**, *274*,
1531–1534.

30 HAJDUK, P.J., BOYD, S., NETTESHEIM, D.,
NIENABER, V., SEVERIN, J., SMITH, R.,
DAVIDSON, D., ROCKWAY, T., and FESIK,
S.W. Identification of novel inhibitors of

urokinase via NMR-based screening. *J. Med. Chem.* **2000**, *43*, 3862–3866.

31 FEJZO, J., LEPRE, C.A., PENG, J.W., BEMIS, G.W., AJAY, MURCKO, M.A., and MOORE, J.M. The SHAPES strategy: an NMR-based approach for lead generation in drug discovery. *Chem. Biol.* **1999**, *6*, 755–769.

32 KUMAR, S. and GUNNARSSON, K. Small molecule drug screening based on surface plasmon resonance. In *Advantages of Drug Discovery Technology*, HARVEY, A.L. (Ed.). Wiley, Chichester, UK, **1998**, 97–114.

33 NIENABER, V.L., RICHARDSON, P.L., KLIGHOFER, V., BOUSKA, J.J., GIRANDA, V.L., and GREER, J. Discovering novel ligands for macromolecules using X-ray crystallographic screening. *Nat. Biotechnol.* **2000**, *18*, 1105–1107.

34 LESUISSE, D., LANGE, G., DEPREZ, P., BENARD, D., SCHOOT, B., DELETTRE, G., MARQUETTE, J.P., BROTO, P., JEAN-BAPTISTE, V., BICHET, P., SARUBBI, E., and MANDINE, E. SAR and X-ray – a new approach combining fragment-based screening and rational drug design: application to the discovery of nanomolar inhibitors of src SH2. *J. Med. Chem.* **2002**, *45*, 2379–2387.

35 STAHL, M., TODOROV, N.P., JAMES, T., MAUSER, H., BÖHM, H.-J., and DEAN, P.M. A validation study on the practical use of automated de novo design. *J. Comput.-Aided Mol. Des.* **2002**, *16*, 459–478.

36 BÖHM, H.-J. and SCHNEIDER, G. (Eds). *Virtual Screening for Bioactive Molecules*. Wiley-VCH, Weinheim, **2000**.

37 BAJORATH, J. Integration of virtual and high throughput screening. *Nat. Rev. Drug Disc.* **2002**, *1*, 882–894.

38 SMITH, A. Screening for drug discovery: the leading question. *Nature* **2002**, *418*, 453–459.

39 OPREA, T.I., DAVIS, A.M., TEAGUE, S.J., and LEESON, P.D. Is there a difference between leads and drugs? a historical perspective. *J. Chem. Inf. Comput. Sci.* **2001**, *41*, 1308–1315.

40 TEAGUE, S.J., DAVIS, A.M., LEESON, P.D., and OPREA, T. The design of leadlike combinatorial libraries. *Angew. Chem., Int. Ed.* **1999**, *38*, 3743–3748.

41 OPREA, T.I. Current trends in lead discovery: Are we looking for the appropriate properties? *J. Comput.-Aided. Mol. Des.* **2002**, *16*, 325–334.

42 HANN, M.M., LEACH, A.R., and HARPER, G. Molecular complexity and its impact on the probability of finding leads for drug discovery. *J. Chem. Inf. Comput. Sci.* **2001**, *41*, 856–864.

43 MURRAY, C.W. and VERDONK, M.L. The consequences of translational and rotational entropy lost by small molecules on binding to proteins. *J. Comput.-Aided Mol. Des.* **2002**, *16*, 741–753.

44 RISHTON, G.M. Nonleadlikeness and leadlikeness in biochemical screening. *Drug Disc. Today* **2003**, *8*, 86–96.

45 BLUNDELL, T.L., JHOTI, H., and ABELL, C. High-throughput crystallography for lead discovery in drug design. *Nat. Rev. Drug Disc.* **2002**, *1*, 45–54.

46 BOEHM, H.J., BOEHRINGER, M., BUR, D., GMUENDER, H., HUBER, W., KLAUS, W., KOSTREWA, D., KUEHNE, H., LUEBBERS, T., MEUNIER-KELLER, N., and MUELLER, F. Novel inhibitors of DNA gyrase: 3D structure based biased needle screening, hit validation by biophysical methods, 3D guided optimization. A promising alternative to random screening. *J. Med. Chem.* **2000**, *43*, 2664–2674.

47 PICKETT, S.D., SHERBORNE, B.S., WILKINSON, T., BENNETT, J., BORKAKOTI, N., BROADHURST, M., HURST, D., KILFORD, I., MCKINNELL, M., and JONES, P.S. Discovery of low molecular weight inhibitors of IMPDH via virtual needle screening. *Bioorg. Med. Chem. Lett.* **2003**, *13*, 1691–1694.

48 LIPINSKI, C.A. Drug-like properties and the causes of poor solubility and poor permeability. *J. Pharm. Toxicol. Methods* **2000**, *44*, 235–249.

49 GREENE, J., KAHN, S., SAVOJ, H., SPRAGUE, P., and TEIG, S. Chemical function queries for 3D database search. *J. Chem. Inf. Comput. Sci.* **1994**, *34*, 1297–1308.

50 SPRAGUE, P.W. Automated chemical hypothesis generation and database searching with CATALYST. *Perspect. Drug. Disc. Des.* **1995**, *3*, 1–20.

51 KUBINYI, H. (Ed.). *3D-QSAR in Drug Design. Theory, Methods and Applications.* ESCOM, Leiden, Netherlands, **1993**.

52 KUBINYI, H., FOLKERS, G., and MARTIN, Y.C. (Eds). *3D-QSAR in Drug Design, Vol. 2.* ESCOM, Dordrecht, Netherlands, **1998**.

53 HANSCH, C. and LEO, A. (Eds) *Exploring QSAR: Fundamentals and Applications in Chemistry and Biology.* American Chemical Society, Washington, DC, **1995**.

54 DEBNATH, A.K. Quantitative structure-activity relationship (QSAR) paradigm – Hansch era to new millennium. *Mini Rev. Med. Chem.* **2001**, *1*, 187–195.

55 KELLOGG, G.E. and SEMUS, S.F. 3D QSAR in modern drug design. In *Modern Methods of Drug Discovery*, HILLISCH, A. and HILGENFELD, R. (Eds). Birkhäuser, Basel, **2003**, 223–241.

56 RCSB Protein Data Bank, from the Research Collaboratory for Structural Bioinformatics, http://www.rcsb.org/pdb/index.html.

57 BERMAN, H.M., WESTBROOK, J., FENG, Z., GILLILAND, G., BHAT, T.N., WEISSIG, H., SHINDYALOV, I.N., and BOURNE, P.E. The protein data bank. *Nucleic Acids Res.* **2000**, *28*, 235–242.

58 BOHACEK, R.S., MCMARTIN, C., and GUIDA, W.C. The art and practice of structure-based drug design: a molecular modeling perspective. *Med. Res. Rev.* **1996**, *16*, 3–50.

59 KUBINYI, H. Structure-based design of enzyme inhibitors and receptor ligands. *Curr. Opin. Drug Disc. Dev.* **1998**, *1*, 4–15.

60 MURCKO, M., CARON, P.R., and CHARIFSON, P.S. Structure-based drug design. *Annu. Rep. Med. Chem.* **1999**, *34*, 297–306.

61 KLEBE, G. Recent developments in structure-based drug design. *J. Mol. Med.* **2000**, *78*, 269–281.

62 BÖHM, H.-J. and STAHL, M. Structure-based library design: molecular modelling merges with combinatorial chemistry. *Curr. Opin. Chem. Biol.* **2000**, *4*, 283–286.

63 MATTOS, C. and RINGE, D. Multiple binding modes. In *3D-QSAR in Drug Design. Theory, Methods and Applications,* KUBINYI, H. (Ed.). ESCOM, Leiden, Netherlands, **1993**, 226–254.

64 MAIGNAN, S., GUILLOTEAU, J.-P., CHOI-SLEDESKI, Y.M., BECKER, M.R., EWING, W.R., PAULS, H.W., SPADA, A.P., and MIKOL, V. Molecular structures of human factor Xa complexed with ketopiperazine inhibitors: preference for a neutral group in the S1 pocket. *J. Med. Chem.* **2003**, *46*, 685–690.

65 TEAGUE, S.J. Implications of protein flexibility for drug discovery. *Nat. Rev. Drug Disc.* **2003**, *2*, 527–541.

66 REYDA, S., SOHN, C., KLEBE, G., RALL, K., ULLMANN, D., JAKUBKE, H.D., and STUBBS, M.T. Reconstructing the binding site of factor Xa in trypsin reveals ligand-induced structural plasticity. *J. Mol. Biol.* **2003**, *325*, 963–977.

67 LADBURY, J.E. Just add water! The effect of water on the specificity of protein-ligand binding sites and its potential application to drug design. *Chem. Biol.* **1996**, *3*, 973–980.

68 STUBBS, M.T., REYDA, S., DULLWEBER, F., MOLLER, M., KLEBE, G., DORSCH, D., MEDERSKI, W.W.K.R., and WURZIGER, H. pH-dependent binding modes observed in trypsin crystals: lessons for structure-based drug design. *ChemBioChem* **2002**, *3*, 246–249.

69 DAVIS, A.M., TEAGUE, S.J., and KLEYWEGT, G.J. Application and limitations of X-ray crystallographic data in structure-based ligand and drug design. *Angew. Chem., Int. Ed. Engl.* **2003**, *42*, 2718–2736.

70 HOOFT, R.W.W., VRIEND, G., SANDER, C., and ABOLA, E.E. Errors in protein structures. *Nature* **1996**, *381*, 272.

71 NISSINK, J.W.M., MURRAY, C., HARTSHORN, M., VERDONK, M.L., COLE, J.C., and TAYLOR, R. A new test set for validating predictions of protein-ligand interaction. *Proteins* **2002**, *49*, 457–471.

72 BABINE, R.E. and BENDER, S.L. Molecular recognition of protein-ligand complexes: applications to drug design. *Chem. Rev.* **1997**, *97*, 1359–1472.

73 GOHLKE, H. and KLEBE, G. Approaches to the description and prediction of the binding affinity of small-molecule ligands to macromolecular receptors. *Angew. Chem., Int. Ed. Engl.* **2002**, *41*, 2644–2676.

74 TAME, J.R.H. Scoring functions: a view from the bench. *J. Comput.-Aided Mol. Des.* **1999**, *13*, 99–108.

75 OPREA, T.I. and MARSHALL, G.R. Receptor-based prediction of binding affinities. In *3D-QSAR in Drug Design*, Vol. 2, KUBINYI, H., FOLKERS, G., and MARTIN, Y.C. (Eds). Kluwer, Dordrecht, Netherlands, **1998**, 35–61.

76 BÖHM, H.J. and STAHL, M. The use of scoring functions in drug discovery applications. In *Reviews in Computational Chemistry*, Vol. 18, LIPKOWITZ, K.B. and BOYD, D.B. (Eds). Wiley-VCH, Weinheim, **2002**, 41–87.

77 GOHLKE, H. and KLEBE, G. Statistical potentials and scoring functions applied to protein-ligand binding. *Curr. Opin. Struct. Biol.* **2001**, *11*, 231–235.

78 CHARIFSON, P.S., CORKEREY, J.J., MURCKO, M.A., and WALTERS, W.P. Consensus scoring: a method for obtaining improved hit rates from docking databases of three-dimensional structures into proteins. *J. Med. Chem.* **1999**, *42*, 5100–5109.

79 CLARK, R.D., STRIZHEV, A., LEONARD, J.M., BLAKE, J.F., and MATTHEW, J.B. Consensus scoring for ligand/protein interactions. *J. Mol. Graph. Mod.* **2002**, *20*, 281–295.

80 GOHLKE, H. and KLEBE, G. DrugScore meets comfa: adaptation of fields for molecular comparison (AFMoC) or how to tailor knowledge-based pair-potentials to a particular protein. *J. Med. Chem.* **2002**, *45*, 4153–4170.

81 GOHLKE, H., HENDLICH, M., and KLEBE, G. Knowledge-based scoring function to predict protein-ligand interactions. *J. Mol. Biol.* **2000**, *295*, 337–356.

82 CRAMER, R.D., PATTERSON, D.E., and BUNCE, J.E. Comparative molecular field analysis (CoMFA). 1. Effect of shape on binding of steroids to carrier proteins. *J. Am. Chem. Soc.* **1988**, *110*, 5959–5967.

83 CLARK, M., CRAMER, R.D., JONES, D.M., PATTERSON, D.E., and SIMEROTH, P.E. Comparative molecular field analysis (CoMFA). 2. Towards its use with

3D-structural databases. *Tetrahedron. Comput. Methods* **1990**, *3*, 47–59.

84 (a) WOLD, S., ALBANO, C., DUNN, W.J., EDLUND, U., ESBENSON, K., GELADI, P., HELLBERG, S., LINDBERG, W., and SJÖSTRÖM, M. Multivariate data analysis in chemistry. In *Chemometrics: Mathematics and Statistics in Chemistry*, KOWALSKI, B. (Ed.). Reidel, Dordrecht, Netherlands, **1984**, 17–95; (b) DUNN, W.J., WOLD, S., EDLUND, U., HELLBERG, S., and GASTEIGER, J. Multivariate structure-activity relationships between data from a battery of biological tests and an ensemble of structural descriptors: the PLS method. *Quant. Struct.-Act. Relat.* **1984**, *3*, 131–137.

85 ORTIZ, A.R., PISABARRO, M.T., GAGO, F., and WADE, R.C. Prediction of drug binding affinities by comparative binding energy analysis. *J. Med. Chem.* **1995**, *38*, 2681–2691.

86 ORTIZ, A.R., PASTOR, M., PALOMER, A., CRUCIANI, G., GAGO, F., and WADE, R.C. Reliability of comparative molecular field analysis models: effects of data scaling and variable selection using a set of human synovial fluid phospholipase A2 inhibitors. *J. Med. Chem.* **1997**, *40*, 1136–1148.

87 PEREZ, C., PASTOR, M., ORTIZ, A.R., and GAGO, F. Comparative binding energy analysis of HIV-1 protease inhibitors: incorporation of solvent effects and validation as a powerful tool in receptor-based drug design. *J. Med. Chem.* **1998**, *41*, 836–852.

88 WANG, T. and WADE, R.C. Comparative binding energy (COMBINE) analysis of influenza neuraminidase-inhibitor complexes. *J. Med. Chem.* **2001**, *44*, 961–971.

89 ROGNAN, D., LAUEMOLLER, S.L., HOLM, A., BUUS, S., and TSCHINKE, V. Predicting binding affinities of protein ligands from three-dimensional models: application to peptide binding to Class I major histocompatibility proteins. *J. Med. Chem.* **1999**, *42*, 4650–4658.

90 MURRAY, C.W., AUTON, T.R., and ELDRIDGE, M.D. Empirical scoring functions. II. The testing of an empirical scoring function for the prediction of ligand-receptor affinities and the use of

Bayesian regression to improve the quality of the model. *J. Comput.-Aided Mol. Des.* **1998**, *12*, 503–519.

91 ELDRIDGE, M.D., MURRAY, C.W., AUTON, T.R., PAOLINI, G.V., and MEE, R.P. Empirical scoring functions: I. The development of a fast empirical scoring function to estimate the binding affinity of ligands in receptor complexes. *J. Comput.-Aided Mol. Des.* **1997**, *11*, 425–445.

92 KLEBE, G., ABRAHAM, U., and MIETZNER, T. Molecular similarity indices in a comparative analysis (CoMSIA) of drug molecules to correlate and predict their biological activity. *J. Med. Chem.* **1994**, *37*, 4130–4146.

93 (a) WOLD, S. Cross-validatory estimation of the number of component in factor and principal component models. *Technometrics* **1978**, *4*, 397–405; (b) CRAMER, R.D., BUNCE, J.D. and PATTERSON, D.E. Crossvalidation, bootstrapping and partial least squares compared with multiple regression in conventional QSAR studies. *Quant. Struct.-Act. Relat.* **1988**, *7*, 18–25.

94 WALLER, C.L., OPREA, T.I., GIOLITTI, A., and MARSHALL, G.R. Three-dimensional QSAR of human immunodeficiency virus (I) protease inhibitors. 1. A CoMFA study employing experimentally-determined alignment rules. *J. Med. Chem.* **1993**, *36*, 4152–4160.

95 OPREA, T.I., WALLER, C.L., and MARSHALL, G.R. Three-dimensional quantitative structure-activity relationship of human immuno-deficiency virus (I) protease inhibitors. 2. Predictive power using limited exploration of alternate binding modes. *J. Med. Chem.* **1994**, *37*, 2206–2215.

96 WEI, D.T., MEADOWS, J.C., and KELLOGG, G.E. Effects of entropy on QSAR equations for HIV-1 protease: 1. Using hydropathic binding descriptors. 2. Unrestrained complex structure optimizations. *Med. Chem. Res.* **1997**, *7*, 259–270.

97 SIPPL, W. Receptor-based 3D QSAR analysis of estrogen receptor ligands – merging the accuracy of

receptor-based alignments with the computational efficiency of ligand-based methods. *J. Comput.-Aided Mol. Des.* **2000**, *14*, 559–572.

98 SIPPL, W., CONTRERAS, J.M., PARROT, I., RIVAL, Y.M., and WERMUTH, C.G. Structure-based 3D QSAR and design of novel acetylcholinesterase inhibitors. *J. Comput.-Aided Mol. Des.* **2001**, *15*, 395–410.

99 MATTER, H., SCHWAB, W., BARBIER, D., BILLEN, G., HAASE, B., NEISES, B., SCHUDOK, M., THORWART, W., BRACHVOGEL, V., LÖNZE, P., and WEITHMANN, K.U. Quantitative structure-activity relationship of human neutrophil collagenase (MMP-8) inhibitors using comparative molecular field and X-Ray structure analysis. *J. Med. Chem.* **1999**, *42*, 1908–1920.

100 MATTER, H. and SCHWAB, W. Affinity and selectivity of matrix metalloproteinase inhibitors: a chemometrical study from the perspective of ligands and proteins. *J. Med. Chem.* **1999**, *42*, 4506–4523.

101 NAUMANN, T. and MATTER, H. Structural classification of protein kinases using 3D molecular interaction field analysis of their ligand binding sites: target family landscapes. *J. Med. Chem.* **2002**, *45*, 2366–2378.

102 MATTER, H., DEFOSSA, E., HEINELT, U., BLOHM, P.M., SCHNEIDER, D., MÜLLER, A., HEROK, S., SCHREUDER, H., LIESUM, A., BRACHVOGEL, V., LÖNZE, P., WALSER, A., AL-OBEIDI, F., and WILDGOOSE, P. Design and quantitative structure-activity relationship of 3-amidinobenzyl-1H-indole-2-carboxamides as potent, non-chiral and selective inhibitors of blood coagulation factor Xa. *J. Med. Chem.* **2002**, *45*, 2749–2769.

103 MATTER, H. Computational approaches towards the quantification of molecular diversity and design of compound libraries. In *Modern Methods of Drug Discovery*, HILLISCH, A. and HILGENFELD, R. (Eds). Birkhäuser, Basel, **2003**, 125–156.

104 SCHUFFENHAUER, A., ZIMMERMANN, J., STOOP, R., VAN DER VYVER, J.J., LECCHINI, S., and JACOBY, E. An ontology for pharmaceutical ligands and its application for in silico screening and library design. *J. Chem. Inf. Comput. Sci.* **2002**, *42*, 947–955.

105 SCHUFFENHAUER, A., FLOERSHEIM, P., ACKLIN, P., and JACOBY, E. Similarity metrics for ligands reflecting the similarity of the target proteins. *J. Chem. Inf. Comput. Sci.* **2003**, *43*, 391–405.

106 WONG, G., KOEHLER, K.F., SKOLNICK, P., GU, Z.-Q., ANANTHAN, S., SCHÖNHOLZER, P., HUNKELER, W., ZHANG, W., and COOK, J.M. Synthetic and computer-assisted analysis of the structural requirements for selective, high-affinity ligand binding to diazepam-insensitive benzodiazepine receptors. *J. Med. Chem.* **1993**, *36*, 1820–1830.

107 MATTER, H., DEFOSSA, E., HEINELT, U., NAUMANN, T., SCHREUDER, H. and WILDGOOSE, P. Combining structure-based design and 3D-QSAR towards the discovery of non-chiral, potent and selective factor Xa inhibitors. In *Rational Approaches to Drug Design, Proceedings of the 13th European Symposium on Quantitative Structure-Activity Relationships*, HÖLTJE, H.-D. and SIPPL, W. (Eds). Prous Science, Barcelona, **2001**, 177–185.

108 KASTENHOLZ, M.A., PASTOR, M., CRUCIANI, G., HAAKSMA, E.E.J., and FOX, T. GRID/CPCA: A new computational tool to design selective ligands. *J. Med. Chem.* **2000**, *43*, 3033–3044.

109 BOEHM, M., STUERZEBECHER, J., and KLEBE, G. Three-dimensional quantitative structure-activity relationship analyses using comparative molecular field analysis and comparative molecular similarity indices analysis to elucidate selectivity differences of inhibitors binding to trypsin, thrombin, and factor Xa. *J. Med. Chem.* **1999**, *42*, 458–477.

110 CHEN, Q., WU, C., MAXWELL, D., KRUDY, G.A., DIXON, R.A.F., and YOU, T.J. A 3D QSAR analysis of in vitro binding affinity and selectivity of 3-isoxazolylsulfonylaminothiophenes as endothelin receptor antagonists. *Quant. Struct.-Act. Relat.* **1999**, *18*, 124–133.

111 BOSTROM, J., BOEHM, M., GUNDERTOFTE, K., and KLEBE, G. A 3D QSAR study on a set of dopamine D4 receptor

antagonists. *J. Chem. Inf. Comput. Sci.* **2003**, *43*, 1020–1027.

112 BALLE, T., ANDERSEN, K., SOBY, K.K., and LILJEFORS, T. $\alpha1$ Adrenoceptor subtype selectivity 3D-QSAR models for a new class of $\alpha1$ adrenoceptor antagonists derived from the novel antipsychotic sertindole. *J. Mol. Graph. Mod.* **2003**, *21*, 523–534.

113 GOODFORD, P. A computational procedure for determining energetically favorable binding sites on biologically important macromolecules. *J. Med. Chem.* **1985**, *28*, 849–857.

114 NISSINK, J.W.M., VERDONK, M.L., and KLEBE, G. Simple knowledge-based descriptors to predict protein-ligand interactions. Methodology and validation. *J. Comput.-Aided Mol. Des.* **2000**, *14*, 787–803.

115 VERDONK, M.L., COLE, J.C., WATSON, P., GILLET, V., and WILLETT, P. SuperStar: improved knowledge-based interaction fields for protein binding sites. *J. Mol. Biol.* **2001**, *307*, 841–859.

116 GOHLKE, H., HENDLICH, M., and KLEBE, G. Predicting binding modes, binding affinities and "hot spots" for protein-ligand complexes using a knowledge-based scoring function. *Perspect. Drug Disc. Dev.* **2000**, *20*, 115–144.

117 WESTERHUIS, J.A., KOURTI, T., and MACGREGOR, J.F. Analysis of multi-block and hierarchical PCA and PLS models. *J. Chemometrics* **1998**, *12*, 301–321.

118 RIDDERSTRÖM, M., ZAMORA, I., FJELLSTRÖM, O., and ANDERSSON, T.B. Analysis of selective regions in the active sites of human cytochromes P450, 2C8, 2C9, 2C18, and 2C19 homology models using GRID/CPCA. *J. Med. Chem.* **2001**, *44*, 4072–4081.

119 WILLIAMS, P.A., COSME, J., SRIDHAR, V., JOHNSON, E.F., and MCREE, D.E. Mammalian microsomal cytochrome P450 monooxygenase: structural adaptations for membrane binding and functional diversity. *Mol. Cell* **2000**, *5*, 121–131.

120 TERP, G.E., CRUCIANI, G., CHRISTENSEN, I.T., and JORGENSEN, F.S. Structural differences of matrix metalloproteinases

with potential implications for inhibitor selectivity examined by the GRID/CPCA approach. *J. Med. Chem.* **2002**, *45*, 2675–2684.

121 MYSHKIN, E. and WANG, B. Chemometrical classification of ephrin ligands and eph kinases using GRID/CPCA approach. *J. Chem. Inf. Comput. Sci.* **2003**, *43*, 1004–1010.

122 KURZ, M., BRACHVOGEL, V., MATTER, H., STENGELIN, S., THÜRING, H., and KRAMER, W. Insights Into the Bile Acid Transportation System the human ileal lipid-binding protein (ILBP) – cholyltaurine complex and its comparison to homologous structures. *Proteins* **2003**, *50*, 312–328.

123 SHERIDAN, R.P., HOLLOWAY, M.K., MCGAUGHEY, G., MOSLEY, R.T., and SINGH, S.B. A simple method for visualizing the differences between related receptor sites. *J. Mol. Graph. Mod.* **2002**, *21*, 71–79.

124 MATTER, H., NAUMANN, T., and PIRARD, B. Target family landscapes to match ligand selectivity with binding site topology in chemical biology. In *EuroQSAR2002: Designing Drugs and Crop Protectants: Processes, Problems and Solutions*, FORD, M., LIVINGSTONE, D., DEARDEN, J., and VAN DE WATERBEEMD, H. (Eds). Blackwell, Oxford, UK, **2003**, 183–185.

125 LAPINSH, M., PRUSIS, P., GUTCAITS, A., LUNDSTEDT, T., and WIKBERG, J.E.S. Development of proteo-chemometrics: a novel technology for the analysis of drug-receptor interactions. *Biochim. Biophys. Acta* **2001**, *1525*, 180–190.

126 LAPINSH, M., PRUSIS, P., MUTULE, I., MUTULIS, F., and WIKBERG, J.E.S. QSAR and proteo-chemometric analysis of the interaction of a series of organic compounds with melanocortin receptor subtypes. *J. Med. Chem.* **2003**, *46*, 2572–2579.

127 LAPINSH, M., PRUSIS, P., LUNDSTEDT, T., and WIKBERG, J.E.S. Proteochemo-metrics modeling of the interaction of amine G-protein coupled receptors with a diverse set of ligands. *Mol. Pharm.* **2002**, *61*, 1465–1475.

128 EKINS, S., WALLER, C.L., SWANN, P.W., CRUCIANI, G., WRIGHTON, S.A., and

WIKEL, J.H. Progress in predicting human ADME parameters in silico. *J. Pharm. Toxicol. Methods* **2000**, *44*, 251–272.

129 LIPINSKI, C.A., LOMBARDO, F., DOMINY, B.W., and FEENEY, P.J. Experimental and computational approaches to estimate solubility and permeability in drug discovery and development settings. *Adv. Drug Deliv. Rev.* **2001**, *46*, 3–26.

130 JORGENSEN, W.L. and DUFFY, E.M. Prediction of drug solubility from structure. *Adv. Drug Deliv. Rev* **2002**, *54*, 355–366.

131 ROSENBERG, M.F., KAMIS, A.B., CALLAGHAN, R., HIGGINS, C.F., and FORD, R.C. Three-dimensional structures of the mammalian multidrug resistance P-glycoprotein demonstrate major conformational changes in the transmembrane domains upon nucleotide binding. *J. Biol. Chem.* **2003**, *278*, 8294–8299.

132 LITMAN, T., DRULEY, T.E., STEIN, W.D., and BATES, S.E. From MDR to MXR: new understanding of multidrug resistance systems, their properties and clinical significance. *Cell. Mol. Life Sci.* **2001**, *58*, 931–959.

133 ZHAO, Y.H., LE, J., ABRAHAM, M.H., HERSEY, A., EDDERSHAW, P.J., LUSCOMBE, C.N., BUTINA, D., BECK, G., SHERBORNE, B., COOPER, I., and PLATTS, J.A. Evaluation of human intestinal absorption data and subsequent derivation of a quantitative structure-activity relationship (QSAR) with the Abraham descriptors. *J. Pharm. Sci.* **2001**, *90*, 749–784.

134 FU, X.C., LIANG, W.Q., and YU, Q.S. Correlation of drug absorption with molecular charge distribution. *Pharmazie* **2001**, *56*, 267–268.

135 AGATONOVICH-KUSTRIN, S., BERESFORD, R., and YUSOF, A.P.M. ANN modeling of the penetration across a polydimethylsiloxane membrane from theoretically derived molecular descriptors. *J. Pharm. Biomed. Anal.* **2001**, *25*, 227–237.

136 NORINDER, U. and ÖSTERBERG, T. Theoretical calculation and prediction of drug transport processes using simple parameters and partial least squares

projections to latent structures (PLS) statistics. The use of electrotopological state indices. *J. Pharm. Sci.* **2001**, *90*, 1076–1085.

137 AGORAM, B., WOLTOSZ, W.S., and BOLGER, M.B. Predicting the impact of physiological and biochemical processes on oral drug bioavailability. *Adv. Drug Deliv. Rev.* **2001**, *90*, S41–S67.

138 NORRIS, D.A., LEESMAN, G.D., SINKO, P.J., and GRASS, G.M. Development of predictive pharmacokinetic simulation models for drug discovery. *J. Control Relat.* **2000**, *65*, 55–62.

139 PARROTT, N. and LAVÉ, T. Prediction of intestinal absorption: comparative assessment of Gastroplus and Idea. *Eur. J. Pharm. Sci.* **2002**, *17*, 51–61.

140 KARIV, I., ROURICK, R.A., KASSEL, D.B., and CHUNG, T.D.Y. Improvement of "hit-to-lead" optimization by integration of in vitro HTS experimental models for early determination of pharmacokinetic properties. *Comb. Chem. High Throughput Screen.* **2002**, *5*, 459–472.

141 BERESFORD, A.P., SELICK, H.E., and TARBIT, M.H. The emerging importance of predictive ADME simulation in drug discovery. *Drug Disc. Today* **2002**, *7*, 109–116.

142 YOSHIDA, F. and TOPLISS, J.G. QSAR model for drug human oral bioavailability. *J. Med. Chem.* **2000**, *43*, 2575–2585.

143 MADAN, A., USUKI, E., BURTON, L.A., OGILVIE, B.W., and PARKINSON, A. In vitro approaches for studying the inhibition of drug-metabolizing enzymes and identifying the drug-metabolizing enzymes responsible for the metabolism of drugs. In *Drug Drug Interaction*, RODRIGUEZ, A.D. (Ed.). Marcel Dekker, New York, **2002**, 217–294.

144 DE GROOT, M.J., ALEX, A.A., and JONES, B.C. Development of a combined protein and pharmacophore model for cytochrome P450 2C9. *J. Med. Chem.* **2002**, *45*, 1983–1993.

145 BIDWELL, L.M., MCMANUS, M., GAEDIGK, A., KAKUTA, Y., NEGISHI, M., PEDERSEN, L., and MARTIN, J.L. Crystal structure of human catecholamine sulfotransferase. *J. Mol. Biol.* **1999**, *293*, 521–530.

146 EKINS, S., DE GROOT, M., and JONES, J.P. Pharmacophore and three-dimensional quantitative structure activity relationship methods for modeling cytochrome P450 active sites. *Drug Metab. Dispos.* **2001**, *29*, 936–944.

147 HIGGINS, L., KORZEKWA, K.R., RAO, S., SHOU, M., and JONES, J.P. An assessment of the reaction energetics for cytochrome P450-mediated reactions. *Arch. Biochem. Biophys.* **2001**, *385*, 220–230.

148 SINGH, S.B., SHEN, L.Q., WALKER, M.J., and SHERIDAN, R.P. A model for predicting likely sites of CYP3A4-mediated metabolism on drug-like molecules. *J. Med. Chem.* **2003**, *46*, 1330–1336.

149 SHEN, M., XIAO, Y., GOLBRAIKH, A., GOMBAR, V.K., and TROPSHA, A. Development and validation of k-nearest-neighbor QSPR models of metabolic stability of drug candidates. *J. Med. Chem.* **2003**, *46*, 3013–3020.

150 LANGOWSKI, J. and LONG, A. Computer systems for the prediction of xenobiotic metabolism. *Adv. Drug Deliv. Rev.* **2002**, *54*, 407–415.

151 GREENE, N. Computer systems for the prediction of toxicity: an update. *Adv. Drug Deliv. Rev.* **2002**, *54*, 417–431.

152 VANDENBERG, J.I., WALKER, B.D., and CAMPBELL, T.J. HERG K+ channels: friend and foe. *Trends Pharmacol. Sci.* **2001**, *22*, 240–246.

153 VAN DE WATERBEEMD, H., SMITH, D.A., BEAUMONT, K., and WALKER, D.K. Property-based design: optimization of drug absorption and pharmacokinetics. *J. Med. Chem.* **2001**, *44*, 1313–1333.

154 VAN DE WATERBEEMD, H. and GIFFORD, E. ADMET in silico modelling: towards prediction paradise? *Nat. Drug Disc.* **2003**, *2*, 192–204.

155 CRUCIANI, G., CRIVORI, P., CARRUPT, P.A., and TESTA, B. Molecular fields in quantitative structure-permeation relationships: the VolSurf approach. *THEOCHEM* **2000**, *503*, 17–30.

156 CRUCIANI, G., PASTOR, M., and GUBA, W. VolSurf: a new tool for the pharmacokinetic optimization of lead compounds. *Eur. J. Pharm. Sci.* **2000**, *11*, S29–S39.

157 GUBA, W. and CRUCIANI, G. Molecular field-derived descriptors for the multivariate modelling of pharmacokinetic data. In *Molecular Modelling and Prediction of Bioactivity, Proceedings of the 12th European Symposium on Quantitative Structure-Activity Relationships (QSAR'98)*, GUNDERTOFTE, K. and JORGENSEN, F.S. (Eds). Plenum Press, New York, **2000**, 89–95.

158 ALIFRANGIS, L.H., CHRISTENSEN, I.T., BERGLUND, A., SANDBERG, M., HOVGAARD, L., and FROKJAER, S. Structure-property model for membrane partitioning of oligopeptides. *J. Med. Chem.* **2000**, *43*, 103–113.

159 CRIVORI, P., CRUCIANI, G., CARRUPT, P.A., and TESTA, B. Predicting blood-brain barrier permeation from three-dimensional molecular structure. *J. Med. Chem.* **2000**, *43*, 2204–2216.

160 PFEIFFER-MAREK, S., MATTER, H., ENGLERT, H., GERLACH, U., KNIEPS, S., SCHUDOK, M., and LEHR, K.H. Using VolSurf descriptors to model pharmacokinetic properties during lead optimization. In *Designing Drugs and Crop Protectants, Proceedings of the 14th European Symposium on Quantitative Structure-Activity Relationship (QSAR 2002)*, FORD, M., LIVINGSTONE, D., DEARDEN, J., and VAN DE WATERBEEMD, H. (Eds). Blackwell Publishers, Oxford, **2003**, 104–105.

161 COLMENAREJO, G., ALVAREZ-PEDRAGLIO, A., and LAVANDERA, J.-L. Cheminformatic models to predict binding affinities to human serum albumin. *J. Med. Chem.* **2001**, *44*, 4370–4378.

162 PETITPAS, I., BHATTACHARYA, A.A., TWINE, S., EAST, M., and CURRY, S. Crystal structure analysis of warfarin binding to human serum albumin. Anatomy of drug site I. *J. Biol. Chem.* **2001**, *276*, 22804–22809.

163 ZHU, C., JIANG, L., CHEN, T.M., and HWANG, K.K. A comparative study of artificial membrane permeability assay for high throughput profiling of drug absorption potential. *Eur. J. Med. Chem.* **2002**, *37*, 399–407.

164 SALMINEN, T., PULLI, A., and TASKINEN, J. Relationship between immobilised artificial membrane chromatographic retention and the brain penetration of structurally diverse drugs. *J. Pharm. Biomed. Anal.* **1997**, *15*, 469–477.

165 IRVINE, J.D., LOCKHART, L.T., CHEONG, J., TOLAN, J.W., SELICK, H.E., and GROVE, J.R. MDCK (Madin-Darby canine kidney) cells: A tool for membrane permeability screening. *J. Pharm. Sci.* **1999**, *88*, 28–33.

166 YAZDANIAN, M., GLYNN, S.L., WRIGHT, J.L., and HAWI, A. Correlating partitioning and Caco-2 cell permeability of structurally diverse small molecular weight compounds. *Pharm. Res.* **1998**, *15*, 1490–1494.

167 DOYLE, D.A., MORAIS CABRAL, J., PFUETZNER, R.A., KUO, A., GULBIS, J.M., COHEN, S.L., CHAIT, B.T., and MACKINNON, R. The structure of the potassium channel: molecular basis of K+ conduction and selectivity. *Science* **1998**, *280*, 69–77.

168 MITCHESON, J.S., CHEN, J., LIN, M., CULBERSON, C., and SANGUINETTI, M.C. A structural basis for drug-induced long QT syndrome. *Proc. Nat. Acad Sci. USA* **2000**, *97*, 12329–12333.

169 PEARLSTEIN, R., VAZ, R., and RAMPE, D. Understanding the structure-activity relationship of the human ether-a-go-go-related gene cardiac K+ channel. A model for bad behavior. *J. Med. Chem.* **2003**, *46*, 2017–2022.

170 ROCHE, O., TRUBE, G., ZUEGGE, J., PFLIMLIN, P., ALANINE, A., and SCHNEIDER, G. A virtual screening method for prediction of the hERG potassium channel liability of compound libraries. *ChemBioChem* **2002**, *3*, 455–459.

171 CAVALLI, A., POLUZZI, E., DE PONTI, F., and RECANATINI, M. Toward a pharmacophore for drugs inducing the long QT syndrome: insights from a CoMFA study of HERG K+ channel blockers. *J. Med. Chem.* **2002**, *45*, 3844–3853.

172 MATTER, H. unpublished results.

173 PEARLSTEIN, R.A., VAZ, R.J., KANG, J., CHEN, X.-L., PREOBRAZHENSKAYA, M., SHCHEKOTIKHIN, A.E., KOROLEV, A.M.,

LYSENKOVA, L.N., MIROSHNIKOVA, O.V., HENDRIX, J., and RAMPE, D. Characterization of HERG potassium channel inhibition using CoMSiA 3D QSAR and homology modeling approaches. *Bioorg. Med. Chem. Lett.* **2003**, *13*, 1829–1835.

174 JIANG, Y., LEE, A., CHEN, J., CADENE, M., CHAIT, B.T., and MACKINNON, R. Crystal structure and mechanism of a calcium-gated potassium channel. *Nature* **2002**, *417*, 515–522.

175 FRIESEN, R.W., DUCHARME, Y., BALL, R.G., BLOUIN, M., BOULET, L., COTE, B., FRENETTE, R., GIRARD, M., GUAY, D., HUANG, Z., JONES, T.R., LALIBERTE, F., LYNCH, J.J., MANCINI, J., MARTINS, E., MASSON, P., MUISE, E., PON, D.J., SIEGL, P.K.S., STYHLER, A., TSOU, N.N., TURNER, M.J., YOUNG, R.N., and GIRARD, Y. Optimization of a tertiary alcohol series of phosphodiesterase-4 (PDE4) inhibitors: structure-activity relationship related to PDE4 inhibition and human ether-a-go-go related gene potassium channel binding affinity. *J. Med. Chem.* **2003**, *46*, 2413–2426.

176 STRESSER, D.M., BLANCHARD, A.P., TURNER, S.D., ERVE, J.C.L., DANDENEAU, A.A., MILLER, V.P., and CRESPI, C.L. Substrate-dependent modulation of CYP3A4 catalytic activity: analysis of 27 test compounds with four fluorometric substrates. *Drug Metab. Dispos.* **2000**, *28*, 1440–1448.

177 LEWIS, D.F.V. Structural characteristics of human P450s involved in drug metabolism: QSARs and lipophilicity profiles. *Toxicology* **2000**, *144*, 197–203.

178 LEWIS, D.F.V. Essential requirements for substrate binding affinity and selectivity toward human CYP2 family enzymes. *Arch. Biochem. Biophys.* **2003**, *409*, 32–44.

179 EKINS, S., BRAVI, G., RING, B.J., GILLESPIE, T.A., GILLESPIE, J.S., VAMDENBRANDEN, M., WRIGHTON, S.A., and WIKEL, J.H. Three-dimensional quantitative structure activity relationship analyses of substrates for CYP2B6. *J. Pharm. Exp. Ther.* **1999**, *288*, 21–29.

180 EKINS, S., BRAVI, G., WIKEL, J., and WRIGHTON, S.A. Three-dimensional quantitative structure activity relationship analysis of cytochrome P-450 3A4 substrates. *J. Pharm. Exp. Ther.* **1999**, *291*, 424–433.

181 EKINS, S., BRAVI, G., BINKLEY, S., GILLESPIE, J.S., RING, B.J., WIKEL, J.H., and WRIGHTON, S.A. Three- and four-dimensional quantitative structure activity relationship analyses of cytochrome P-450 3A4 inhibitors. *J. Pharm. Exp. Ther.* **1999**, *290*, 429–438.

182 EKINS, S., BRAVI, G., BINKLEY, S., GILLESPIE, J.S., RING, B.J., WIKEL, J.H., and WRIGHTON, S.A. Three- and four-dimensional-quantitative structure activity relationship (3D/4D-QSAR) analyses of CYP2C9 inhibitors. *Drug Metab. Dispos.* **2000**, *28*, 994–1002.

183 EKINS, S., BRAVI, G., BINKLEY, S., GILLESPIE, J.S., RING, B.J., WIKEL, J.H., and WRIGHTON, S.A. Three and four dimensional-quantitative structure activity relationship (3D/4D-QSAR) analyses of CYP2D6 inhibitors. *Pharmacogenetics* **1999**, *9*, 477–489.

184 RAO, S., AOYAMA, R., SCHRAG, M., TRAGER, W.F., RETTIE, A., and JONES, J.F. A Refined 3-dimensional QSAR of cytochrome P450 2C9: computational predictions of drug interactions. *J. Med. Chem.* **2000**, *43*, 2789–2796.

185 AFZELIUS, L., MASIMIREMBWA, C.M., KARLEN, A., ANDERSSON, T.B., and ZAMORA, I. Discriminant and quantitative PLS analysis of competitive CYP2C9 inhibitors versus non-inhibitors using alignment independent GRIND descriptors. *J. Comput.-Aided Mol. Des.* **2002**, *16*, 443–458.

186 PASTOR, M., CRUCIANI, G., MCLAY, I., PICKETT, S., and CLEMENTI, S. GRid-INdependent descriptors (GRIND): a novel class of alignment-independent three-dimensional molecular descriptors. *J. Med. Chem.* **2000**, *43*, 3233–3243.

187 DE GROOT, M.J., ACKLAND, M.J., HORNE, V.A., ALEX, A.A., and JONES, B.C. Novel approach to predicting P450-mediated drug metabolism: development of a combined protein and pharmacophore model for CYP2D6. *J. Med. Chem.* **1999**, *42*, 1515–1524.

188 DE GROOT, M.J., ACKLAND, M.J., HORNE, V.A., ALEX, A.A., and JONES, B.C. A novel approach to predicting P450 mediated drug metabolism. CYP2D6 catalyzed N-dealkylation reactions and qualitative metabolite predictions using a combined protein and pharmacophore model for CYP2D6. *J. Med. Chem.* **1999**, *42*, 4062–4070.

189 ZAMORA, I., AFZELIUS, L., and CRUCIANI, G. Predicting drug metabolism: a site of metabolism prediction tool applied to the cytochrome P450 2C9. *J. Med. Chem.* **2003**, *46*, 2313–2324.

190 WESTER, M.R., JOHNSON, E.F., MARQUES-SOARES, C., DIJOLS, S., DANSETTE, P.M., MANSUY, D., and STOUT, C.D. Structure of mammalian cytochrome P450 2C5 complexed with diclofenac at 2.1 A resolution: evidence for an induced fit model of substrate binding. *Biochem.* **2003**, *42*, 9335–9345.

191 WESTER, M.R., JOHNSON, E.F., MARQUES-SOARES, C., DANSETTE, P.M., MANSUY, D., and STOUT, C.D. Structure of a substrate complex of mammalian cytochrome P450 2C5 at 2.3 a resolution: evidence for multiple substrate binding modes. *Biochem.* **2003**, *42*, 6370–6379.

192 WILLIAMS, P.A., COSME, J., WARD, A., ANGOVE, H.C., VINKOVIC, D.M., and JHOTI, H. Crystal structure of human cytochrome P450 2C9 with bound warfarin. *Nature* **2003**, *424*, 464–468.

193 HELLBERG, S., ERIKSSON, L., JONSSON, J., LINDGREN, F., SJÖSTRÖM, M., SKAGERBERG, B., WOLD, S., and ANDREWS, P. Minimum analog peptide sets (MAPS) for quantitative structure-activity relationships. *Int. J. Pept. Prot. Res.* **1991**, *37*, 414–424.

194 ATTERWILL, C.K. and WING, M.G. In vitro preclinical lead optimization technologies (PLOTs) in pharmaceutical development. *Toxicol. Lett.* **2002**, *127*, 143–151.

195 LINUSSON, A., GOTTFRIES, J., OLSSON, T., OERNSKOV, E., FOLESTAD, S., NORDEN, B., and WOLD, S. Statistical molecular design, parallel synthesis, and biological evaluation of a library of thrombin inhibitors. *J. Med. Chem.* **2001**, *44*, 3424–3439.

196 ERIKSSON, U.G., BREDBERG, U., GISLEN, K., JOHANSSON, L.C., FRISON, L., AHNOFF, M., and GUSTAFSSON, D. Pharmacokinetics and pharmaco-dynamics of ximelagatran, a novel oral direct thrombin inhibitor, in young healthy male subjects. *Eur. J. Clin. Pharmacol.* **2003**, *59*, 35–43.

197 SUGANO, K., YOSHIDA, S., TAKAKU, M., HARAMURA, M., SAITOH, R., NABUCHI, Y., and USHIO, H. Quantitative structure-intestinal permeability relationship of benzamidine analogue thrombin inhibitor. *Bioorg. Med. Chem. Lett.* **2000**, *10*, 1939–1942.

198 BURGEY, C.S., ROBINSON, K.A., LYLE, T.A., SANDERSON, P.E.J., LEWIS, S.D., LUCAS, B.J., KRUEGER, J.A., SINGH, R., MILLER-STEIN, C., WHITE, R.B., WONG, B., LYLE, E.A., WILLIAMS, P.D., COBURN, C.A., DORSEY, B.D., BARROW, J.C., STRANIERI, M.T., HOLAHAN, M.A., SITKO, G.R., COOK, J.J., MCMASTERS, D.R., MCDONOUGH, C.M., SANDERS, W.M., WALLACE, A.A., CLAYTON, F.C., BOHN, D., LEONARD, Y.M., DETWILER, T.J., LYNCH, J.J., YAN, Y., CHEN, Z., KUO, L., GARDELL, S.J., SHAFER, J.A., and VACCA, J.P. Metabolism-directed optimization of 3-aminopyrazinone acetamide thrombin inhibitors. Development of an orally bioavailable series containing P1 and P3 pyridines. *J. Med. Chem.* **2003**, *46*, 461–473.

199 HASEGAWA, K., SHINDOH, H., SHIRATORI, Y., OHTSUKA, T., AOKI, Y., ICHIHARA, S., HORII, I., and SHIMMA, N. Cassette dosing approach and quantitative structure-pharmacokinetic relationship study of antifungal N-myristoyl-transferase inhibitors. *J. Chem. Inf. Comput. Sci.* **2002**, *42*, 968–975.

200 MCKENNA, J.M., HALLEY, F., SOUNESS, J.E., MCLAY, I.M., PICKETT, S.D., COLLIS, A.J., PAGE, K., and AHMED, I. An algorithm-directed two component library synthesized via solid-phase methodology yielding potent and orally bioavailable p38 MAP kinase inhibitors. *J. Med. Chem.* **2002**, *45*, 2173–2184.

201 PICKETT, S.D., MCLAY, I.M., and CLARK, D.E. Enhancing the hit-to-lead properties of lead optimization libraries.

J. Chem. Inf. Comput. Sci. **2000**, *40*, 263–272.

202 CHOI-SLEDESKI, Y.M., KEARNEY, R., POLI, G., PAULS, H., GARDNER, C., GONG, Y., BECKER, M., DAVIS, R., SPADA, A., LIANG, G., CHU, V., BROWN, K., COLLUSSI, D., LEADLEY, R., REBELLO, S., MOXEY, P., MORGAN, S., BENTLEY, R., KASIEWSKI, C., MAIGNAN, S., GUILLOTEAU, J.P., and MIKOL, V. Discovery of an orally efficacious inhibitor of coagulation factor Xa which incorporates a neutral P1 ligand. *J. Med. Chem.* **2003**, *46*, 681–684.

203 TUCKER, T.J., BRADY, S.F., LUMMA, W.C., LEWIS, S.D., GARDELL, S.J., NAYLOR-OLSEN, A.M., YAN, Y., SISKO, J.T., STAUFFER, K.J., LUCAS, B.J., LYNCH, J.J., COOK, J.J., STRANIERI, M.T., HOLAHAN, M.A., LYLE, E.A., BASKIN, E.P., CHEN, I.W., DANCHECK, K.B., KRUEGER, J.A., COOPER, C.M., and VACCA, J.P. Design and synthesis of a series of potent and orally bioavailable noncovalent thrombin inhibitors that utilize nonbasic groups in the P1 position. *J. Med. Chem.* **1998**, *41*, 310–319.

204 MATTER, H., SCHUDOK, M., SCHWAB, W., THORWART, W., BARBIER, D., BILLEN, G., HAASE B., NEISES, B., WEITHMANN, K.U., and WOLLMANN, T. Tetrahydroisoquinoline-3-carboxylate based matrix metalloproteinase inhibitors: design, synthesis and structure-activity relationship. *Bioorg. Med. Chem.* **2002**, *10*, 3529–3544.

205 BARINGHAUS, K.H., MATTER, H., STENGELIN, S., and KRAMER, W. Substrate specificity of the ileal and the hepatic Na+/bile acid cotransporters of the rabbit. II. A reliable 3D QSAR pharmacophore model for the ileal Na+/bile acid cotransporter. *J. Lipid Res.* **1999**, *40*, 2158–2168.

15

Chemoinformatic Tools for Library Design and the Hit-to-Lead Process: A User's Perspective

Robert Alan Goodnow, Jr., Paul Gillespie, and Konrad Bleicher

15.1
The Need for Leads: The Sources of Leads and the Challenge to Find Them

The modern drug discovery process requires methods to regularly generate high-quality, druglike leads for various biological targets in order to function efficiently and for eventual success. The limited rate of success for progressing a small molecule with activity for a biological target through preclinical and clinical research to the launch of a new drug makes it necessary for most companies to work on multiple projects simultaneously. There are a few published studies with respect to preclinical attrition rates when going from target validation to clinical candidate selection, but they are certainly severely limiting [1]. Rough estimates have been published that only 1 in 10 000 compounds screened will ever reach the selection stage as a clinical candidate [2, 3]. Estimates of rates of successful drug launch from the clinical candidate stage are between 10 and 20%. In terms of time, estimates range from 10 to 15 years for discovery and development before launch of a new drug. The way in which research and development organizations are taking a family of lead molecules into clinical development is one of two reasons cited for higher success rates in delivering compounds for launch; selecting compounds for clinical studies according to stricter, more aggressive quality criteria is also cited as a factor that differentiates organizations with higher rates of successful drug launch [4]. The attrition of compounds and biological targets during drug discovery is not only disappointing but is also very costly. DiMasi and colleagues have recently published a study in which they estimate that the costs from discovery to launch for each drug would reach on average US$403 million; capitalization of these costs is estimated to reach over US$800 million [5].

With the obstacles of limited success rates of long-term and costly discovery efforts, it is little wonder that pharmaceutical organizations endeavor to develop more efficient strategies and processes at every step of the drug discovery process (Figure 15.1). Of the many strategies, some organizations have chosen to focus on generating greater numbers of high-quality lead molecules for which there may be fewer potential liabilities when transforming these molecules into drugs. It stands to reason that if corporate compound collections can be populated with diverse sets of molecules with good druglike properties, the transformation of these molecules to drugs will be faster and more efficient. Efficient chemoinformatic processes, which provide good leads from hit-to-lead efforts, add to the overall efficiency of drug discovery. The need

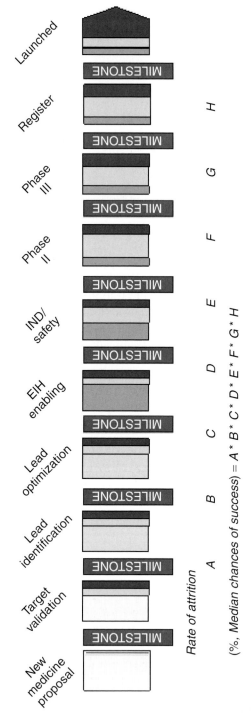

Fig. 15.1 Cumulative attrition rates in the drug discovery process milestones. Decreasing the attrition rate for any of these milestones has significant potential impact for improving the overall efficacy of launching drugs to market.

for good leads is particularly acute since the mapping of the human genome and developments in genomic and genetic approaches to new target identification [6]. Some have characterized the state of pharmaceutical research as being "target rich, but lead poor" [7].

Many drug leads are identified by assaying natural products, historical compound collections, combinatorial chemistry libraries, compounds synthesized according to knowledge-based approaches such as study of literature and patents, and purchase of commercially available single compounds. Compounds obtained in this way are often assayed in HTS, the predominant means to identify potential leads in a random manner. Depending on the biological target, the assay of hundreds of thousands of single compounds usually produces some number of primary hits. The process of primary hit filtration and assessment is known as hit-to-lead [8–11] (Section 15.7). Experience has shown that of the many primary hits identified from a high-throughput screen, only a handful are likely to advance to the resource-intensive lead optimization phase.

For targets without leads, a lead generation approach is employed if work on such a target is to continue. Such approaches usually require the construction of small-molecule libraries in order to maximize the chances of finding compounds of some level of minimal potency necessary to identify a *bona fide* lead useful for further optimization. Lead generation library design should emphasize design quality of the individual molecules over their numbers. This topic has been discussed extensively [12]. It is along this line of thought that the application of combinatorial chemistry has evolved from an emphasis on the synthesis of large numbers of diverse molecules to the synthesis of sets of compounds focused on protein families or specific targets. Chemoinformatic tools provide rational, objective ways to focus libraries to biological targets while assuring the creation of druglike compounds. Application of the proper chemoinformatic tool to the task at hand requires awareness of options and limitations for various applications. It is the intent of this chapter to demonstrate how evaluation and application of some chemoinformatic tools can make a positive impact on lead finding for drug discovery. Examples of the use and interpretation of calculated molecular properties in lead identification suggest ways in which these tools and processes can be improved. Lead generation and hit-to-lead activities at Roche Basel have been described [13, 14].

15.2
Property Predictions

In lead identification, the goal is to focus as quickly as possible on druglike compounds. If properties of compounds can be reliably predicted, then costs and timelines can be significantly reduced. This is true for real compounds and is especially true for virtual compounds, wherein we can avoid the synthesis of compounds with poorly predicted properties. Many publications have appeared in this area, describing algorithms for the prediction of a number of properties, including solubility, druglikeness, absorption, metabolism, oral bioavailability, various toxicities, frequent hitters and, most common of all, activity against a given biological target. In this section, we describe the state of the art in the literature with regard to solubility, druglikeness, and frequent hitters, and we also describe our experimental observations on these properties and some of the available tools.

15.3
Prediction of Solubility

Solubility is a very important requirement for a drug. While there are examples of drugs with low solubility, these tend to be compounds where very low doses are required. On the other hand, there are several cases where low solubility can become a serious hurdle:

- Compounds with low solubility and high permeability (BCS Class II [15]). For these compounds, oral bioavailability may be limited by solubility, and significant formulation efforts may be required before acceptable properties are obtained.
- Compounds with low solubility and low permeability (BCS Class IV [15]). These compounds tend to have poor bioavailability, and even where formulation can help, variable exposure can be seen and safety liabilities may result.
- Some *in vitro* ADME and safety assays (e.g., microsomal stability and hERG blockade) may not be possible with poorly soluble compounds.
- Compounds with low solubility are not appropriate for IV administration.

Over the last ten years or so, a trend toward increasing molecular weight and hydrophobicity has been observed in compounds prepared for biological screening. There has been a corresponding decrease in the solubility of many of these compounds with the result that poor solubility is becoming an issue for many programs. Lipinski has noted that the changes in molecular weight and hydrophobicity trends correspond to changes in the screening paradigm at several companies around 1990, where the screening throughput increased considerably, and it became possible to screen compounds with low thermodynamic solubility, which previously would not have been screened [16].

Having a computational tool to predict solubility of compounds would be of great value. Much work has been done in this area, with a variety of prediction tools reported. These have been well reviewed recently by Livingstone [17], so only a few points will be covered here.

Several computational methods require the user to input experimental parameters in order to make predictions. Examples include Yalkowsky's general solvation equation [18], which requires melting points, and Abraham's method [19], which requires five experimentally derived parameters. These methods are not useful for cases before the compounds are synthesized (e.g., in library design or virtual screening), and so they will not be considered further here.

In addition to methods that rely on experimental inputs, there are prediction tools that are based on structure alone. A number of different techniques have been used to correlate a variety of structure-derived descriptors with observed solubilities, including linear regression [20–25] and neural networks [26–36]. Interestingly, despite the diversity in approaches used to model the solubility data, the sources of solubility data are not very diverse; many publications in this area use data from the AQUASOL Database of Aqueous Solubility maintained by Professor Samuel H. Yalkowsky at the University of Arizona, or data from the Physical/Chemical Property Database (PHYSPROP) from the Syracuse Research Corporation [37]. The PHYSPROP data also come ultimately from Professor Yalkowsky.

We have studied the performance of several prediction methods to see how well in-house thermodynamic solubility measurements could be predicted. Among the prediction methods we studied were Huuskonen's method [26], ACD/Solubility DB [38], Meylan's method [21] as implemented in QMPRPlus, and the SimulationsPlus solubility prediction as implemented in QMPRPlus [39]. In general, we found the predictions to

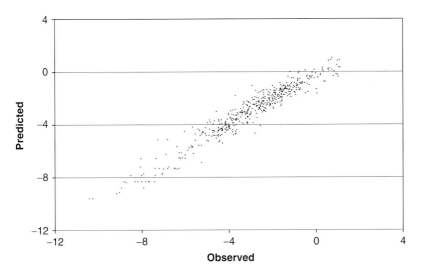

Fig. 15.2 Predicted versus observed solubility values for 493 compounds in the Gasteiger dataset [33]; $r^2 = 0.93$.

Tab. 15.1 r^2 values for QMPRPlus solubility predictions for various datasets

Dataset	n	r^2 for QMPRPlus predictions
Abraham [19]	662	0.85
Chen [24]	267	0.69
Yalkowsky [27]	430	0.90
Jurs [28]	399	0.78
Gasteiger [33]	493	0.93
Yaffe [35]	517	0.92
Roche	1526	0.17

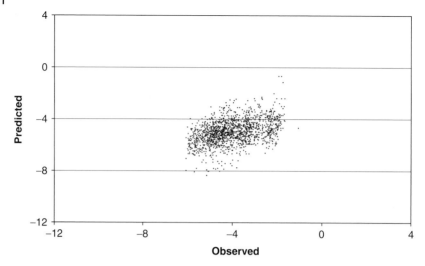

Fig. 15.3 Predicted versus observed solubility values for 1526 compounds in the Roche dataset; $r^2 = 0.17$.

be poor, with the SimulationsPlus prediction being generally the best. Table 15.1 shows the results for QMPRPlus solubility predictions on six datasets from the literature and for a set of compounds tested at Roche. The table shows that QMPRPlus performs extremely well for the literature data (see also Figure 15.2), with r^2 values ranging from 0.69 to 0.92, but that the predictions for the in-house compounds are poor, with an r^2 value of 0.17 (see also Figure 15.3).

In investigating why the predictions are much poorer for our compounds than for literature datasets, we considered the following possibilities:

- QMPRPlus predicts intrinsic solubility (the solubility of the unionized species) while we are measuring the actual solubility at pH 6.5.
- The training set for the prediction from QMPRPlus is not representative of the sorts of compounds for which we are measuring solubility.
- The purity and/or crystallinity of our samples may not be sufficiently good to generate accurate experimental solubility data. Lipinski has noted that the physical form of the first batch synthesized is often very different from the compound studied later in development [16]. For example, the compound may be amorphous, or if crystalline, it may not be in the most stable crystalline form. It may also contain traces of solvent or of trifluoroacetic acid.

There are two pieces of evidence suggesting that ionizable groups are responsible for at least a part of the problem. First, the measured solubilities are generally higher than the predicted values, and charged species are expected to be more soluble than the unionized species. Second, most of the compounds for which the predictions are poorest are alkylamines (206 of 316 compounds where the measured solubility is at least 100-fold greater than the predicted solubility). On the other hand, even if we remove alkylamines and carboxylic acids from our dataset, we find that the predictions are still poor: the r^2 value improves only to 0.22. This suggests that the difference between intrinsic and measured solubility does not explain all of the differences between predicted and measured solubilities.

The QMPRPlus solubility training set (from Syracuse Research Corporation [37]) is not in the public domain, so we cannot assess directly how similar the training set compounds are to our test compounds. We can approach this question indirectly by looking at the number of compounds flagged by QMPRPlus as having extrapolated solubility predictions. These compounds have descriptor values outside the range of the training set, and so are likely to be unlike the compounds in the training set. We compared the number of extrapolations for our compounds with six datasets from the literature. We found that the number of extrapolations is significantly higher for our dataset, suggesting that our compounds are less similar to the QMPRPlus training set than the literature compounds (Table 15.2).

We can also look at other literature datasets to gain an idea of how similar our compounds are to compounds for which QMPRPlus gives very good predictions. We have looked at four simple descriptors: molecular weight, topological polar surface area [40], chemical complexity [41], and rotatable bond count, using John Bradshaw's

Tab. 15.2 Numbers of compounds flagged as extrapolations in QMPRPlus solubility predictions

Dataset	n	Flagged	Percentage flagged
Abraham [19]	662	2	0
Chen [24]	267	35	13
Yalkowsky [27]	430	13	3
Jurs [28]	399	18	5
Gasteiger [33]	493	17	3
Yaffe [35]	517	4	1
Roche	1526	569	37

Tab. 15.3 Median values for molecular weight, polar surface area, chemical complexity, and rotatable bond count for literature datasets and for a set of Roche compounds

Dataset	MW	TPSA	Complexity	Rotatable bonds	n
Abraham [19]	135	20	162	1	662
Chen [24]	237	63	339	2	267
Yalkowsky [27]	181	27	223	2	430
Jurs [28]	167	43	222	2	399
Gasteiger [33]	175	26	216	2	493
Yaffe [35]	118	0	96	1	517
Roche	441	84	648	7	1526

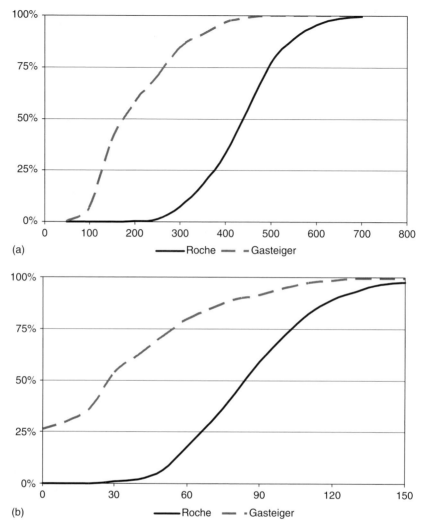

Fig. 15.4 (a) Cumulative molecular weight distributions of compounds in the Gasteiger [33] and Roche datasets. (b) Cumulative topological polar surface area ($Å^2$) distributions of compounds in the Gasteiger [33] and Roche datasets. (c) Cumulative chemical complexity distributions of compounds in the Gasteiger [33] and Roche datasets. (d) Cumulative rotatable bond count distributions of compounds in the Gasteiger [33] and Roche datasets.

SMARTS expression[a] [42]. From Table 15.3, it is clear that the compounds in our test set are significantly different in size, polar surface area and complexity, compared to the kinds of molecules that are often used in the generation of algorithms for the prediction of solubility.

---

[a] John Bradshaw's SMARTS expression for rotatable bond count is [!$(*#*)&!D1]-&!@[!$(*#*)&!D1]

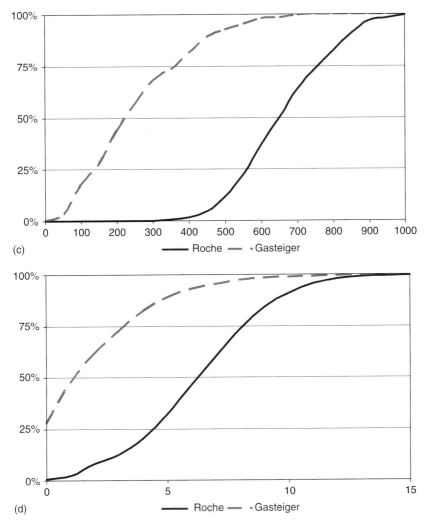

(c)

(d)

Fig. 15.4 (Continued).

The differences can also be seen in Figure 15.4, which compares cumulative distributions of these properties between the Gasteiger and Roche datasets. Many of the literature compounds are very simple, with low molecular weight, few polar atoms, and few functional groups. They have often been included in solubility datasets because they are well characterized and because accurate solubility data are available for them, rather than because they are druglike. The inclusion of many such simple compounds in a training set for a solubility prediction tool may focus the tool on an area of chemistry space that is not well populated with druglike molecules and may make the tool less useful for the prediction of the solubility

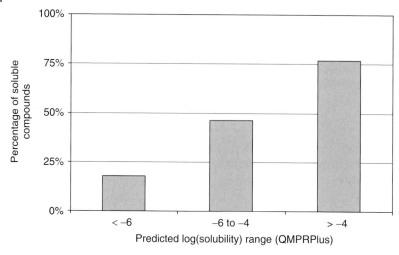

Fig. 15.5   Percentage of soluble compounds as a function of QMPRPlus solubility predictions.

for leadlike and druglike compounds. We think it is likely that a prediction tool trained with data from more druglike compounds would give better predictions for our compounds.

Even though we have seen that the solubility predictions are poor for individual compounds, there is still a question over whether such predictions could nevertheless be useful in library design. We believe that they can. Figure 15.5 shows that QMPRPlus gets the general trend correct: it shows that the proportion of soluble compounds increases with an increase in the solubility predicted by QMPRPlus (where a soluble compound is defined to have intrinsic solubility greater than 10^{-4} M).

15.4
Druglikeness

Many papers have been published over the last few years covering computational approaches to distinguishing between drugs and nondrugs. These have been reviewed [43–48]. For us, three questions stand out in considering the utility of such tools.

15.4.1
Are There Differences between Drugs and Nondrugs?

To be able to distinguish between drugs and nondrugs, we must accept that there is a fundamental difference between drugs and nondrugs. At one level, this is obviously true – we can easily distinguish between drugs and substances such as acetyl chloride, ethyl acetate, Styrofoam, paint, and lead. The latter substances fall into the group called "far-from-drugs" by Ajay and colleagues [49], and they can be distinguished from drugs and "close-to-drugs" using rules-based systems such as those described by Sadowski and Kubinyi [50], and other groups. Such rules are simple to encode and allow for the "rapid elimination of swill" or REOS [43].

A more difficult question is whether it is possible to separate drugs from substances without obvious flags. We have to ask if there really are fundamental differences between these two groups of compounds. There may not be – the list of drugs changes over time, with new chemical entities being approved, and marketed drugs being withdrawn occasionally as new safety issues are recognized. Proudfoot has pointed out that the pharmacopeia of the last fifty or a hundred years looks very different from the molecules that have been approved over the last decade [51], and Gershell and Atkins have listed several drugs that have been withdrawn recently [52]. In addition, drug approval is in part an accident of history – two compounds with similar structures, properties, and activities do not have an equal chance of becoming drugs: the first-found compound is more likely to be developed. As a result, druglikeness is clearly different from other fundamental properties of compounds such as melting point, lipophilicity, intrinsic solubility, or oral bioavailability, which do not change over time, and do not depend on when the compound is discovered. Using a larger pool of "druglike" compounds (e.g., WDI or MDDR compounds) rather than drugs in the training set overcomes issues related to changes in the pharmacopeia over time, but it introduces a new issue. By including "close-to-drugs" in the training set, we run the risk of making the prediction tool less accurate in the most useful function it could perform in preclinical research: distinguishing between truly druglike compounds and "close-to-drugs".

15.4.2
Is the Problem Tractable within a Single Program?

As anyone with experience in drug discovery and development will know, there are many hurdles on the road from target identification to drug approval. The following are some examples: the drug must be able to bind to a protein (usually a receptor or an enzyme) with sufficient potency; the drug must have acceptable solubility; often, the drug should have good oral bioavailability; the drug should have an appropriate half-life in humans, which will often be commensurate with once-a-day dosing; the drug should not inhibit the major human isoforms of the cytochrome P450 enzymes; the drug should have a good safety profile; and, in particular, should not cause problems in the areas of genotoxicity (e.g., MNT and Ames tests), phototoxicity, phospholipidosis, and cardiac safety (e.g., at the hERG channel).

A reliable druglikeness predictor would give high prediction scores only to compounds that have satisfactory properties based on all of these criteria, or in other words, there would be few false positives. There has been considerable effort expended over the last 10 to 20 years in modeling individual components of this process, including solubility [18–36], ADME properties [53–70], and toxicities [71–88]. Individually, each of these predictions has false positives and false negatives, so it is difficult to expect that a computational model, which implicitly includes all of these factors, will give very reliable predictions.

15.4.3
Do We Have a Training Set that Will Allow Us to Address the Issue?

Most prediction tools for druglikeness use the MDL® Available Chemicals Directory (ACD) [89] as a list of nondrugs and one of three databases as a list of drugs: MDL® Drug Data Report (MDDR) [89], MDL® Comprehensive Medicinal Chemistry (CMC) [89], or Derwent World Drug Index (WDI) [90].

ACD is a collection of about 250 000 commercially available chemicals, and includes a very broad range of substances including solvents, inorganic materials, polymers, peptides, proteins, pesticides, drugs, and organic reagents. Generally, a rules-based system such as REOS is used to remove the "far-from-drugs" group, yielding a set of approximately 175 000 [50] "nondrugs". We have found that this set contains about 900 drugs (where drugs are defined as compounds appearing in the FDA Electronic Orange Book [91] or Goodman and Gilman's *The Pharmacological Basis of Therapeutics* [92]). Most publications in this area further prune the filtered ACD list to remove compounds that are also found in MDDR or WDI, but as Oprea points out [93], we should be aware that ACD *per se* is not a drug-free list.

MDDR is a database derived principally from the patent literature, journals, and meetings. It contains the structures of approximately 130 000 compounds of which approximately 1000 are launched drugs. WDI is a Derwent database of approximately 73 000 marketed drugs and development compounds, drawn from journals, scientific meetings, and approved name lists. CMC is a list of about 8400 compounds taken from the Drug Compendium in Pergamon's Comprehensive Medicinal Chemistry [94] and also from the United States Approved Names list.

By comparing the proportion of drugs in ACD and MDDR, we can see that the assumption that ACD is a set of nondrugs and that MDDR is a set of drugs is not completely accurate: ACD contains quite a few drugs and MDDR contains many nondrugs. In addition, we should bear in mind that many of the compounds in ACD are building blocks with lower complexity than many drugs, and so we should be aware of the possibility that using ACD compounds as a nondruglike set may introduce a bias toward higher druglike predictions for compounds with higher molecular weight. The differences in complexity between ACD compounds and various sets of druglike compounds have been shown very clearly [93].

15.4.4
Approaches to the Prediction of Druglikeness

Several different approaches have been used to generate druglikeness prediction tools. The simplest approach is to identify a range of descriptor values for druglike compounds. The best known example of this approach is the Lipinski rules [16], which predict poor absorption for compounds exceeding cutoffs for molecular weight, lipophilicity, and counts of hydrogen bond donors and acceptors. This approach has the beauty of simplicity, which makes it easy to implement and understand, and the results are generally in accord with the experiences of most medicinal chemists, although as Lipinski points out, there are exceptions. Oprea has pointed out that the Lipinski rules do not distinguish druglike from nondruglike compounds: 90% of ACD complies with the rule-of-five [93]. However, the Lipinski rules are still of great value in highlighting the region of chemistry space where compounds with acceptable absorption properties are more likely to be found. Several other publications have outlined descriptor ranges for druglike compounds [93, 95–103].

Another approach is to generate an artificial neural network [49, 50, 104, 105] or decision tree [106] to distinguish between lists of "drugs" and "nondrugs". Neural

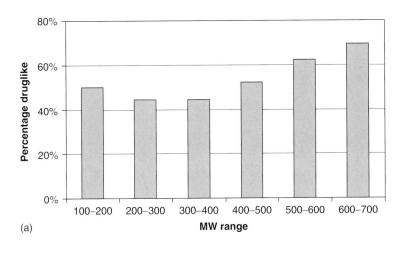

(a)

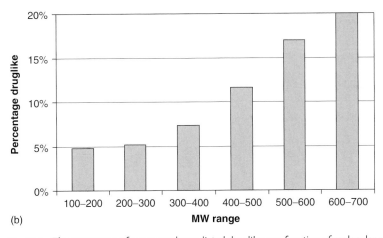

(b)

Fig. 15.6 The percentage of compounds predicted druglike as a function of molecular weight.

networks suffer from a disadvantage in terms of the interpretability of the model: it is not possible to examine the "rules" that lead to one compound receiving a high score and another a low score. This is not a problem in filtering (e.g., in virtual screening, or in targeting a list of compounds for acquisition), but it can be a problem in library design. The Sadowski–Kubinyi paper clearly illustrates that overtraining is not an issue in this neural network (NN) model; other methods (e.g., PLS) show similar trends. Other approaches include the prediction of druglikeness based on the calculated biological activities of a compound [107] or based on the intuition of medicinal chemists [108].

At Roche, we have implemented a neural network to predict druglikeness using the dataset kindly provided by Kubinyi and Sadowski [50]. The complexity of the system means that few general rules can be deduced – we have only observed that halogens and nitriles are strongly disfavored and that there is a preference for compounds with higher molecular weights. In an analysis of a collection of 525 000 commercially available screening compounds from a variety of suppliers, we found that the percentage of compounds predicted to be druglike (using Sadowski and Kubinyi's cutoff of 0.3 to define druglike compounds) is reasonably constant (range: 44–52%) in the molecular weight range 200 to 500 but rises to 70% for compounds in the molecular weight range 600 to 700 (see Figure 15.6a). This means that compounds with higher molecular weights are predicted to have a higher chance of being druglike, which is not expected from the Lipinski rules, which suggest that such compounds are more likely to have poorer absorption properties. The unexpectedly higher percentage of heavier compounds predicted to be druglike is even more pronounced when a higher cutoff of the predicted value is used (see Figure 15.6b).

15.5
Frequent Hitters

Some compounds appear again and again in hit lists, showing activity against a wide range of targets. Some of these compounds are promiscuous binders that interact with a number of different proteins; some are reactive; some interfere with the assay readout, either because they are fluorescent or because they bind to one of the proteins used in the assay (e.g., many carboxylic acids exhibit weak binding to nonspecific antiphosphoprotein antibodies, and biotin derivatives bind to streptavidin); and some form large aggregates that have enzyme inhibitory activity. Frequent hitters have been recognized by several groups [109–112], and a prediction tool has been developed to identify them [113].

Figure 15.7 shows an analysis of the selectivity of 2094 recent hits in high-throughput screens at Roche. Of these, almost half were specific for the screen in which they were identified as a hit – there were no other screens where these compounds had been titrated in a dose-response assay. However, 8% of the compounds had been tested in dose-response assays for more than five projects. These compounds are frequent hitters, and the most commonly occurring substructures are shown in Figure 15.8. Several of these compounds either are electrophilic, or can give electrophilic species on oxidation. The nitrobenzoxadiazoles are known fluorescent compounds. The oxindoles

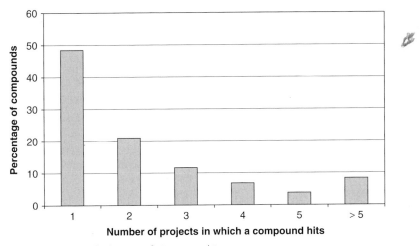

Fig. 15.7   Analysis of selectivity of 2094 recent hits.

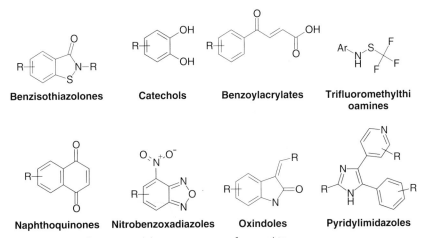

Fig. 15.8   Commonly occurring substructures among frequent hitters.

and pyridylimidazoles are an interesting class. Although examples of each of these classes are potent inhibitors of a variety of kinases, medicinal chemists appear reluctant to abandon them, preferring to use the term "privileged structure" borrowed from the GPCR world [114] rather than calling them frequent hitters.

15.6
Identification of a Lead Series

The goal of the hit-to-lead process is to develop a hit or series of hits to the point where significant resources can be committed for lead optimization. Occasionally, a hit from high-throughput screening will have the properties required, but it is more common for compounds to go through a hit-to-lead process before there is confidence that a reasonable series has been identified.

The definition of a lead series varies from company to company, from department to department within a company, and even from project to project. Figure 15.9 shows some of the generic properties that are expected from most lead series. Lead compounds should show reproducible activity in both a primary and secondary assay, preferably with a different readout. The activity will generally be in the low micromolar range, although for some targets, much more potent activity would be required. The requirement for plausible SAR in the series gives us confidence that we understand which areas of the molecule can be modified in the lead optimization phase. The requirement for selectivity is project dependent: sometimes there is a closely related protein where activity would be undesirable. If we cannot show in the hit-to-lead phase that we have a reasonable expectation of achieving selectivity against such related targets, then the series would not proceed.

The chemical attributes of a lead series include knowing the chemical structure and purity of the hit(s). There are often cases where the activity of a screening sample cannot be confirmed when the compound is resynthesized and this can sometimes be traced to impurities in the original sample, and there are cases where the structure of the hit was not assigned correctly. In order to justify the resources required for lead optimization, it is important to know the structure of the lead. Synthetic tractability can

Activity
- Reproducible concentration-dependent activity
- Active in secondary assay/cell-based assay
- Generally, IC50 < 10 μM
- Plausible SAR in the series
- Binding kinetics analyzed
- Selectivity assessed

Chemistry
- Known structure and purity
- Stability assessed
- Synthetically tractable
- Series is patentable
- Chemistry space for optimization

ADME/Tox
- Solubility measured
- LogD measured
- Permeability measured
- Protein binding measured or predicted
- Metabolic liabilities measured or predicted
- Pharmacokinetics assessed
- Toxicity alerts predicted
- hERG liability assessed

Fig. 15.9 General characteristics of a lead series.

be an issue for hits with great structural complexity, for example, some hits derived from natural products. In some cases, it can be difficult to resynthesize the hits and to explore structure-activity relationships. Compounds amenable to high-throughput chemistry approaches are desirable. Clearly, it is important to have opportunities to secure patent protection before investing significantly in a series, and there must be reasonable chemistry space for optimization.

The ADME and toxicity requirements at the lead series stage are project- and therapeutic area dependent. For example, a compound intended for oral delivery will have pharmacokinetic requirements that are different from those of a compound intended for intravenous administration, and a drug that will be taken every day for the lifetime of the patient will have toxicity concerns that are different from those of a drug that will only be taken once. Generally, at the lead series stage, our goal is to have identified the ADMET liabilities of the series within the requirements of the project, and to have demonstrated a reasonable likelihood that any problem(s) can be resolved. If, for example, solubility is identified as a limitation of the series, and every modification to the compound to improve solubility results in an unacceptable loss of potency, then the series will be rejected. It is increasingly necessary to provide some evidence of the potential for good *in vivo* pharmacokinetic properties for a lead series.

15.7
The Hit-to-lead Process

Our approach to the hit-to-lead process is shown in Figure 15.10. It is shown as a linear process, but several of the steps may require iterations.

The process starts with a high-throughput screening campaign, which identifies a set of primary hits that have generally been assayed once at a single concentration. These compounds are then tested in triplicate, again at a single concentration. Confirmed hits, compounds that test positive in at least two of the three confirmation assays, are then titrated, and compounds with activity in the desired range are validated hits. Depending on the project, selectivity assays may be required before the declaration of validated hits.

Validated hits then enter hit-to-lead activities. The first step is data analysis and prioritization. Typically, the hit rate in a high-throughput screen is of the order of 0.1%, although higher hits rates of the order of 1% are sometimes encountered. If the screening collection is of the order of 500 000 compounds, then the primary hit rate will generally be in the range of 500 to 5000 compounds. Given typical rates of attrition in the confirmation and validation stages, screening campaigns can often deliver several hundred validated hits, and the first task is to prioritize among these.

15.7.1
Prioritization of Hits

The first step in prioritization is to ensure that the validated hits do not contain any undesired substructures. This can be achieved using a set of rules such as

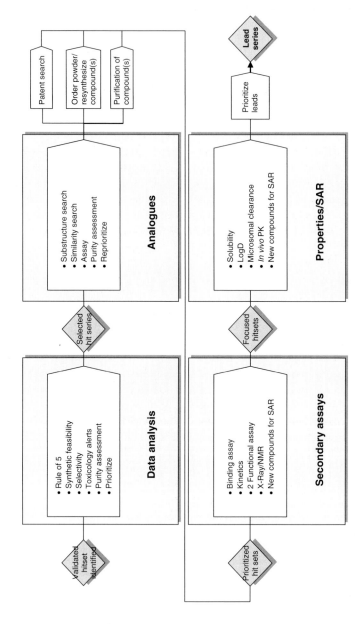

Fig. 15.10 Hit-to-lead workflow.

the REOS system used at Vertex [43]. The next step is to assemble data for broad consultation within the medicinal chemistry community. These data include the structure and activity of the compounds, their selectivity (dose-response data from other projects), purity data, predicted toxicity flags from DEREK [86], and molecular weight, CLOGP, and counts of hydrogen bond donors and acceptors for comparison with the Lipinski rules [16]. The data are assembled in a Microsoft Excel spreadsheet and then made available to medicinal chemists through the intranet. The chemists can vote for or against compounds and make comments about their suitability for further evaluation. We have found this process very valuable in accessing a huge reservoir of medicinal chemistry expertise accumulated over decades of experience. It is also valuable in helping to educate newer medicinal chemists and in initiating discussions.

The votes are tallied and presented at a meeting where a number of compounds are prioritized. We have noted, as have Lipinksi [16] and Takaoka and coworkers [108], that chemists fall into very different schools of thought. One group is very focused on prioritizing the compounds and pays close attention to flags; the other group sees potential in every compound and can point out ways in which problems can be overcome. One other observation is that chemists tend to use relative rather than absolute values in judging compounds: we judge compounds relative to the other hits on the list, rather than using consistent standards. We have to be careful to guard against further evaluation of compounds of low quality just because there was nothing better identified in the screen.

We have analyzed a dataset of the voting results for 1412 validated hits over 13 screens. We normalized the votes to take account of the number of people participating (range 13–20), and scaled the votes to the range 0 to 1, where 0 indicates that all voters voted against a compound and 1 indicates that all voters voted in favor. The results are shown in Figure 15.11.

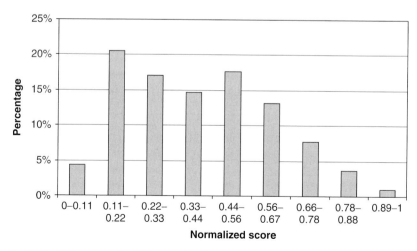

Fig. 15.11 Voting scores for 1412 screening hits.

Overall, we have found that more compounds are rejected than accepted by medicinal chemists. If we split the compounds into three groups by popularity (using cutoffs in the normalized score of 0.33 and 0.66), we find that the compounds with high, medium, and low scores account for 13, 45, and 42% of the dataset, respectively.

Table 15.4 is an analysis of the molecular properties of the three groups of compounds, patterned after the table of Blake [103]. This table and Figure 15.12 show that compounds with high values of molecular weight, lipophilicity, polar surface area, and rotatable bonds are efficiently removed by our prioritization process. Interestingly, if we look at the mean number of some functional groups in the dataset, we see that compounds with alkylamines are somewhat preferred for, while compounds containing the carboxylic acid functionality are strongly disfavored.

Tab. 15.4 Calculated molecular properties of hits (after Blake [103])

Property[a]	Top-ranked group	Middle-ranked group	Bottom-ranked group
Entries	180	637	595
KOWCLOGP	3.65	3.95	3.62
% KOW_CLOGP > 5	18.3	28.1	27.6
% Rule-of-five violations	19.4	30.9	39.8
TPSA	64.4	71.0	77.6
% PSA > 200	0	0	4.7
MW	338	368	374
% MW > 500	1.1	9.9	19.2
Rotatable bonds	4	5	5
% Rotatable bonds > 10	5.0	10.0	13.4
HBA	5	5	5
HBD	1	1	1
Alkyl amines (mean)	0.16	0.15	0.08
Carboxylates (mean)	0.03	0.13	0.16
Halogens (mean)	0.63	0.84	0.73

a)   The average value of each descriptor is the median value unless otherwise stated.

Tab. 15.5 Factors influencing chemists' prioritization of screening hits

More important factors	Less important factors
CLOGP	CLOGP
Molecular weight	Molecular weight
Number of active analogs	Predicted druglikeness
Patentability	Predicted solubility
Potency	Predicted toxicity
Selectivity	Selectivity
Synthetic tractability	
Toxicity	

Table 15.5 shows the factors that influence the prioritization of hits by medicinal chemists. Three factors (molecular weight, CLOGP, and selectivity) appear on both sides of the table, as both more important and less important factors in the prioritization. At first, this appears contradictory but in fact it reflects the different schools of thought mentioned above, where some chemists use these factors to filter the hits, while others view problems in these areas as something that can be fixed at a later time. In our experience, most chemists do filter out compounds with extreme values of molecular weight and hydrophobicity.

Two of the other factors that are important to many chemists are patentability and the number of active compounds in the set of validated hits. Current chemoinformatics tools are not well suited to predict these factors.

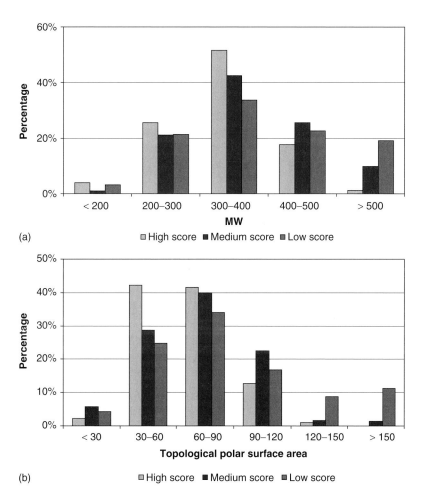

Fig. 15.12 (a) Molecular weight distribution of screening hits. (b) Topological polar surface area [40] distribution of screening hits. (c) Barone complexity [41] distribution of screening hits. (d) Bradshaw rotatable bond [42] distribution of screening hits.

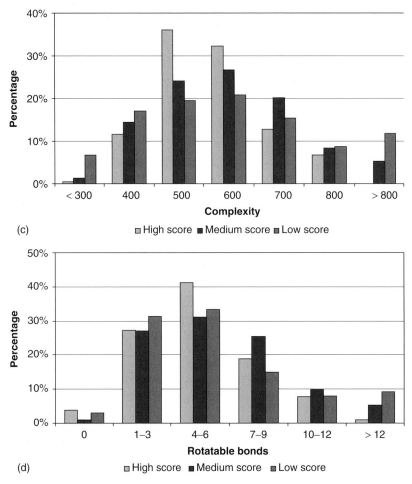

(c)

(d)

Fig. 15.12 *(Continued)*.

15.7.2
Identification of Analogs

The next step is to identify additional analogs of the prioritized hits for screening. These compounds may be available externally. There are many suppliers of compounds for screening, and ChemNavigator, Inc. [115], ChemBridge Corporation [116], Specs [117], and Maybridge plc [118] have convenient websites at which suitable analogs can be identified by substructure and similarity searching. We also search for compounds available internally. We have found that it is useful to rescreen compounds that tested negative during the high-throughput screen because there are a number of false negatives in every screen. Furthermore, there are often additional compounds available that were not screened in the primary screen. These compounds may have been acquired after the last distribution of screening master plates, or they may be available in quantities too small to allow for global distribution for high-throughput screening. Analogs are screened, and the results often give very valuable SAR information, allowing us to make preliminary judgments about the likely size of chemistry space available for optimization. With the analog screening information in hand, we can then reprioritize the remaining hit series.

15.7.3
Additional Assays

Secondary assays depend on the project. Where the primary screen was a cell-based assay, the secondary assay may be a radioligand competition binding assay. In other cases, such as where the primary screen was a biochemical assay, the secondary assay may be a cellular assay, and may be functional or mechanistic. One of the issues that may arise at this stage is that compounds with reasonable activity in the primary assay may not show activity in the secondary assay. There can be a number of reasons for this, including insufficient potency, inability of the compound to get into cells, or a higher intracellular concentration of the natural ligand (e.g., ATP) if the inhibitor is a competitive inhibitor. It is often necessary at this stage to prepare additional compounds in the series to get compounds of sufficient potency and/or permeability so that cellular activity can be demonstrated.

Several high-throughput *in vitro* ADME assays help to triage compounds. Assessments of solubility, partition coefficient, permeability (in either a Caco2 assay [119] or in the PAMPA format [120]), and an assessment of metabolism in human liver microsomes give a good first glimpse of possible issues. If problems are identified, then further analogs can be prepared to see if the problem is tractable or if the series should be dropped.

Before the lead series milestone, selectivity is generally determined against a panel of up to 50 related proteins in the case of kinase and GPCR targets. This is done either in-house or with the help of an external partner (e.g., MDS Panlabs [121] or Cerep [122]), and it gives confidence that the series is either selective or that it will be possible to build selectivity into the series during the lead optimization process. Also, before the lead series milestone, the hERG liability of the series is assessed, again either in-house or with the help of an external partner (e.g., Zenas Technologies L.L.C. [123]).

Throughout the period of the hit-to-lead process, additional analogs may be prepared to improve understanding of the SAR of the series to give an impression of whether chemistry space in the series is limited and to show that analogs can be prepared in a timely manner.

If hits with good properties were selected at the outset, we can hope to arrive at the end of the hit-to-lead process with a series of compounds meeting our lead series criteria, where issues have been flagged and appear tractable. These compounds then provide the launching point for the next phase, lead optimization.

15.8
Leads from Libraries: General Principles, Practical Considerations

Developments in high-throughput chemistry methodology and automated synthesis and purification technology have created an environment in which it is possible to seriously consider small-molecule libraries as a source of leads for drug discovery. The development of chemistry methods for high-throughput synthesis has progressed regularly during the last ten years. A survey of high-throughput chemistry (HTC) methods has been serially published by Dolle [124–127]. Developing concurrently are chemoinformatic tools and processes for guiding library designs to the synthesis of molecules useful for drug discovery. The early enthusiasm for creating large libraries of compounds without regard to compound purity or design quality has given way to more stringent expectations for this approach in drug discovery [128]. Instead of focusing on libraries of larger numbers, many efforts are aimed at increasing the medicinal chemistry focus and expertise encoded in libraries toward specific targets and creating this focus with a smaller number of compounds [129]. Further, the application of chemoinformatic tools to create well designed, information-rich arrays of small molecules directed toward specific targets has added value to the molecules that are found to be active. Many of these strategies have been clearly summarized in a review of computational design strategies for combinatorial libraries [130].

The discovery of lead compounds by screening of small-molecule libraries has been summarized in two reviews [131, 132]. Publications cited in these references collectively make the case that lead finding through HTS of corporate libraries is one of several successful approaches to delivering clinical candidates. The use of parallel and combinatorial chemistry methods is often described as a technique to enhance the potency, selectivity, and druglike properties of initial HTS hits. In most of these publications, however, the library design practices are not described in detail. It is likely that R groups (also referred to as diversity reagents) are often selected both on the basis of the availability of reagents and on assumptions about SAR around the central template.

When designing small-molecule libraries, we have found that several general aspects must be treated, often separately, with a variety of chemoinformatic tools. Specifically, the selection of templates or cores and the R groups are the result of comparison of multiple options. Given the virtually infinite size of chemistry space, it is not practical to consider all possible options and then select the single best one. Rather, efficiency in lead discovery requires the convenience to iteratively compare multiple calculated properties among currently feasible options among available templates and reagents. Also, visualization of these calculated properties in a summary format in order to facilitate selection of R groups is necessary. The most readily adopted chemoinformatic tools are simple to use by nonexperts, provide rapid and easily interpretable visualizations, and contrast good options with less attractive options. Web-based tools are particularly popular.

Much has been written and reviewed on design practices for libraries, particularly for larger libraries [133–142]. Larger libraries (greater than 5000 members) require substantial planning, materials, and time for execution and delivery. Often, such libraries are created with the intent of adding to corporate screening collections without targeting specific targets. To this end, the design concepts of druglikeness and diversity are centrally important to exemplify new template ideas through compound library synthesis. Given the time and effort necessary for such libraries, computational methods have been published to optimize diversity and druglike properties while minimizing the number of reagents necessary. For example, Pickett and coworkers described a method called MOGA for efficient optimization library size and configuration while addressing issues of diversity, reagent cost, and druglike properties [143]. It was shown that increasing the number of objectives for optimization results in a corresponding increase in the number of solutions for reagent selection to the extent that no single solution was apparent.

While sophisticated chemoinformatic tools for library design provide a rationale for reagent selection, the quality of library design is ultimately determined by aspects that are beyond the control of chemoinformatic processes: template designs relevant to drug discovery, availability of diversity reagents, and feasibility of high-throughput chemistry methods. The number of readily available reagents is limited. Only a fraction of the large number of reagents in ACD are available in multigram quantities at costs of US$100 per gram or less. Often arraying diverse functional groups around a template by means of a particular chemistry will be limited by chemical compatibility. For example, one will not find isocyanate reagents containing a free primary amine. When it becomes necessary to include such functional group combinations, it is necessary to develop protecting group schemes, which add cost and complexity to the library design and synthesis process. Thus, for many commercially available reagents, the diversity of readily available reagents appropriate for inclusion into a library design at a reasonable cost is often quite limited.

Given these limitations, it is not surprising that many organizations have coupled building block synthesis efforts to their library building efforts. To guide the basic design of building blocks for use in combinatorial chemistry libraries, there are reports of analyzing the basic structural elements found in drugs [144, 145] and isosteric replacement programs [146]. At this level of consideration, the number of theoretically possible solutions to the building blocks again becomes large and beyond tractability. As a result, practical solutions for building block and library designs usually incorporate some aspect of biological targeting and expertise in medicinal chemistry.

When following up on a potential lead identified from HTS, smaller libraries are often preferred. Such libraries must be made, purified, and assayed rapidly to complete the hit-to-lead process. Often, questions concerning preliminary SAR are encoded into these hit-to-lead libraries. We have found that chemoinformatic analysis of these smaller, more focused library designs, particularly those synthesized in a hit-to-lead exploration, must clearly convey options for selecting among a limited number of compounds for rapid synthesis and assay. Design for druglikeness or leadlikeness is clearly and rigorously guided by the benchmarks of the hit-to-lead process. The design process must also be convenient such that design and analysis are routine. Finally, we have found that the quality of library designs is fundamentally determined not only by careful selection of diversity reagents according to chemoinformatic analysis but also by the quality and druglikeness of the core idea. Consideration of information about the

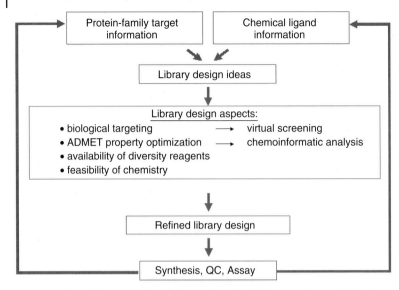

Fig. 15.13 General process outline and considerations for design of targeted libraries.

biological target can add value to the design process. For example, an understanding of the site of activity can guide the molecule design emphasis even at the early stage of lead generation. An outline of the general library design process is diagrammed in Figure 15.13.

We have established several processes for collecting and archiving lead generation library proposals from Roche chemists at all sites. This leverages the collective medicinal chemistry thinking about quality lead generation ideas. A database of these proposals serves as a repository of chemistry-ready templates for creating virtual libraries. Analysis of these libraries by various chemoinformatic tools highlights lead generation designs that are more druglike than ideas proposed on the basis of reported combinatorial chemistry.

15.9
Druglikeness in Small-molecule Libraries

Druglikeness has been extensively reviewed, both in Section 15.4 and in the references cited therein. The first chemoinformatic analysis required is to choose R groups that allow us to stay within the boundaries of druglikeness as most clearly stated by the Lipinski rule-of-five [16]. We have taken the approach to add upon these basic rules other information derived from chemoinformatic calculations to distinguish an optimal set of diversity reagents among those available for synthesis. Additional calculated information includes rotatable bond count and predictions of solubility, polar surface area, blood-brain barrier penetrability, hERG activity, and protein binding (*vide infra*).

The use of *in silico* screening to add value to the discovery of druglike molecules has been well reviewed [56]. Prediction of ADME properties for virtual library compounds is of particular interest. Reasons for attrition of drugs in clinical development have been analyzed by Kennedy, who highlighted inappropriate ADMET properties as responsible for 40% of the failures in drug development [147]. Kennedy's analysis has been reconsidered by excluding anti-infectives from this analysis, which seemed to overemphasize drug failures due to poor ADME properties. This time, only 7% of failures in development were due to poor ADME properties [148]. Although one may be tempted to conclude that ADME properties are therefore not a potential problem, these numbers do not take into account the unreported number of molecules that cannot be advanced beyond *in vitro* profiling because of poor pharmacokinetic behavior. We have observed many potent molecules that cannot be progressed to the stage of clinical candidate selection because of poor pharmacokinetic properties. Often, ADME assays have clearly indicated the problems that were encountered in *in vivo* experiments. Therefore, we see a great potential value for accurate ADME prediction tools and consider their application as very useful in the earliest stages of lead identification.

Calculation of ADME predictions is now routine and often high throughput. However, unlike many published studies in which calculated properties are validated with experimental data for a diverse collection of molecules, for virtual small-molecule libraries, there are usually no physical data to validate the predictions. Often, the molecules that are the subject of calculation are dissimilar to those molecules used to develop prediction tools. As a result, one is usually looking for trends in the prediction as a function of the selection of specific diversity reagents around a common core or scaffold. Thus, it is important to consider predicted molecular properties with care and to interpret the results with the proper level of expectation.

15.10
Data Reduction and Viewing for Virtual Library Design

We have found that the manner in which chemoinformatic information is displayed is important for interpretation and decision making about library designs. We find that property displays with respect to a particular diversity reagent are superior to that of the whole library; it is easier to improve a library design by removal of a specific diversity reagent. The advantage of designing libraries based on calculated properties of potential products as opposed to the reagents has been commented upon [149, 150]. In the analysis process developed here, the calculated properties of the products can be separated and displayed for each R group. In this way, one can choose to include an R group for its impact on the virtual products' calculated properties. Many library design systems do not facilitate the ready viewing of structures by chemists. We have found that design software that provides a simple list without showing the structures of the R groups is less compelling and less often employed than software that facilitates a medicinal chemist's interaction and application of experience and insight.

It has been proposed that a sparse matrix provides superior coverage to the selection of apparently optimal diversity reagents. However, the final collection of molecules that comprise a library at assay may be quite different from that designed at the outset. This is due to failures of the chemistry and/or purification of final products. Therefore, one must include some redundancy into the reagent set; further, the design must take into account the overall impact of a reagent on the aggregate of molecules in which it will be represented. With these considerations in mind, it would seem that an acceptable library design could be approximated based on construction of an optimal set of diversity reagents. To this end, we have developed a web-based reagent sorting and viewing system (Figure 15.14) [151]. The calculated molecular properties are displayed as histograms associated with each diversity reagent. In this way, it is possible to develop a sense of the best diversity reagents to select among a given set based on a series of calculated properties. We find this a useful complement to many of the reported design tools useful for designing libraries.

15.11
Druglikeness

The challenges of interpreting druglikeness calculations for single molecules are treated in detail in Section 15.4.1. We have tried to use optimization of predicted druglikeness scores (Section 15.4) as a factor in library design, but we have found that poor scores often result from the presence of a substructure that is not present in the drug training set. For example, the presence of a trifluoromethyl group or a dichloroaryl group lowers the druglikeness score substantially (see Figure 15.15). Although these fragments are not frequently found in drugs, they are present in useful hits and leads. A lead optimization effort would likely replace such fragments with more acceptable isosteres. As a result, although we use druglikeness scoring as a means to review in summary format the diversity reagents that we are selecting, we use other chemoinformatic metrics to make critical decisions for library design.

15.12
Complexity and Andrews' Binding Energy

Perhaps a useful variant of the druglikeness calculation for library designs is comparison of the Andrews' binding energy [152] and the Barone complexity scores [41]. Andrews' binding energy calculations reflect the maximum binding potential of molecules in an ideal receptor setting. When interpreting Andrews' binding energy, it should be noted that Andrews' binding energy calculations contain a significant bias toward small and charged molecules, as discussed elsewhere in this book [153]. It is of interest to compare Andrews' binding energy for a set of validated hits for which the binding affinities (lower micromolar) are known approximately (Figure 15.16). In this comparison, molecular complexity is also considered. In reporting another method for calculating molecular complexity, Hann provides evidence that simpler compounds are more likely to be hits from a screening process [99]. Therefore, a better lead may be one that is simple yet still has substantial potency. It has been noted that lead optimization often results in the

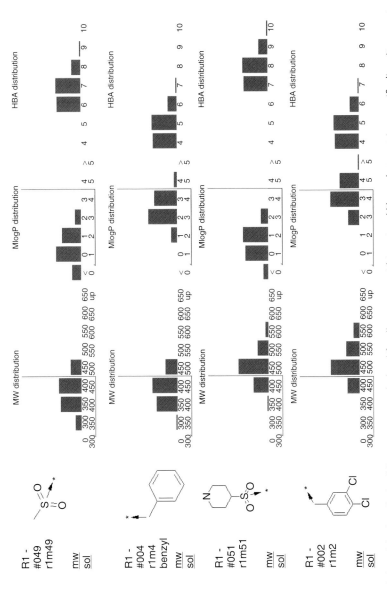

Fig. 15.14 Web-based display of rule-of-five parameters calculated for all structures within a virtual library that contain a specific diversity reagent. The complete virtual library is composed of 2400 members derived from an array of 40 × 60 reagents.

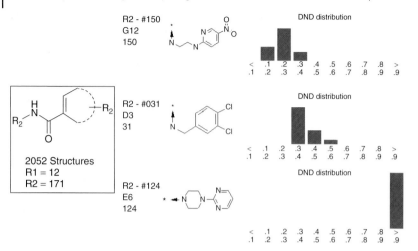

Fig. 15.15 Selected examples of the distribution of predicted druglikeness scores (DND: drug neural network) for a library of 2052 compounds. As described in Section 15.4, druglikeness scoring was performed with a neural network to predict druglikeness using the dataset kindly provided by Kubinyi and Sadowski. Scores close to 1.0 predict druglikeness; scores approaching zero predict nondruglikeness.

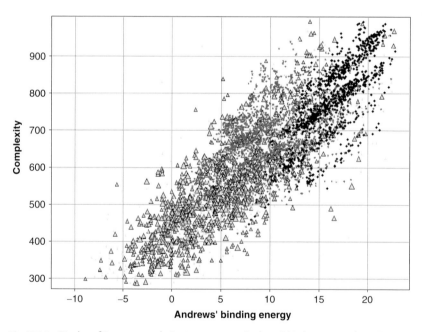

Fig. 15.16 Display of Barone complexity scores versus Andrews' binding energy for 1400 Roche validated HTS hits (triangles) and for two library designs of 2400 (circles) and 1100 (diamonds) virtual structures. The size of the triangles correlates with the rated attractiveness of a group of medicinal chemists as described in Section 15.7.1.

evolution of more complex molecules in order to enhance potency and selectivity. High-throughput chemistry and, particularly, combinatorial chemistry synthesis based on multicomponent condensation reactions tend to be overly complex and functionalized. Hits that are appealing to chemists as candidates for further optimization are often simple structures without excessive functionalizations; this is due to the likelihood that complexity will increase during lead optimization [154]. A convenient way to gauge the potential attractiveness of a virtual library design is to compare molecular complexity scores of the virtual compounds to those calculated for hits that have been rated for their potential lead attractiveness by medicinal chemists.

Figure 15.16 shows the display of calculated complexity versus Andrews' binding energy for a virtual library in comparison to a selection of Roche validated hits. From this display, one concludes that the virtual library design represented by circles is similar to some of the more complex structures in the HTS hit set; however, the synthesized library design represented by diamonds is composed mostly of structures more complex than the molecules in the validated hit set. The simpler virtual library design is an expansion of a validated lead series and is more likely to provide useful hits. Although hits have come from the more complex design, they have been difficult to progress to *bona fide* leads. Library designs that provide excessively complex hits will have less impact in lead generation than those that provide simpler structures. Conversely, the lower ranges of Andrews' binding energy scoring are also potentially useful for avoiding the synthesis of uncharged compounds, which lack sufficient functionality for any high-affinity binding interaction.

15.13
Solubility

The acceptable solubility of structures generated in a HTC library generation approach is a frequent problem [155, 156]. The poor correlation of various solubility predictions has been discussed in Section 15.3. It would seem that increased validation of the prediction and vetting of the experimentally measured datasets is a method for improvement. However, we have found when analyzing virtual libraries that the trends in predicted solubility seem generally within expectations and are a useful method to guide diversity reagent selection toward more soluble molecules (see Figure 15.17). Where prediction has indicated potential for insolubility and such compounds were synthesized, we often encountered problems with such library designs. We find that solubility prediction routines are most useful for eliminating those R groups that are predicted to create highly insoluble compounds in combination with any other reagents.

15.14
Polar Surface Area

Good library designs include considerations for creating molecules that are likely to have good passive cellular permeability and intestinal absorption [157]. Polar surface

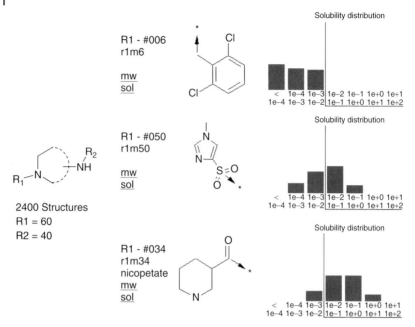

Fig. 15.17 Selected examples of the distribution of predicted solubilities for diversity reagents within a 2400-member library built around a common core. Solubilities were calculated using the simulations Plus QMPRPlus solubility prediction tools and expressed in mg mL^{-1}. Significant number of aqueous solubility predictions below 10 μg mL^{-1} was considered sufficient criteria for deselection of that diversity reagent.

area (PSA) is a metric that has been correlated to intestinal absorption [158], Caco-2 monolayer penetration [159, 160] and blood-brain barrier passage [161]. In these studies, 140 Å^2 has been designated as an upper limit beyond which most studied molecules do not have good passive cellular permeability [162]. Compounds with PSA scores less than 140 Å^2 had generally favorable human fractional absorption. For molecules whose site of action is expected to be the CNS, PSA scores below 100 Å^2 predominate. In another study, two sets of well-absorbed and poorly absorbed compounds were compared in terms of AlogP [49] and PSA scores. In this comparison, [163] most of the well-absorbed molecules cluster below approximately 120 Å^2. We have used the Ertl method for fast PSA scoring [40] in analyzing our diversity reagent selection, particularly in the case of libraries intended for CNS targets (see Figure 15.18).

15.15
Number of Rotatable Bonds

Veber and coauthors have analyzed more than 1000 drug candidates and have shown that compounds that have fewer rotatable bonds and tPSA scores less than 140 Å^2

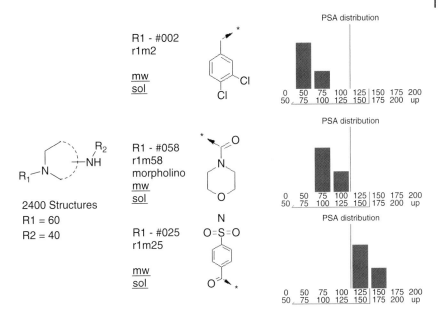

Fig. 15.18 Selected examples of the distribution of predicted PSA scores for diversity reagents within a 2400-member library built around a common core. Topological PSA scores were calculated according to the method published by Ertl and expressed in Å². Significant number of predicted scores greater than 140 Å² is sufficient criteria for deselection of that diversity reagent.

tend to have greater oral bioavailability [101]. It was shown that when the molecular weight is above 450, variance in oral bioavailability could be correlated with the variance in the number of rotatable bonds. In library designs, the number of rotatable bonds is a central feature of the template and functionality for linking diversity reagents. However, the quality of the designs can be further optimized by the selection of diversity reagents that do not add excessively to the total count of rotatable bonds. Although an apparently simple metric to calculate, different algorithms count rotatable bonds in different ways. It is necessary to standardize and benchmark one's rule set before eliminating diversity reagents that contribute to an excessive number of rotatable bonds [164].

15.16
hERG Channel Binding

The ability of a molecule to bind to the potassium channel, hERG (human ether-a-go-go related gene), is a serious pharmacological concern and may lead to the failure to progress an otherwise active and druglike molecule during the lead optimization process. Cardiac QT interval prolongation has been associated to some extent with

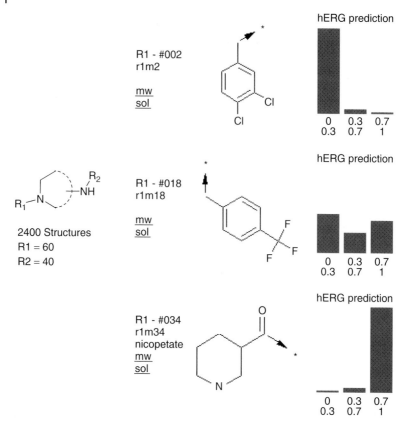

Fig. 15.19 Examples of the distribution of predicted hERG binding likelihood for diversity reagents within a 2400-member library built around a common core. Binding to the hERG channel was predicted according to the method reported by Roche scientists [77]. Predicted scores of less than 0.3 are flagged as potential for concern. Scores above 0.7 are not flagged as potential hERG binders. There is no basis for interpretation for scores between 0.7 and 0.3.

hERG channel binding. There are reports of the withdrawal of marketed drugs that were found to have hERG-related cardiac safety issues [52]. Predictions of hERG channel binding [77] are also considered at times when designing libraries, particularly in the case of following up on molecules for which there has been an issue with these properties. We have found many results in terms of library design that are difficult to interpret. This is likely due to the fact that the hERG channel training sets are small and composed of molecules that are often dissimilar to the structures in the virtual library. As larger hERG datasets for diverse chemical structures are developed, we anticipate enhanced reliability of predictions. At this time, it is of interest to test the capability of predictions tools for hERG by visualization of library designs. Examples of diversity reagents highlighted by these methods are shown in Figure 15.19.

15.17
Chemoinformatic Analysis of the Predicted Hansch Substituent Constants of the Diversity Reagents for Design of Vector Exploration Libraries

When targeting library designs to a biological target, filtering R groups for appropriate druglike parameters must be dynamically balanced with other bases for selection. Having selected a template or core with defined vectors at which diversity elements are to be introduced, we need a systematic means for reagent selection. This is particularly true in cases in which we lack biostructural information to guide the selection process. Selection of reagents based on Craig [165] and Topliss [166] plots is a standard practice in medicinal chemistry, but the number of substituents for which data are available is limited. Further, these plots organize only part of the interactions that one may wish to consider when sorting reagents for inclusion into a library. Significant forces of molecular recognition of small molecules by proteins include steric effects (*MR*, molar refractivity), inductive and resonance components (*F*, field parameter and *R*, resonance parameter), and finally the hydrophobic substituent constant, π. Colleagues at Roche have developed an artificial neural network to correlate substituent constants with topological descriptors for these four parameters [167]. The ability to predict such parameters is ideally suited to a chemoinformatic guided library design approach. A web interface has been developed for these tools requiring the user to enter SMILES [168] strings representing the diversity element as it is attached to a central core. For convenient viewing, the calculated properties can be viewed in a two-dimensional plot by summing field and resonance parameters (Figure 15.20), the two primary components that comprise the Hammett constant σ [169]. Reagents are then selected by sampling various sectors of the space created by arraying the predicted parameters as described; sampling is also done according to *MR* calculations. Accordingly, reagents that have a range of values for electronic, hydrophobic, and molecular size descriptors are selected. The chemical diversity represented in this way is then more easily interpreted when hits are detected from libraries designed in this manner.

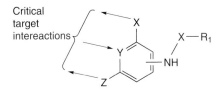

Critical target intereactions

Fig. 15.20 Exploration of the effect of adding diverse substituents to particular vectors arrayed around a central template presumed critical to a binding interaction.

When following up on validated hits from libraries designed according to R group selection by predicted Hansch parameters, it is convenient and rational to select other R groups in a similar chemistry space or sector. Notable is the inclusion of linking group functionality into the Hansch parameter prediction and how different linkers can change the predicted parameter (Figure 15.21). This level of detailed sorting of characterization according to Hansch parameters goes far beyond that reported in the scientific literature. It is useful to consider the display in terms of standard ranges for these parameters (Table 15.6). Diversity reagents are then selected on the basis

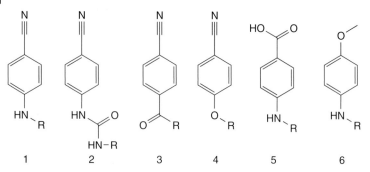

Fig. 15.21 Calculation of Hansch parameters for three diversity reagent examples with different linkage chemistry.

Tab. 15.6 Average ranges for Hansch parameters

Parameter	Average ranges
π (lipophilicity)	−2 to 2
MR (molar refractivity)	0 to 5
F (field effect)	−0.1 to 0.7
R (resonance effect)	−0.2 to 0.3
F + R	−0.3 to 1

Tab. 15.7 Hansch parameters for R Groups in Figure 15.21

R group	π	MR	F	R
1	0.69	3.512	0.123	−0.633
2	0.432	4.495	0.255	−0.193
3	0.478	3.579	0.288	0.114
4	0.624	3.345	0.241	−0.34
5	0.185	3.64	0.123	−0.633
6	1.055	3.747	0.123	−0.633

of sampling. Positive values for $F + R$ usually represent electron-withdrawing groups and negative $F + R$ values, electron-donating groups (see Figures 15.22, 15.23, and Table 15.7).

15.18
Targeting Libraries by Virtual Screening

The application of computational algorithms for compound filtering and clustering is routinely used to eliminate undesired structures on the basis of chemically reactive functionalities, predicted liabilities (e.g., frequent hitters, hERG, cyp450 and so forth), or druglike properties. Computational methods are also used to group compounds on

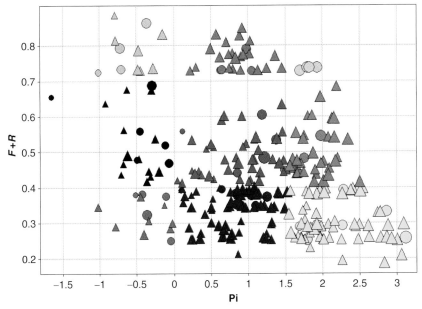

Fig. 15.22 Display of predicted Hansch parameters of the sum of inductive and resonance constants $(F + R)$ versus lipophilicity (π) for 302 commercially available carboxylic acids, acid chlorides, sulphonyl chlorides and isocyanates. Symbol size corresponds to larger molecular size (MR). Circles represent the selected R groups and triangles represent the unselected compounds. Colors convey the segmentation of the diversity reagents into nine sectors for selection.

the basis of certain similarity/diversity criteria [170]. This is most often applied in the context of compound purchasing to ensure the quality of molecules to be brought in and also to prevent duplication of topologically similar structures already represented in the compound inventory.

Virtual screening tools are also useful to preselect compounds from the corporate collection for directed or biased screening efforts [171, 172]. This is usually applied when a high-throughput screening assay is not available or when the throughput is not sufficiently high. Many retrospective screening analyses have shown the validity of such an approach where compounds were predicted to be active and increased hit rates were observed. To a lesser extent, prospective experiments have shown that biased screening efforts can lead to novel hits where an HTS campaign did not deliver reasonable structures [173]. The integration of virtual screening and HTS seems to be a logical consequence of the maturation of both disciplines but is often still regarded as being competitive rather then complementary and therefore not as intensively applied in many drug discovery companies as expected [174].

For the generation of targeted libraries, similarity algorithms are useful for computationally screening virtual libraries. Since the number of structures available from such libraries is beyond practical limits, the tools and strategies to be applied for computational screening often differ from the ones used for physically available

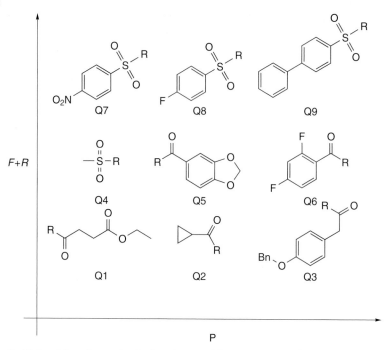

Fig. 15.23 Selected examples for the selection of diversity reagents of 302 choices according to nine sectors selected according to Hansch parameter predictions.

compound collections. Usually, a cascade of virtual screening algorithms ranging from 1-D to 3-D tools is applied to reduce the large number of theoretically available compounds to some 100 to 1000 predicted actives [175]. While in the first instance, target-unrelated properties are used for filtering out undesired structures, 2-D and 3-D ligand/target information comes into play once druglike molecules have to be prioritized in terms of their similarity/diversity. Since either biostructural or ligand information is the basis for biasing compound collections toward particular targets, the molecules can be scored on a predefined basis. "Cherry picking" of molecular structures out of a virtual chemical library generates compound collections that are supposed to show higher hit rates compared to random selection. Retrospective screenings of databases showed that increased hit rates can be obtained for certain target families by using computational algorithms to enrich subsets of those by sorting the compounds in terms of their similarity to particular seed structures or by target information [176].

15.19
Combinatorial Design Based on Biostructural Information

For the prioritization of virtual compound libraries based on biostructural information, usually a fixed target structure is assumed and the energetically minimized conformers of the potential ligands are generated and scored by various binding energy-prediction methods [177]. Several computational algorithms for structure-based *de novo* design have been described, among which LUDI [178], BUILDER [179], DOCK [180], GOLD [181], and FlexX [182] are some of the most prominent software packages. These virtual screening tools identify potential ligand–target interaction points within the predefined receptor-binding pocket. For this developing area of chemoinformatics in drug discovery, many excellent papers have been published. Many of these strategies have been clearly summarized in a review of computational design strategies for combinatorial libraries [130]; virtual screening algorithms and success stories are discussed in Chapter 4 [183].

For the hit-to-lead and lead optimization phases, such tools are highly valuable when the chemical space is limited by preselecting templates and building blocks based on already available information. For example, novel and highly active small-molecule thrombin inhibitors were identified using LUDI [184, 185] as a ligand design tool; in this study, elements and related fragments of known thrombin ligands were used as building blocks [184].

There has been considerable discussion on the validity of docking and scoring functions in structure-based design because of the complex issues involved such as ligand orientation, water participation, or flexibility of the target protein itself [186–190]. As long as the virtual structure-based screening procedure identifies compound proposals that can be realized and tested in a short period of time, the accuracy lacking in the scoring functions can be compensated and finally validated by the biophysical screening experiment itself.

15.20
Ligand-based Combinatorial Design: The RADDAR Approach

For many targets in pharmaceutical research, there is little or no biostructural information available that would allow the application of a docking and scoring strategy. Therefore, information about the ligand is the only source available for design of a novel compound series for targets such as ion channels, membrane-embedded receptors, or transporters. The most widely applied strategies are based on 3-D pharmacophore models that are generated from hypotheses based on a set of compounds showing activities for a particular target [191]. Enumeration of the minimized conformers, which are the basis for 3-D pharmacophore screening, is still time consuming, making extensive 3-D pharmacophore screening of large databases with the currently available software and hardware tools inconvenient.

Virtual screening on a 2-D basis is far more efficient in terms of computing time, although the information content of the resulting virtual hits is far less sophisticated than for a 3-D pharmacophore search. It is in this setting that combinatorial chemistry approaches are applied most effectively. The many interactions possible with a compound library based on virtual hits compensates for the inherent fuzziness of any prediction tool. Examples have been described where the 2-D structure of a single compound was sufficient to generate *de novo* compound proposals resulting in active compounds for targets such as ion channels [192] and G-protein-coupled receptors [193].

Whatever be the tool that is used for virtual screening, tight integration of chemistry and library design with the computational methods is critical. There is no point in virtually screening chemical databases of structures that are synthetically not feasible. The realization of these proposals in a parallel or even combinatorial fashion is of utmost importance. This is still of major concern for most virtual screening algorithms that produce artificial compounds where the chemist is challenged to synthesize compound arrays based on a computational proposal.

In an ideal setting, the chemist must propose libraries based on evaluated chemistry and the availability as well as exclusivity of templates and building blocks. Computational screening of such virtual libraries might not necessarily lead to the most promising proposals, but the fast realization and testing of the ideas allows a quick assessment of the compounds computationally identified and chemically generated. Feeding back this information and either adapting the computational prediction tools, the information used for virtual screening (ligands, pharmacophore model, X-ray data and so forth), or the library proposals themselves lead to a dynamic process. This process has been well established at Roche and is called RADDAR (Roche Adaptive Drug Design and Refinement) (see Figure 15.24).

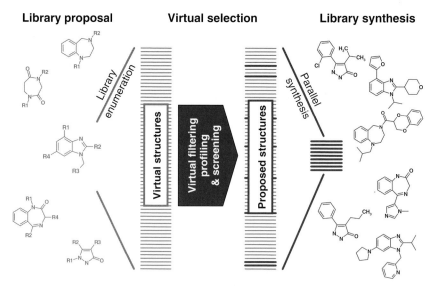

Fig. 15.24 RADDAR: a nearly infinite number of chemical structures can be generated by computationally enumerating combinatorial library proposals. "Cherry picking" of interesting candidates (proposed structures) based on defined computational algorithms allows the chemist to synthesize the relevant subsets of theoretically accessible molecules.

15.21
Virtual Screening of Small-molecule Library with Peptide-derived Pharmacophores

The identification of ligands, and particularly agonists, for peptide GPCRs is often a significant challenge. An understanding of the SAR for endogenous peptides can provide a basis for virtual screening. We have developed a peptide-based SAR for a peptide agonist of the prolactin-releasing peptide hormone receptor (PrRP-R), a GPCR potentially related to feeding and satiety [194]. HTS of the corporate collection failed to identify chemical leads for further elaboration. In order to develop ideas about potential feasibility of small-molecule ligands, a virtual screening approach was attempted. A virtual library of approximately 75 000 unique structures was enumerated based on several hundred small-molecule templates. A database of conformational ensembles of these 75 000 unique structures was generated with Catalyst [195] using the fast mode for conformer generation. Since the peptide agonist pharmacophore contains a charged residue, it was important to enumerate and generate conformers for charged virtual structures. Catalyst was also used to perform mapping analysis of this library and several follow-up virtual libraries against a four-point pharmacophore model derived from the peptide SAR studies. An approach similar to lead identification via virtual screening of peptide-based pharmacophores for urotensin II has been reported [196].

It was important to view the results of these virtual screens in the context of calculated druglike properties. The virtual library generated from template lib00711 was the highest scoring proposal with acceptable molecular weight and PSA score distribution (Figure 15.25, third row left). Other templates scored higher against the three-point pharmacophore model but had high molecular weight and PSA score distributions. One increases the chances of creating active and druglike structures for subsequent optimization. The simultaneous viewing and prioritization was performed using a web-based template display system. It is possible to view the types of molecules that give high overlap scores and then inspect the fitted models. Target tractability in terms of small-molecule therapeutics is readily assessed in this way based on the types of structures that are highlighted. The template that provides the best pharmacophore fitness scoring along with acceptable druglike properties, is selected for further study and design. In this manner, the generation of virtual structures and evaluation of pharmacophore fitness were iterated until 10% hit rate in four-point pharmacophore fitness was achieved for a given set of virtual library compounds. Within the evaluation of hundreds of different template ideas, only a few templates appropriately displayed the necessary pharmacophore elements; of those few, only one design (lib00711), a proposal based on a GPCR targeting premise, also possessed acceptable druglike properties. Presumably, with larger libraries based on other templates, a greater number of feasible ideas would be identified.

This process is an interesting way to test notions about privileged substructures having particular affinity for a member of a target protein family [197]. In this virtual screen, although there were many so-called "GPCR-targeted designs", very few showed any affinity for this specific GPCR target pharmacophore. The privileged structure concept is a strategy that has been employed to bias molecular designs for a target family [198]. While the concept of a privileged structure helps to organize thinking

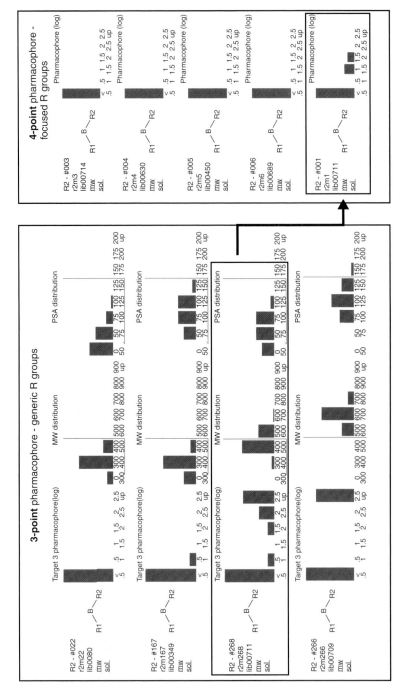

Fig. 15.25 Partial display of virtual screen of conformational ensemble of ~75 000 structure library based on multiple templates. A display of fitness scores to a three-point pharmacophore model provides initial signs of similarity of a library design to that model. Subsequent virtual screening against a more stringent four-point pharmacophore model further highlights potentially useful library designs.

about new template ideas, it is a more general, less stringent way of targeting than pharmacophore model fitting.

15.22
Chemoinformatic Tools and Strategies to Visualize Active Libraries

Identifying leads from the HTS of small-molecule libraries can provide advantages over leads detected as active single molecules. A great deal of SAR information is available when active and inactive hits share a common core or template. This process begins with the examination of the active molecules within the context of the focused library as a whole. An analysis of what was made and available for assay is facilitated by breaking the library molecules into a central core and diversity sets. Usually, this analysis reveals holes in the library design due to limitations in library size or failure in the synthesis chemistry. This deconstruction of the library molecules provides a convenient means to visualize and summarize trends in the SAR [199]. An example below illustrates this process. Such visualization provides a guide to subsequent synthesis for hit-to-lead evaluation of the active molecule.

A focused library was designed based on a cocrystal structure of a kinase with a small-molecule inhibitor with moderate activity. In the library design, 1160 compounds were theoretically possible, but owing to limitations in the synthesis and purification of the crude products before assay, only 826 were obtained. After assay, four submicromolar actives were identified as inhibitors. It was convenient to visualize this library by fragmenting the library around a central core and examining which R groups were present and active (Figure 15.26). The gaps in the diversity of the compounds are clear in this Figure. Such gaps highlight the need to design in some level of redundancy in library design in anticipation of a number of synthesis failures. Obtaining only four active molecules as primary hits within a set of some eight hundred compounds is not untypical even when libraries are designed using detailed structural information. Three of the four active compounds contain the same diversity element for the A substituent. This sort of analysis and data display is helpful to highlight the size and diversity of library coverage necessary in order to enhance the chances of finding a hit within a set of so many inactive compounds. Analysis of the library in terms of active and inactive R groups as opposed to single compounds also facilitates follow-up library synthesis. Follow-up library design and synthesis based on this information have provided a substantially higher percentage of hits of submicromolar inhibitor potency.

15.23
Visualization of Library Designs during Hit-to-lead Efforts

There are several advantages in finding leads from focused libraries. Usually, one must be cautious about interpreting negative HTS data; one cannot necessarily construct a negative SAR from compounds that are inactive during an HTS campaign without some reassay verification. This is often due to concerns about compound purity and integrity. Interpretation of negative data is also not advised as single compounds do not

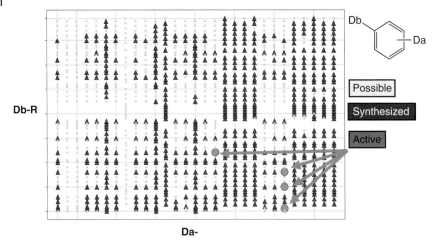

Fig. 15.26 Visualization of the two sets of diversity elements arrayed around a single core. Triangles indicate inactive molecules in a kinase assay; squares indicate synthesis failures; circles indicate active kinase inhibitors within this set.

present the small systematic variations around a common theme usually necessary to begin the construction of a preliminary SAR. Focused libraries, however, are usually released to screening at the same or nearly the same time. These libraries will likely have similar levels of purity and integrity (if quality control practices are in place) and have many small systematic variations around a common theme. We have found that validated hits from focused libraries provide some validation as to the size of libraries necessary in order to find an initial sign of activity for a particular library design.

Hit-to-lead practices with library-derived actives are rapidly directed to follow up on holes in the library that may have been created because of reagents that were not selected or were unavailable. Follow-up libraries are also designed by selecting R groups based on SAR data from screening. During hit-to-lead campaigns, design strategies shift from molecular diversity to dense coverage of an SAR chemical space. A preliminary SAR is possible with the deconvolution of the library into R groups and comparison of active and inactive molecules containing those R groups. Visualization of several simple chemoinformatic calculations can provide a qualitative sense of likely liabilities when attempting to develop hits into leads.

For example, in Figure 15.27, two plots of MlogP versus molecular weight are shown for two different library designs composed of greater than 1000 members each. Each library provided compounds that were active as inhibitors at lower micromolar concentrations against separate enzymes. In the case represented by the left plot (a), from a library of approximately 1400 compounds, 12 lower micromolar compounds were identified. These compounds were found in a central and favorable (leadlike) locale of the MlogP versus MW plot. After identifying several liabilities of these compounds and providing clear strategies for a solution of those liabilities, it was possible to identify these compounds as a lead series for further optimization.

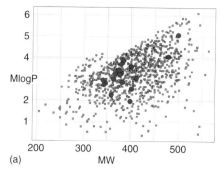

(a)

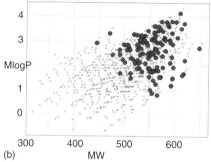

(b)

Fig. 15.27 Visualization of two libraries.
(a) Library of ~1400 compounds for which a lead series was successfully identified after hit-to-lead efforts. (b) Library of ~1100 compounds for which lead identification/ lead generation efforts failed owing to undruglike properties of the library design. Larger circles indicate the enzyme active inhibitors; circles indicate the enzyme inactive inhibitors.

In another case of 1100 compounds, 4 lower micromolar actives were found (Figure 15.27b). Subsequent hit-to-lead efforts were facilitated based on an analysis of R groups contributing to enzyme inhibition. A substantial number of active compounds were quickly synthesized, but most of the active compounds were high in molecular weight and fairly lipophilic toward the upper limits of druglikeness. Instead of binding to smaller, less lipophilic compounds, the enzyme bound to higher, more lipophilic compounds of similar structure. Subsequent hit-to-lead work highlighted several problems with these structures including low solubility and lower cellular permeability. The prediction of these properties for this series was generally consistent with the problems encountered during the hit-to-lead efforts. On the basis of these and other data, this hit series was discontinued. These results provide some insight into the type of ligand necessary to effect enzyme inhibition for this template. It is possible that the enzyme pocket is large, requiring larger, potentially undruglike molecules for efficient inhibition. It would be necessary to test other templates for inhibition in order to find potent inhibitors, which are also smaller and less lipophilic, before concluding that the target is not tractable in small-molecule therapeutics.

15.24
Summary and Outlook for Chemoinformatically Driven Lead Generation

In generating leads from HTS of corporate compound collections and from focused libraries, there are clearly multiple tools and strategies; this situation likely reflects rapid development in this field. We have found that tools currently available for the prediction of solubility and druglikeness have not reached the stage where they can be used to filter hits from screening. Despite this, such tools do capture general trends that can be very useful in library design. Prediction of other features that will make good drugs, such as low hERG binding affinity, will likely evolve as databases of this information for diverse structural classes grow. The evolution of better docking and scoring functions will be complemented by repeated efforts to validate these tools with experimental data. Deriving leads from combinatorial chemistry libraries is an approach that many believe is providing useful leads. Chemoinformatics for lead generation will only add to the precision of such library designs. As lead generation will continue to be a high priority for drug discovery, well-established medicinal chemistry concepts and insight into small-molecule-target interactions coupled with creative organic synthesis will provide a field of ideas to be ordered and prioritized by chemoinformatic methods for lead generation. Establishing effective processes in this aspect of drug discovery will provide a basis for success in tackling the challenges of chemogenomics, the emerging new paradigm in drug discovery.

Acknowledgements

We are pleased to acknowledge many helpful discussions with our colleagues in the computational chemistry groups at Roche-Basel and Roche Nutley. We are indebted to these members of our organization for creating new computational tools for convenient in-house use. We are particularly indebted to Dr. Sung-Sau So, Dr. Hongmao Sun, John Feinberg, Dr. Gisbert Schneider, Dr. Martin Stahl, Dr. Manfred Kansy, Dr. Achim Grenz, Dr. Michael Dillon, and Dr. Hans-Joachim Böhm. Professor Tudor Oprea is also gratefully acknowledged for his insightful editorial comments and suggestions.

References

1 BOLTEN, B.M. and DEGREGORIO, T. Trends in development cycles. *Nat. Rev. Drug Disc.* **2002**, *1*, 335–336.

2 OPREA, T.I. Current trends in lead discovery: Are we looking for the appropriate properties? *J. Comput.-Aided Mol. Des.* **2002**, *16*, 325–334.

3 BOYD, D.B. In *Rational Molecular Design in Drug Research*, LILJEFORS, T., JORGENSEN, F.S., and KROGSGAARD-LARSEN, P. (Eds). Munksgaard, Copenhagen, **1998**, 15–29.

4 Parexel's Pharmaceutical R&D Statistical Sourcebook 2001, High Performance Drug Discovery: Speed to Value, **2001**, 139–140.

5 DEMASI, J.A., HANSEN, R.W., and GRABOWSKI, H.G. The price of innovation: new estimates of drug development costs. *J. Health Econ.* **2003**, *22*, 151–185.

6 HOPKINS, A.L. and GROOM, C.R. Opinion: the druggable genome. *Nat. Rev. Drug Disc.* **2002**, *1*, 727–730.

7 BAILEY, D. and BROWN, D. High throughput chemistry and structure-based design: survival of the smartest. *Drug Disc. Today* **2001**, *6*, 57–58.

8 MICHNE, W.F. Hit-to-lead chemistry: a key element in new lead generation. *Pharm. News* **1996**, *3*, 19–21.

9 CAMPBELL, S. and DIFFERDING, E. Drug discovery strategies: from leads to drugs. *Chim. Nouv.* **2001**, *19*, 3347–3352.

10 TROPSHA, A. and REYNOLDS, C.H. Designing focused libraries for drug discovery: hit to lead to drug. *J. Mol. Graph. Model.* **2002**, *20*, 427–428.

11 BAXTER, A.D. and LOCKEY, P.M. 'Hit' to 'lead' and 'lead' to 'candidate' optimisation using multi-parametric principles. *Drug Disc. World* **2001**, *2*, 9–15.

12 TEAGUE, S.J., DAVIS, A.M., LEESON, P.D., and OPREA, T. The design of leadlike combinatorial libraries. *Angew. Chem., Int. Ed. Engl.* **1999**, *38*, 3743–3748.

13 BLEICHER, K.H., BOHM, H.J., MULLER, K., and ALANINE, A.I. A guide to drug discovery: hit and lead generation: beyond high throughput screening. *Nat. Rev. Drug Disc.* **2003**, *2*, 369–378.

14 ALANINE, A., NETTEKOVEN, M., ROBERTS, E., and THOMAS, A.W. Lead generation – enhancing the success of drug discovery by investing in the hit to lead process. *Comb. Chem. High Throughput Screen.* **2003**, *6*, 51–66.

15 AMIDON, G.L., LENNERÄS, H., SHAH, V.P., and CRISON, J.R. A theoretical basis for a biopharmaceutic drug classification: the correlation of in vitro drug product dissolution and in vivo bioavailablity. *Pharm. Res.* **1995**, *12*, 413–420.

16 LIPINSKI, C.A., LOMBARDO, F., DOMINY, B.W., and FEENEY, P.J. Experimental and computational approaches to estimate solubility and permeability in drug discovery and development settings. *Adv. Drug Deliv. Rev.* **1997**, *23*, 3–25.

17 LIVINGSTONE, D.J. Theoretical property predictions. *Curr. Top. Med. Chem.* **2003**, *3*, 1171–1192.

18 RAN, Y. and YALKOWSKY, S.H. Prediction of drug solubility by the general solubility equation (GSE). *J. Chem. Inf. Comput. Sci.* **2001**, *41*, 354–357.

19 ABRAHAM, M.H. and LE, J. The correlation and prediction of the solubility of compounds in water using an amended solvation energy relationship. *J. Pharm. Sci.* **1999**, *88*, 868–880.

20 JORGENSEN, W.L. and DUFFY, E.M. Prediction of drug solubility from structure. *Adv. Drug Deliv. Rev.* **2002**, *54*, 355–366.

21 MEYLAN, W.M., HOWARD, P.H., and BOETHLING, R.S. Improved method for estimating water solubility from octanol/water partition coefficient. *Environ. Toxicol. Chem.* **1996**, *15*, 100–106.

22 GAO, H., SHANMUGASUNDARAM, V., and LEE, P. Estimation of aqueous solubility of organic compounds with QSPR approach. *Pharm. Res.* **2002**, *19*, 497–503.

23 BUTINA, D. and GOLA, J.M.R. Modeling aqueous solubility. *J. Chem. Inf. Comput. Sci.* **2003**, *43*, 837–841.

24 CHEN, X.-Q., CHO, S.J., LI, Y., and VENKATESH, S. Prediction of aqueous solubility of organic compounds using a quantitative structure-property relationship. *J. Pharm. Sci.* **2002**, *9*, 1838–1852.

25 CHENG, A. and MERZ, K.M. Jr. Prediction of aqueous solubility of a diverse set of compounds using quantitative structure-property relationships. *J. Med. Chem.* **2003**, *46*, 3572–3580.

26 HUUSKONEN, J. Estimation of aqueous solubility for a diverse set of organic compounds based on molecular topology. *J. Chem. Inf. Comput. Sci.* **2000**, *40*, 773–777.

27 RAN, Y., JAIN, N., and YALKOWSKY, S.H. Prediction of aqueous solubility of organic compounds by the general solubility equation (GSE). *J. Chem. Inf. Comput. Sci.* **2001**, *41*, 1208–1217.

28 MCELROY, N.R. and JURS, P.C. Prediction of aqueous solubility of heteroatom-containing organic compounds from molecular structure. *J. Chem. Inf. Comput. Sci.* **2001**, *41*, 1237–1247.

29 TETKO, I.V., TANCHUK, V.Yu., KASHEVA, T.N., and VILLA, A.E.P. Estimation of Aqueous solubility of chemical compounds using E-state indices. *J. Chem. Inf. Comput. Sci.* **2001**, *41*, 1488–1493.

30 BRUNEAU, P. Search for predictive generic model of aqueous solubility using Bayesian neural nets. *J. Chem. Inf. Comput. Sci.* **2001**, *41*, 1605–1616.

31 LIU, R. and SO, S.-S. Development of quantitative structure-property relationship models for early ADME evaluation in drug discovery. 1. Aqueous solubility. *J. Chem. Inf. Comput. Sci.* **2001**, *41*, 1633–1639.

32 ENGKVIST, O. and WREDE, P. High throughput, in silico prediction of aqueous solubility based on one- and two-dimensional descriptors. *J. Chem. Inf. Comput. Sci.* **2002**, *42*, 1247–1249.

33 YAN, A. and GASTEIGER, J. Prediction of aqueous solubility of organic compounds based on a 3D structure representation. *J. Chem. Inf. Comput. Sci.* **2003**, *43*, 429–434.

34 WEGNER, J.K. and ZELL, A. Prediction of aqueous solubility and partition coefficient optimized by a genetic algorithm based descriptor selection method. *J. Chem. Inf. Comput. Sci.* **2003**, *43*, 1077–1084.

35 YAFFE, D., COHEN, Y., ESPINOSA, G., ARENAS, A., and GIRALT, F. A fuzzy ARTMAP based on quantitative structure-property relationships (QSPRs) for predicting aqueous solubility of organic compounds. *J. Chem. Inf. Comput. Sci.* **2001**, *41*, 1177–1207.

36 MANALLACK, D.T., TEHAN, B.G., GANCIA, E., HUDSON, B.D., FORD, M.G., LIVINGSTONE, F.J., WHITLEY, D.C., and PITT, W.R. A consensus neural network-based technique for discriminating soluble and poorly soluble compounds. *J. Chem. Inf. Comput. Sci.* **2003**, *43*, 674–679.

37 Syracuse Research Corporation, 6225 Running Ridge Road, North Syracuse, NY 13212.

38 Advanced Chemistry Development, Inc., 90 Adelaide Street West, Suite 600, Toronto, Ontario M5H 3V9, Canada.

39 Simulations Plus, Inc., 1220 W. Avenue, Lancaster, CA 93534.

40 ERTL, P., ROHDE, B., and SELZER, P. Fast calculation of molecular polar surface area as a sum of fragment-based contributions and its application to the prediction of drug transport properties. *J. Med. Chem.* **2000**, *43*, 3714–3717.

41 BARONE, R. and CHANON, M. A new and simple approach to chemical complexity.

Application to the synthesis of natural products. *J. Chem. Inf. Comput. Sci.* **2001**, *41*, 269–272.

42 BRADSHAW, J. MUG '01 – 15th Daylight User Group Meeting. March 6, **2001**. Website: http://www.daylight.com/ meetings/mug01/Bradshaw/History/.

43 WALTERS, W.P., AJAY, and MURCKO, M.A. Recognizing molecules with drug-like properties. *Curr. Opin. Chem. Biol.* **1999**, *3*, 384–387.

44 CLARK, D.E. and PICKETT, S.D. Computational methods for the prediction of "Drug-Likeness". *Drug Disc. Today* **2000**, *5*, 49–58.

45 BLAKE, J.F. Chemo-informatics – predicting the physicochemical properties of "Drug-like" molecules. *Curr. Opin. Biotech.* **2000**, *11*, 104–107.

46 AJAY. Predicting drug-likeness: Why and How? *Curr. Top. Med. Chem.* **2002**, *2*, 1273–1286.

47 WALTERS, W.P. and MURCKO, M.A. Prediction of 'Drug-likeness'. *Adv. Drug Deliv. Rev.* **2002**, *54*, 255–271.

48 EGAN, W.J., WALTERS, W.P., and MURCKO, M.A. Guiding molecules towards drug-likeness. *Curr. Opin. Drug Disc. Dev.* **2002**, *5*, 540–549.

49 GHOSE, A.K., VISWANADHAN, V.N., and WENDOLOSKI, J.J. Prediction of hydrophobic (Lipophilic) properties of small organic molecules using fragmental methods: an analysis of ALOGP and CLOGP methods. *J. Phys. Chem. A* **1998**, *102*, 3762–3772.

50 SADOWSKI, J. and KUBINYI, H. A scoring scheme for discriminating between drugs and nondrugs. *J. Med. Chem.* **1998**, *41*, 3325–3329.

51 PROUDFOOT, J.R. Exploitation of HTS lead structures: lessons from the successes. Abstracts of Papers, *225th ACS National Meeting*. New Orleans, LA, March 23–27, **2003**, MEDI-193.

52 GERSHELL, L.J. and ATKINS, J.H. Timeline: a brief history of novel drug discovery technologies. *Nat. Rev. Drug Disc.* **2003**, *2*, 321–327.

53 ANDREWS, C.W., BENNETT, L., and YU, L.X. Predicting human oral bioavailability of a compound: development of a novel quantitative structure-bioavailability relationship. *Pharm. Res.* **2000**, *17*, 639–644.

54 STENBERG, P., BERGSTRÖM, C.A.S., LUTHMAN, K., and ARTURSSON, P. Theoretical predictions of drug absorption in drug discovery and development. *Clin. Pharmcokinet.* **2002**, *41*, 877–899.

55 CLARK, D.E. and GROOTENHUIS, P.D. Progress in computational methods for the prediction of ADMET properties. *Curr. Opin. Drug. Disc. Dev.* **2002**, *5*, 382–390.

56 VAN DE WATERBEEMD, H. and GIFFORD, E. ADMET in silico modelling: towards prediction paradise? *Nat. Rev. Drug Disc.* **2003**, *2*, 192–204.

57 MASIMIREMBWA, C.M., RIDDERSTROEM, M., ZAMORA, I., and ANDERSSON, T.B. Combining pharmacophore and protein modeling to predict CYP450 inhibitors and substrates. *Methods Enzymol.* **2002**, *357(Cytochrome P450, Part C)*, 133–144.

58 WESSEL, M.D. and MENTE, S. ADME by computer. *Annu. Rep. Med. Chem.* **2001**, *36*, 257–266.

59 YAN, Z. and CALDWELL, G.W. Metabolism profiling, and cytochrome P450 inhibition & induction in drug discovery. *Curr. Top. Med. Chem.* **2001**, *1*, 403–425.

60 SUSNOW, R.G. and DIXON, S.L. Use of robust classification techniques for the prediction of human cytochrome P450 2D6 inhibition. *J. Chem. Inf. Comput. Sci.* **2003**, *43*, 1308–1315.

61 KOROLEV, D., BALAKIN, K.V., NIKOLSKY, Y., KIRILLOV, E., IVANENKOV, Y.A., SAVCHUK, N.P., IVASHCHENKO, A.A., and NIKOLSKAYA, T. Modeling of human cytochrome P450-mediated drug metabolism using unsupervised machine learning approach. *J. Med. Chem.* **2003**, *46*, 3631–3643.

62 SANGHVI, T., NI, N., MAYERSOHN, M., and YALKOWSKY, S.H. Predicting passive intestinal absoprtion using a single parameter. *QSAR Comb. Sci.* **2003**, *22*, 247–257.

63 MANGA, N., DUFFY, J.C., ROWE, P.H., and CRONIN, M.T.D. A hierarchical QSAR model for urinary excretion of drugs in humans as a predictive tool for biotransformation. *QSAR Comb. Sci.* **2003**, *22*, 263–273.

64 HOU, T. and XU, X. ADME evaluation in drug discovery. *J. Mol. Model.* **2002**, *8*, 337–349.

65 NIWA, T. Using general regression and probabilistic neural networks to predict human intestinal absorption with topological descriptors derived from two-dimensional chemical structures. *J. Chem. Inf. Comput. Sci.* **2003**, *43*, 113–119.

66 KLOPMAN, G., STEFAN, L.R., and SAIAKHOV, R.D. ADME Evaluation 2. A computer model for the prediction of intestinal absorption in humans. *Eur. J. Pharm. Sci.* **2002**, *17*, 253–263.

67 BURTON, P.S., GOODWIN, J.T., VIDMAR, T.J., and AMORE, B.M. Predicting drug absorption: How nature made it a difficult problem. *J. Pharm. Exp. Ther.* **2002**, *303*, 889–895.

68 ZHAO, Y.H., LE, J., ABRAHAM, M.H., HERSEY, A., EDDERSHAW, P.J., LUSCOMBE, C.N., BOUTINA, D., BECK, G., SHERBORNE, B., COOPER, I., and PLATIS, J.A. Evaluation of human intestinal absorption data and subsequent derivation of a quantitative structure-activity relationship (QSAR) with the Abraham descriptors. *J. Pharm. Sci.* **2001**, *90*, 749–784.

69 SHEN, M., XIAO, Y., GOLBRAIKH, A., GOMBAR, V.K., and TROPSHA, A. Development and validation of k-nearest-neighbor QSPR models of metabolic stability of drug candidates. *J. Med. Chem.* **2003**, *46*, 3013–3020.

70 TURNER, J.V., GLASS, B.D., and AGATANOVIC-KUSTRIN, S. Prediction of drug bioavailability based on molecular structure. *Anal. Chim. Acta* **2003**, *485*, 90–102.

71 BENFENATI, E. and GINI, G. Computational predictive programs (Expert Systems) in toxicology. *Toxicology* **1997**, *119*, 213–225.

72 CRONIN, M.T.D. Computer-aided prediction of drug toxicity in high throughput screening. *Pharm. Pharmacol. Commun.* **1998**, *4*, 157–163.

73 MATTHEWS, E.J. and CONTRERA, J.F. A new highly specific method for predicting the carcinogenic potential of pharmaceuticals in rodents using enhanced MCASE QSAR-ES software. *Reg. Tox. Pharm.* **1998**, *28*, 242–264.

74 ROSENKRANZ, H.S., CUNNINGHAM, A.R., ZHANG, Y.P., CLAYCAMP, H.G., MACINA, O.T., SUSSMAN, N.B., GRANT, S.G., and KLOPMAN, G. Development, characterization and application of predictive-toxicology models. *SAR QSAR Env. Res.* **1999**, *10*, 277–298.

75 GREENE, N., JUDSON, P.N., LANGOWSKI, J.J., and MARCHANT, C.A. Knowledge-based expert systems for toxicity and metabolism prediction: DEREK, StAR and METEOR. *SAR QSAR Env. Res.* **1999**, *10*, 299–314.

76 JOHNSON, D.E. and WOLFGANG, G.H.I. Assessing the potential toxicity of new pharmaceuticals. *Curr. Top. Med. Chem.* **2001**, *1*, 233–245.

77 ROCHE, O., TRUBE, G., ZUEGGE, J., PFLIMLIN, P., ALANINE, A., and SCHNEIDER, G. A virtual screening method for prediction of the hERG potassium channel liability of compound libraries. *ChemBioChem* **2002**, *3*, 455–459.

78 BARRATT, M.D. and RODFORD, R.A. The computational prediction of toxicity. *Curr. Opin. Chem. Biol.* **2001**, *5*, 383–388.

79 ADMANS, G., TAKAHASHI, Y., BAN, S., KATO, H., ABE, H., and HANAI, S. Artificial neural network for predicting the toxicity of organic molecules. *Bull. Chem. Soc. Japan* **2001**, *74*, 2451–2461.

80 GREENE, N. Computer systems for the prediction of toxicity: an update. *Adv. Drug Del. Rev* **2002**, *54*, 417–431.

81 FENG, J., LURATI, L., OUYANG, H., ROBINSON, T., WANG, Y., YUAN, S., and YOUNG, S.S. Predictive toxicology: benchmarking molecular descriptors and statistical methods. *J. Chem. Inf. Comput. Sci.* **2003**, *43*, 1463–1470.

82 CAVALLI, A., POLUZZI, E., DE PONTI, F., and RECANATINI, M. Toward a pharmacophore for drugs inducing the long QT syndrome: insights from a CoMFA study of HERG K+ channel blockers. *J. Med. Chem.* **2002**, *45*, 3844–3853.

83 KESERU, G.M. Prediction of hERG potassium channel affinity by traditional and hologram qSAR methods. *Bioorg. Med. Chem. Lett.* **2003**, *13*, 2773–2775.

84 DOBLER, M., LILL, M.A., and VEDANI, A. From crystal structures and their analysis to the in silico prediction of toxic phenomena. *Helv. Chim. Acta* **2003**, *86*, 1554–1568.

85 MUSKAL, S.M., JHA, S.K., KISHORE, M.P., and TYAGI, P. A simple and readily integratable approach to toxicity prediction. *J. Chem. Infor. Comput. Sci.* **2003**, *43*, 1673–1678.

86 LHASA Limited, Department of Chemistry, University of Leeds, Leeds, LS2 9JT, UK.

87 MultiCASE Inc., 23811 Chagrin Blvd. Ste 305, Beachwood, OH 44122.

88 Accelrys Inc., 9685 Scranton Road, San Diego, CA 92121.

89 MDL Information Systems, Inc., 14600 Catalina Street, San Leandro, CA 94577.

90 Derwent World Drug Index, Derwent Information, 14 Great Queen Street, London WC2B 5DF, UK.

91 Website: http://www.fda.gov/cder/ob/default.htm.

92 HARDMAN, J.G., LIMBIRD, L.E., MOLINOFF, P.B., RUDDON, R.W., and GOODMAN GILMAN, A. (Eds.) *Goodman and Gilman's: The Pharmacological Basis of Therapeutics.* McGraw Hill, New York, **1996**.

93 OPREA, T.I. Property distribution of drug-related chemical databases. *J. Comput.-Aided Mol. Des.* **2000**, *14*, 251–264.

94 HANSCH, C., SAMMES, P.G., and TAYLOR, J.B. (Eds.) *Comprehensive Medicinal Chemistry: The Rational Design, Mechanistic Study, and the Therapeutic Application of Chemical Compounds.* Pergamon Press, Oxford, **1990**.

95 PALM, K., STENBERG, P., LUTHMAN, K., and ARTURSSON, P. Polar molecular surface properties predict the intestinal absorption of drugs in humans. *Pharm. Res.* **1997**, *14*, 568–571.

96 GHOSE, A.K., VISWANADHAN, V.N., and WENDOLOSKI, J.J. A knowledge-based approach in designing combinatorial or medicinal chemistry libraries for drug discovery. 1. A qualitative and quantitative characterization of known

drug databases. *J. Comb. Chem.* **1999**, *1*, 55–68.

97 Xu, J. and Stevenson, J. Drug-like index: a new approach to measure drug-like compounds and their diversity. *J. Chem. Inf. Comput. Sci.* **2000**, *40*, 1177–1187.

98 Muegge, I., Heald, S.L., and Brittelli, D. Simple selection criteria for drug-like chemical matter. *J. Med. Chem.* **2001**, *44*, 1841–1846.

99 Hann, M.M., Leach, A.R., and Harper, G. Molecular complexity and its impact on the probability of finding leads for drug discovery. *J. Chem. Inf. Comput. Sci.* **2001**, *41*, 856–864.

100 Muegge, I. Pharmacophore features of potential drugs. *Chem. Eur. Journal* **2002**, *8*, 1976–1981.

101 Veber, D.F., Johnson, S.R., Cheng, H.-Y., Smith, B.R., Ward, K.W., and Kopple, K.D. Molecular properties that influence the oral bioavailability of drug candidates. *J. Med. Chem.* **2002**, *45*, 2615–2623.

102 Feher, M. and Schmidt, J.M. Property distributions: differences between drugs, natural products, and molecules from combinatorial chemistry. *J. Chem. Inf. Comput. Sci.* **2003**, *43*, 218–227.

103 Blake, J.F. Examination of the computed molecular properties of compounds selected for clinical development. *BioTechniques* **2003**, *34*, S16–S20.

104 Frimurer, T.M., Bywater, R., Nærum, L., Lauritsen, L.N., and Brunak, S. Improving the odds in discriminating "Drug-like" from "Non Drug-like" compounds. *J. Chem. Inf. Comput. Sci.* **2000**, *40*, 1315–1324.

105 Murcia-Soler, M., Pérez-Giménez, F., García-March, F.J., Salabert-Salvador, M.T., Díaz-Villanueva, W., and Castro-Bleda, M.J. Drugs and nondrugs: an effective discrimination with topological methods and artificial neural networks. *J. Chem. Inf. Comput. Sci.* **2003**, *43*, 1688–1702.

106 Wagener, M. and Van Geerestein, V.J. Potential drugs and nondrugs: prediction and identification of important structural features. *J. Chem. Inf. Comput. Sci.* **2000**, *40*, 280–292.

107 Anzali, S., Barnickel, G., Cezanne, B., Krug, M., Filimonov, D., and Poroikov, V. Discriminating between drugs and nondrugs by prediction of activity spectra for substances (PASS). *J. Med. Chem.* **2001**, *44*, 3325–3329.

108 Takaoka, Y., Endo, Y., Yamanobe, S., Kakinuma, H., Okubo, T., Shimazaki, Y., Ota, T., and Sumiya, S. Development of a method for evaluating drug-likeness and ease of synthesis using a data det in which compounds are assigned scores based on chemists' intuition. *J. Chem. Inf. Comput. Sci.* **2003**, *43*, 1269–1275.

109 Rishton, G.M. Reactive compounds and in vitro false positives in HTS. *Drug Disc. Today* **1997**, *2*, 382–384.

110 McGovern, S.L., Caselli, E., Grigorieff, N., and Shoichet, B.K. A common mechanism underlying promiscuous inhibitors from virtual and high throughput screening. *J. Med. Chem.* **2002**, *45*, 1712–1722.

111 McGovern, S.L. and Shoichet, B.K. Kinase inhibitors: not just for kinases anymore. *J. Med. Chem.* **2003**, *46*, 1478–1483.

112 Rishton, G.M. Nonleadlikeness and leadlikeness in biochemical screening. *Drug Disc. Today* **2003**, *8*, 86–96.

113 Roche, O., Schneider, P., Zuegge, J., Guba, W., Kansy, M., Alanine, A., Bleicher, K., Danel, F., Gutknecht, E., Rogers-Evans, M., Neidhart, W., Stalder, H., Dillon, M., Sjoegren, E., Fotouhi, N., Gillespie, P., Goodnow, R., Harris, W., Jones, P., Taniguchi, M., Tsujii, S., von der Saal, W., Zimmermann, G., and Schneider, G. Development of a virtual screening method for identification of "Frequent Hitters" in compound libraries. *J. Med. Chem.* **2002**, *45*, 137–142.

114 Evans, B.E., Rittle, K.E., Bock, M.G., DiPardo, R.M., Freidinger, R.M., Whitter, W.L., Lundell, G.F., Veber, D.F., Anderson, P.S., Chang, R.S.L., Lotti, V.J., Cerino, D.J., Chen, T.B., Kling, P.J., Kunkel, K.A., Springer, J.P., and Hirshfield, J. Methods for drug discovery: development of potent, selective, orally effective cholecystokinin antagonists. *J. Med. Chem.* **1988**, *31*, 2235–2246.

115 ChemNavigator, Inc., 6126 Nancy Ridge Drive, Suite 117, San Diego, CA 92121.

Website: http://www.chemnavigator.com.

116 ChemBridge Corporation, 16981 Via Tazon, Suite G, San Diego, CA 92127. Website: http://www.hit2lead.com.

117 Specs, Fleminglaan 16, 2289 CP Rijswijk, The Netherlands. Website: http://www.specs.net.

118 Maybridge plc, Trevillett, Tintagel, Cornwall PL34 OHW, UK. Website: http://www.maybridge.com.

119 ARTURSSON, P., PALM, K., and LUTHMAN, K. Caco-2 monolayers in experimental and theoretical predictions of drug transport. *Adv. Drug Deliv. Rev* 2001, *46*, 27–43.

120 KANSY, M., SENNER, F., and GUBERNATOR, K. Physicochemical high throughput screening: parallel artificial membrane permeation assay in the description of passive absorption processes. *J. Med. Chem.* 1998, *41*, 1007–1010.

121 MDS-Panlabs International, 11804 North Creek Parkway South, Bothell, WA 98011.

122 Cerep, 128 Rue Danton, BP 50601 92506 Rueil-Malmaison, France.

123 Zenas Technologies L.L.C., 5896 Fleur de Lis Drive, New Orleans, LA70124.

124 DOLLE, R.E. Comprehensive survey of combinatorial library synthesis: 2001. *J. Comb. Chem.* 2002, *4*, 369–418.

125 DOLLE, R.E. Comprehensive survey of combinatorial library synthesis: 2000. *J. Comb. Chem.* 2001, *3*, 477–517.

126 DOLLE, R.E. Comprehensive survey of combinatorial library synthesis: 1999. *J. Comb. Chem.* 2000, *2*, 383–433.

127 DOLLE, R.E. and NELSON, K.H. Jr. Comprehensive survey of combinatorial library synthesis: 1998. *J. Comb. Chem.* 1999, *1*, 235–282.

128 MERRITT, A.T. and GERRITZ, S.W. Combinatorial chemistry rumors of the demise of combinatorial chemistry are greatly exaggerated. *Curr. Opin. Chem. Biol.* 2003, *7*, 305–307.

129 VALLER, M.J. and GREEN, D. Diversity screening versus focussed screening in drug discovery. *Drug Disc. Today* 2000, *5*, 286–293.

130 ROSE, S. and STEVENS, A. Computational design strategies for combinatorial libraries. *Curr. Opin. Chem. Biol.* 2003, *7*, 331–339.

131 GOLEBIOWSKI, A., KLOPFENSTEIN, S.R., and PORTLOCK, D.E. Lead compounds discovered from libraries: Part 2. *Curr. Opin. Chem. Biol.* 2003, *7*, 308–325.

132 GOLEBIOWSKI, A., KLOPFENSTEIN, S.R., and PORTLOCK, D.E. Lead compounds discovered from libraries. *Curr. Opin. Chem. Biol.* 2001, *5*, 273–284.

133 GHOSE, A.K. and VISWANADHAN, V.N. (Eds) *Combinatorial Library Design and Evaluation – Principles, Software Tools and Applications in Drug Discovery.* Marcel Dekker, New York, 2001.

134 GOOD, A.C. and LEWIS, R.A. New methodology for profiling combinatorial libraries and screening sets: cleaning up the design process with HARPick. *J. Med. Chem.* 1997, *40*, 3926–3936.

135 MARTIN, E.J. and CRITCHLOW, R.E. Beyond mere diversity: tailoring combinatorial libraries for drug discovery. *J. Comb. Chem.* 1999, *1*, 32–45.

136 GILLET, V.J., WILLETT, P., BRADSHAW, J., and GREEN, D. Selecting combinatorial libraries to optimize diversity and physical properties. *J. Chem. Inf. Comput. Sci.* 1999, *39*, 169–177.

137 BROWN, R.D., HASSAN, M., and WALDMAN, M. Combinatorial library design for diversity, cost efficiency, and drug-like character. *J. Mol. Graph. Model.* 2000, *18*, 427–437.

138 GILLET, V.J., KHATIB, W., WILLETT, P., FLEMING, P.J., and GREEN, D.V.S. Combinatorial library design using a multiobjective genetic algorithm. *J. Chem. Inf. Comput. Sci.* 2002, *42*, 375–385.

139 GILLET, V.J., WILLETT, P., FLEMING, P.J., and GREEN, D.V.S. Designing focused libraries using MoSELECT. *J. Mol. Graph. Model.* 2002, *20*, 491–498.

140 AGRAFIOTIS, D.K. Multiobjective optimization of combinatorial libraries. *IBM J. Res. Dev.* 2001, *45*, 545–566.

141 PEARLMAN, R.S. and SMITH, K.M. Novel software tools for chemical diversity. *Perspect. Drug Disc. Des.* 1998, *9–11*, 339–353.

142 MARTIN, Y.C. Diverse viewpoints on computational aspects of molecular

diversity. *J. Comb. Chem.* **2001**, *3*, 231–250.

143 WRIGHT, T., GILLET, V.J., GREEN, D.V.S., and PICKETT, S.D. Optimizing the size and configuration of combinatorial libraries. *J. Chem. Inf. Comput. Sci.* **2003**, *43*, 381–390.

144 BEMIS, G.W. and MURCKO, M.A. Properties of known drugs. 2. Side chains. *J. Med. Chem.* **1999**, *42*, 5095–5099.

145 BEMIS, G.W. and MURCKO, M.A. The properties of known drugs. 1. Molecular frameworks. *J. Med. Chem.* **1996**, *39*, 2887–2893.

146 ERTL, P. Cheminformatics analysis of organic substituents: identification of the most common substituents, calculation of substituent properties, and automatic identification of drug-like bioisosteric groups. *J. Chem. Inf. Comput. Sci.* **2003**, *43*, 374–380.

147 KENNEDY, T. Managing the drug discovery/development interface. *Drug Disc. Today* **1997**, *2*, 436–444.

148 KUBINYI, H. Opinion: Drug research: myths, hype and reality. *Nat. Rev. Drug Disc.* **2003**, *2*, 665–668.

149 PEARLMAN R.S. and SMITH K.M. Combinatorial library design in the new century. Abstracts of Papers, *221st ACS National Meeting*. San Diego, CA, April 1–5, **2001**, BTEC-039.

150 JAMOIS, E.A., HASSAN, M., and WALDMAN, M. Evaluation of reagent-based and product-based strategies in the design of combinatorial library subsets. *J. Chem. Inf. Comput. Sci.* **2000**, *40*, 63–70.

151 FEINBERG, J. and GOODNOW, R.A. unpublished method.

152 ANDREWS, P.R., CRAIK, D.J., and MARTIN, J.L. Functional group contributions to drug-receptor interactions. *J. Med. Chem.* **1984**, *27*, 1648–1657.

153 OPREA, T.I. Chemoinformatics in lead discovery. In *Chemoinformatics in Drug Discovery*, OPREA, T.I. (Ed.). Wiley-VCH, Weinheim, 2004, 25–41.

154 OPREA, T.I., DAVIS, A.M., TEAGUE, S.J., and LEESON, P.D. Is there a difference between leads and drugs? A historical perspective. *J. Chem. Inf. Comput. Sci.* **2001**, *41*, 1308–1315.

155 LIPINSKI, C.A. Integration of physicochemical property considerations into the design of combinatorial libraries. *Pharm. News* **2002**, *9*, 195–202.

156 LIPINSKI, C.A. Avoiding investment in doomed drugs. Is poor solubility an industry wide problem? *Curr. Drug Disc.* **2001**, 17–19.

157 NORINDER, U., ÖSTERBERG, T., and ARTURSSON, P. Theoretical calculation and prediction of Caco-2 cell permeability using MolSurf parametrization and PLS statistics. *Pharm. Res.* **1997**, *14*, 1786–1791.

158 STENBERG, P., LUTHMAN, K., ELLENS, H., LEE, C.P., SMITH, P.L., LAGO, A., ELLIOTT, J.D., and ARTURSSON, P. Prediction of the intestinal absorption of endothelin receptor antagonists using three theoretical methods of increasing complexity. *Pharm. Res.* **1999**, *16*, 1520–1526.

159 PALM, K., LUTHMAN, K., UNGELL, A.-L., STRANDLUND, G., and ARTURSSON, P. Correlation of drug absorption with molecular surface properties. *J. Pharm. Sci.* **1996**, *85*, 32–39.

160 STENBERG, P., LUTHMAN, K., and ARTURSSON, P. Prediction of membrane permeability to peptides from calculated dynamic molecular surface properties. *Pharm. Res.* **1999**, *16*, 205–212.

161 CLARK, D.E. Rapid calculation of polar molecular surface area and its application to the prediction of transport phenomena. 2. Prediction of blood-brain barrier penetration. *J. Pharm. Sci.* **1999**, *88*, 815–821.

162 PALM, K., STENBERG, P., LUTHMAN, K., and ARTURSSON, P. Polar molecular surface properties predict the intestinal absorption of drugs in humans. *Pharm. Res.* **1997**, *14*, 568–571.

163 EGAN, W.J., MERZ, K.M. Jr., and BALDWIN, J.J. Prediction of drug absorption using multivariate statistics. *J. Med. Chem.* **2000**, *43*, 3867–3877.

164 ERICKSON, J.A., JALAIE, M., ROBERTSON, D.H., LEWIS, R.A., and VIETH, M. Lessons in molecular recognition: the effects of ligand and protein flexibility on molecular docking accuracy. *J. Med. Chem.* **2004**, *47*, 45–55.

165 CRAIG, P.N. Interdependence between physical parameters and selection of substituent groups for correlation studies. *J. Med. Chem.* **1971**, *14*, 680–684.

166 TOPLISS, J.G. Utilization of operational schemes for analog synthesis in drug design. *J. Med. Chem.* **1972**, *15*, 1006–1011.

167 CHIU, T.-L. and SO, S.-S. Development of neural network QSPR models for Hansch substituent constants. 1. Method and validations. *J. Chem. Inf. Comput Sci.* **2003**, *43*, 147–153.

168 WEININGER, D. SMILES, a chemical language and information system. 1. Introduction to methodology and encoding rules. *J. Chem. Inf. Comput. Sci.* **1988**, *28*, 31–36.

169 HANSCH, C. and LEO, A. *Exploring QSAR, Fundamentals and Applications in Chemistry and Biology.* ACS Professional Reference Book, ACS, Washington, DC, **1995**, Chapter 1.

170 BÖHM, H.J. and SCHNEIDER, G. (Eds) *Virtual Screening for Bioactive Molecules.* Wiley-VCH, Weinheim, **2000**.

171 TOLEDO-SHERMAN, L.M. and CHEN, D. High throughput virtual screening for drug discovery in parallel. *Curr. Opin. Drug Disc. Dev.* **2002**, *5*, 414–421.

172 DILLER, D.J. and MERZ, K.M. Jr. High throughput docking for library design and library prioritization. *Proteins: Struct., Funct. Genet.* **2001**, *43*, 113–124.

173 BOEHM, H.J., BOEHRINGER, M., BUR, D., GMUENDER, H., HUBER, W., KALUS, W., KOSTREWA, D., KUEHNE, H., LUEBBERS, T., MEUNIER-KELLER, N., and MUELLER, F. Novel inhibitors of DNA gyrase: 3D structure based biased needle screening, hit validation by biophysical methods, and 3D guided optimization. A promising alternative to random screening. *J. Med. Chem.* **2000**, *43*, 2664–2674.

174 BAJORATH, J. Integration of virtual and high throughput screening. *Nat. Rev. Drug Disc.* **2002**, *1*, 882–894.

175 WALTERS, W.P., STAHL, M.T., and MURCKO, M.A. Virtual screening – an overview. *Drug Disc. Today* **1998**, *3*, 160–178.

176 BISSANTZ, C., FOLKERS, G., and ROGNAN, D. Protein-based virtual screening of chemical databases. 1. Evaluation of different docking/scoring combinations. *J. Med. Chem.* **2000**, *43*, 4759–4767.

177 ABAGYAN, R. and TOTROV, M. High throughput docking for lead generation. *Curr. Opin. Chem. Biol.* **2001**, *5*, 375–382.

178 BÖHM, H.J. The computer program LUDI: a new method for the de novo design of enzyme inhibitors. *J. Comput.-Aided Mol. Des.* **1992**, *6*, 61–78.

179 ROE, D.C. and KUNTZ, I.D. BUILDER v.2: improving the chemistry of a de novo design strategy. *J. Comput.-Aided Mol. Des.* **1995**, *9*, 269–282.

180 KUNTZ, I.D., BLANEY, J.M., OATLEY, S.J., LANGRIDGE, R., and FERRIN, T.E. A geometric approach to macro-molecule-ligand interactions. *J. Mol. Biol.* **1982**, *161*, 269–288.

181 JONES, G., WILLETT, P., GLEN, R.C., LEACH, A.R., and TAYLOR, R. Development and validation of a genetic algorithm for flexible docking. *J. Mol. Biol.* **1997**, *267*, 727–748.

182 RAREY, M., KRAMER, B., LENGAUER, T., and KLEBE, G. A fast flexible docking method using an incremental construction algorithm. *J. Mol. Biol.* **1996**, *261*, 470–489.

183 RAREY, M., LEMMEN, C., and MATTER, H. Algorithmic engines in virtual screening. In *Chemoinformatics in Drug Discovery*, OPREA, T.I. (Ed.). Wiley-VCH, Weinheim, **2004**, 59–115.

184 BÖHM, H.-J., BANNER, D.W., and WEBER, L. Combinatorial docking and combinatorial chemistry: design of potent non-peptide thrombin inhibitors. *J. Comput.-Aided Mol. Des.* **1999**, *13*, 51–56.

185 BOEHM, H.J. LUDI: rule-based automatic design of new substituents for enzyme inhibitor leads. *J. Comput.-Aided Mol. Des.* **1992**, *6*, 593–606.

186 WANG, R., LU, Y., and WANG, S. Comparative evaluation of 11 scoring functions for molecular docking. *J. Med. Chem.* **2003**, *46*, 2287–2303.

187 SCHULZ-GASCH, T. and STAHL, M. Binding site characteristics in structure-based virtual screening: evaluation of current docking tools. *J. Mol. Model.* **2003**, *9*, 47–57.

188 Böhm, H.-J. and Stahl, M. The use of scoring functions in drug discovery applications. *Rev. Comput. Chem.* **2002**, *18*, 41–87.

189 Muegge, I. and Rarey, M. Small molecule docking and scoring. *Rev. Comput. Chem.* **2001**, *17*, 1–60.

190 Stahl, M. and Rarey, M. Detailed analysis of scoring functions for virtual screening. *J. Med. Chem.* **2001**, *44*, 1035–1042.

191 Mason, J.S., Good, A.C., and Martin, E.J. 3-D Pharmacophores in drug discovery. *Curr. Pharm. Des.* **2001**, *7*, 567–597.

192 Schneider, G., Clement-chomienne, O., Hilfiger, L., Schneider, P., Kirsch, S., Bohm, H.-J., and Neidhart, W. Virtual screening for bioactive molecules by evolutionary de novo design. *Angew. Chem., Int. Ed. Engl.* **2000**, *39*, 4130–4133.

193 Schneider, G., Neidhart, W., and Adam, G. Integrating virtual screening methods to the quest for novel membrane protein ligands. *Curr. Med. Chem. CNS Agents* **2001**, *1*, 99–112.

194 Danho, W., Swistok, J., Khan, W., Truitt, T., Kurylko, G., Fry, D., Greeley, D., Sun, H., Dvorozniak, M., Mackie, G., Spence, C., and Goodnow, Jr, R. Structure-activity relationship and bioactive conformation of prolactin releasing peptides (PrRPs), ligand for a potential obesity target. In *Proceedings of the 18th American Peptide Symposium*, Boston, July 19–23, Chorev, M. and Sawyer, R.K. (Eds). **2003**.

195 Catalyst Tutorial, Release 4.6, Accelrys Inc., 9685 Scranton Road, San Diego, CA 92121.

196 Flohr, S., Kurz, M., Kostenis, E., Brkovich, A., Fournier, A., and Klabunde, T. Identification of nonpeptidic urotensin II receptor antagonists by virtual screening based on a pharmacophore model derived from structure-activity relationships and nuclear magnetic resonance studies on urotensin II. *J. Med. Chem.* **2002**, *45*, 1799–1805.

197 Patchett, A. and Nargund, R.P. Privileged structures – an update. *Annu. Rev. Med. Chem.* **2000**, *35*, 289–298.

198 Mason, J.S., Morize, I., Menard, P.R., Cheney, D.L., Hulme, C., and Labaudiniere, R.F. New 4-point pharmacophore method for molecular similarity and diversity applications: overview of the method and applications, including a novel approach to the design of combinatorial libraries containing privileged substructures. *J. Med. Chem.* **1999**, *42*, 3251–3264.

199 A. Smellie, unpublished method.

16

Application of Predictive QSAR Models to Database Mining

Alexander Tropsha

16.1
Introduction

The field of Quantitative Structure-Activity Relationships (QSAR), as an integral part of computer-aided drug design and discovery, is experiencing one of the most exciting periods in its history. Modern QSAR approaches are characterized by the use of multiple descriptors of chemical structure combined with the application of both linear and nonlinear optimization approaches and a strong emphasis on rigorous model validation to afford robust and predictive QSAR models. The most important recent developments in the field concur with a substantial increase in the size of experimental datasets available for the analysis and an increased application of QSAR models as virtual screening tools to discover biologically active molecules in chemical databases and/or virtual chemical libraries. The latter focus differs substantially from the traditional emphasis on developing so-called explanatory QSAR models characterized by high statistical significance but only as applied to training sets of molecules with known chemical structure and biological activity.

Recent trends in both 2-D and 3-D QSAR studies have focused on the development of optimal QSAR models through variable selection. This procedure selects only a subset of available chemical descriptors – those that are most meaningful and statistically significant in terms of correlation with biological activity. The optimum selection of variables is achieved by combining stochastic search methods with correlation methods such as MLR, partial least squares (PLS) analysis, or artificial neural networks (ANN) (e.g., [1–4]). More specifically, these methods employ generalized simulated annealing [1], genetic algorithms, [2] or evolutionary algorithms [3] as the stochastic optimization tool. It has been demonstrated that such methods combined with various chemometric tools effectively improve QSAR models, compared to those without variable selection.

Quantitative Structure-Activity Relationship models are used increasingly in chemical data mining and combinatorial library design [5, 6]. For example, three-dimensional (3-D) stereoelectronic pharmacophore based on QSAR modeling was used recently to search the National Cancer Institute Repository of Small Molecules [7] to find new leads for inhibiting HIV type 1 reverse transcriptase at the nonnucleoside binding site [8]. A descriptor pharmacophore concept was introduced by us recently [9] on the basis of variable selection QSAR: the descriptor pharmacophore is defined as a subset of

Chemoinformatics in Drug Discovery. Edited by Tudor I. Oprea
Copyright © 2004 WILEY-VCH Verlag GmbH & Co. KGaA, Weinheim
ISBN: 3-527-30753-2

molecular descriptors that afford the most statistically significant QSAR model. It has been demonstrated that chemical similarity searches using descriptor pharmacophores, as opposed to using all descriptors, afford more efficient mining of chemical databases or virtual libraries to discover compounds with the desired biological activity [9, 10].

Many reviews discussing different QSAR-modeling methodologies have been published (e.g., see [11, 12] and references therein). This chapter concentrates on the utility of QSAR models for virtual screening of chemical databases or virtual libraries to discover bioactive compounds of potential pharmaceutical significance. This approach to virtual screening is less common than structure-based methods utilizing docking and scoring (for recent reviews, see [13–15]). However, as we discuss and illustrate in this chapter, rigorously validated QSAR models with statistically proven predictive power afford their application as reliable *virtual screening* tools for database mining or chemical library design.

16.2
Building Predictive QSAR Models: The Importance of Validation

The process of QSAR model development is typically divided into three steps: data preparation, data analysis, and model validation. The implementation of these steps is generally determined by the researchers' interests, experience, and software availability. The resulting models are then frequently employed, at least in theory, to design new molecules based on chemical features or trends found statistically significant with respect to underlying biological activity. For instance, the popular 3-D QSAR approach, CoMFA (Comparative Molecular Field Analysis), [16] makes suggestions regarding steric or electronic modifications of the training set compounds that are likely to increase their activity.

The first stage includes the selection of a dataset for QSAR studies and the calculation of molecular descriptors. The second stage deals with the selection of a statistical data analysis and correlation technique, either linear or nonlinear such as PLS or ANN. Many different algorithms and computer software are available for this purpose; in all approaches, descriptors serve as independent variables and biological activities serve as dependent variables.

Typically, the final part of QSAR model development is the model validation [17, 18], when the predictive power of the model is tested on an independent set of compounds. In essence, predictive power is one of the most important characteristics of QSAR models. It can be defined as the ability of a model to predict accurately the target property (e.g., biological activity) of compounds that were not used for model development. The typical problem of QSAR modeling is that at the time of the model development a researcher only has, essentially, training set molecules, so predictive ability can only be characterized by statistical characteristics of the training set model and not by true external validation.

Most of the QSAR-modeling methods implement the leave-one-out (LOO) (or leave-some-out) cross-validation procedure. The outcome from this procedure is a cross-validated correlation coefficient q^2, which is calculated according to the following formula:

$$q^2 = 1 - \frac{\sum (y_i - \hat{y}_i)^2}{\sum (y_i - \bar{y})^2},$$

(1)

where y_i, $\hat{y}_i$, and $\bar{y}$ are the actual, estimated by LOO cross-validation procedure, and the average activities, respectively. The summations in (1) are performed over all compounds that are used to build a model (training set). Frequently, q^2 is used as a criterion of both robustness and predictive ability of the model. Many authors consider a high q^2 (for instance, $q^2 > 0.5$) as an indicator or even as the ultimate proof of the high predictive power of a QSAR model. They do not test the models for their ability to predict the activity of compounds of an external test set (i.e., compounds that have not been used in the QSAR model development). For instance, recent publications [19–22] provide several examples in which the authors claimed that their models had high predictive ability without even validating them with an external test set. Some authors validate their models using only one or two compounds that were not used in QSAR model development [23, 24], and still claim that their models are highly predictive.

A widely used approach to establish model robustness is the randomization of response [25] (i.e., in our case of activities). It consists of repeating the calculation procedure with randomized activities and subsequent probability assessments of the resultant statistics. Frequently, it is used along with the cross validation. Sometimes, models based on the randomized data have high q^2 values, which can be explained by a chance correlation or structural redundancy [26]. If all QSAR models obtained in the Y-randomization test have relatively high values for both R^2 and LOO q^2, it implies that an acceptable QSAR model cannot be obtained for the given dataset by the current modeling method.

The following example illustrates the danger of developing a QSAR model not subjected to the Y-randomization test. Recently, a CoMFA study of 16 antagonists of the dopamine D_2 receptor [27] reported q^2 values exceeding 0.9 for the training set, and a test set containing three compounds only produced an R^2 of 0.99. The technique used differed from standard CoMFA methodology in that the conformation of each compound was individually adjusted based upon the magnitude of prediction error. This process was repeated until the model could no longer be improved, after which the CoMFA columns that conflicted with the experimental results were eliminated from the model. This resulted in the generation of models containing only 3 to 7% of the total field information produced within CoMFA (68–147 columns used out of 2112–2376). With this large number of descriptors, there will be a small subset, just by mere chance, whose variance correlates with the target property for this small training set. A significant bulk of the paper was then devoted to the discussion of the resulting CoMFA contour plots in terms of their guidance for the future design of D_2 antagonists. However, a simple Y-randomization test (Oloff and Tropsha, unpublished) demonstrates that with a similar technique, many models with acceptable values of LOO q^2 could be obtained for the same dataset with a randomly created **Y**-vector. Clearly, there was no real structure-activity relationship and, consequently, any interpretation of this or any other QSAR model created in this fashion is spurious.

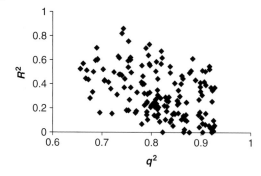

Fig. 16.1 Beware of q^2! External R^2 (for the test set) shows no correlation with the "predictive" LOO q^2 (for the training set). (Adapted from Ref. [17]).

Thus, it is still uncommon to test QSAR models (characterized by a reasonably high q^2) for their ability to predict accurately biological activities of compounds not included in the training set. In contrast to such expectations, it has been shown that if a test set with known values of biological activities is available for prediction, there exists no correlation between the LOO cross-validated q^2 and the correlation coefficient R^2 between the predicted and observed activities for the test set (Figure 16.1). In our experience [17, 28], this phenomenon is characteristic of many datasets and is independent of the descriptor types and optimization techniques used to develop training set models. In a recent review, we emphasized the importance of external validation in developing reliable models [18].

As was suggested in several recent publications by both our colleagues [28–31] and by us [18, 32], the only way to ensure the high predictive power of a QSAR model is to demonstrate a significant correlation between predicted and observed activities of compounds for a validation (test) set, which was not employed in model development. We have shown [18, 29] that various *commonly accepted* statistical characteristics of QSAR models derived for a training set are *insufficient* to establish and estimate the predictive power of QSAR models. We emphasize that external validation must be made, in fact, a mandatory part of model development. One can achieve this goal by dividing the experimental SAR dataset into a training set and a test set, which are used for model development and validation, respectively. We believe that special approaches should be used to select a training set to ensure the highest significance and predictive power of QSAR models [17, 33]. Our recent reviews and publications describe several algorithms that can be employed for such division [18, 29, 33].

In order to estimate the true predictive power of a QSAR model, one needs to compare the predicted and observed activities of a sufficiently large external test set of compounds that were not used in the model development. One convenient parameter is an external q^2 defined as follows (similar to Eq. (1) for the training set):

$$q_{\text{ext}}^2 = 1 - \frac{\sum\limits_{i=1}^{\text{test}} (y_i - \hat{y}_i)^2}{\sum\limits_{i=1}^{\text{test}} (y_i - \bar{y}_{\text{tr}})^2} \tag{2}$$

where y_i and $\hat{y}_i$ are the measured and predicted (over the test set) values, respectively, of the dependent variable and $\bar{y}_{tr}$ is the averaged value of the dependent variable for the training set; the summations run over all compounds in the test set. Certainly, this formula is only meaningful when $\bar{y}_{tr}$ does not differ significantly from the similar value for the test set [34]. In principle, given the entire collection of compounds with known structure and activity, there is no specific reason to select one particular group of compounds as a training (or test) set; thus, the division of the dataset into multiple training and test sets [33] or an interchangeable definition of these sets [35] is recommended.

The use of the following statistical characteristics of the test set was also recommended [33]: (1) correlation coefficient R^2 between the predicted and observed activities, (2) coefficients of determination (predicted versus observed activities R_0^2, and observed versus predicted activities $R_0'^2$), (3) slopes k and k' of the regression lines through the origin. Thus, we consider a QSAR model predictive if the following conditions are satisfied [33]:

$$q^2 > 0.5 \tag{3}$$

$$R^2 > 0.6 \tag{4}$$

$$\frac{(R^2 - R_0^2)}{R^2} < 0.1 \quad \text{or} \quad \frac{(R^2 - R_0'^2)}{R^2} < 0.1 \tag{5}$$

$$0.85 \leq k \leq 1.15 \quad \text{or} \quad 0.85 \leq k' \leq 1.15 \tag{6}$$

We have demonstrated [29, 33] that all of the above criteria are indeed necessary to adequately assess the predictive ability of a QSAR model.

16.3
Defining Model Applicability Domain

It needs to be emphasized that no matter how robust, significant, and validated a QSAR may be, it cannot be expected to reliably predict the modeled property for the entire universe of chemicals. Therefore, before a QSAR model is put into use for screening chemicals, its domain of application must be defined and predictions for only those chemicals that fall in this domain should be considered reliable. Some approaches that aid in defining the applicability domain are described below.

Extent of Extrapolation   For a regression-like QSAR, a simple measure of a chemical being too far from the applicability domain of the model is its leverage, h_i [36], which is defined as

$$h_i = x_i^T (X^T X)^{-1} x_i \quad (i = 1, \ldots, n) \tag{7}$$

where x_i is the descriptor row vector of the query compound, and X is the $n \times k - 1$ matrix of k model descriptor values for n training set compounds. The superscript T refers to the transpose of the matrix/vector. The warning leverage h^* is, generally, fixed at $3k/n$, where n is the number of training compounds and k is the number of model parameters. A leverage greater than the warning leverage h^* means that the predicted response is the result of substantial extrapolation of the model and, therefore, may not be reliable [37, 38].

Effective Prediction Domain   Similarly, for regression-like models, especially when the model descriptors are significantly correlated, Mandel [39] proposed the formulation of effective prediction domain (EPD). It has been demonstrated, with examples, that a regression model is justified inside and on the periphery of the EPD. Clearly, if a compound is determined to be too far from the EPD, its prediction from the model should not be considered reliable.

Residual Standard Deviation   Another important approach that can be used to evaluate the applicability domain is the degree-of-fit method developed originally by Lindberg et al. [40] and modified recently by Cho et al. [6]. According to the original method, the predicted y values are considered to be reliable if the following condition is met:

$$s^2 < s_a^2(E_x)F \tag{8}$$

where s^2 is the residual standard deviation (RSD) of descriptor values generated for a test compound, $s_a^2(E_x)$ is the RSD of the $\mathbf{X}$ matrix after dimensions (components) a, and F is the F-statistic at the probability level α and $(p - a)/2$ and $(p - a)(n - a - 1)/2$ degrees of freedom. The RSD of descriptor values generated for a test compound is calculated using the following equation:

$$s^2 = \|e\|/(p - a) \tag{9}$$

where p is the number of x-variables, a is the number of components, and $\|e\|$ is the sum of squared residuals e_i expressed as

$$e_i = x_i - x_i\mathbf{BB}' \tag{10}$$

where x_i is the i-th x-variable, and $\mathbf{B}$ and $\mathbf{B}'$ represent the weight matrix and transposed weight matrix of x-variables, respectively. Since the lowest possible value of F is 1.00 at $\alpha = 0.10$ (when both degrees of freedom are equal to infinity), we decided to replace F with the degree-of-fit factor f to simplify the above condition. Thus, the modified degree-of-fit condition [6] is as follows: predicted y values are considered to be reliable if

$$s^2 < s_a^2(E_x)f \tag{11}$$

Similarity Distance   In the case of a nonlinear method such as the k Nearest Neighbor (kNN) QSAR [41], since the models are based on chemical similarity calculations, a large similarity distance could signal query compounds that are too dissimilar to the training set compounds. We have proposed [41] a cutoff value, $\mathbf{D}_c$ (Eq. 12), that defines a similarity distance threshold for external compounds.

$$\mathbf{D}_c = \mathbf{Z}\sigma + \gamma \tag{12}$$

Here γ is the average and σ is the standard deviation of the Euclidean distances of the k nearest neighbors of each compound in the training set in the chemical descriptor space, and $\mathbf{Z}$ is an empirical parameter to control the significance level, with the default value of 0.5. If the distance from an external compound to its nearest neighbor in the training set is above $\mathbf{D}_c$, we label its prediction unreliable.

16.4
Validated QSAR Modeling as an Empirical Data-modeling Approach: Combinatorial QSAR

In QSAR modeling, the input data can be represented as $N \times M$ matrices where N rows correspond to chemical compounds and M columns contain values of independently calculated chemical descriptors except for the first column, which contains the values of the biological activity (or some target property, e.g., compound toxicity). In the statistical data analysis sense, the descriptors are viewed as independent variables, and the target property is regarded as a dependent variable. The task of statistical data modeling as applied to experimental datasets, represented in the form of the above matrix, is to establish an empirical relationship (linear or nonlinear) between the independent variables and the dependent property that is characterized by the high value of the correlation coefficient between actual and predicted values of the biological activity (reviewed in [11]).

In recent years, our group has concentrated on the development of the general principles of validated QSAR modeling [18]. As discussed above, we argue that, in order to be reliable and predictive, QSAR models should (1) be statistically significant and robust, (2) be validated by making accurate predictions for external datasets that were not used in the model development, and (3) have their application boundaries defined. We believe that QSAR modeling is an empirical, exploratory research in which the models with the best validated predictive power should be sought by combinatorial exploration of various combinations of statistical data-modeling techniques and different types of chemical descriptors [42].

Our extensive experience with QSAR modeling led to the following strategy that incorporates several critical steps [18]. (1) Data preparation, which includes (*i*) collection and curation of the target property data; (*ii*) calculation of molecular descriptors for chemicals with acceptable target properties; and (*iii*) merging of the property and descriptor values in a manageable SAR database. We emphasize that the quality of biological data is of utmost importance; a recent review [43] provides an example of a significant decrease in the external predictive power of the CoMFA QSAR model for HIV protease inhibitors when the model was reevaluated using correct binding affinity data. (2) Model generation, which implies establishing statistically significant relationships between target property and descriptor values using linear or nonlinear modeling techniques. (3) Model validation, which implies quantitative assessment of the model robustness and its predictive power based on rational division of the original dataset into the training and test sets [29, 33]. (4) Definition of the application domain of the model in the space of chemical descriptors used in deriving the model [18, 33].

Most recently, we have realized the importance of both the combinatorial approach to QSAR modeling and the use of consensus prediction of activities for external compounds by averaging the predicted activity values resulting from all validated models [42]. This strategy is driven by the concept that if an implicit structure-activity relationship exists for a given dataset, it can be formally manifested via a variety of QSAR models utilizing different descriptors and optimization protocols. We believe that multiple alternative QSAR models should be developed (as opposed to a single model using some favorite QSAR method) for each dataset. There are several popular commercial and noncommercial software packages that provide users with various descriptor types and data-modeling capabilities. However, practically every package employs only one (or a few) type of descriptors (see [11] for the discussion of various 2-D and 3-D descriptors) and, typically, a single (or a few) molecular modeling technique(s). For instance, popular QSAR-modeling packages such as CoMFA [16], Catalyst [44], binary QSAR as implemented by the CCG [45], CASE [46], and TopCAT [47] provide a relatively easy-to-use interface and allow users to build single models with (if successful) reasonable internal accuracy typically characterized by q^2. However, as emphasized in the previous section, the training set modeling alone is insufficient to afford models with validated predictive power; therefore, the QSAR model development process has to be modified to incorporate an independent model validation and applicability domain definition [18, 33]. Since QSAR modeling is relatively fast (and in principle, can be completely automated), these alternative models could be explored simultaneously while making predictions for external datasets. Consensus predictions of biological activity for novel compounds based on several QSAR models, especially when predictions converge, provide more confidence in the activity estimates and better justification for the experimental validation of these compounds. Finally, we emphasize the use of molecular descriptors (and descriptor pharmacophores) of chemical structures as opposed to chemical fragments, which allows building QSAR models for more chemically diverse molecules.

Our combinatorial approach to QSAR model building is summarized on the workflow diagram (Figure 16.2). To achieve QSAR models of the highest internal, and most importantly, external accuracy, the combinatorial QSAR approach explores all possible combinations of various descriptor types and optimization methods along with external model validation. The need to develop and employ the combinatorial QSAR approach is dictated by our experience in QSAR modeling, which suggests, as mentioned above, that QSAR is a highly experimental area of statistical data modeling, where it is impossible to decide *a priori* as to which particular QSAR-modeling method will prove most successful. Every particular combination of descriptor sets and optimization techniques is likely to capture certain unique aspects of the structure-activity relationship. Since our ultimate goal is to use the resulting models in database mining to discover *diverse* biologically active molecules, application of different combination of modeling techniques and descriptor sets shall increase our chances for success.

16.5
Validated QSAR Models as Virtual Screening Tools

Although combinatorial chemistry and HTS have offered medicinal chemists a much broader range of possibilities for lead discovery and optimization, the number of chemical compounds that can be reasonably synthesized, which is sometimes called "virtual chemistry space", is still far beyond today's capability of chemical synthesis and biological assay. Therefore, medicinal chemists continue to face the same problem as before: which compounds should be chosen for the next round of synthesis and testing? For chemoinformaticians, the task is to develop and utilize various computer programs to evaluate a very large number of chemical compounds and recommend the most promising ones for bench medicinal chemists. This process can be called *virtual*

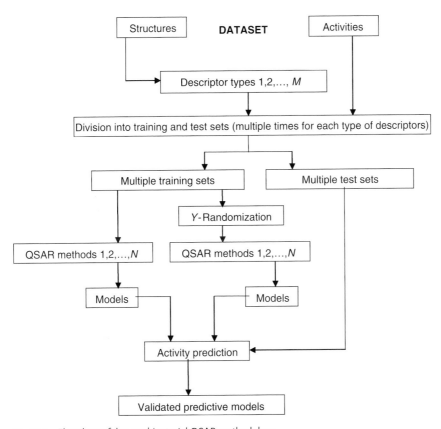

Fig. 16.2   Flowchart of the combinatorial QSAR methodology.

screening or chemical database searching [48, 49]. A large number of computational methods exist for virtual screening, but which one is chosen will depend on the information available and the task at hand in practice.

Database mining associated with pharmacophore identification is a common and efficient approach for lead compounds discovery. Pharmacophore identification refers to the computational way of identifying the essential 3-D structural features and configurations that are responsible for the biological activity of a series of compounds. Once a pharmacophore model has been developed for a particular set of biologically active molecules, it can be used to search databases of 3-D structures with the aim of finding new, structurally different lead molecules with the desired biological activity, thus playing an increasingly important role in this phase of drug development projects [50].

An obvious parallel can be established between the search for pharmacophoric elements that are mainly responsible for the specificity of drug action and the identification of descriptors contributing the most to the correlation with biological activity. The selection of specific pharmacophoric features responsible for biological activity is directly analogous to the selection of specific chemical descriptors that contribute the most to an explanatory QSAR model. Recently, we have demonstrated that kNN-QSAR models can be used in database mining, that is, finding molecular structures that are similar in their activity to the probe molecules, or even predicting the activities for the compounds in a database [5, 9]. First, a preconstructed QSAR model can be used as a means of screening compounds from existing databases (or virtual libraries) for high-predicted biological activity. Alternatively, variables selected by QSAR optimization can be used for similarity searches to improve the performance of the database mining methods (Figure 16.3).

The advantage of using QSAR models for database mining is that it affords not only the compounds selection but also quantitative prediction of their activity. For illustration, we shall discuss our recent success in developing validated predictive models of anticonvulsants [51] and their application to the discovery of novel potent compounds by the means of database mining [10].

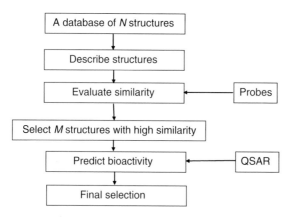

Fig. 16.3 Flowchart of database mining using QSAR models.

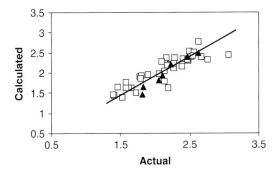

Fig. 16.4 The best kNN model for the training (squares) and test (triangles) sets of anticonvulsants ($q^2 = 0.79$, $R^2 = 0.90$).

Initially, we had applied kNN and SA-PLS QSAR approaches to a dataset of 48 chemically diverse functionalized amino acids with anticonvulsant activity that were synthesized previously, and successful QSAR models of FAA anticonvulsants were developed (Figure 16.4) [51]. Both methods utilized multiple descriptors such as molecular connectivity indices or atom pair descriptors, which are derived from two-dimensional molecular topology. QSAR models with high internal accuracy were generated, with leave-one-out cross-validated R^2 (q^2) values ranging between 0.6 and 0.8. The q^2 values for the actual dataset were significantly higher than those obtained for the same dataset with randomly shuffled activity values, indicating that models were statistically significant. The original dataset was further divided into several training and test sets, and highly predictive models providing q^2 values greater than 0.5 for the training sets and R^2 values greater than 0.6 for the test sets (Figure 16.4).

In the second stage of this process, we had applied the validated QSAR models and descriptor pharmacophore concepts (cf. Figure 16.3) to mine available chemical databases for new lead FAA anticonvulsant agents. Two databases had been thoroughly explored: the National Cancer Institute [7] and the Maybridge [52] databases, including 237 771 and 55 273 chemical structures, respectively. Database mining was performed independently using 10 individual QSAR models that had been extensively validated using several criteria of robustness and accuracy. Descriptor pharmacophores were also used to calculate the distance matrix for similarity analysis. Each individual QSAR model selected a number of database mining hits independently, and the consensus hits (i.e., those selected by all models) were further explored experimentally for their anticonvulsant activity.

Experimental Validation I As a result of computational screening of the NCI database, 27 compounds were selected as potential anticonvulsant agents and submitted to our experimental collaborators. Of these 27 compounds, our collaborators selected two (compounds **C4** and **C6** in Table 16.1) for synthesis and evaluation; their choice was based on the ease of synthesis and the fact that these two compounds featured a fully modified amino group, which would not be expected to be active, on the basis of prior experience. An interesting feature of these compounds was the presence of a terminal Carbobenzyloxy (Cbz) group, a feature that was absent in any of the training set compounds [51]. Several additional compounds, which were close analogs of these two, were either taken from the literature (**C1**) or designed in our collaborator's laboratory. In total, seven compounds were resynthesized and sent to the NIH for the MES test (a standard test for the anticonvulsant activity, which was used for the training set compounds as well). The chemical structures of all compounds and the results of testing are shown in Table 16.1. All seven compounds were tested in parallel on both mice and rats. Each compound was tested in three doses, 300, 100, and 30 mg kg^{-1} in mice, and only in one dose, 30 mg kg^{-1}, in rats. The biological results indicate that upon initial and secondary screening, *five out of seven compounds tested showed anticonvulsant activity with* ED$_{50}$ *less than* 100 mg kg^{-1}, *which is considered promising by the* NIH *standard* (Table 16.1). Interestingly, all seven compounds were also found to be very active in the same tests performed on rats (a complete set of experimental data on rats for the training set was not available, and therefore no QSAR models for rats were built).

Experimental Validation II   Mining of the Maybridge database yielded two compounds that were synthesized and sent to NIH for the MES anticonvulsant test. *Very* promising and exciting results have also been obtained. One of the compounds (**C9**) shows moderate anticonvulsant activity of ED$_{50}$ between 30 and 100 mg kg^{-1} (in mice), while the other (**C8**) is a *very* potent anticonvulsant agent with ED$_{50}$ of 18 mg kg^{-1} in mice (ip). In summary, both compounds were found to be very active in both mice and rats (Table 16.2).

Figure 16.5 summarizes our approach to using validated QSAR models for virtual screening as applied to the anticonvulsant dataset. It presents a practical example of the drug discovery workflow that can be generalized for any dataset in which sufficient data to develop reliable QSAR models is available.

These results certainly appear very promising and reassuring in terms of our computational strategies, which emphasize that rigorous validation of QSAR models as well as conservative extrapolation are responsible for a very high hit rate (Tables 16.1 and 16.2). However, additional developments of methodology are certainly required to improve model accuracy since quantitative agreement between actual and predicted anticonvulsant activity is not excellent.

It is important to note that *none* of the compounds identified in external databases as potent anticonvulsants and validated experimentally belongs to the same class of FAA molecules as the training set. This observation is very stimulating because it underscores the power of our methodology to identify potent anticonvulsants of novel chemical classes as compared to the training set compounds, which is one of the most important goals of virtual screening.

Tab. 16.1 Results of anticonvulsant activity testing from anticonvulsant screening project at the National Institutes of Health. Data for most promising compounds are in bold

Cpd structure (hydrogen depleted graphs)	ID No.	mp[c]	Mice (ip)[a]		Rats (po)[b]		Predicted mice (ip)[f]
			MES,[d] ED$_{50}$ (mg kg^{-1})	Tox,[e] TD$_{50}$ (mg kg^{-1})	MES,[d] ED$_{50}$ (mg kg^{-1})	Tox,[e] TD$_{50}$ (mg kg^{-1})	MES ED$_{50}$ (mg kg^{-1})[g]
	C1	–	>30, <100	>100, <300	<30	>100 (ip)	64.3
	C2	99–100	>100, <300	>100, <300	52 [1.0] (28–78)	>500	35.0
	C3	57–58	**43 [0.25] (41–46)**	133 [0.25] (118–142)	<30	>30	41.3
	C4	40–41	**52 [0.25][g] (51–53)**	91 [0.25] (76–114)	~30	>30	17.0

(continued overleaf)

Tab. 16.1 (*Continued*)

Cpd structure (hydrogen depleted graphs)	ID No.	mp[c]	Mice (ip)[a]		Rats (po)[b]		Predicted mice (ip)[f]
			MES,[d] ED50 (mg kg⁻¹)	Tox,[e] TD50 (mg kg⁻¹)	MES,[d] ED50 (mg kg⁻¹)	Tox,[e] TD50 (mg kg⁻¹)	MES ED50 (mg kg⁻¹)[g]
	C5	76–77	86 [0.25] (79–95)	141 [0.25] (126–161)	<30	>30	22.4
	C6	63–64	>100, <300	>100, <300	~30	>30	12.1
	C7	Oil	74 [0.25] (69–79)	126 [0.25] (111–147)	<30	>30	27.7

a) The compounds were administered intraperitoneally. ED50 and TD50 values are in milligrams per kilogram. Numbers in parenthesis are 95% confidence intervals.
b) The dose effect data was obtained at the "time of peak effect" (indicated in hours in the brackets).
c) The compounds were administered orally.
c) Melting points (°C) are uncorrected.
d) MES = maximal electroshock seizure test.
e) Tox = neurologic toxicity determined from rotorod test.
f) [10, 49]
g) ED50 = 52 mg kg⁻¹ after 0.25 h upon ip administration in rats (3/4 animals protected).

Tab. 16.2 Results of anticonvulsant activity testing from anticonvulsant screening project at the National Institutes of Health (Maybridge database)

Cpd structure	NIH ADD No.	mp	Mice (ip)		Rats (po)		Predicted mice (ip)
			MES, ED_{50} $(mg\,kg^{-1})$	Tox, TD_{50} $(mg\,kg^{-1})$	MES, ED_{50} $(mg\,kg^{-1})$	Tox, TD_{50} $(mg\,kg^{-1})$	ED_{50} $(mg\,kg^{-1})$
	C8	48–49	18 [0.25] (13–24)	50 [0.25] (34–69)	11.2	>500	37.5
	C9	128–129	>30, <100	>100, <300	~30	>30	29.6

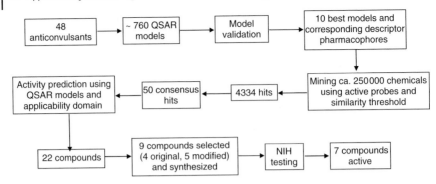

Fig. 16.5 Computer-aided drug discovery workflow based on combination of QSAR modeling and consensus database mining as applied to the discovery of novel anticonvulsants [10]. The workflow emphasizes the importance of model validation and applicability domain in ensuring high hit rates as a result of database mining with predictive QSAR models.

16.6
Conclusions and Outlook

In this chapter, we have reviewed modeling approaches for the development of the externally validated and predictive QSAR models that can be employed in mining chemical databases or virtual compound libraries for novel bioactive compounds. We stress that any approach to QSAR model development should be rigorous and comprehensive. We have discussed several criteria of model robustness, placing particular emphasis on the external predictive power of the model. We have demonstrated that the traditional LOO q^2 is the insufficient statistical characteristic of model predictive power; although low values of q^2 almost certainly indicate that the model is not expected to be predictive externally, high values of this parameter alone also provide no guarantee that the underlying models can be used for the reliable external prediction. Our recommended scheme to validated QSAR model development includes the following steps. (1) The initial stage of any modeling method is to achieve the highest value of cross-validated R^2 (q^2) using the leave-one-out cross-validation method. (2) In order to test the predictive ability of the final model(s), we use external prediction R^2. In this process, a fraction of compounds that are not used to develop the initial models are used, and the activity of these compounds is then predicted from the resulting model(s), yielding the external prediction R^2. (3) The robustness of a model is tested by comparing it with models derived for random datasets. The latter are generated by multiple random shuffling of target properties (i.e., biological activities) of the original dataset. Only if the difference between the q^2 values of the original and random datasets is sufficiently large (0.3 and higher) is the original model considered acceptable. (4) The externally validated models can then be applied to mine external databases or virtual libraries using applicability domain to exclude compounds that are too dissimilar to the training set to make any reliable prediction of their activity. We have demonstrated that the application of these principles to the development of the predictive QSAR models of anticonvulsant agents, [49] and their application to database mining led to the discovery of novel anticonvulsant agents [10] (Tables 16.1 and 16.2 and Figure 16.5).

This chapter emphasizes the emerging shift from more traditional use of QSAR modeling for lead optimization toward application of these models for lead discovery in chemical databases or virtual libraries. This shift is dictated by the rapidly growing sizes and diversity of both experimental datasets with known biological activity that is amenable to QSAR modeling and inexpensive chemical libraries and databases that are available for virtual screening and experimental validation, as well as by continuing exploration of sophisticated data-modeling techniques for building QSAR models in multidimensional spaces of chemical descriptors. We conclude that continued emphasis on the robustness and *external* predictive power as well as applicability domain as the most important parameters of QSAR models shall increase their practical use as reliable virtual screening tools for pharmaceutical lead identification.

References

1 ROGERS, D. HOPFINGER, A.J. Application of genetic function approximation to quantitative structure-activity relationships and quantitative structure-property relationships. *J. Chem. Inf. Comput. Sci.* **1994**, *34*, 854–866.

2 KUBINYI, H. Variable selection in QSAR studies. I. An evolutionary algorithm. *Quantum Struct.-Act. Relat.* **1994**, *13*, 285–294.

3 SO, S.S. KARPLUS, M. Evolutionary optimization in quantitative structure-activity relationship: an application of genetic neural networks. *J. Med. Chem.* **1996**, *39*, 1521–1530.

4 SO, S.S. KARPLUS, M. Genetic neural networks for quantitative structure-activity relationships: improvements and application of benzodiazepine affinity for benzodiazepine/GABAA receptors. *J. Med. Chem.* **1996**, *39*, 5246–5256.

5 TROPSHA, A., CHO, S.J., ZHENG, W. "New Tricks for an Old Dog": development and application of novel QSAR methods for rational design of combinatorial chemical libraries and database mining. In *Rational Drug Design: Novel Methodology and Practical Applications*, ACS Symposium Series Vol. 719, PARRILL, A.L., REDDY, M.R. (Eds). **1999**, 198–211.

6 CHO, S.J., ZHENG, W., TROPSHA, A. Rational combinatorial library design. 2. Rational design of targeted combinatorial peptide libraries using chemical similarity probe and the inverse QSAR approaches. *J. Chem. Inf. Comput. Sci.* **1998**, *38*, 259–268.

7 NCI. http://dtp.nci.nih.gov/docs/ 3d_database/structural_information/ smiles_strings.html, **2004**.

8 GUSSIO, R., PATTABIRAMAN, N., KELLOGG, G.E., ZAHAREVITZ, D.W. Use of 3D QSAR methodology for data mining the National Cancer Institute Repository of Small Molecules: application to HIV-1 reverse transcriptase inhibition. *Methods* **1998**, *14*, 255–263.

9 TROPSHA, A. ZHENG, W. Identification of the descriptor pharmacophores using variable selection QSAR: applications to database mining. *Curr. Pharm. Des.* **2001**, *7*, 599–612.

10 SHEN, M., BEGUIN, C., GOLBRAIKH, A., STABLES, J.P., KOHN, H., TROPSHA, A. Application of predictive QSAR models to database mining: identification and experimental validation of novel anticonvulsant compounds. *J. Med. Chem.* **2004**, *47*, 2356–2364.

11 TROPSHA, A. Recent trends in quantitative structure-activity relationships.

In *Burger's Medicinal Chemistry and Drug Discovery*, Vol. 1, ABRAHAM, D. (Ed.). John Wiley & Sons, New York, **2003**, 49–77.

12 OPREA, T.I. 3D-QSAR modeling in drug design. In *Computational Medicinal Chemistry and Drug Discovery*, TOLLENAERE, J., DE WINTER, H., LANGENAEKER, W., BULTINCK, P. (Eds). Marcel Dekker, New York, **2004**, 571–616.

13 WONG, C.F. MCCAMMON, J.A. Protein flexibility and computer sided drug design. *Annu. Rev. Pharmacol. Toxicol.* **2003**, *43*, 31–45.

14 TAYLOR, R.D., JEWSBURY, P.J., ESSEX, J.W. A review of protein-small molecule docking methods. *J. Comput.-Aided Mol. Des.* **2002**, *16*, 151–166.

15 MUEGGE, I. Selection criteria for drug-like compounds. *Med. Res. Rev.* **2003**, *23*, 302–321.

16 CRAMER III, R.D., PATTERSON, D.E., BUNCE, J.D. Recent advances in comparative molecular field analysis (CoMFA). *Prog. Clin. Biol. Res.* **1989**, *291*, 161–165.

17 GOLBRAIKH, A. TROPSHA, A. Beware of q2! *J. Mol. Graph. Model.* **2002**, *20*, 269–276.

18 TROPSHA, A., GRAMATICA, P., GOMBAR, V.K. The importance of being earnest: validation is the absolute essential for successful application and interpretation of QSPR models. *Quantum Struct.-Act. Relat. Combust. Sci.* **2003**, *22*, 69–77.

19 GIRONES, X., GALLEGOS, A., CARBO-DORCA, R. Modeling antimalarial activity: application of kinetic energy density quantum similarity measures as descriptors in QSAR. *J. Chem. Inf. Comput. Sci.* **2000**, *40*, 1400–1407.

20 BORDAS, B., KOMIVES, T., SZANTO, Z., LOPATA, A. Comparative three-dimensional quantitative structure-activity relationship study of safeners and herbicides. *J. Agric. Food Chem.* **2000**, *48*, 926–931.

21 FAN, Y., SHI, L.M., KOHN, K.W., POMMIER, Y., WEINSTEIN, J.N. Quantitative structure-antitumor activity relationships of camptothecin analogues: cluster analysis and genetic algorithm-based studies. *J.. Med. Chem.* **2001**, *44*, 3254–3263.

22 SUZUKI, T., IDE, K., ISHIDA, M., SHAPIRO, S. Classification of environmental estrogens by physicochemical properties using principal component analysis and hierarchical cluster analysis. *J. Chem. Inf. Comput. Sci.* **2001**, *41*, 718–726.

23 RECANATINI, M., CAVALLI, A., BELLUTI, F., PIAZZI, L., RAMPA, A., BISI, A., GOBBI, S., VALENTI, P., ANDRISANO, V., BARTOLINI, M., CAVRINI, V. SAR of 9-amino-1,2,3,4-tetrahydroacridine-based acetylcholinesterase inhibitors: synthesis, enzyme inhibitory activity, QSAR, and structure-based CoMFA of tacrine analogues. *J. Med. Chem.* **2000**, *43*, 2007–2018.

24 MORON, J.A., CAMPILLO, M., PEREZ, V., UNZETA, M., PARDO, L. Molecular determinants of MAO selectivity in a series of indolylmethylamine derivatives: biological activities, 3D-QSAR/CoMFA analysis, and computational simulation of ligand recognition. *J. Med. Chem.* **2000**, *43*, 1684–1691.

25 WOLD, S. ERIKSSON, L. Statistical validation of QSAR results. In *Chemometrics Methods in Molecular Design*, WATERBEEMD, H. v. d. (Ed.). Wiley-VCH, Weinheim, **1995**, 309–318.

26 CLARK, R.D., SPROUS, D.G., LEONARD, J.M. Validating models based on large dataset. In *Rational Approaches to Drug Design, Proceedings of the 13th European Symposium on Quantitative Structure-Activity Relationship*, Aug 27 – Sept 1, HÖLTJE, H.-D. SIPPL, W. (Eds). Prous Science, Düsseldorf, **2001**, 475–485.

27 WILCOX, R.E., HUANG, W.H., BRUSNIAK, M.Y., WILCOX, D.M., PEARLMAN, R.S., TEETER, M.M., DURAND, C.J., WIENS, B.L., NEVE, K.A. CoMFA-based prediction of agonist affinities at recombinant wild type versus serine to alanine point mutated D2 dopamine receptors. *J. Med. Chem.* **2000**, *43*, 3005–3019.

28 KUBINYI, H., HAMPRECHT, F.A., MIETZNER, T. Three-dimensional quantitative similarity-activity relationships (3D QSiAR) from SEAL similarity matrices. *J. Med. Chem.* **1998**, *41*, 2553–2564.

29 NOVELLINO, E., FATTORUSSO, C., GRECO, G. Use of comparative molecular field analysis and cluster analysis in series design. *Pharm. Acta Helv.* **1995**, *70*, 149–154.

30 NORINDER, U. Single and domain made variable selection in 3D QSAR applications. *J. Chemometrics* **1996**, *10*, 95–105.

31 ZEFIROV, N.S. PALYULIN, V.A. QSAR for boiling points of "small" sulfides. Are the "high-quality structure-property-activity regressions" the real high quality QSAR models? *J. Chem. Inf. Comput. Sci.* **2001**, *41*, 1022–1027.

32 GOLBRAIKH, A. TROPSHA, A. Predictive QSAR modeling based on diversity sampling of experimental datasets for the training and test set selection. *J. Comput.-Aided Mol. Des.* **2002**, *16*, 357–369.

33 GOLBRAIKH, A., SHEN, M., XIAO, Z., XIAO, Y.D., LEE, K.H., TROPSHA, A. Rational selection of training and test sets for the development of validated QSAR models. *J. Comput.-Aided Mol. Des.* **2003**, *17*, 241–253.

34 OPREA, T.I. GARCIA, A.E. Three-dimensional quantitative structure-activity relationships of steroid aromatase inhibitors. *J. Comput.-Aided Mol. Des.* **1996**, *10*, 186–200.

35 OPREA, T.I. Rapid estimation of hydrophobicity for virtual combinatorial library analysis. *SAR QSAR Environ. Res.* **2001**, *12*, 129–141.

36 ATKINSON, A.C. *Plots, Transformations and Regression*. Clarendon Press, Oxford, UK, **1985**, 129–141.

37 GRAMATICA, P. PAPA, E. QSAR modeling of bioconcentration factor by theoretical molecular descriptors. *Quantum Struct.-Act. Relat.* **2003**, *22*, 374–385.

38 GRAMATICA, P., PILUTTI, P., PAPA, E. QSAR prediction of ozone tropospheric degradation. *Quant. Struct.-Act. Relat.* **2003**, *22*, 364–373.

39 MANDEL, J. The regression analysis of collinear data. *J. Res. Natl. Bur. Stand.* **1985**, *90*, 465–476.

40 LINDBERG, W., PERSSON, J.-A., WOLD, S. Partial least-squares method for spectrofluorimetric analysis of mixtures of humic acid and ligninsulfonate. *Anal. Chem.* **1983**, *55*, 643–648.

41 ZHENG, W. TROPSHA, A. Novel variable selection quantitative structure – property relationship approach based on the k-nearest-neighbor principle.

J. Chem. Inf. Comput. Sci. **2000**, *40*, 185–194.

42 KOVATCHEVA, A., GOLBRAIKH, A., OLOFF, S., XIAO, Y.D., ZHENG, W., WOLSCHANN, P., BUCHBAUER, G., TROPSHA, A. Combinatorial QSAR of ambergris fragrance compounds. *J. Chem. Inf. Comput. Sci.* **2004**, *44*, 582–595.

43 OPREA, T.I., ZAMORA, I., SVENSSON, P. Qvo Vadis, Scoring Functions? Toward an Integrated Pharmacokinetic and Binding Affinity Prediction Framework. In *Combinatorial Library Design and Evaluation for Drug Design*, GHOSE, A., VISWANADHAN, V.N., (Ed.). Marcel Dekker, New York, **2001**, 233–266.

44 *Catalyst. 4.9*. Accelrys, San Diego, CA, **2003**.

45 *Binary-QSAR*. Chemical Computing Group, Montreal, Canada, **1998**.

46 ROSENKRANZ, H.S. KLOPMAN, G. CASE, the computer-automated structure evaluation system, as an alternative to extensive animal testing. *Toxicol. Ind. Health* **1988**, *4*, 533–540.

47 ENSLEIN, K., GOMBAR, V., BLAKE, B. Use of SAR in computer-assisted prediction of carcinogenicity and mutagenicity of chemicals by the TOPCAT program. *Mutat. Res.* **1994**, *305*, 47–61.

48 OPREA, T.I. MATTER, H. Integrating virtual screening in lead discovery. *Curr. Opin. Chem. Biol.* **2004**, *8*, 349–358.

49 RAREY, M., LEMMEN, C., MATTER, H. Algorithmic engines in virtual screening. In *Chemoinformatics in Drug Discovery*, OPREA, T.I. (Ed.). Wiley-VCH, Weinheim, **2004**, 59–115.

50 GÜNER, O. (Ed.) *Pharmacophore Perception, Development, and Use in Drug Design*. International University Line, La Jolla, CA, **2000**.

51 SHEN, M., LETIRAN, A., XIAO, Y., GOLBRAIKH, A., KOHN, H., TROPSHA, A. Quantitative structure-activity relationship analysis of functionalized amino acid anticonvulsant agents using k nearest neighbor and simulated annealing PLS methods. *J. Med. Chem.* **2002**, *45*, 2811–2823.

52 Maybridge, http://www.daylight.com/ products/databases/Maybridge.html, **2004**.

17
Drug Discovery in Academia – A Case Study

Donald J. Abraham

17.1
Introduction

Discovering a new therapeutic agent and shepherding it through the chain of events that result in an approved new drug application (NDA) by the FDA is an arduous quest by even the largest and most successful pharmaceutical companies. While the process is primarily the purview of industry, the past few decades have seen an increase in academic discoveries being commercialized. Our goal is to provide readers with both the science and technology transfer issues involved with drug discovery in an academic setting and the process involved with linking the intellectual property with business. It is our hope that the take-home lessons listed at the end of this article will help other academicians who are interested in seeing their discoveries translate to therapeutic agents. The chapter summarizes the role of various components that play a role in translating basic research to industry. While not all of these introductory topics are germane for every case, they are meant to provide an overview of what an academician might encounter if she/he wishes to pursue the development of a promising agent beyond publication.

17.2
Linking the University with Business and Drug Discovery

17.2.1
Start-up Companies

The first spin-off company arising from academia is believed to have been created by Horace Darwin, the youngest son of Charles Darwin, in 1881. The result was the Cambridge Scientific Instrument Company, supplying Cambridge University's research laboratories [1]. Drug discovery in an academic setting has played an increasingly important role during the last three decades. This was not true during the first three quarters of the twentieth century. There are only isolated examples, during this period, of drugs that were discovered in Universities that eventually became marketed therapeutic agents. The role model provided by Genentech, founded in 1976 by venture capitalist

Chemoinformatics in Drug Discovery. Edited by Tudor I. Oprea
Copyright © 2004 WILEY-VCH Verlag GmbH & Co. KGaA, Weinheim
ISBN: 3-527-30753-2

Robert A. Swanson and biochemist Dr Herbert W. Boyer, and the changes in state and federal laws (the 1980 Bayh-Dole Act) paved the way for universities and academicians to participate in fostering business alliances and creating new companies.

Start-up companies emerge now from academic discoveries worldwide. The largest numbers of companies are located in clusters. California (San Francisco/San Diego) has the highest number of start-ups followed by Massachusetts (Boston Area), with North Carolina not far behind. Other notable clusters are found in Europe, that is, Cambridge in the United Kingdom, and Rhineland, Rein-Neckar, Berlin-Brandenburg and Munich in Germany (www.biojapan.de/features/sumi.ppt). Mehta has published a summary of the number of start-ups initiated by university technology-transfer processes. There were over 200 a year from 1994 to 1996, 300 a year from 1997 to 1999 and over 454 in the year 2000 [2]. Clearly, universities are playing an increasing role in the biotechnology revolution.

17.2.2
Licensing

The Bayh-Doyle Act gave patent rights to universities that arose from research supported by federal funding with the stipulation that any royalties be shared with the inventors. University intellectual property offices were established during this period to pursue licensing opportunities for professorial (and research group) inventions more effectively. The Bayh-Dole policy was then expanded by most universities to include nonfederally funded research. According to the Association of University Technology Managers (AUTM), between the years 1980 and 2000, over 3300 university spinout companies were formed in the United States and Canada, including 454 in the year 2000 alone [3]. The AUTM report finds that since 1980, academic licensing has resulted in at least 3376 new companies. The total sponsored research expenditures for 190 reporting academic institutions were $29.5 billion, up 10%, sponsored research expenditures funded by federal sources were $18.1 billion, up 8%, and the total 2002 sponsored research expenditures funded by industry were $2.7 billion, with no increase over 2001. Invention disclosures (13 032), new US patent applications (5545) and patents issued (3764) were also record highs. The continuing interest of academia in commercialization of research discoveries suggests that even more future industry leaders will come from university laboratories. The once-proud fiscal output of academic athletic programs can be dwarfed by comparison, but, unfortunately, the athletic programs still have a much higher publicity position in universities.

Even with this technology-transfer success, the learning curve for university licensing offices with biotechnology companies has apparently lacked expertise and know-how in structuring deals with businesses. Edwards, Murray and Yu recently reviewed university licensing to biotechnology companies for the past 25 years [4]. They concluded that universities often neglected important economic aspects in licensing agreements with biotechnology firms. The Alliance Database used in the study sampled 119 research institutions and all known commercial alliances made by 122 biotechnology companies where 36 instances revealed both full upstream and downstream economic terms. One major conclusion was that only one license in five provided for maintenance fees, while milestone payments were even more neglected.

Universities and colleges are limited in their ability to narrowly focus on the complex and difficult task of discovering and developing a new therapeutic agent from the bench to clinical trials. The reason for this is obvious. The long-standing role of the academy has been directed to scholarship in all of its forms and not to following a discovery to its commercial endpoint. Besides, academic institutions, unlike industry, cannot afford to set up expensively tailor-made laboratories and employ a multitude of high-salaried individuals who span the diversity needed. Another obvious difference is that university professors have the autonomy that industrial scientists rarely obtain unless they are well proven. Massive screening of large combinatorial chemical libraries is more the purview of industry, obviously due to the cost of expensive robots and an integrated cadre of high-paid specialists. And, as drug discovery professionals well know, scientific advances and new therapeutic agents arise more often than one would believe by serendipity from an unexpected finding [5].

17.3
Research Parks

Biotechnology research parks located near universities have also emerged as a growing trend since the 1980s. These parks were developed for a duality of reasons:

1. to attract businesses that might benefit from the alliances with academia and provide a source of well-educated and trained workforce;
2. the parks could enable faculty to be entrepreneurs, starting companies while retaining faculty status.

In 2002, the Association of University Research Parks (AURP) contracted with Association Research Inc. (ARI) to develop a profile of US and Canadian Research Parks [6]. The ARI sent out questionnaires to 195 entities believed to be operating in research parks and received 87 written responses with 79 yielding sufficient information to be stored in a database. Research parks such as the Research Triangle Park occupy vast tracks of land (7000 acres), while others such as the University City Science Center (16 acres) and Audubon (3 acres) are relatively compact. The average size of a research park in the survey was 628 acres and the median 180 acres. Employment ranged from 10 000 to 42 000 for 62 research parks.

17.4
Conflict of Interest Issues for Academicians

During the 1980s and early 1990s, there were little or no official university regulations at most institutions regarding faculty participation in or sitting on boards of start-up companies they founded. Much of the concerns since the middle 1990s arose from clinical trials being conducted in medical schools by faculty with financial interests in the sponsors of the trials. In one celebrated case, after a young patient died following a gene-transfer procedure, it was discovered that there was massive underreporting to the NIH of adverse events of related gene-transfer experiments in settings where investigators and, in some cases, institutions held significant financial positions. In 1988, Congress passed new regulations concerning individual conflict of interest in federally sponsored research. According to David Korn, senior vice president for Biomedical and Health Sciences Research at the Association of American Medical Colleges, mandating the federal law proved much easier than its implementation, which took seven years [7]. During the years since the federal law for government-sponsored research was passed, a plethora of new university rules and regulations and others that govern almost all faculty involvement with industry interactions have appeared at our institution. Unfortunately, draconian regulations have evolved to such an extent that even the most basic of research efforts, far removed from clinical trials, are scrutinized for all industry- and even government-sponsored studies. Faculty conflict-of-interest committees have taken center stage in academia, placing increased restrictive provisions in place to assure that faculty ownership and/or interactions with companies are in compliance with state and federal laws. It is not clear whether such committees have members with extensive experience in these areas and this brings to mind the old adage, "the blind leading the blind". As Max Perutz (whose Medical Research Council Laboratory of Molecular Biology fostered 12 Nobel Laureates) mentions in the preface of his book, *I Wish I Had Made You Angrier* [8], "... because creativity in science as in the arts, cannot be organized [. . . .] but hierarchical organizations, inflexible bureaucratic rules, and mountains of futile paper work can kill it".

In academia today, the appearance of a possible conflict of interest is enough to derail efforts to conduct industrial research in academic laboratories as well as to inhibit start-up company formation while remaining in the university. This is especially true if the faculty member has a grant from her/his company to perform research in her/his laboratory, even in the nonclinically related research areas. In this writer's opinion, universities have come full circle, separating any real entrepreneurship from the academy. The future might look brighter if David Korn's thinking on conflict of interest can be successfully implemented. "Conflicts of interest are ubiquitous and inevitable in academic life, and the challenge for academic medicine is not to eradicate them, which is fanciful and would be inimical to public policy goals, but to recognize them and manage them sensibly, effectively, and in a manner that can withstand public scrutiny" [7].

Another issue of concern has been the tension created by university intellectual property management versus the role of universities to discover basic knowledge. One of the most contentious issues now being debated is whether the role of a university is compromised by an imbalance or a shift toward goal-oriented research and the creation of wealth [2]. There are those who see the universities losing the altruistic nature of education and the pursuit of knowledge for knowledge's sake. In this regard, it should be remembered that even the pursuit of knowledge for knowledge's sake is *not* primarily supported by the university but by federal government funding agencies such as NIH and NSF.

17.5
Drug Discovery in Academia

Traditional drug discovery research in academia has focused on developing new methodologies, uncovering new agonists or antagonists and isolating or cloning new targets. Methodologies employed in universities for drug discovery or method development include, but are not limited to, QSAR, structure-based drug design, molecular modeling, computational chemistry, combinatorial chemistry coupled with high-throughput screening, proteomics and natural products synthesis and screening. For comprehensive reviews in these areas, see these chapters published in the sixth edition of Burger's Medicinal Chemistry and Drug Discovery [9]. The more rational the approach employed by academic researchers, the greater the chance to obtain NIH funding. For this primary reason, high-throughput screening in academia is not in vogue, even if well funded by industrial grants, since screening without advancing theory can be viewed as lacking in high-quality scholarship.

17.5.1
Clinical Trials in Academia

The availability of medical schools to conduct clinical trials on new drug entities funded by pharmaceutical companies has been ongoing from the early years of the twentieth century until the present day. Indeed, a university's research budget can be handsomely increased through such contracts. Advocates also point to the importance of the university-published findings of clinical trials in the search for new therapeutic agents. Some clinical findings can produce information that will provide a basis for new drug discovery, although this is rare.

The NIH and the pharmaceutical industry have formed a new partnership to overcome barriers to early-phase clinical trials. The Secretary of Health and Human Services, Tommy Thompson, announced on July 9, 2003 grant awards for six cancer centers involved in a unique public–private partnership. Five pharmaceutical companies together with NCI put forth a total of $5.7 million for this partnership. Institutions receiving the funding include: Massachusetts General Hospital; University of Colorado Health Sciences Center; Washington University, St. Louis; University of Pittsburgh Cancer Institute; University of California, Davis Cancer Center; and Ohio State University Comprehensive Cancer Center (http://www.bms.com/news/other/data/pf_other_news_3855.html).

17.5.1.1 Molecular Modeling and Computational Chemistry

Academia has been rich in producing theoretical computational methodology that underpins molecular modeling. The following software arose from universities or private and publicly funded institutes: AMBER [10], INSIGHT [11, 12], CHARMM [13], SYBYL [14], GRID [15], DOCK [16] and HINT (Hydropathic INTeractions) [39]. All except AMBER were commercialized.

17.5.1.2 **Structure-based Drug Design**

Paul Ehrlich, in the early twentieth century, proposed that drugs interacted with receptors in a way similar to how a key fits a lock [17].[a] The scientists' great dream of viewing the structure of a biological lock and key in three dimensions became a reality when Max Perutz solved the phase problem for determining the structure of large biological molecules at atomic resolution using X-ray Crystallography [18–20]. The first suggestion to use X-ray crystallography of biological molecules for drug design purposes came from academia and was published in 1974 [21]. Structure-based design is perhaps the royalty of methodologies in drug discovery since it is the most rational approach and one suited in present-day academia for federal funding proposals. A Google search using *structure-based drug design* shows about 1 300 000 hits as of April 2004.

Structure-based design as conducted in industry and academia is a cyclic process. One proceeds from an initial active compound whose complex with the protein or target receptor is determined. This is usually followed by the design of a better binding molecule with an increase in biological activity. A new structure determination of the complex is then made and the cycle continued until a potential clinical candidate is obtained. The cyclic process refines each stage of discovery. The key to success for designing a clinical candidate, in our experience, involved selection of a structural scaffold for the initial lead molecule that possessed a low-toxicity profile, since toxicity had derailed successful drug discovery more than any other step in the process. When a plateau is reached in biological testing, one of the candidates can be forwarded for *in vivo* evaluation in animals.

17.6
Case Study: The Discovery and Development of Allosteric Effectors of Hemoglobin

> *"Paul Ehrlich, the father of modern drug discovery, stated that scientists need the four German G's: Geschicht (Skill), Geduld (Patience), Geld (Money) and Glück (Luck)"*[8]. The four German G's were all evident in this case study.

Our research group and associated colleagues have had the good fortune of having an allosteric effector of hemoglobin, RSR13, discovered in our laboratories, proceed to an NDA submission, which is being reviewed by the FDA as a radiation sensitizing agent for the treatment of breast cancer metastasis to brain. We have had a second agent proceed to a phase-one clinical trial for the treatment of sickle-cell anemia. The editor thought that a history of the discovery, design and development of a molecule that we have seen translated to clinical trials might provide readers with a case study that puts flesh on the preceding topics.

Seeing a chemical agent proceed through the maze required for FDA approval as a new drug entity makes one truly appreciative of the extensive expertise required at every stage. It is impossible to credit all of those involved in this 12-year quest. One thing is certain. Without collaborative skills at every level, a strong perseverance to continue

---

[a] http://www.chemheritage.org/EducationalServices/pharm/chemo/readings/ehrlich/pabio.htm.

despite the odds, a lot of money and some good fortune, RSR 13 would not have reached the NDA stage.

17.6.1
Geduld (Patience)

I started my academic career working in the cancer chemotherapeutics arena.[b] A switch was made to focus on sickle-cell anemia in 1975. The reason for the switch was straightforward. My long-standing interest in drug discovery has always been directed toward structure-based drug design [21]. In the early- to mid-1970s, hemoglobin was the only large molecule drug target whose structure was determined at atomic resolution. At about the same time, another group headed by Peter Goodford at Burroughs Wellcome, United Kingdom, had the same idea. Goodford's group was the first to develop a structure-based antisickling agent that reached clinical trials (BW12C, compound 1a) [22, 23]. Later, we confirmed the proposed BW12C binding site; however, BW12C interactions with the protein were different from those proposed [24]. BW12C was subsequently dropped from further study because of a short half-life and unfavorable route of administration for a chronic disease.

BW12C

Our own efforts languished from four failed NIH grants. At that time, it was not clear to reviewers that structure-based drug design using the native adult hemoglobin (HbA) coordinates would be productive. It took five years, with four failed attempts, to win the first NIH grant and another decade plus to get an agent into clinical trials.

17.6.2
Glück (Luck)

After four failed NIH grants and without funds, our group dwindled to only my oldest son William, who was in high school and volunteered to work for the summer. Serendipity and an unexpected ally saved the day. The ally was the goods delivery person

---

[b] When the author was ten years old he watched his maternal grandfather painfully die of cancer. He told his mother that one day he wanted to discover a drug to treat cancer. Thinking that MDs discovered new medicines, he enrolled in premedicine only to find that his real interest was in organic chemistry. After switching majors, he was very pleased to discover that chemists usually discover new drugs.

in Salk Hall at the University of Pittsburgh, Mr Robert Heflin. One day, while we were riding together in the elevator, Bob asked why I looked so distraught, and I told him it was due to the fact that I had lost all research funds since the NIH did not believe in my idea to design an antisickling agent from the structure of hemoglobin. Mr Heflin told me that he could help, and I almost said no thanks, thinking it not very likely. However, his sincerity was so appreciated that I took him to my office, where a student and I had put together a lab-quip plastic model of HbA composed of 5000 atoms. I told Mr Heflin to pretend that the giant globe-shaped tetramer was the moon and the potential drug was a space ship that had to land near the mutant binding site. Bob responded with a big smile on his face and repeated he could help. I asked how, and he responded that he knew the famous first baseman for the Pittsburgh Pirates, Mr Willie Stargell. Mr Stargell was heading a group of black athletes in the National Baseball league funding sickle-cell anemia research. Mr Stargell provided the University of Pittsburgh, where my laboratories were at that time, $18 000 for our first sickle-cell anemia funding for the following few years. This seed money revived our research efforts to provide enough results and experience to get our first NIH grant to study sickle-cell anemia, an RFP (Request for Proposal) from the NIH for sickle cell structure based drug design.

17.6.3
Geschick (Skill)

Another problem our group had was to make the transition from small-molecule- to large-molecule crystallography. At that time (1970–1975), only a few centers in the world were equipped to determine the structure of a large molecule using X-ray crystallography. At that time, groups normally had to spend a decade or more to complete a single structure, and the process was very labor intensive. It was clear after five years (1975–1980) that my small crystallography laboratory in neither the Pharmacy School[c] nor the well-equipped Department of Crystallography at the University of Pittsburgh (under Prof. George Alan Jeffrey and famous for small molecule structure determination) had the equipment and skills to progress rapidly with solving potential drug complexes with hemoglobin.

While reading a *Nature* commentary one afternoon, a plan of action crystallized. The author had written that if you want to get hit by lightening you must go where it storms. It was clear that if we were to make rapid progress and learn hemoglobin crystallography, there was only one place to go, Cambridge, United Kingdom in Max Perutz's laboratory at the Medical Research Council Laboratory of Molecular Biology (MRCLMB). Fortunately, a symposium was being held in honor of Max's 65th birthday at Airlee House in Virginia (1980), only a four-hour drive from Pittsburgh. With some nervousness, I approached Max during a poster session with my ideas, and to my great surprise, he immediately invited me to join him in Cambridge, United Kingdom at the Medical Research Council Laboratory of Molecular Biology.

---

[c] My dark room for developing X-ray films was in my coat closet, and to take cold temperature diffraction of crystals grown at 4 °C, I had to wait until winter to open the window.

Mr Robert Heflin again came to our rescue to link me up with funds to study in Cambridge via Ms. Ruth White, the then director of the Sickle-Cell Anemia Society in Pittsburgh. In one day, Ms. White arranged for me to get a Heinz Foundation grant to travel and work in Perutz's laboratory.

I worked with Max Perutz and colleagues at the MRCLMB for the next eight years during periodic visits (1980–1988). During 16 trips to the MRCLMB, we slowly worked out the structural parameters for a therapeutic agent that would bind strongly to hemoglobin. The overall template for binding to specific Hb sites that prevent polymerization of HbS was found to have an aromatic halogen ring connected to a polar side chain. This information enabled our group in Pittsburgh to design and synthesize or select several potent antisickling agents but all failed as potential therapeutic agents because of red cell deformability and/or *in vivo* toxicity at the high *in vivo* dosage needed to interact with the approximately 650 g of HbS in sickle-cell-anemia patients. I asked Prof. Don Witiak, a colleague from Ohio State University at that time, if there was a known drug with the halogen aromatic polar template that could be given in high doses. He told us yes, the antilipidemic drug Clofibrate that had clofibric acid (CFA) as the active component. It was given to humans in 2-gram-per-day quantities.

We immediately tested CFA and found that it exhibited a strong antigelling activity [25, 26]. Just as important was the fact that CFA, when administered orally as the ethyl ester clofibrate (Clofibrate), could be given in high doses (2 g day^{-1}). X-ray analyses of cocrystallized Hb and CFA showed two binding sites, one strong and one very weak, separated by several angstroms [27].

Since CFA might be a candidate to treat sickle-cell anemia, we decided to determine what effect CFA might have on the oxygen binding properties of hemoglobin solutions. Many antigelling agents left-shift the oxygen binding curve, producing a high-affinity oxy-Hb relaxed (R) state that does not incorporate into the polymerization of tense (T) state deoxy-HbS. It was a surprise when the antigelling CFA, which inhibits sickle-cell Hb polymerization, was found to shift the Hb oxygen binding curve toward the right, that is, toward T-state Hb in a manner similar to that of the natural *in vivo* allosteric effector 2,3-diphosphoglcerate (2,3-DPG) [25]. An agent that can produce an *in vivo* right shift in the oxygen binding curve had been long sought as a potential to treat human hypoxic conditions such as stroke and angina. It was obvious that 2,3-DPG, however,

2,3 DPG

Clofibric Acid
CFA

could not be used as a therapeutic agent since its five negative charges prevent its transport across the hydrophobic red cell membrane.

The CFA binding site [25, 28] was far removed from the 2,3-DPG site at the surface of the β subunits [29] (compare Figures 17.1 and 17.2c).

The determination of the CFA binding site on Hb was the first report of a tense state (deoxy state) non-2,3-DPG binding site. However, our excitement for using clofibrate to treat sickle-cell anemia was quickly dashed when we checked the literature and discovered that CFA binds by well over 90% to serum proteins, regardless of dose. Later, we confirmed the serum protein binding in whole blood using oxygen equilibrium analyses that indicated that very little CFA was transported into erythrocytes in quantities sufficient to interact with hemoglobin.

The shift in the oxygen binding curve in the direction opposite to that desired to treat sickle-cell anemia opened new fields of potential treatment for ischemic disease states such as stroke and angina mentioned above, as well as use for enhancing radiation of tumors, during transplant surgery to keep vital organs oxygenated, or even to greatly extend the shelf life of stored blood since aged blood is not effective due to 2,3-DPG depletion. When we published our CFA results [25], the first clinically oriented researchers to use the clofibrate discovery (to the surprise of Max Perutz and I) were the radiation oncologists who were testing a theory that radiation treatment of hypoxic tumors was enhanced if oxygen levels were increased [30, 31].

Perutz and Poyart tested another antilipidemic agent, bezafibrate (BZF), and found it much more effective than CFA [32]. It was my task on a subsequent visit to the MRCLMB to determine the binding site for bezafibrate, and I found that it linked both the high- and low-occupancy CFA sites as well as a new type of hydrogen bond between Asn108β and the halogenated phenyl ring of bezafibrate [27, 28]. Perutz gave a lecture at Harvard and pointed out the new type of hydrogen bond, and this stimulated Burley

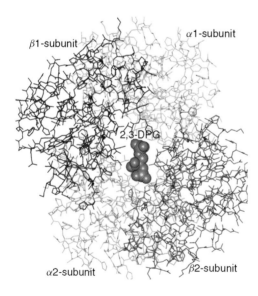

β1-subunit

α1-subunit

2,3-DPG

α2-subunit

β2-subunit

Fig. 17.1 View of the 2,3-DPG binding site at the mouth of the cleft of deoxy hemoglobin [29].

and Petsko to look for this general type of hydrogen bond in protein structure [33, 34]. This is one of the few examples where drug binding has helped elucidate and understand protein structure.

Lalezari and Lalezari synthesized urea derivatives of bezafibrate, and, with Perutz, determined the binding site of the most potent derivatives to be the same as that we had discovered for bezafibrate [35]. Although all of these compounds were extremely potent, they were not suitable clinical candidates on account of being hampered again by serum protein binding [36, 37].

Lelazari Series   Bezafibrate

From this point, our laboratory used a more classical structure-activity approach to find the best clinical candidate. To get a structure-activity spread, we synthesized a series of bis-phenyl fibrate analogs, making every type of substitution in the three-atom linking chain between the aromatic rings ($NH-CO-CH_2$) that shortened the four-atom bezafibrate linking chain [37]. To our surprise and great delight, one class (RSR) produced a large shift of the oxygen binding curve of hemoglobin in the presence of whole blood, something that either we or the others had not observed previously with any other classes of molecules. The two most active molecules were RSR 4, the 3,5-dichloro derivative, and RSR 13, the 3,5-dimethyl derivative. RSR are the initials of Ram S. Randad, the excellent postdoctoral researcher who synthesized this class of molecules.[d] Another postdoctoral researcher, Ahmed A. Mehanna returned for a second stay in our laboratory and played a significant role in overseeing the synthesis for all classes of molecules being made in our group. Seeing the RSR 4 and RSR 13 results in whole blood compared to the results for the Lalezari molecules and bezafibrate discussed below was the first time I was confident enough to believe that we had a discovery that might result in a real therapeutic agent [37].

---

[d] In our research group, all new compounds are coded with the initials of the person who synthesizes them. Ram Randad's initials appear now routinely in the chemical literature attached to widely studied RSR 13 allosteric effector.

RSR series

MM series

RSR 13 R1,3 =CH3; R2 = H   MM 30 R1,3 = Cl; R2 = H

RSR 56 R1,3 = OCH3; R2 = H   MM 25 R1,R3 = H; R2 = Cl

The substitutions that reversed the amide bond, the derivatives of graduate student Mona A. Mahran's (the MM) series were all weaker allosteric effectors. We spent the next few years sorting out the molecular reason for the behavior of this class of molecules.

X-ray crystallographic analyses alone could not sort out the reason for differences between the RSR, MM and other weak acting series. Figure 17.2a is a stereo diagram showing the overlap of four allosteric effectors that bind at the same deoxy-Hb site but differ in their allosteric potency. Only small differences in atomic positions are apparent when comparing the strong RSR molecules with the MM molecules.

The computational program HINT[e] (Hydropathic INTeractions), developed in our laboratories [39], however, provided invaluable information fingering the important interactions between the strong and weak acting allosteric effectors. The HINT analyses revealed that the amide linkage between the two aromatic rings of the compounds must be orientated such that the carbonyl oxygen forms a hydrogen bond with the side-chain amine of αLys99 (Figure 17.2b) [28, 36, 40]. Three other important interactions were found. The first was the water-mediated hydrogen bonds between the effector molecule and the protein, the most important occurring between the effector's terminal carboxylate and the side-chain guanidinium moiety of residue αArg141. The second, a hydrophobic interaction, involves a methyl or halogen substituent on the effector's terminal aromatic ring and a hydrophobic groove created by Hb residues Phe36α, Lys99α, Leu100α, His103α and Asn108β. The third, a hydrogen bond, is formed between the side-chain amide nitrogen of Asn108β and the electron cloud of the effector's terminal aromatic ring [36, 40, 41]. I first observed the new hydrogen bond

---

[e] Glen Kellogg came into my group as a postdoctoral researcher and wrote the code for HINT on the basis of ideas published by Al Leo and myself [38]. Professor Kellogg deserves credit for the development and use of HINT in industrial and academic laboratories worldwide.

while contouring the Hb binding site of bezafibrate[f] [27, 32]. As mentioned above, Burley and Petsko subsequently pointed out that this type of hydrogen bond existed in a number of proteins [33, 34]. Perutz and Levitt estimated this bond to be about 3 kcal mol^{-1}, much stronger than we originally thought [42].

Dr. Gajanan Joshi of our group next measured, with painstaking experiments, the binding constants of over thirty allosteric effectors and compared them with the number of binding sites and found that all the constants agreed with the number of crystallographic binding sites found [43]. The degree of right shift in the oxygen binding

(a)

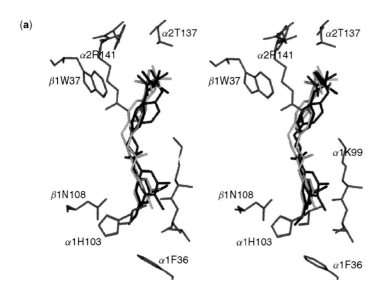

Fig. 17.2 Stereoview of allosteric binding site in deoxy hemoglobin. A similar compound environment is observed at the symmetry-related site, not shown here. **(a)** Overlap of four right-shifting allosteric effectors of hemoglobin: (RSR13, yellow), (RSR56, black), (MM30, red), and (MM25, cyan). The four effectors bind at the same site in deoxy hemoglobin. The stronger acting RSR compounds differ from the much weaker MM compounds by reversal of the amide bond located between the two phenyl rings. As a result, in both RSR13 and RSR56, the carbonyl oxygen faces and makes a key hydrogen bonding interaction with the amine of αLys99. In contrast, the carbonyl oxygen of the MM compounds is oriented away from the αLys99 amine. The αLys99 interaction with the RSR compounds appears to be critical in the allosteric differences. **(b)** Detailed interactions between RSR13 and hemoglobin, showing key hydrogen bonding interactions that help constrain the T-state and explain the allosteric nature of this compound and those of other related compounds. **(c)** The RSR 13 binding site. Compare the difference with the 2,3-DPG binding site in Figure 17.1 at the base of the β subunits. Two molecules of RSR 13 bind symmetrically around the two fold axis from the top of the subunits down the central water cavity to the αβ subunit interfaces (around 20 angstroms).

---

[f] I called Max Perutz at home that evening. It was around 8:00 p.m. and he immediately walked to the MRCLMB to see this exciting interaction. Max had a childlike enthusiasm for scientific discoveries till the end of his life.

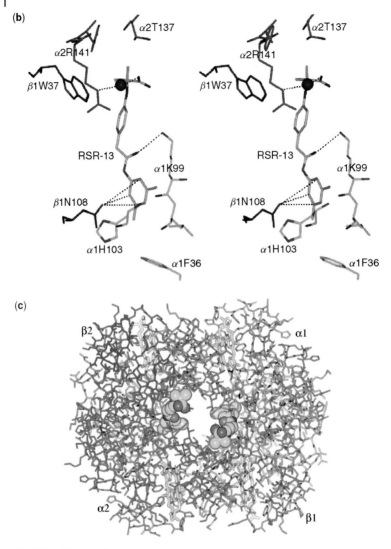

(b)

(c)

Fig. 17.2 *(Continued)*.

curve produced by these compounds was not solely related to their binding constant, providing a structural basis for E. J. Ariens' theory of intrinsic activity. These studies enabled us to get information directly related to the atomic mechanism of action and reasons for the allosteric shift. The conclusions from our structural studies provide a near-complete understanding for the observed structure–activity results.

1. All of these derivatives bind and overlap in the bezafibrate binding site. Direct analysis from the crystal structures did not reveal the reasons for the observed maximum activity.

2. HINT clearly diagnosed that the amide bond oxygen must face and form hydrogen bond with Lys99α.
3. The interactions between the allosteric effectors and hemoglobin add hydrogen bonds to the Hb tense state, similar to DPG, and therefore stabilize that state, resulting in an increased delivery of oxygen.

$$HbO_2(R\ state) + Allosteric\ Effector \longleftrightarrow Hb(T\ state) \cdot Allosteric\ Effector + O_2$$

4. A single substitution on the terminal aromatic ring at the 3 (meta) position of the terminal aromatic ring in the RSR Series oriented the substituted group away from the alpha helix containing Lys99α (G helix) and toward the sterically more accessible Hb central water cavity. The mono-substituted phenyl derivatives were all much weaker allosteric effectors than the 3,5-di- or 3,4,5-tri-substituted molecules. However, the di- or tri- substitutions at 3- and 5-, or 3,4,5-positions forced one methyl group into the G helix and we believe that it acts like a screwdriver wedged into a gear, preventing it from undergoing the allosteric transition to the R state because of restriction of movement of the G helix.
5. The extra binding affinity by the addition of the second or third substitution was not in proportion to the degree of shift in the oxygen binding curve. There were some derivatives with lower activity having binding constants equal or close to the most active analogs. The difference for the increased activity was the entropic placement of the second substituent into the G helix, hindering its movement during the allosteric transition to the tense state [41].

17.6.4
Geld (Money)

17.6.4.1   Advancing a Lead Compound to Clinical Trials

Preclinical Studies at Virginia Commonwealth University's Medical School on the Medical College of Virginia Campus
Figures 17.3a (university units) and 17.3b (individuals) show the scheme we used to link the university together to forward the new molecule to phase-one clinical trial. We were fortunate to link a number of Virginia Commonwealth University (VCU) medical professors on the Medical College of Virginia campus to test RSR 13 in the preclinical evaluation for its efficacy in treating hypoxic diseases. The first radiation oncology animal investigations were performed by Prof. Rupert Schmidt-Ullrich's group [45, 46] and the first investigations of RSR 13 for potential use in stroke were performed by Prof. Hermes Kontos' [47] and Prof. Ross Bullock's groups [48]. Other VCU professors and their colleagues looked at RSR 13 to counter hypoxia in brain injury and for surgical uses. Professors Albert Munson and Jean Meade (VCU Department of Pharmacology and Toxicology) performed the first toxicology evaluations that were key to the initial venture capitalization. Linking well-funded university professors together across disciplines who wanted to study RSR 13 in their areas of specialty resulted in numerous publications indicating an excellent prognosis for efficacy and safety *in vivo*.

(a)

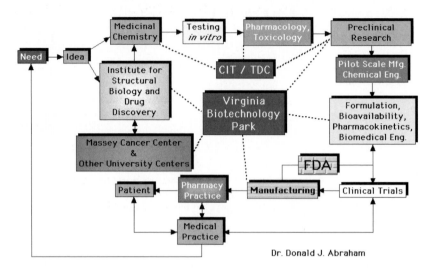

Institute for Structural Biology and Drug Discovery
Virginia Commonwealth University

Dr. Donald J. Abraham

(b)

Institute for Structural Biology and Drug Discovery
Virginia Commonwealth University

Dr. Donald J. Abraham

Fig. 17.3

Surgical Blood Loss: Development Comparison

	Research	Preclinical	IND	Phase I	Phase II
Somatogen					
Project	Genetically linked Hb with Hb Presbyterian substitution (Asn108β->Lys)				
Cum. Spending (million $)		$18.7	$18.7	$75.1	$113.6
Valuation (million $)		$30.0	$102.0	$122.0	$445.0
Date		7/89 - 6/91	6/91	2/92 - 8/93	9/93 - 9/95
Duration / Cum. Time (mos.)		24/48	7/55	19/74	24/98
Northfield					
Project	Glutaraldehyde crosslinked multimer Hb				
Cum. Spending (million $)				$36.4	$56.1
Valuation (million $)				$90.0	$231.0
Date				87 - 2/94	3/94 - 9/95
Duration / Cum. Time (mos.)				93/98	16/114
Alliance Pharmaceuticals					
Project	Perfluorocarbon				
Cum. Spending (million $)				$48.3	$73.7
Valuation (million $)				$297.0	$305.0
Date				2/92 - 5/93	6/94 - 6/95
Duration / Cum. Time (mos.)				15/unk	12/unk
Baxter					
Project	Crosslinked Hb with diasperins				
Allos Pharmaceuticals					
Project	Allosteric effectors				
Cum. Spending (million $)			$2.2	$2.5	
Valuation (million $)			$8.5	$8.5	
Date			6/95	7/95 - 9/95	
Duration / Cum. Time (mos.)			12	3/15	

Fig. 17.4 The comparison of the costs of several blood substitutes companies to develop an oxygen carrying molecule vs. the Allos Therapeutics Inc. allosteric effector RSR 13 to an IND and phase one clinical trial.

Allos Therapeutics Inc. was then able to proceed through a phase-one study on the basis of the basic and preclinical studies for only $2 million. This is perhaps a record for a new drug. The following graph prepared by the former Allos CEO, Stephen Hoffman, compared the cost of development of RSR 13 to the blood substitute companies that also sought to increase oxygen delivery *in vivo* (Figure 17.4).

The Role of the Virginia Biotechnology Research Park in Our Drug Discovery Efforts

Advancing any new agent from the laboratory bench to clinical trials is rare. Advancing it through an academic network is even rarer. The advent of the Virginia Biotechnology Research Park (VBRP) adjacent to the Medical College of Virginia campus of Virginia Commonwealth University turned out to be a key player in the process toward commercialization of the first hemoglobin allosteric effector RSR 13. VCU President

Fig. 17.3 (a) The linked departments and institutes at VCU that were associated with moving both Hemoglobin allosteric effectors from the bench to phase one clinical trials. The CIT box is the State of Virginia Center for Innovative Technology that held the original VCU patents and matched funds with industry to develop new drugs. The Institute for Structural Biology and Drug Discovery is located in the Virginia Biotechnology Research Park. (b) VCU professors in the Institute for Structural Biology and Drug Discovery and professors from the associated departments in Figure 17.1(a), who also performed translational research that performed basic and/or clinical research in advancing RSR 13 and (vanillin) an antisickling molecule to clinical trials.

Eugene Trani and also the Chairman of the Board for the VBRP encouraged faculty to initiate biotechnology companies for the research park. Several faculty-initiated companies sprang to life, including what was first named HemoTech Inc. and later named Allos Therapeutics, a company I was encouraged to found by the VCU President Eugene Trani's office. Mr James Farinholt, President Trani's lead person for business development at VCU at that time, helped tremendously in sorting out the details needed to tie the university and state with any future funding by venture groups or pharmaceutical houses.

The name chosen for Allos Therapeutics Inc. was taken from the scientific terminology for allosteric equilibriums modulated by allosteric effectors. Allosteric proteins regulate some of the most important pathways for life. Allos' initial focus was to discover and develop allosteric effectors as therapeutic agents. Having associated with a number of large pharmaceutical houses in drug discovery efforts toward finding an antisickling drug, it was clear that nothing was advanced through the complicated maze of a large pharmaceutical house's drug development teams without a champion overseeing every phase. We had no such champion, and the concept of an allosteric effector as a drug was a hindrance. Few, if any, in the drug discovery world understood the theory of allosteric regulators and their potential as a new type of drug action.

On the other hand, small start-up companies focused intensely and entirely on what they were created to do, with the incentive that early and even later employees would be well rewarded if the venture was successful. It was clear to me, in this case, that if the translation from basic and preclinical research to clinical trials were to occur, our best chance was to initiate a company. For the most part, big pharmaceutical houses have stayed clear of allosteric effectors as drugs. We had the enormous advantage that far more is known about hemoglobin as an allosteric protein than any other.

Venture Capital and New Companies

The best discoveries can lay dormant without recognition of their importance by the investigator(s). Medicinal chemists involved in drug discovery in academia must consider patenting as well as publishing. Publishing the results of a new compound without patenting assures that it will not be translated to a new drug. Publications and accompanying patents are not just necessary but critical to attract venture groups who can provide start-up funds for a new company.

HemoTech Inc. went without major funding for several years and no one seemed interested. The director of the VBRP had contact with a Wall Street Journal reporter who was interested in early stage discoveries. The Wall Street Journal published an article [49] on the same day we published the first *in vivo* results combined with the crystallographic binding site for RSR 13 and RSR 4 [36]. This drew much attention from the venture capital groups and large pharmaceutical houses. While Ortho Biotech, a Johnson and Johnson subsidiary, wished to license the technology, Johnson and Johnson's venture outlet decided that a start-up company would be better, and a conglomerate of Venture groups with them, called Med Vest, provided funds for the toxicology and phase-one trial for RSR 13. The due diligence performed by venture groups can be much more extensive than imagined. After the company was funded, I was provided with the due diligence, including letters of approval and extensive review of our basic science studies by colleagues in a host of academic institutions. The process made NIH reviews for research grants look trivial, which those of us who apply routinely for NIH funding know is not the case.

Essential was the selection of Stephen Hoffman, MD, Ph.D., as the CEO for HemoTech Inc.[g] Stephen, whose Ph.D. was in Organic Chemistry, had both a firm hand on the science and the medical aspects. Dr Hoffman's experience as the scientific founder of Somatogen, the first blood substitute company to go public, provided him with a firm understanding of the chemistry and physiology of oxygen delivery. Stephen was also able to link our new company with a venture investment group headed by Mr James Daverman of Marquette Ventures, who had invested in Somatogen and therefore had considerable knowledge in the hemoglobin field. Fortunately for us, Mr Daverman was a member of the Medvest Inc. conglomerate venture group, who funded Allos Therapeutics Inc. Overall, the understanding of allosteric regulation processes and their potential by RSR 13 investors was not only a surprise but also made all the difference in their decision to provide the first two million dollars.

Initial Toxicology and Venture Capital Funding   The choice of which allosteric effector to forward to clinical trials was made intuitively. A crucial decision had to be made since the potential venture capital group, Med Vest, headed by hematologist turned entrepreneur, Dr Geoffrey Brooke, would pay for only one toxicity study before a decision was made on the commitment of the first two million dollars to fund our new company. I had to chose from the top two candidates, either the weaker acting 3,5-dimethyl derivative (RSR 13) or the more potent 3,5-dichloro RSR 4 molecule. Both molecules completely overlapped in their binding sites with only the longer bond lengths for the chlorine atoms. We had considerable experience testing halogenated aromatic acids in our sickle-cell assay and almost all were unsuitable due to toxicity issues. I also believed that, in general, in drug discovery, alkane derivatives had less toxicity issues than the corresponding halogen derivatives. Therefore I chose the 3,5-disubstituted methyl groups (RSR 13) over the 3,5-disubstituted chlorines (RSR 4). The whole advance of this project would be determined from the choice made. Fortunately, this choice of RSR 13 was well founded. The dichloro derivative RSR 4 tended to lyse erythrocyte membranes and was not suitable for *in vivo* study at the doses needed. To our further surprise, the dimethyl derivative, RSR 13, was even more active *in vivo* in animal studies than RSR 4, opposite of the *in vitro* tests.

Synthetic Chemistry and Purification   Both the two-step synthetic scheme we had used for RSR 13 and the method for purification of the sodium salt of RSR 13 had to be altered for large-scale preparations to be used in clinical trials. The normal two-step synthesis involved a standard carbene reaction run in chloroform.

---

[g] The name was later changed to Allos Therapeutics Inc.

RSR 13 via carbene reaction

Because of the environmental restrictions on the use of chloroform in industrial chemistry, an alternate synthesis was devised by the scale-up chemists. It involved a seemingly prohibitive step, at least as professors instruct students in organic chemistry: an SN_2 reaction on a tertiary carbon.

RSR 13 Industrial synthesis

The other change that needed to be made in the synthesis of RSR 13 for *in vivo* administration was the method of purification. RSR 13 is used *in vivo* as the sodium salt. I prepared the first batch for *in vivo* toxicology by triturating RSR 13 sodium salt with acetone to remove any vestiges of water. However, the first industrial scale-up procedure called for crystallization of the salt from ethanol–water. The ethanol–water crystals were not as soluble as the acetone triturated method and could not be formulated at a reasonable volume. We performed the crystal structure determination of the ethanol–water crystals and found that it was a heptahydrate (Figure 17.5) [50]. The problem for large-scale production of RSR 13 was solved eventually by the industrial producers of RSR 13.

A recent Chemical and Engineering (C & E) News article highlighted the outsourcing of RSR 13 by Allos Therapeutics Inc. and three other pharmaceutical companies of different sizes. In particular, Allos Therapeutics Inc. contracted Hovione, a Portuguese pharmaceutical chemical company to synthesize RSR 13 under cGMP conditions for clinical trials. While the synthesis was not difficult, Douglas G. Johnson, VP for manufacturing at Allos reported that an unsuspected impurity profile showed up upon the scale-up and could only be remedied by equipment changes. Hovione were committed to this project and agreed to purchase the special equipment. The C & E News article concludes that good service leads to repeat business, and Hovione remained committed to Allos and purchased the needed equipment [50a].

Allos Therapeutics Inc. and Clinical Trials with a Submission of an NDA

The phase-one safety study was performed by Prof. Jurgen Venitz in our Department of Pharmaceutics in the School of Pharmacy. Jurgen is one of the best-qualified phase-one

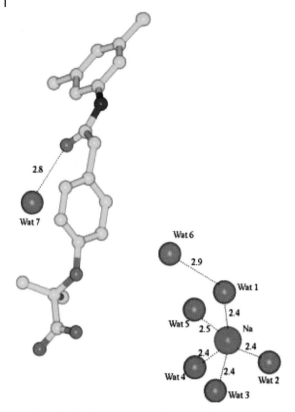

Fig. 17.5 RSR 13 Heptahydrate Sodium Salt. The sodium atom is in light blue, water atoms are large red spheres, carbons yellow, nitrogen dark blue, and oxygens red. This particular RSR 13 salt was not soluble enough for use in clinical trials.

clinical trials scientists in the country. Being convinced that RSR 13 was indeed safe, I volunteered to be the first subject, but was not allowed to be enrolled in the study because of my age and the fact that RSR 13 came from my laboratories. I was allowed to be present at the first injection of RSR 13 in humans. It was a nervous time, even with the firm belief that safety issues with RSR 13 should not be an issue.[h] And it was not.

CEO, Stephen Hoffman and VP for clinical studies, Michael Gerber of Allos Therapeutics guided RSR 13 through a series of phase-one and phase-two clinical trials for radiation treatment of brain tumors and for potential use in cardiopulmonary bypass surgery [46, 51–55]. Considering the cost of running a phase-three clinical trial, only one was possible. The very positive phase-two results for use of RSR 13 to treat metastatic brain cancer provided the impetus for selecting that indication for a phase-three trial.

---

[h] The author felt like he was watching the troops, in this case the first medical students who took RSR 13, go ashore at the battle of Normandy while being safely sequestered on a ship watching the invasion and praying that no one would be injured.

Stephen Hoffman as CEO of Allos raised about $40 million of private funding before he and the current CEO, Michael Hart, did a tremendous job of taking the company public and raising $90 million. The public offering just made the window by a few dozen hours for an IPO, as the window quickly closed after President Clinton announced that the human genome sequence could not be patented. This news sent the biotechnology stock market plummeting.

The total sum of $140 million and subsequent private funding of around $12 million enabled the completion of the phase-three clinical trials with about enough remaining to conduct a second phase-three trial if the FDA did not approve the current NDA. The total amount spent if the RSR 13 is approved is greatly reduced compared to the estimated current average cost of over half a billion dollars for a new therapeutic agent.

The phase-three results were unblinded in March 2003 and overall, for all types of metastatic brain cancer, were reported as nonpositive, showing the drug to have efficacy but not at the statistical level set at the beginning of the trial by Allos Therapeutics Inc. for a successful phase-three trial. The subset analysis for the different types of metastatic cancer to brain, however, showed that breast cancer patients had a 100% increase in survival. The 500 plus patient trial also demonstrated RSR 13 to be extremely safe, confirming our early design feature of using a known antilipidemic drug, clofibrate, as a substructure. An NDA for RSR 13 with trade name *efaproxaril* was reviewed by the FDA on a fast track review for use in treatment of metastases of breast cancer to brain. On June 2, 2004, the FDA gave an approvable letter to market efaproxaril if the ongoing phase-three trial, called ENRICH (Enhancing Whole Brain Radiation Therapy in Patients with Breast Cancer and Hypoxic Brain Metastases), designed specifically for the use of efaproxaril to treat metastatic breast cancer to brain, is successful. The ENRICH phase three trial will seek to enroll approximately 360 patients at up to 50 cancer centers across North/South America and Europe. In the letter, the FDA stated, "if the study shows effectiveness in this population (increased survival) using the prespecified analysis, and the study is otherwise satisfactory, we believe it would, together with the subset result in RT-009, support approval" (http://biz.yahoo.com/prnews/040602/law025_1.html).

On August 3, 2004, the FDA gave efaproxaril orphan drug status, which provides seven years of patent protection after approval if the Enrich trial is successful (http://biz.yahoo.com/prnews/040803/latu114_1.html). Orphan drug status also qualifies the product for possible funding to support clinical trials, study design assistance from the FDA during development, and other financial incentives and tax benefits.

Just the Beginning

Normally, the body only uses about 25% of Hb's oxygen for normal biological processes. The potential magnitude of having a therapeutic agent that delivers stored oxygen cannot be underestimated. To date, no such agent has been available to study the physiology of numerous *in vivo* pathways that might be regulated with an increase or decrease in oxygen pressure.

It is my opinion that if RSR 13 is approved, its potential use in medicine could be far beyond the radiation treatment of tumors. RSR 13 is the first molecule with human efficacy to safely shift the oxygen levels to tissues. RSR 13 has a short half-life with IV infusion and therefore will be limited to acute uses. The potential acute use for cardiovascular areas might prove to be the largest target. RSR 13 is also a potential drug that could be used illegally in sports to increase VO_2 max, the volume and velocity of oxygen released to tissue during exercise. Apparently, RSR 13 was credited for a disqualification of a bike rider in the Italian version of the Tour de France, the 2001 Giro d'Italia, see http://www.totalbike.com/news/article/556/ and http://perso.wanadoo.fr/jc.auriol/histoires.htm. Allos Therapeutics Inc. has been working with the Olympics committee on how to detect any illegal uses of RSR 13.

Another Allosteric Effector for Treatment of Sickle-cell Disease

In the 1990s, we advanced another Hb allosteric effector to clinical trials, vanillin [56]. Dr. Martin Safo of our group has recently extended our search for a nontoxic aldehyde food substance as a potential antisickling agent. The latest molecule 5-HMF (5-hydroxymethyl-2-furfural) just completed in vivo tests in sickle-cell transgenic mice with outstanding results [57]. We hope our experience with RSR 13 will aid us in forwarding 5-HMF through clinical trails.

Take Home Messages

1. Collaborative efforts across academic units were a vital key for successful translation from the bench to the bedside.
2. Use of a known low-toxicity scaffold (drug) for building molecular specificity was the most important advance, permitting us to overcome toxicity issues due to the large doses required to treat almost a pound and a half of the in vivo receptor (hemoglobin).
3. When in doubt as to which molecule to forward as a clinical candidate, consider metabolism and toxicity profiles as well as biological activity.
4. Having a dedicated start-up company to champion moving a molecule through all the steps to an NDA made the difference as large pharmaceutical companies are less likely to champion a compound from academia.
5. Serendipity still continues to play its important role in drug discovery. All four German Gs of Paul Ehrlich [8] that scientists need were evident in our case study.

Acknowledgments

The author acknowledges the financial support for the Hb basic research and preclinical studies from our long-standing NIH grant: 5 RO1HL32793, NIH grant NO1-HB-1-301, Allos Therapeutics Inc. grants, The H.J. Heinz Company, The Willie Stargell Foundation, and The C. Usher Foundation and The European Molecular Biology Organization (EMBO) and the Max Perutz Fund for short term fellowships to visit the Medical Research Council Laboratory of Molecular Biology. The author also wishes to thank the help of Mrs Ruth White and the Sickle-Cell Society of Pittsburgh, the Deans and Administrators at the University of Pittsburgh and Virginia Commonwealth University for the freedom to pursue this work as needed and for financial contributions. Last, but not the least, we thank our many colleagues and collaborators in Europe and the United States, who put their shoulder to the till to provide the strong science contributions that built a solid foundation for advancing two agents to clinical trials. A very special thanks to my wife Nancy L. Abraham, who also provided me the freedom and support to spend whatever time was needed at our universities or abroad to bring this work to fruition.

References

1 CONTRERAS, J., EAVIS, K, NEWELL, S. *The Dizzying Rise of University Spinouts.* Internet 36, 25. **2002**. Tornado Insider.

2 MEHTA, S. The emerging role of academia in commercializing innovation. *Nat. Biotechnol.* **2004**, *22*, 21–24.

3 PRESSMAN, L. (Ed.). *AUTM Licensing Survey FY 2002.* The Association of University Technology Managers Inc. **2002**.

4 EDWARDS, M.G., MURRAY F., YU, R. Value creation and sharing among universities, biotechnology and pharma. *Nat. Biotechnol.* **2003**, *21*, 618–624.

5 ROBERTS, R.M. *Serendipity, Accidental Discoveries in Science.* John Wiley & Sons, New York, **1989**.

6 DONNELLY, B.E. CLAYMAN, B.P. *Research Parks in Profile,* Vol. 1. **2003**.

7 KORN, D. *Industry, Academia, Investigator: Managing the Relationships.* Office of Research Virginia Commonwealth University, **2002**.

8 PERUTZ, M.F. *I Wish I Would Have Made You Angrier Earlier: Essays on Science, Scientists, and Humanity.* Cold Spring Harbor Laboratory Press, Plainview, NY, **1998**.

9 ABRAHAM, D.J. (Ed.). History of quantitative structure activity relationships. In *Burger's Medicinal Chemistry and Drug Discovery,* Volume 3. John Wiley & Sons, Hoboken, NJ, **2003**, 1–48. Chapters by SELASSIE, C.D.; TROPSHA, A. Recent trends in quantitative structure-activity relationships, 49–76; MARSHALL, G.R. BEUSEN, D.D. Molecular modeling in drug design, 77–168;

KOLLMAN, P.A. CASE, D.A. Drug-target binding forces: advances in force field approaches, 169–185; MASON, J.S. PICKETT, S.D. Combinatorial library design, Molecular similarity and diversity applications, 187–242; MUEGGE, I. ENYEDY, I. Virtual screening, 243–279; HARDY, L.W., SAFO, M.K., ABRAHAM, D.J. Structure-based drug design, 417–469; BUSS, A.D., COX, B., WAIGH, R.D. Natural products as leads for new pharmaceuticals, 847–900; chapter by MITSCHER, L.A. DUTTA, A. Combinational chemistry and multiple parallel synthesis, Volume 2, 1–36.

10 WEINER, S.J., KOLLMAN, P.A., CASE, D.A., SINGH, U.C., GHIO, C., ALAGONA, G., PROFETA, S., WEINER, P.J. A new force field for molecular mechanics simulation of nucleic acids and proteins. *J. Am. Chem. Soc.* **1984**, *106*, 765.

11 HAGLER, A.T. MOULT, J. Computer simulation of the solvent structure around biological macromolecules. *Nature* **1978**, *272*, 222–226.

12 HAGLER, A.T., HULER, E., LIFSON, S. Energy functions for peptides and proteins. I. Derivation of a consistent force field including the hydrogen bond from amide crystals. *J. Am. Chem. Soc.* **1974**, *96*, 5319–5327.

13 BROOKS, B.R., BRUCCOLERI, R.E., OLAFSON, B.D., STATES, D.J., SWAMINATHAN, S., KARPLUS, M. CHARMM: a program for macromolecular energy, minimization, and dynamics calculations. *J. Comput. Chem.* **1983**, *4*, 187–217.

14 TAYLOR, J.S., GARRETT, D.S., COHRS, M.P. Solution-state structure of the Dewar pyrimidinone photoproduct of thymidylyl-(3'-5')-thymidine. *Biochemistry* **1988**, *27*, 7206–7215.

15 WADE, R.C. GOODFORD, P.J. The role of hydrogen-bonds in drug binding. *Prog. Clin. Biol. Res.* **1989**, *289*, 433–444.

16 SHOICHET, B.K., STROUD, R.M., SANTI, D.V., KUNTZ, I.D., PERRY, K.M. Structure-based discovery of inhibitors of thymidylate synthase. *Science* **1993**, *259*, 1445–1450.

17 ERLICH, P. In *Collected Papers of Paul Erlich*, HIMMELWEIT, F (Ed.). Pergamon, London, **1957**.

18 KENDREW, J.C. PERUTZ, M.F. X-ray studies of compounds of biological interest. *Annu. Rev. Biochem.* **1957**, *26*, 327–372.

19 LADNER, R.C., HEIDNER, E.J., PERUTZ, M.F. The structure of horse methaemoglobin at 2-0 A resolution. *J. Mol. Biol.* **1977**, *114*, 385–414.

20 PERUTZ, M.F. Stereochemistry of cooperative effects in haemoglobin. *Nature* **1970**, *228*, 726–739.

21 ABRAHAM, D.J. The potential role of single crystal X-ray diffraction in medicinal chemistry. *Intra Sci. Chem. Rep.* **1974**, *8*(4), 1–9.

22 FITZHARRIS, P., MCLEAN, A.E., SPARKS, R.G., WEATHERLEY, B.C., WHITE, R.D., WOOTTON, R. The effects in volunteers of BW12C, a compound designed to left-shift the blood-oxygen saturation curve. *Br. J. Clin. Pharmacol.* **1985**, *19*, 471–481.

23 BEDDELL, C.R., GOODFORD, P.J., KNEEN, G., WHITE, R.D., WILKINSON, S., WOOTTON, R. Substituted benzaldehydes designed to increase the oxygen affinity of human haemoglobin and inhibit the sickling of sickle erythrocytes. *Br. J. Pharmacol.* **1984** *82*, 397–407.

24 WIREKO, F.C. ABRAHAM, D.J. X-ray diffraction study of the binding of the antisickling agent 12C79 to human hemoglobin. *Proc. Natl. Acad. Sci. U. S. A.* **1991**, *88*, 2209–2211.

25 ABRAHAM, D.J., PERUTZ, M.F., PHILLIPS, S.E. Physiological and X-ray studies of potential antisickling agents. *Proc. Natl. Acad. Sci. U.S.A.* **1983**, *80*, 324–328.

26 KENNEDY, P.E., WILLIAMS, F.L., ABRAHAM, D.J. Design, synthesis, and testing of potential antisickling agents. 3. Ethacrynic acid. *J. Med. Chem.* **1984**, *27*, 103–105.

27 PERUTZ, M.F., FERMI, G., ABRAHAM, D.J., POYART, C., BURSAUX, E. Hemoglobin as a receptor of drugs and peptides: X-ray studies of the stereochemistry of binding. *J. Am. Chem. Soc.* **1986** *108*, 1064–1978.

28 MEHANNA, A.S. ABRAHAM, D.J. Comparison of crystal and solution hemoglobin binding of selected antigelling agents and allosteric modifiers. *Biochemistry* **1990**, *29*, 3944–3952.

29 ARNONE, A. X-ray diffraction study of binding of 2,3-diphosphoglycerate to human deoxyhaemoglobin. *Nature* **1972**, *237*, 146–149.

30 HIRST, D.G., WOOD, P.J., SCHWARTZ, H.C. The modification of hemoglobin affinity for oxygen and tumor radiosensitivity by antilipidemic drugs. *Radiat. Res.* **1987**, *112*, 164–172.

31 HIRST, D.G. WOOD, P.J. Chlorophenoxy acetic acid derivatives as hemoglobin modifiers and tumor radiosensitizers. *Int. J. Radiat. Oncol. Biol. Phys.* **1989**, *16*, 1183–1186.

32 PERUTZ, M.F. POYART, C. Bezafibrate lowers oxygen affinity of haemoglobin. *Lancet* **1983**, *2*, 881–882.

33 BURLEY, S.K. PETSKO, G.A. Aromatic-aromatic interaction: a mechanism of protein structure stabilization. *Science* **1985**, *229*, 23–28.

34 BURLEY, S.K. PETSKO, G.A. Amino-aromatic interactions in proteins. *FEBS Lett.* **1986**, *203*, 139–143.

35 LALEZARI, I., LALEZARI, P., POYART, C., MARDEN, M., KISTER, J., BOHN, B., FERMI, G., and PERUTZ, M.F. New effectors of human hemoglobin: structure and function. *Biochemistry* **1990**, *29*, 1515–1523.

36 ABRAHAM, D.J., WIREKO, F.C., RANDAD, R.S., POYART, C., KISTER, J., BOHN, B., LIARD, J.F., and KUNERT, M.P. Allosteric modifiers of hemoglobin: 2-[4-[[(3,5-disubstituted anilino)-carbonyl]methyl]phenoxy]-2-methylpropionic acid derivatives that lower the oxygen affinity of hemoglobin

in red cell suspensions, in whole blood, and in vivo in rats. *Biochemistry* **1992**, *31*, 9141–9149.

37 RANDAD R.S., MAHRAN, M.A., MEHANNA, A.S., ABRAHAM, D.J. Allosteric modifiers of hemoglobin. 1. Design, synthesis, testing, and structure-allosteric activity relationship of novel hemoglobin oxygen affinity decreasing agents. *J. Med. Chem.* **1991**, *34*, 752–757.

38 ABRAHAM, D.J. LEO, A.J. Extension of the fragment method to calculate amino acid zwitterion and side chain partition coefficients. *Proteins* **1987**, *2*, 130–152.

39 KELLOGG, G.E., SEMUS, S.F., and ABRAHAM, D.J. HINT: a new method of empirical hydrophobic field calculation for CoMFA. *J. Comput.-Aided Mol. Des.* **1991**, *5*, 545–552.

40 WIREKO, F.C., KELLOGG, G.E., and ABRAHAM, D.J. Allosteric modifiers of hemoglobin. 2. Crystallographically determined binding sites and hydrophobic binding/interaction analysis of novel hemoglobin oxygen effectors. *J. Med. Chem.* **1991**, *34*, 758–767.

41 SAFO, M.K., MOURE, C.M., BURNETT, J.C., JOSHI, G.S., and ABRAHAM, D.J. High-resolution crystal structure of deoxy hemoglobin complexed with a potent allosteric effector. *Protein Sci.* **2001**, *10*, 951–957.

42 LEVITT, M., PERUTZ and M.F. Aromatic rings act as hydrogen bond acceptors. *J. Mol. Biol.* **1988**, *201*, 751–754.

43 ABRAHAM, D.J., KISTER, J., JOSHI, G.S., MARDEN, M.C., and POYART, C. Intrinsic activity at the molecular level: E.J.Ariens' concept visualized. *J. Mol. Biol.* **1995**, *248*, 845–855.

44 ABRAHAM, D.J., SAFO, M.K., BOYIRI, T., DANSO-DANQUAH, R.E., KISTER, J., and POYART, C. How allosteric effectors can bind to the same protein residue and produce opposite shifts in the allosteric equilibrium. *Biochemistry* **1995**, *34*, 15006–15020.

45 KHANDELWAL, S.R., RANDAD, R.S., LIN, P.S., and MENG, H., PITTMAN, R.N., KONTOS, H.A., CHOI, S.C., ABRAHAM, D.J., and SCHMIDT-ULLRICH, R. Enhanced oxygenation in vivo by allosteric inhibitors of hemoglobin saturation. *Am. J. Physiol.* **1993**, *265*, H1450–H1453.

46 KHANDELWAL, S.R., KAVANAGH, B.D., LIN, P.S., TRUONG, Q.T., LU, J., ABRAHAM, D.J., and SCHMIDT-ULLRICH, R.K. RSR13, an allosteric effector of haemoglobin, and carbogen radiosensitize FSAII and SCCVII tumours in C3H mice. *Br. J. Cancer* **1999**, *79*, 814–820.

47 WEI, E.P., RANDAD, R.S., LEVASSEUR, J.E., ABRAHAM, D.J., and KONTOS, H.A. Effect of local change in O2 saturation of hemoglobin on cerebral vasodilation from hypoxia and hypotension. *Am. J. Physiol.* **1993**, *265*, H1439–H1443.

48 DOPPENBERG, E.M., WATSON, J.C., BULLOCK, R., GERBER, M.J., ZAUNER, A., and ABRAHAM, D.J. The rationale for, and effects of oxygen delivery enhancement to ischemic brain in a feline model of human stroke. *Ann. N. Y. Acad. Sci.* **1997**, *825*, 241–257.

48a DOPPENBERG, E.M., RICE, M.R., ALESSANDRI, B., QIAN, Y., DI, X., and BULLOCK, R. Reducing hemoglobin oxygen affinity does not increase hydroxyl radicals after acute subdural hematoma in the rat. *J. Neurotrauma* **1999**, *16*, 123–133.

49 BISHOP, J.E. Technology and health: new chemicals boost blood's ability to deliver oxygen, researchers report. *The Wall Street J.*, Sept. 29, **1992**.

50 SAFO, M.K. and ABRAHAM, D.J. Crystal Structure of Rsr 13 Sodium Salt Hepta-Hydrate. **2004**, Unpublished data.

50a McCOY, M. Pharma outsourcing; A look at four pharmaceutical outsourcing relationships shows that good service leads to repeated business. Chemical and Engineering News, 2004, April 5.

51 KILGORE, K.S., SHWARTZ, C.F., GALLAGHER, M.A., STEFFEN, R.P., MOSCA, R.S., and BOLLING, S.F. RSR13, a synthetic allosteric modifier of hemoglobin, improves myocardial recovery following hypothermic cardiopulmonary bypass. *Circulation* **1999**, *100*, 11351–11356.

52 KLEINBERG, L., GROSSMAN, S.A., PIANTADOSI, S., PEARLMAN, J., ENGELHARD, H., LESSER, G., RUFFER, J., and GERBER, M. Phase I trial to determine the safety, pharmacodynamics, and pharmacokinetics of RSR13, a novel radioenhancer, in newly diagnosed

glioblastoma multiforme. *J. Clin. Oncol.* **1999**, *17*, 2593–2603.

53 KLEINBERG, L., GROSSMAN, S.A., CARSON, K., LESSER, G., O'NEILL, A., PEARLMAN, J., PHILLIPS, P., HERMAN, T., and GERBER, M. Survival of patients with newly diagnosed glioblastoma multiforme treated with RSR13 and radiotherapy: results of a phase II new approaches to brain tumor therapy CNS consortium safety and efficacy study. *J. Clin. Oncol.* **2002**, *20*, 3149–3155.

54 SHAW, E., SCOTT, C., SUH, J., KADISH, S., STEA, B., HACKMAN, J., PEARLMAN, A., MURRAY, K., GASPAR, L., MEHTA, M., CURRAN, W., and GERBER, M. RSR13 plus cranial radiation therapy in patients with brain metastases: comparison with the Radiation Therapy Oncology Group Recursive Partioning Analysis Brain Metastases Database. *J. Clin. Oncol.* **2003**, *21*, 2364–2371.

55 Notes: Department of Radiation Oncology, Wake Forest University School of Medicine, Winston-Salem, NC 27157-1030, USA eshaw@wfubmceduFAU – SHAW, EDWARD WHAR, J.A., GERBER, M., VENITZ, J., and BALIGA, N. Allosteric modification of oxygen delivery by hemoglobin *Anesth. Analg.* **2001**, *92*, 615–620.

56 ABRAHAM, D.J., MEHANNA, A.S., WIREKO, F.C., WHITNEY, J., THOMAS, R.P., and ORRINGER, E.P. Vanillin, a potential agent for the treatment of sickle cell anemia. *Blood* **1991**, *77*, 1334–1341.

57 SAFO, M.K., ABDULMALIK, O., DANSO-DANQUAH, R., BURNETT, J.C., NOKURI, S., JOSHI, G.S., MUSAYEV, F.N., ASAKURA, T., and ABRAHAM, D.J. Structural basis for the potent antisickling effect of a novel class of five-membered heterocyclic aldehydic compounds. *J. Med. Chem.*, **2004**. Web Release Date: 6 August, 2004; (Article)DOI: 10.1021/jm0498001.

Subject Index

a

ABE *see* Andrew's binding energy
academic licensing 456
ACAT models 348
ACD *see* Available Chemicals Directory
ACD software
– name 230
– solubility 385
ACE *see* angiotensine-converting enzyme
active site induced feature 135
active site mapping 204
active-analog approach 8
activity landscapes 318
ADAM 73
ADME(T) 15, 17, 27, 33, 45, 90, 144, 179,
    306, 397
– chemotype 361
– prediction 338, 407
– properties 347, 357
affinity prediction 12
AFMoC 67
aggregation 290
Alfuzosin 292
alignments 68, 89, 123, 292
Allos Therapeutics 471
allosteric regulators 472
allosteric site 18
Almond 355
AMBER 459
AMOEBA 15
AMSOL 96
anagyrine 233
Andrew's binding energy (ABE) 32, 408
angiotensin-II
– antagonists 98
– receptor 99
angiotensin-converting enzyme (ACE) 10
– inhibitor 1
ANN *see* artificial neural network 393
anticonvulsants 447

b

BCR-ABL tyrosine kinase 94
BCUT 82, 153, 160, 177 f., 325, 330
– diversity analysis 168
Beilstein 223, 304
benzodiazepine 16
– receptor 343
bezafibrate 464, 468
BFGS update 70
bile-acid resorption 364
bioactivity distribution 225
bioavailability 348, 361, 413
bioisosteres 290
bioisosteric analogs 295
bioisosterism 295
BioPrint 119, 128, 224
bitstring representations 88
Bleep 66
blood-brain barrier penetrability 406
Boehringer Mannheim wall chart 253
boosting methods 75
BUILDER 419

antitargets 334, 338, 343, 388
Apache 251
apomorphine 7
AQUASOL 385
aqueous solubility 283, 347, 383, 403
– prediction 384, 389, 411
artificial network 393
artificial neural network 437
atorvastatin 245
attrition 48
attrition rate 333
AUR-STORE 224
auto-correlation vector 82
AutoDOCK 15
Available Chemical Directory (ACD) 32,
    277, 287, 320, 392

Chemoinformatics in Drug Discovery. Edited by Tudor I. Oprea
Copyright © 2004 WILEY-VCH Verlag GmbH & Co. KGaA, Weinheim
ISBN: 3-527-30753-2

building blocks 300
Bupivacaine 34
butaclamol 7

c

Cabinet 237, 241 ff., 247, 251
– library 252
Caco-2 assay 403
Caco-2 cells 348, 353
Cambridge Structural Database (CSD) 4, 65
cannabinoid CB1/CB2 receptor 304
canonical SMILES 283
captopril 4
carbonic anhydrase II 94
cardiac T-type Ca^{2+} channel 98
carisoprodol 233
CASE 444
CASP competitions 13
Catalyst 54, 98, 125, 421, 444
CATS 82, 98
CDK-2 180, 342
– inhibitors 341
Cerep 403
Cerivastatin 34
CERN 248
cGMP 475
charge-transfer 64
CHARMM 459
checkpoint kinase-1 282
ChemBank 224
ChemBioBase 224
ChemBridge 402
ChemGPS 131
Chemical Abstracts Service (CAS) 1
chemical descriptors 437
Chemical Diversity Labs 310
chemical diversity space 17
chemical similarity searches 438
chemistry space 146, 318, 325, 329, 445
– descriptors 330
ChemNavigator 402
ChemoSoft 288, 294
chemotypes 26, 53, 131, 292
– chemistry space 357
– in databases 223
ChemScore 341
ChemSpace 88, 98
ChemTree 153
chlorpromazine 7
citric acid 260
CK2α 92
classification 75
ClassPharmer 152

cleavage rules 291
clinical research 25
clinical trials 477
– in academia 459
clique detection algorithms 77
Clofibrate 463
clofibric acid 463
clogP 29, 34, 35, 53, 144, 153, 155, 156, 164, 176, 179, 227, 235, 307, 329, 399
clustering algorithm 320
CMC *see* Comprehensive Medicinal Chemistry
CombiDock 89
CombiLibMaker 178
combinatorial chemistry 16, 17, 87, 317
– libraries 176
Combinatorial Chemistry Consortium 178
combinatorial libraries, target-specific 309
combinatorial library design, product-based 205
COMBINE 341
CoMFA *see* Comparative Molecular Field Analysis
Comparative Molecular Field Analysis (CoMFA) 11, 68, 75, 123, 340, 342 f., 438, 443 f.
Common Object Request Broker 243
ComPharm 119, 123, 130
complexity 337, 387, 392, 408
component specification language 206
compound
– acquisition 321, 327
– attrition 320
– collections 328, 381
– filters 47, 324
– selection 320, 325
Comprehensive Medicinal Chemistry (CMC) 223
computer-aided drug design 6, 199
CoMSIA 68, 342
CONCORD 2, 70, 178
conflict of interest 457 ff.
conformational flexibility 67, 72, 78, 85
connection table 276
consensus scoring 91
contact matrix 13
CORBA *see* Common Object Request Broker
CORINA 70, 274, 282
Cox-2
– inhibitors 10
– reductase 10
– *see* also cyclooxygenase-II
CPCA 344, 347
C-QSAR 237
cross validation 76, 438

CSD *see* Cambridge Structural Database
cyclooxygenase-II (Cox2) 119
cytochrome P450 306, 345, 348, 355, 391
– CYP3A4 90
– CYP2D6 90
– enzymes 17
– isozymes 37

d

DAS *see* Distributed Annotation System
data integration 241
database integration
– federation 244
– mining 438, 446
– unification 244
Daylight 176, 279
– CIS 54
– fingerprint 45, 187
– HTTP toolkit 251 f.
DBCROSS 3
DBMAKER 2
de novo design 87
– active site mapping 204
– component specification language 206
– computer-aided 199
– database property indexing 203
– random sampling 204
– structure-based 419
– unwanted structures removal 207
– use of in-house intellectual property 202
– versus lead optimization 200
decision
– process 32
– site 161
– tree 75, 393
decision-making process 26
degree-of-fit method 442
DEREK 399
design
– D-optimal 178, 190
– ligand-based 181
– structure-based 190
– target-focused 182
Dexfenfluramine 34
Diclofenac 131
2,3-diphosphoglycerate 5, 463, 469
2,4-diaminopyrimidine 96
dictionary of Chinese medicine 255 f.
Digital Object Identifier (DOI) 226
dihydrofolate reductase (DHFR) 35
diligence 472
dipole 63
directed tweak technique 71

DISCO 125
discontinued drugs 34
dissimilarity 123
distance geometry 72
Distributed Annotation System (DAS) 243
diverse solutions (DVS) 163, 178, 325, 330
diversity 17, 88, 145, 175, 205
– analysis 131
– design of experiments 190
– optimization with MoSELECT 188
– optimization with stochastic methods 186
– receptor-relevant subspaces 184
– simultaneous optimization 185
DANN gyrase 97
DOCK 15, 72, 73, 77, 87, 91, 94, 274, 419, 459
docking 61, 64, 72, 89, 306
– placement of hydrogens 278
DOCKIT 73
document databases 245
DOI *see* Digital Object Identifier
dopamine D_2 receptor 295 f., 439
– applicability domain 441
– binary 444
– combinatorial 444, 445
– consensus prediction 444
– data preparation 443
– experimental validation 447 ff.
– model generation 443
– model robustness 444
– model validation 444
– Y-randomization 451
D-optimal design 178
drug approval 391
Drug Data Report 392
drug discovery 455
– attrition rate 333, 381
– in academia 459
– value chain 333, 335
Drug Index 392
druglike 44, 48, 63, 90, 127, 179, 291, 336, 381, 389, 392, 404
– chemical space 33
– properties 4
– property space 33
druglikeness 310, 323, 383, 390 ff., 406
– prediction 408
Drugmatrix 224
DrugScore 66, 340, 345, 347
duplicate removal 276
DVS *see* Diverse Solutions

e

ECN *see* Enzyme Commission Number
efaproxaril 477
effective prediction domain 442
EGFR 237
embedding 72
Empath – metabolic pathway chart 253
enalapril 1, 10
ENRICH 477
ensemble 293
enthalpy 211
– of binding 12
enthalpy-entropy compensation 66
entropy 63, 66, 211
– of binding 12, 18
enzyme commission codebook 254
Enzyme Commission Number (ECN) 247, 253
Eph kinases 345
estrogen receptor-α 94
EUDOC 96
exclusion criteria 324
Extensible Markup Language (XML) 242

f

factor Xa 97, 342 f., 363
false negatives 391, 402
false positives 27, 148, 151, 290, 391
farnesyltransferase 96, 98
fast Fourier transform 74
FBPA *see* Fuzzy Bipolar Pharmacophore Autocorrellograms
feature trees 76, 83, 89
filtering 156
filters 327, 330
fingerprints 74, 176
– Barnard 88
– hashed 82
– minifingerprints 82
– pharmacophore 120
flavin monooxygenase 349
FlexE 278
flexible bonds *see* rotatable bonds
FlexS 71
FlexX 71, 87, 89, 94, 97, 273, 277, 419
FlexX-C 89
FlexX-Pharm 72
flurbiprofen 131, 134 f.
focused/diverse library design 183
focused library 182
fold predictions 13
force fields 67
FORM 249
formal charges 274

Fourier transform 73
frequent hitters 27, 90, 290, 383, 394 f.
– *see also* promiscuous inhibitors
Fuzzy Bipolar Pharmacophore Autocorrellograms (FBPA) 121, 132
fuzzy matching 82

g

GaP space (Gridding and Partitioning) 46, 53
GASP 125
GastroPlus 348
GB/SA 18
Genentech 455
genetic algorithm-driven procedure 124
genetic algorithms 86
geometric hashing 80
Giro d'Italia 478
glycogen synthase kinase-3 98
GOLD 72, 96, 419
Google 241, 266, 460
GPCR 18, 179, 291, 345
– annotator 224
– ligands 293, 308, 310
GPCR Annotator 224
GPCR-based library 99
G-protein-coupled receptor *see* GPCR
graph matching 77
graph theory 6
GRID 68, 344 f., 350, 459
GRID/CPCA 344
GSK-3 98
guanine 272

h

Hammerhead 87
Hammett constant 415
Hansch parameters 415, 417
hashcodes 153, 159, 276
hashing 79
hemoglobin 5, 15, 18, 463
– allosteric effector 460 ff.
hERG 37, 90, 290, 384, 391, 406, 413
high-throughput screening *see* HTS
HINT 68, 342, 459, 468f.
HipHop 125, 130
hit-to-lead 395, 397, 403, 423
– libraries 405
– work flow 398
HIV protease 342, 443
HIV-1 protease 98
HMG-CoA reductase 10, 257
Hodgkin score 133
Hovione 475

5-HT2c ligands 99
HTML 248, 251
HTS 27, 317
– actives rescue 169
– compound prioritization 149
– confirmation rate 152
– cost 143
– false negatives 146
– false positives 147, 165, 168
– peptide hydrolase target 162
– protein kinase target 167
– protein-protein interaction 168
– *see also* high throughput screening
HTTP *see* Hyper Text Transfer Protocol
Httpd 251
hydrogen
– acceptors 153
– bonding 278
– bonds 64, 464, 468
– donors 153
– placement in docking studies 278
hydrophobic
– effect 64, 118
– interactions 337, 468
– substituent constant 415
hydrophobicity 384
Hyper Text Transfer Protocol (HTTP)
 241, 248, 251
hypervalent nitrogen 280
HypoGen 125

i

ICM 94
iDEA 348
IL-8 289
indoline/cytochrome C peroxidase 273
informative design 180
inosine 5'-monophosphate dehydroenase
 97
INSIGHT 459
intellectual property 25, 201, 289, 300,
 310, 455
– management 458
intestinal absorption 352 f.
ionization 271

k

kinase 44, 224
– inhibitor 65, 91
kNN 447
knowledge databases 181
knowledge-based scoring functions 66
Kohonen maps 298, 307

l

lead
– discovery 17, 26, 28, 43, 404
– generation 15, 406
– identificaton 383
optimization
– – by structure-based design 339
– – compound profile 334
– – criteria 337
– series 396
leadlike 32, 34, 43, 48, 144, 152, 337, 390,
 405, 424
– properties 396
Leadscope 153
learning
– active 76
– supervised 74
– unsupervised 74
least-squares fitting 7
leatherface 279
Levinthal paradox 13
Levomethadyl 34
libraries 87
library
– design 287, 404 f., 407, 438
– target-focused 182
LibraryDesigner 178, 183
ligand-based design 181
LIGPLOT 254
linear descriptors 81
linear optimization 69
Lipitor 245, 256
lipophilicity 350
– *see also* hydrophobicity
"lock and key" hypothesis 117
logD 348
logD$_{7.4}$ 37
logP 29, 210, 337, 348
long QT syndrome 354, 413
– *see also* hERG
– *see also* QT prolongation
LPAM-1 289
LUDI 87, 89, 96, 419

m

MACCS 176
machine learning 74, 118
magic bullet 1
manganese 6
Material Safety Data Sheets (MSDS)
 246
matrix metalloproteinase 342

maximum
– common subgraph 77
– dissimilarity 319
– dissimilarity algorithm 327
Maybridge 402, 447
MDCK cells 353
MDDR *see* MDL Drug Data Report
MDL Drug Data Report (MDDR) 32 f.,
    179, 181, 223, 277, 291, 324
– database 99
MDR1 348
MDS Panlabs 403
Merck Index 228, 233
META 349
MetabolExpert 349
metabolic instability 361
Metabolite 349
MetaSite 355
Meteor 349
Metropolis Monte Carlo 85
MIME 249
MlogP 424
MLR 437
MMP-3 343
MMP-8 364
MODDE 190
MOE 157, 161
molar refractivity 415
MOLCAD 341
Molconn-Z 187
molecular
– complexity 34, 52, 151
– descriptors 307
– design, statistical 360
– diversity 317
– fingerprints 320
– flexibility 71
– hash codes 276
– properties 306
– recognition 16, 49, 63, 121, 415
– replacement 74
– similarity 317
– volume 6, 8
– weight 228, 307, 324, 329, 337, 384, 387,
    392, 399, 413, 421, 424
MolSurf 353
MSDS *see* Material Safety Data Sheets
multidimensional optimization 335, 357,
    360
multiple binding modes 18, 25, 63
multipole representation 13
muscarinic receptor 99

n

N-arylpiperazine 295, 297
National Cancer Institute 447
NCl database 98
Netscape 256
neural networks 75, 298
neuropeptide Y1 289
new drug application 455
Newton methods 70
N-myristoyltransferase 363
nondrugs 33
nonlinear optimization 69
nonpolar surface area 227
NP-hard problems 77
number
– of hydrogen bond donors 235
– of hydrogen bonds 144, 324, 399
– of rings 29, 34, 337
– of rotable bonds 12, 92, 144, 156, 176,
    324, 329, 387, 400

o

Occam's razor 18
octanol/water partition coefficient 210,
    324, 329
Olanzapine 290
Omeprazole 34
optimal diversity 407
optimization 396, 408
– multidimensional 335
oral bioavailability 363, 383
orphan drug status 477
overfitting 76

p

P1 purinergic receptor 98
P38 MAP kinase 363
PACK 214
PAMPA 348, 353, 403
Pareto ranking 188
partial least squares analysis (PLS) 12,
    210, 340, 343, 350, 394, 437
patenting 2, 472
pattern recognition 79, 128
PCA *see* principal component analysis
PDB 5, 65, 66, 134, 214, 246, 272, 277
PDR *see* Physician's Desk Reference
pentaazacrown scaffold 6
peripheral structural motifs 296
permeability 28, 403
permute 274
pharmacodynamics 25
pharmacokinetic properties 350

pharmacokinetics 25, 361
pharmacophore 7, 43, 53, 62, 94, 98, 179, 205, 419, 421, 427, 446
– field intensities 124
– filter 101
– hypothesis 125
– keys 45
– points 46
– searches 306
– similarity scoring 132
pharmacophoric
– pattern 8
– points 72
phosphodiesterase 4 (PDE4) 35
– inhibitors 355
phosphonate 50
phosphoramidate 50
phospho-ribosyl transferase 272
Photo ARKive 255 f.
Physician's Desk Reference (PDR) 246
PHYSPROP 385
PICCOLO 186
Pittsburgh Pirates 462
Plackett-Burman design 190
PLAID *see* Protein-Ligand Accessibility and Interaction Diagram
Planet 253 f.
plasmepsin II 96
PLS *see* partial least squares analysis
PLUMS 185
PMF 66
polar surface area (PSA) 29, 34, 144, 153, 164, 175, 227, 235, 307, 348, 363, 387, 400, 411 f., 421
pose clustering 80
potency 404
potentials of mean force 127
prediction of binding affinity 63
Principal Component Analysis (PCA) 75, 119, 307, 344
privileged
– scaffolds 289
– structural motifs 205
– structure 288, 395, 421
– substructures 290
product-based
– filtering 205
– selection 288
PRO-LIGAND 87
promiscuous
– binders 290, 394
– inhibitors 147, 151
property index 203
PRO-SELECT 97

Protein Data Bank *see* PDB
Protein Data Base *see* PDB
Protein-Ligand Accessibility and Interaction Diagram (PLAID) 254
protein-ligand complexes 65, 277
protein-protein interaction 168
proteochemometrics 345
protonation 63, 65
– states 274
Prous Science 293
ProVal 13
PrRP-R 421
PSA *see* polar surface area
pseudochirality of pyramidal nitrogens 274
purinergic receptor 98
PyMOL 256
Pyridostigmine 34
pyrimidine trione 272

q

QMPRPlus 385
QSAR *see* Quantitative Structure-Activity Relationships
QT prolongation 349
– *see also* hERG
– *see also* long QT syndrome
Quantitative Structure-Activity Relationships (QSAR) 5, 7, 26, 35, 75, 118, 127, 364, 437
– 3D 68, 75, 123, 438
– predictive power 440
– Y-randomization 439
quasi-Newton methods 70
QXP 96

r

RACHEL 15, 201, 217
RADDAR approach 419 ff.
random sampling 204
Rapid Overlay of Chemical Shapes (ROCS) 171
rational drug design 287 f.
reactive compounds 156
reagent-based selection 288
RECAP 55, 291, 304
redox chemistry 6
REG (SciFinder) 246
REOS 306, 391
retinoic acid receptor-α 94
retrosynthetic
– analysis 283, 291
– fragments 296
reverse transcriptase 437
ring hashcodes 159, 163, 167
ring-pattern classes 331

RO5 *see* rule-of-five
ROCS *see* Rapid Overlay of Chemical Shapes
rotatable bonds 29, 34, 53, 235, 307, 412
– as building block separators 202
RSR 13 465, 467, 473
rule-of-five 29, 33 f., 48, 90, 179, 227, 235,
    324, 336, 363, 392, 399
rule-of-three 235

s

S-adenosylmethionine 8
SA-PLS 447
SAR *see* Structure-Activity Relationship
SAR Navigator 153
SC-558 135
scaffold hopping 59, 98, 283
SciFinder 223, 228
scoring 64
scoring function 13, 66, 209, 278
– focused 211
– limitations 100
– targeted 213
search engines 241
SELECT 187
selectivity 334, 404
Sequence Retrieval System (SRS) 242
serendipity 457, 461
serum albumin 350
server-server communications 250, 265
sickle-cell anemia 460
Sickle-Cell Anemia Society 463
signal transduction 288
similarity 17, 62, 69, 81, 84, 94, 118, 122,
    305, 317, 330, 417
– Carbo 84
– ComPharm 133
– Cosine 81
– distance 442
– Hodgkin 84
– Tanimoto 81, 295, 327
simplex 69
simulated annealing 85
SimulationsPlus 385
simultaneous optimization 185
SLIDE 73
SLN 271
SMARTS 118, 271, 279, 388
SMILES 178, 247, 253, 255, 279, 415
social evolution 267
solid-phase synthesis 16
solubility 28
solvation models 18
Somatogen 473

Specs 402
Spresi 223
SPROUT 87
SRS *see* Sequence Retrieval System
statistical molecular design 203
stereochemistry 227
stochastic methods 186
stromelysin 272
structural
– exclusion criteria 319
– keys 81
– motif 295
Structure-Activity Relationship (SAR)
    4, 67, 171
structure-based
– design 190, 460
– drug design 15
subgraph
– matching 77
– maximum common 77
substructure filters 330
superoxide dismutase 6
SuperStar 344
support vector machines (SVMs) 75
surface
– area 6, 12
– complementarity 291
SVMs *see* support vector machines
SwissProt 226
SYBYL 459

t

Tachykinin NK1 296
Tanimoto similarity 81, 295, 327
target function 213
targeted scoring functions 67
tautomer conversion 28
tautomerism 271
– 2-hydroxypyridine 280
– imidazole 275
– pyrazoles 281
– 2-pyridones 280
– tetrazole 275
tautomerization 63, 65
TCP/IP 246
– port numbers 266
thermodynamics 211
thermolysin 50
thrombin 96, 214, 343 f., 347, 360, 363
thyroid hormone receptor 95, 98
thyrotropin releasing hormone (TRH) 9
TOPAS 88
TopCAT 444

topological descriptor 119
topology-based similarity 77
toxicity prediction 16
toxicology 25
traditional chinese medicines 255 f.
TRH *see* thyrotropin relasing hormone
trivalent nitrogen – configuration 272, 276
tRNA-guanine transglycosylase 94
trypsin 344

u

undesirable
– compounds 145
– fragments 90
– functional groups 324
– physicochemical properties 145
– substructures 92
undesired structures 199, 207, 416
– removal 207
– *see also* substructure filters
UNITY 176
urotensin II 421
– receptor 99

v

VALIDATE 12, 210
vanillin 478
variable selction 437
vascular cell adhesion molecule-1 99
vasopressin V1A/V2 296
venture capital 472 f.
virtual combinatorial chemistry 205
virtual libraries 288
– GPCR agents 309
virtual screening 26, 43, 59, 73, 175, 205,
    271, 417, 420 f., 437

– ADMET filters 350
– databases 281
– ligand-based 61
– pharmacophore-based 62, 117
– similarity-based 62
– strategies 90, 91
– structure-based 61, 94
– work flow 90
Virtual Screening of Virtual Libraries (VSVL)
    54
visualization 298, 423
VLA-4 289
VolSurf 350, 353, 359, 361, 364

w

Wall Street Journal 472
water solubility 227
WDI *see* World Drug Index
web robots 266
WOMBAT 29, 31, 35, 223 ff., 253, 255
– errors from literature 231
– human versus rat 234
– quality control 228
– target classe profile 226
World Drug Index (WDI) 23, 29, 223, 246,
    254 f.

x

XML *see* Extensible Markup Language

y

Y-randomization 439

z

Zipfian distribution 46